ESSENTIALS OF BASIC MATHEMATICS

THIRD EDITION

ESSENTIALS OF BASIC MATHEMATICS

THIRD EDITION

ALLYN J. WASHINGTON
Dutchess Community College

SAMUEL H. PLOTKIN
Montgomery County Community College

CAROLYN E. EDMOND

THE BENJAMIN/CUMMINGS PUBLISHING COMPANY, INC.
Menlo Park, California   Reading, Massachusetts
London   Amsterdam   Don Mills, Ontario   Sydney

Sponsoring Editor: Susan A. Newman
Production Editor: Greg Hubit
Coordinator: Margaret Moore
Cover Designer: Henry Brever
Cover Photographer: George Fry
Book Designer: Janet Bollow

Library of Congress Cataloging in Publication Data

Washington, Allyn J
Essentials of basic mathematics.

Includes index.
1. Mathematics—1961– I. Plotkin, Samuel H., joint author. II. Edmond, Carolyn E., joint author. III. Title.
QA39.2.W373 1981 512′.1 80-27335
ISBN 0-8053-9529-6

ABCDEFGHIJ-DO-8321

The Benjamin/Cummings Publishing Company, Inc.
2727 Sand Hill Road
Menlo Park, California 94025

PREFACE

PURPOSE OF THIS BOOK

This text was written for those students who need to develop an understanding of the fundamental concepts and techniques of elementary mathematics. Algebra, basic topics from geometry, and a brief introduction to trigonometry are included in the text, which may be used by those students who require a review of these topics or by those who have not previously studied them. Also, since most of the topics normally included in intermediate algebra are developed, the text may be used in courses covering algebra, or algebra and geometry, including courses intended as background for further work in mathematics.

MATHEMATICAL CONTENT

The approach in the development of the material is informal and intuitive. Discussions of the basic concepts are intended to give the student an understanding of the material, how it is developed, and how more complex material develops from it. More detail has been included in those topics which experience has shown are more difficult for students to grasp, although topics which normally cause less difficulty are also developed in sufficient detail. Special cases are noted but not overemphasized.

Arithmetic Review For courses in which a review of topics from arithmetic is necessary, the material in Appendix A is appropriate. Also, this material may be used for purposes of reference and review by any student using this text.

Sets Set terminology and notation are introduced and used where it is deemed appropriate.

Geometry Chapter 13 covers geometry; it includes material on basic figures, with which the student is probably reasonably familiar, in addition to discussions of similar triangles and solid geometric figures and a brief introduction to reasoning and proof in mathematics.

EXAMPLES AND EXERCISES

Word problems appear in most sets of worked examples and exercises. The continual solving of such problems accustoms the student to them. The common approach of devoting a few sections to such problems and essentially avoiding them elsewhere often allows the student to avoid them and never develop the necessary technique for their solution. For this reason, after some of the simplest forms have been introduced, Chapter 7 is devoted to the interpretation and setting up of such problems.

Numerous worked examples and exercises are included. There is a set of exercises at the end of each section, and each chapter includes a summary section with a set of review exercises. There are many drill exercises in the section and chapter review exercise sets to help the student develop techniques. Also, many exercises illustrate the use of mathematics in areas such as physics, chemistry, astronomy, engineering, business, biology, social science, and the humanities. Therefore, this material may be used as a basic algebra-geometry text for students regardless of their intended major.

NEW FEATURES IN THIS EDITION

The third edition includes all the basic features of the first two editions. Most sections have been somewhat rewritten to include additional explanatory material and new examples and exercises.

The new features in this edition are the following:

1. *Inequalities:* Additional material on inequalities is included. In Section 3-4 inequalities which include equalities as well as the representation of inequalities on a number line are covered. Section 10-6 covers the graphing of inequalities.
2. *Ratio and proportion:* Section 3-5 is now a separate section on ratio and proportion.

3. *Factoring:* Additional types of factoring are now included. Section 5-5 includes factoring of the sum or difference of cubes and factoring by grouping.
4. *Functions and graphs:* The chapter on functions and graphs (Chapter 10) is now presented before the material on logarithms so that a more general treatment of logarithms may be covered. Section 10-7 now covers the slope of a line, and Section 10-8 introduces the concept of an inverse function.
5. *Logarithms:* Chapter 11 now covers exponential and logarithmic functions. The general definition and properties of logarithms are now presented, rather than only the specialized use of logarithms as a calculational device. Also, natural logarithms are introduced.
6. *Solid geometry:* The coverage of solid geometry figures (formerly section 13-5) has been split into sections (now 13-5 and 13-6).
7. *Metric units:* Many more exercises involving metric units are included. Also, Appendix B provides an introduction to the use of the metric (SI) system.
8. *Calculators:* The advent of the use of calculators has influenced several changes. The change in the coverage of logarithms is one example of this. Also, many exercises lend themselves to calculator solution, particularly those of Chapter 14 on trigonometry. The section on finding square roots which was included in the first two editions has been deleted. Although the use of the calculator is not required, it is assumed that many students will use one.
9. *Examples and Exercises:* There are now over 4000 exercises, an increase of about 40%, and nearly 500 worked examples, an increase of over 25%.

ACKNOWLEDGMENTS

The authors wish to acknowledge the help and suggestions given them by those who have used the first two editions. In particular, we wish to thank Martin Johnson of Montgomery County Community College for his numerous valuable comments and suggestions. Also, we wish to thank Robert Borque of Southern Maine Vocational Technical Institute and Jerry Wethington of Mesa College for their many valuable suggestions in their reviews of the manuscript. Also, we wish to thank the other members of the Mathematics Departments at both Dutchess Community College and Montgomery County Community College for their many helpful suggestions. The assistance, cooperation, and diligent efforts of the Benjamin/Cummings staff during the production of this text are also greatly appreciated.

A.J.W.
S.H.P.
C.E.E.

CONTENTS

1
NUMBERS

1-1 THE SIGNIFICANCE OF MATHEMATICS

Mankind's endeavor to learn about the environment is manifested in many of the events of today. Artificial satellites used for scientific research and communications circle the earth; spacecraft have transmitted important information about other planets back to earth; men have made explorations on the surface of the moon; computers, with unbelievable speed, perform tasks which previously took endless hours or were prohibitively complicated; diseases which once crippled many are now almost unknown in many areas. These are obvious and impressive examples of progress. Much progress has been made also in geological studies of the earth, in psychological studies of the mind, in analyses of elections and voting trends, in weather forecasting, in determining new sources of natural resources, and in numerous studies related to social problems.

One of the basic reasons for the tremendous development of certain fields of knowledge is that it has been possible to describe phenomena in these fields mathematically. In particular, physics, chemistry, astronomy, and biology, which have strong mathematical bases, have made rapid advances. However, mathematics is used to some degree in almost all fields of knowledge. Today, many advanced topics of mathematics are important in the development of the computer, which, in turn, is having a strong impact on the scientific and business worlds. Advanced topics in the calculus are used in the analysis of music. Geometric properties have played an important role in art. Statistics is used extensively in business, psychology, political science, and many other fields.

Obviously, in this text, we cannot hope to develop the mathematics necessary to solve most of the problems which arise in nature. However, the material presented here is the foundation for the more advanced mathematics which can be applied to the solution of such problems. For those of you who will go on with your study of mathematics, all the topics presented here are of importance for your later work. Thus, you should strive to understand these topics as they are presented.

Mathematics can be used directly in solving many kinds of problems. Some examples of the problems that can be solved by the mathematics covered in this text are: (1) determining interest received from financial investments; (2) making indirect measurements of distances which cannot otherwise be measured; (3) determining the dimensions of maps and art reproductions; (4) making binary calculations (which are used in many computers); (5) constructing graphs for economics and psychology. Numerous other illustrations of the uses of mathematics are given throughout the text.

There are still other benefits to be derived from this study of mathematics. The more mathematics a person knows, the more appreciation he or she has of today's developments. Also, studying mathematics for itself has rewards. One sees the development of a major field of study with a very logical structure. While the practical applications of certain topics may not be immediately obvious, history has shown us that many mathematical topics once considered as curiosities later became very important in applications. This is true, for example, of much of the mathematics which is applicable to computers.

We hope that you will proceed with a feeling for the usefulness of mathematics and not with apprehension of its difficulties. Some of the topics presented will probably be challenging–however, they should be met and readily conquered if the material is given sufficient attention.

Exercises 1-1

1. Skim through the articles of a principal newspaper such as today's *New York Times* and note the number of items which mention or imply the use of mathematics or are about developments affected by mathematics.
2. Go to your college library and look through at least two references from each of the following areas: physics, engineering, chemistry, biology, geology, psychology, economics, business, art, music, and literature. What conclusion do you draw with respect to the mathematical content of the references?
3. Make a list of 15 words which you use with familiarity and which are related to mathematics. (Do not include any names of numbers.)
4. List 5 possible situations in everyday life where the use of some form of mathematics arises.

1-2 POSITIVE AND NEGATIVE NUMBERS

We shall being our study of mathematics by considering the various kinds of numbers that we shall be using. This chapter will be concerned with the basic properties of these numbers and the basic operations of addition, subtraction, multiplication, and division. It is assumed that the reader is familiar with the basic operations of arithmetic. In this section, we shall consider the meanings of positive and negative numbers, and the operations on these numbers will be taken up in the following sections.

In the development of the basic operations in arithmetic, the subtraction of one number from a smaller number is not considered. In fact, the whole numbers and fractions that are employed in arithmetic would provide no answer for such a problem. It is therefore necessary to introduce a new kind of number, the **negative number.** Negative numbers are employed extensively in algebra and also have applications in numerous practical situations. For example, they may be used to represent temperatures below zero or monetary accounts with deficits.

A negative number is a number less than zero. The symbols for addition $(+)$ and subtraction $(-)$ are also used to indicate positive and negative numbers, respectively. For example, $+5$ (positive 5) is a **positive number** (one which is greater than zero), and -5 (negative 5) is a negative number. Since the positive numbers correspond directly to the numbers that are used in arithmetic, there is no danger of confusion if the positive sign before a positive number is omitted. However, we will never omit the negative sign when designating a negative number.

In this discussion, we see that the $+$ sign and the $-$ sign are used in two senses. One is to indicate addition or subtraction, and the other is to designate a positive or a negative number. Consider Example A.

Example A The expression $3 + 6$ means "add the number 6 to the number 3." Here, the plus sign indicates addition, and the numbers are **unsigned numbers,** *meaning that they are positive numbers.*

The expression $(+3) + (+6)$ means "add the number $+6$ to the number $+3$." Here, the middle plus sign indicates addition, whereas the other plus signs denote positive numbers. Of course, the answer here would be the same as the answer to $3 + 6$.

The expression $3 - 6$ means "subtract the number 6 from the number 3." Here, the minus sign indicates subtraction, and the numbers are unsigned.

The expression $(+3) - (+6)$ means "subtract the number $+6$ from the number $+3$." Note that the plus signs designate positive numbers.

The expression $(+3) - (-6)$ means "subtract the number -6 from the number $+3$." Here, the first minus sign indicates subtraction, whereas the second designates a negative number.

The method of performing these operations will be discussed in the following section.

Positive and negative numbers can be designated on a scale as shown in Fig. 1-1. *On a horizontal line, we choose a point which we call the* **origin,** *and here we locate the number 0 (zero); we locate positive numbers on this line to the right of 0 and the negative numbers to the left of 0.* As we see, every positive number has a corresponding negative number on the opposite side of the origin. Therefore, we can consider the negative numbers as being the **opposites** of the corresponding positive numbers. In the same way, the positive numbers are opposites of the negative numbers. The number zero at the origin is neither positive nor negative.

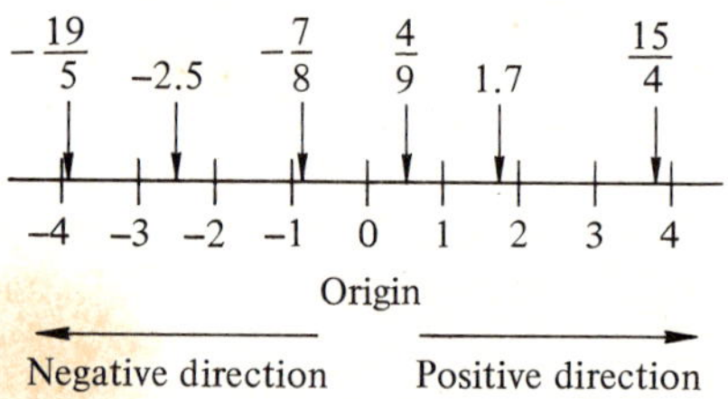

Figure 1-1

When numbers are marked + or −, they are often called **signed** (or **directed**) **numbers** to indicate their opposite qualities. Since positive numbers are directed to the right of the origin, it follows that numbers increase from left to right. Therefore, *any number on the scale is smaller than all numbers on its right. The symbols > and < mean* "**greater than**" *and* "**less than,**" *respectively.* Their use is shown in Example B.

Example B The expression $2 > 1$ means "the number 2 is greater than the number 1." It also means that, on the scale in Fig. 1-1, we should find 2 to the right of 1.

The expression $0 < 4$ means "0 is less than 4." This means that 0 is to the left of 4.

The expression $-3 < 0$ means "-3 is less than 0." The number -3 lies to the left of 0 on the scale.

The expression $-3 > -5$ means "-3 is greater than -5." The number -3 is to the right of the number -5 on the scale.

While the positive direction is conveniently taken to the right of the origin on the number scale, we may select any direction as the positive direction so long as we take the opposite direction to be negative. Consider the illustrations in Example C.

Example C If we say that the positive direction is up, we then take the negative direction to be down. East may be positive if west is made negative, or north may be positive if south is made negative.

If a temperature above zero were called positive, a temperature below zero would be called negative.

If a counterclockwise movement were called positive, a clockwise movement would be negative.

The value of a number without its sign is called its **absolute value.** That is, if we disregarded the signs of $+5$ and -5, the value would be the same. This is equivalent to saying that the distances from the origin to the points $+5$ and -5 on the number scale are equal. The absolute value of a number is indicated by the symbol $|\ |$. The number $+5$ does not equal the number -5, but $|+5| = |-5|$. Absolute values are used in performing algebraic operations with signed numbers. Consider Example D.

Example D The absolute value of $+8$ is 8. The absolute value of -2 is 2. The absolute value of $-\frac{3}{2}$ is $\frac{3}{2}$. We may write these equalities as $|+8| = 8$, $|-2| = 2$ and $|-\frac{3}{2}| = \frac{3}{2}$.

In applications, numbers are often used to represent some type of measurement and therefore have certain units of measurement associated with them. *Such numbers are referred to as* **denominate numbers.** Both the traditional English units and metric units are used in this text. A discussion of these units, and the symbols which are used, is found in Appendix B.

Example E To indicate that an object is ten feet long, we represent the length as 10 ft.

To indicate that the speed of a projectile is 150 meters per second, we represent the speed as 150 m/s. (Note the use of s for second. We use s rather than sec, which is also common.)

Exercises 1-2

In Exercises 1 through 8, state the meanings of the given expressions, as in Example A.

1. $2 + 7$ **2.** $8 - 5$ **3.** $(+5) + (+9)$ **4.** $(+6) - (+2)$

5. $(-4) + (-3)$ **6.** $(+7) - (-3)$ **7.** $(-5) - (+2)$ **8.** $(-1) - (-5)$

In Exercises 9 through 12, determine the opposite of each of the numbers given.

9. $+7, -7$ **10.** $-4, 4$ **11.** $\frac{1}{2}, -\frac{1}{3}$ **12.** $-0.4, +2.6$

In Exercises 13 through 16, locate the approximate positions of the given numbers on a number scale such as that shown in Fig. 1-1.

13. $4, -4$ **14.** $0.5, -0.5$ **15.** $-\frac{13}{4}, +\frac{13}{22}$ **16.** $-2.3, +3.2$

In Exercises 17 through 20, determine which of the given numbers is the largest.

17. $4, -5, -1$ **18.** $-2, -5, 0$ **19.** $-6, -2, -4$ **20.** $-1, -2, -3$

In Exercises 21 through 32, insert the proper sign, > or < or =, between the given numbers

21. $6 \quad 2$ **22.** $8 \quad -3$ **23.** $0 \quad 4$ **24.** $-1 \quad -5$

25. $-3 \quad -7$ **26.** $-3 \quad 0$ **27.** $\frac{2}{3} \quad \frac{1}{2}$ **28.** $\frac{4}{5} \quad \frac{9}{8}$

29. $0.8 \quad 1.2$ **30.** $2.3 \quad -2.4$ **31.** $|-9| \quad |9|$ **32.** $|-3| \quad -|-3|$

In Exercises 33 through 36, find the absolute value of each of the given numbers.

33. $+6, -6$ **34.** $-9, +9$ **35.** $-\frac{6}{7}, \frac{8}{5}$ **36.** $2.4, -0.1$

In Exercises 37 through 46, answer the given questions involving positive and negative numbers.

37. If a football team gains 2 yd on a given play and this gain is represented by $+2$, determine the number which represents a loss of 4 yd.

38. If a deposit of \$4 is represented as $+4$, determine the signed number which represents a bank withdrawal of \$150.

39. The latitude of Springfield, Massachusetts, is about 42° north. If this is represented as $+42$, what is the latitude of Montevideo, Uruguay? (Consult a map.)

40. If the distance above ground level is represented as a positive number, what signed number represents the level of the bottom of a well 100 m deep?

41. If $+30$ represents the score a pinochle player gets for successfully completing a bid, what signed number is used to denote that he was set at a bid of 35?

42. The great Pyramid in Egypt was built about 2600 B.C. Assuming that years designated as A.D. are considered as positive, state this date as a signed number.

43. If the amount of money a person owes is represented as a negative number, which person, one whose financial position is -10 or one whose financial position is -20, is in the better financial position?

44. If the number of years to elapse between a future date and now is represented as a positive number, what number would be used to represent the number of years that have elapsed since the United States first put a man on the moon (1969)?

45. Note the way in which the financial section of a newspaper shows that a given stock gained or lost value during the day.

46. Use the proper sign, > or <, to show which temperature, $-30°F$ or $-5°F$, is the higher.

1-3 ADDITION AND SUBTRACTION OF POSITIVE AND NEGATIVE NUMBERS

In the previous section, we discussed the meaning and the designation of positive and negative numbers. In this section, we shall show how the normal operations of addition and subtraction are performed with positive and negative numbers. The following sections will take up the multiplication and division of positive and negative numbers.

We have noted that positive numbers are equal to the equivalent unsigned numbers encountered in arithmetic. Therefore, in the addition of positive numbers, the **sum** (*the result of addition*) is found in the same manner as we find the sum of unsigned numbers. This procedure is illustrated in Example A.

Example A Since $4 + 5 = 9$, we have $(+4) + (+5) = +9$. Thus, the sum of $+4$ and $+5$ is $+9$.

In the same way, since $3 + 12 = 15$, we have $(+3) + (+12) = +15$.

When adding two numbers, we often make use of a very important property of addition—the **commutative law of addition,** *which states that the order in which numbers are added does not matter.* For example, since $3 + 5 = 8$ and $5 + 3 = 8$, we have $3 + 5 = 5 + 3$. We shall often find it convenient to rearrange the order in which numbers are added.

Another important property of addition is the **associative law of addition,** *which states that the sum of three numbers taken in a certain order may be found either by grouping the first two together or by grouping the second two together.* This is illustrated in Example B.

Example B Since $2 + 3 = 5$ and $5 + 6 = 11$, we have $(2 + 3) + 6 = 11$. Also, since $3 + 6 = 9$ and $2 + 9 = 11$, we have $2 + (3 + 6) = 11$. Thus, $(2 + 3) + 6 = 2 + (3 + 6)$ is an example of the associative law of addition. The associative law effectively forms the basis of finding sums of more than two numbers.

Let us now consider the addition of two negative numbers. If we consider a negative number as representing a loss, we can think of adding one negative number to another as adding one loss to another. When we add one loss to another, our result is a still greater loss. Thus, we add the absolute values and designate the sum as negative.

Example C From the above considerations, we have $(-4) + (-5) = -(4 + 5) = -9$. Here we have added the absolute value of -4 and the absolute value of -5 in the expression $(4 + 5)$.

In the same way, $(-3) + (-12) = -(3 + 12) = -15$.

When considering the addition of two numbers of unlike signs, we may think of combining a gain with a loss. If the gain is greater than the loss, we have a net gain; conversely, if the loss is greater than the gain, we have a net loss. Thus, the number of smaller absolute value is subtracted from the one of greater absolute value, with the result taking on the sign of the number of greater absolute value.

Example D

$$(+5) + (-4) = +(5 - 4) = +1$$
$$(-5) + (+4) = -(5 - 4) = -1$$
$$(-3) + (+12) = +(12 - 3) = +9$$
$$(+3) + (-12) = -(12 - 3) = -9$$

To summarize the processes used when adding positive and negative numbers, we

add numbers of like signs by adding the absolute values of the numbers and giving the sum the common sign. To add two numbers with unlike signs, we find the difference of their absolute values and give the difference the sign of the number of larger absolute value.

Example E

$$(+8) + (+2) = +(8 + 2) = +10$$
$$(-8) + (-2) = -(8 + 2) = -10$$
$$(+8) + (-2) = +(8 - 2) = +6$$
$$(-8) + (+2) = -(8 - 2) = -6$$

To subtract positive and negative numbers, we use the same definition for subtraction that we use in arithmetic. That is, *we find a number* (*the* **difference**) *which, when added to the number being subtracted* (*the* **subtrahend**), *will give the other number* (*the* **minuend**).

Example F

$$(+7) - (+3) = +4 \quad \text{since} \quad (+4) + (+3) = +7$$
$$(+3) - (+7) = -4 \quad \text{since} \quad (-4) + (+7) = +3$$
$$(-7) - (+3) = -10 \quad \text{since} \quad (-10) + (+3) = -7$$
$$(+3) - (-7) = +10 \quad \text{since} \quad (+10) + (-7) = +3$$
$$(-7) - (-3) = -4 \quad \text{since} \quad (-4) + (-3) = -7$$
$$(-3) - (-7) = +4 \quad \text{since} \quad (+4) + (-7) = -3$$

If we examine the results of Example F, we note that the same result would be obtained in each case if we change the operation to addition *and* changed the sign of each of the numbers being subtracted.

Example G

$(+7) - (+3) = +4$ and $(+7) + (-3) = +4$

$(+3) - (+7) = -4$ and $(+3) + (-7) = -4$

$(-7) - (+3) = -10$ and $(-7) + (-3) = -10$

$(+3) - (-7) = +10$ and $(+3) + (+7) = +10$

$(-7) - (-3) = -4$ and $(-7) + (+3) = -4$

$(-3) - (-7) = +4$ and $(-3) + (+7) = +4$

There are several very important conclusions to be drawn from the illustrations of Example G. The first is that

to subtract one signed number from another, we may change the sign of the subtrahend and then add it to the minuend, using the rules for the addition of signed numbers.

Also, from the first two illustrations, we see that

subtracting a positive number is equivalent to adding a negative number of the same absolute value.

Finally, the last three illustrations point out that

subtracting a negative number is equivalent to adding a positive number of the same absolute value.

This last point may be thought of as removing (subtracting) a loss (a negative number), which results in a gain.

Example H

$(+8) - (+5) = (+8) + (-5) = +3$

$(+5) - (+8) = (+5) + (-8) = -3$

$(+8) - (-5) = (+8) + (+5) = +13$

$(+5) - (-8) = (+5) + (+8) = +13$

Thus, we see that $-(+5)$ and $+(-5)$ are the same, and that $-(-5)$ and $+(+5)$ are the same.

In general, when working with signed numbers, we will find it convenient to change the operations and the numbers such that the result is obtained by the addition and subtraction of positive numbers (written as unsigned numbers). When doing this, we must remember that adding a negative number is equivalent to subtracting a positive number, and subtracting a negative number is equivalent to adding a positive number. Consider the illustrations in Example I.

Example I

$$(+5) + (+2) = 5 + 2 = 7$$

$$(+5) + (-2) = 5 - 2 = 3$$

$$(+5) - (+2) = 5 - 2 = 3$$

$$(+2) - (+5) = 2 - 5 = -3$$

$$(-5) + (+2) = -5 + 2 = -3$$

$$(-5) - (-2) = -5 + 2 = -3$$

$$(-4) + (-2) - (-7) = -4 - 2 + 7 = -6 + 7 = 1$$

$$(-5) - (-2) - (+9) = -5 + 2 - 9 = -3 - 9 = -12$$

Note that, when a sign is not denoted before a signed number, we assume that it may be treated as if a + sign preceded it. That is, we treat (-4) as $+(-4)$, and $(+4)$ as $+(+4)$.

At this point we note that we did not mention commutativity in connection with subtraction. *Subtraction is not commutative*, since the order in which numbers are subtracted *does* matter.

Example J Considering the numbers $+2$ and $+4$, we know that $(+2) - (+4) = -2$ and $(+4) - (+2) = +2$. Therefore, $(+2) - (+4)$ *does not equal* $(+4) - (+2)$; this is written as $(+2) - (+4) \neq (+4) - (+2)$. Thus, we see that subtraction is not commutative.

Exercises 1-3

In Exercises 1 through 4, state whether the commutative law of addition or the associative law of addition is illustrated.

1. $7 + 9 = 9 + 7$

2. $3 + (8 + 4) = (3 + 8) + 4$

3. $(8 + 1) + 6 = 8 + (1 + 6)$

4. $2 + (3 + 4) = (3 + 4) + 2$

In Exercises 5 through 36, perform the indicated operations on the given positive and negative numbers.

5. $(+2)+(+9)$
6. $(+3)+(+8)$
7. $(-6)+(-9)$
8. $(-6)+(-5)$
9. $(-3)+(+6)$
10. $(-1)+(+8)$
11. $(+2)+(-10)$
12. $(+3)+(-11)$
13. $(+12)-(+5)$
14. $(+14)-(+6)$
15. $(+7)-(+10)$
16. $(+2)-(+15)$
17. $(-6)-(+4)$
18. $(-3)-(+8)$
19. $(+7)-(-9)$
20. $(+14)-(-2)$
21. $(-6)-(-7)$
22. $(-9)-(-5)$
23. $(+1)+(-5)+(-2)$
24. $(+6)+(+5)+(-3)$
25. $(+2)+(-8)-(+2)$
26. $(-4)-(+8)+(-9)$
27. $(+5)-(-3)-(+7)$
28. $(-7)-(-1)-(+6)$
29. $(-7)-(-15)+(-2)$
30. $(-6)+(-13)-(-11)$
31. $(-9)-(-5)+(-8)-(+5)$
32. $(-7)-(-5)+(-4)-(+6)$
33. $(+3)-(-7)-(+9)+(-3)$
34. $(+6)+(-11)-(-5)-(+8)$
35. $(-6)-(+9)-(-12)+(-14)-(-1)$
36. $(-5)+(+6)-(+3)-(-14)+(-3)$

In Exercises 37 through 42, set up the given problems in terms of the addition or subtraction of signed numbers, and then perform the operations. (The operation of addition or subtraction that should be used in setting up the problem is indicated in each case.)

37. The temperature of an object is -10°F. (a) If the temperature of the object increases (use addition) by 5°F, what is the resulting temperature? (b) If the temperature of the object decreases (use subtraction) by 5°F, what is the resulting temperature?

38. While watching a rocket launching, an observer notes that it is "blastoff minus 10 minutes." Assuming no delays, how long has the rocket been in the air 30 min later? (Use subtraction.)

39. An object 25 m above ground level falls to the bottom of a 30-m well. How far does it fall? (Use subtraction.)

40. A person has a debt of \$17. Another person owes him \$11. What is the sum (use addition) of these "assets" to him?

41. If a velocity is considered negative when an object is moving downward and positive when it is moving upward, what is the change in velocity of a ball moving at 60 ft/s just before it hits the floor and just after it hits the floor? (Use subtraction.)

42. In 4 successive years, the population of a certain city changed as follows: increased by 5000; increased by 2000; decreased by 3000; increased by 1000. Find the sum (use addition) of these changes.

In Exercises 43 and 44, answer the questions as directed.

43. Using the numbers -2 and -7, show that subtraction is not commutative. (See Example J.)

44. What rule of English grammar is equivalent to subtracting a negative number in mathematics?

1-4 MULTIPLICATION OF POSITIVE AND NEGATIVE NUMBERS

In developing the principles used in the multiplication of positive and negative numbers, we wish to be able to preserve the definitions which hold for the multiplication of numbers in arithmetic. Therefore, as in arithmetic, the multiplication of signed numbers is the process of repeated addition, where equal numbers are added together a specified number of times. *The numbers to be multiplied are called* **factors,** *and the result is called the* **product.**

Example A In multiplying 4 by 3, which we can denote by 4×3 or $4 \cdot 3$, we mean

$$4 \times 3 = 3 + 3 + 3 + 3 = 12$$

In the same way, we have

$$3 \times 4 = 4 + 4 + 4 = 12$$

Here, 3 and 4 are the factors and 12 is the product.

In Example A, we see the basic meaning of the multiplication of two numbers. Also, a very important property of multiplication is demonstrated. This property is the **commutative law of multiplication,** *which states that the order of multiplication does not matter*. In Example A, this law is illustrated by noting that $4 \times 3 = 3 \times 4$. We recall that a similar property was true in the case of the addition of two numbers.

Another property of multiplication is the **associative law of multiplication,** *which states that the product of three numbers taken in a certain order may be found either by grouping the first two together or by grouping the second two together*. Again, this is similar to the associative law of addition. Consider the illustrations in Example B.

Example B Since $2 \times 3 = 6$ and $6 \times 4 = 24$, we have $(2 \times 3) \times 4 = 24$. Also, since $3 \times 4 = 12$ and $2 \times 12 = 24$, we have $2 \times (3 \times 4) = 24$. Thus, $(2 \times 3) \times 4 = 2 \times (3 \times 4)$ is an example of the associative law of multiplication. This law forms the basis of the multiplication of more than two numbers.

By the use of the principles that we have established for the addition of signed numbers and the commutative and associative laws of multiplication, we can now establish the principles used for the multiplication of these positive and negative numbers. Consider Example C.

Example C Since $3 \times 4 = 12$ and since $3 = +3$, $4 = +4$, and $12 = +12$, we have $(+3) \times (+4) = +12$.

Since $3 \times (-4)$ means $(-4) + (-4) + (-4)$, since $(-4) + (-4) + (-4) = -12$, and since $3 = +3$, we have $(+3) \times (-4) = -12$.

Also, since multiplication is commutative, we have $(-4) \times (+3) = -12$.

We see from Example C that the product of two positive numbers is positive. Also, we note that the product of a positive number and a negative number is negative. Hence, we see that

changing the sign of one factor changes the sign of the product.

Applying this principle, we may determine the product of two negative numbers.

Example D From Example C, we have seen that

$$(-4) \times (+3) = -12$$

Applying the principle that changing the sign of one factor changes the sign of the product, we change $+3$ to -3 and -12 to $+12$. This leads to

$$(-4) \times (-3) = +12$$

Thus, we conclude that

the product of two negative numbers is positive.

This last illustration may be looked upon as subtracting -3 four times or $-(-3) - (-3) - (-3) - (-3) = +12$. We may also think of this as negating four \$3 debts, which would result in a gain of \$12. Note that this example is consistent with the fact that subtracting a negative number is equivalent to adding a positive number of the same absolute value.

Continuing our consideration of the multiplication of two negative numbers, let us multiply -3 by -2. Setting up the multiplication in the standard vertical manner, we have

$$\begin{array}{r} -3 \\ \times\ -2 \\ \hline \end{array}$$

Suppose we think of the lower minus sign as a direction sign. The 2 indicates the number of times the lower minus sign directs operation on the -3. Thus, the -2 says "take away a -3 two times", and taking away a -3 two times means becoming positive by 6.

From the above discussion, we can state the following results:

A. ***If two factors have like signs, their product is positive.***
B. ***If two factors have unlike signs, their product is negative.***

If there are three or more factors, we can use the associative and commutative laws and multiply two factors at a time, in any order, observing each time the sign of the product. We can then conclude:

A. ***The product is positive if all factors are positive or if there is an even number of negative factors.***
B. ***The product is negative if there is an odd number of negative factors.***

If any factor is zero, it follows that the product is zero because zero times any number is zero.

Example E

$$(+4) \times (+7) = +28 \qquad (+4) \times (-7) = -28$$
$$(-4) \times (-7) = +28 \qquad (-4) \times (+7) = -28$$
$$(+2) \times (+3) \times (+5) = +30 \qquad \text{(no negative factors)}$$
$$(-2) \times (+3) \times (+5) = -30 \qquad \text{(one negative factor)}$$
$$(-2) \times (-3) \times (+5) = +30 \qquad \text{(two negative factors)}$$
$$(-2) \times (-3) \times (-5) = -30 \qquad \text{(three negative factors)}$$
$$(-1) \times (-2) \times (-3) \times (-5) = +30 \qquad \text{(four negative factors)}$$
$$(-1) \times (0) \times (-3) \times (-5) = 0 \qquad \text{(zero is a factor)}$$

If a given number appears as a factor in a product more than once, it is customary to write the number only once, *indicating the number of times it would appear as a factor by writing an* **exponent** *with the number.* An exponent is a small number to the right and slightly above the number.

Example F Instead of writing 5×5, we write 5^2. Here, the 2 is the exponent. The expression 5^2 is read as "5 squared" or "5 to the second power."

Rather than writing $(-2) \times (-2) \times (-2)$, we write $(-2)^3$. Here, the 3 is the exponent, and $(-2)^3$ is read as "negative 2 cubed" or "negative 2 to the third power."

The product $3 \times 3 \times 3 \times 3 \times 3$ is written as 3^5. The 5 is the exponent, and 3^5 is read as "the fifth power of 3."

We must note the use of parentheses in raising a signed number to a power. The meaning of $(-2)^3$ is indicated in Example F. However, -2^3 means the negative of 2^3. Since parentheses are not placed around -2, only the 2 is being raised to the third power. Thus, $-2^3 = -(2^3)$.

From our earlier discussion of the multiplication of signed numbers, we can see that a positive number raised to a power will have a positive result.

However, *a negative number will have a positive result if the power is an even number, and it will have a negative result if the power is odd.* This situation is due to the fact that the exponent indicates the number of factors of the number.

Example G

$(+3)^2 = +9$ since $(+3)^2 = (+3) \times (+3)$

$(+4)^3 = +64$ since $(+4)^3 = (+4) \times (+4) \times (+4)$

$(-3)^2 = +9$ since $(-3)^2 = (-3) \times (-3)$

$-3^2 = -9$ since $-3^2 = -(3^2) = -(3 \times 3)$

$(-3)^3 = -27$ since $(-3)^3 = (-3) \times (-3) \times (-3)$

$-3^3 = -27$ since $-3^3 = -(3^3) = -(3 \times 3 \times 3)$

$(-2)^4 = +16$ since $(-2)^4 = (-2) \times (-2) \times (-2) \times (-2)$

$-2^4 = -16$ since $-2^4 = -(2^4) = -(2 \times 2 \times 2 \times 2)$

Also,

$$(-1)^{20} = +1 \qquad -1^{20} = -1 \qquad (-1)^{21} = -1 \qquad -1^{21} = -1$$

In any product where the factors are signed numbers, the multiplication of these signed numbers may be indicated by writing them, in their parentheses, adjacent to each other. This eliminates the need for using the $\times$ or $\cdot$ as signs of multiplication.

Example H We may write $(-3) \times (+4)$ as $(-3)(+4)$ because we realize that this form indicates the multiplication of -3 and $+4$.

Therefore, $(-2)(-3)(+4) = +24$, since

$$(-2)(-3)(+4) = (-2) \times (-3) \times (+4)$$

Exercises 1-4

In Exercises 1 through 4, state whether the commutative law of multiplication or the associative law of multiplication is illustrated.

1. $3 \times (4 \times 5) = (3 \times 4) \times 5$

2. $5 \times 9 = 9 \times 5$

3. $(3 \times 2) \times 7 = 7 \times (3 \times 2)$

4. $(3 \times 2) \times 7 = 3 \times (2 \times 7)$

In Exercises 5 through 32, find the indicated products.

5. $(+8) \times (+10)$

6. $(+12) \times (+5)$

7. $(+7)(-9)$

8. $(-4)(+11)$

9. $(-7) \times (+12)$

10. $(+15) \times (-4)$

11. $(-2) \times (-15)$

12. $(-9) \times (-8)$

13. $(+4) \times (-2) \times (+7)$

14. $(-3) \times (+6) \times (+5)$

15. $(-10)(-3)(+8)$

16. $(-12)(+4)(-10)$

17. $(+2) \times (-5) \times (-9) \times (+4)$

18. $(-7) \times (-6) \times (+10) \times (-1)$

19. $(+8)(0)(-4)(-1)$

20. $(-1)(-3)(-5)(-2)$

21. $(+3)(-1)(+4)(-7)(-2)$

22. $(-5)(+3)(-2)(0)(-9)$

23. $(+6)^2; (-6)^2$

24. $(+5)^3; (-5)^3$

25. $(-7)^2; -7^2$

26. $-9^2; (-9)^2$

27. $(-4)^5$

28. $(-4)^6$

29. $(-1)^{16}; -1^{16}$

30. $(-1)^{19}; -1^{19}$

31. $(-2)^7; -2^7$

32. $-2^8; (-2)^8$

In Exercises 33 through 36, set up the given problems in terms of the multiplication of signed numbers, and then perform the operations.

33. Metal expands when heated and contracts when cooled. The length of a particular metal rod changes by 2 mm (millimeters) for each 1°C change of temperature. What is the total change in length of the rod as the temperature changes from 0°C to −25°C?

34. The price of a stock fell $3 each day for 4 consecutive days. Find the net change in the price of the stock during this period.

35. At a certain time, an airplane had been descending at the rate of 1000 ft/min for 7 min. How much greater was its altitude 7 min earlier?

36. A gambler lost money in the first day of gambling. On the second day she gambled for four hours and managed to reduce her losses by $10 each hour. What was the change in her monetary holdings during the second day?

1-5 DIVISION OF POSITIVE AND NEGATIVE NUMBERS

The next operation we shall consider is that of division of positive and negative numbers. Again, we wish to make our definitions consistent with those of arithmetic, where division is the process of determining how many times one number is contained in another. Here, *the number being divided is called the* **dividend,** *the number being divided into the dividend is the* **divisor,** *and the result is the* **quotient.**

Just as we define subtraction in terms of addition, we may also define division in terms of multiplication. When dividing 8 by 2, for example, we ask ourselves "2 times what number equals 8?" Since $2 \times 4 = 8$, we know that $8 \div 2 = 4$, where the symbol $\div$ means "divided by." Thus, we see that the product of the quotient and divisor equals the dividend. Due to this relationship of multiplication and division, they are known as **inverse operations.** Each may be thought of as "undoing" the other. In the same way, addition and subtraction are inverse operations.

Since multiplication and division are inverse operations, we can establish the principles used in the division of signed numbers from those we have already established for the multiplication of signed numbers. The basis for these principles is seen in Example A.

Example A

$$(+15) \div (+5) = +3 \quad \text{since} \quad (+5) \times (+3) = +15$$
$$(+15) \div (-5) = -3 \quad \text{since} \quad (-5) \times (-3) = +15$$
$$(-15) \div (+5) = -3 \quad \text{since} \quad (+5) \times (-3) = -15$$
$$(-15) \div (-5) = +3 \quad \text{since} \quad (-5) \times (+3) = -15$$

We see from the illustrations of Example A that

the quotient of two numbers of like signs is positive and the quotient of two numbers of unlike signs is negative.

Thus, we see that the division of two signed numbers follows the same principles as the multiplication of two signed numbers.

The division of one number by another is very often expressed in the form of a fraction. *The dividend becomes the* **numerator** *of the fraction and the divisor is the* **denominator** *of the fraction.*

Example B $(+21) \div (-3)$ in fractional form is $\frac{+21}{-3}$. Thus,

$$\frac{+21}{-3} = -7 \qquad \frac{-21}{+3} = -7 \quad \text{and} \quad \frac{-21}{-3} = +7$$

If either the numerator or the denominator (or both) of a fraction is the product of two or more signed numbers, this product may be found and then the value of the fraction can be determined by the appropriate division. However, since the same basic principles apply for the multiplication and the division of signed numbers, we need only determine the number of negative factors of the numerator and denominator combined. *If the total number of negative factors is odd, the quotient is negative. Otherwise, it is positive.* The reduction of the fraction to lowest terms then follows as with any arithmetic fraction.

Factors common to both numerator and denominator can be divided out.

Example C

$$\frac{(-2)(+12)}{-4} = +\frac{(2)(12)}{4} = \frac{(1)(12)}{2} = \frac{6}{1} = 6$$

The sign of the result is positive because the original fraction has two negative factors (-2 and -4) in the numerator and denominator. The resulting fraction is then simplified by dividing out two factors of 2 from both the numerator and the denominator. Since the resulting denominator is 1, it does not have to be written.

In a similar manner, we have

$$\frac{(-3)(+6)(-5)}{(-1)(+10)} = -\frac{(3)(6)(5)}{(1)(10)} = -\frac{(3)(6)(1)}{(1)(2)} = -\frac{(3)(3)(1)}{(1)(1)} = -9$$

The result is negative since the original fraction had three negative factors. The fraction was then simplified by dividing out a factor of 5 and a factor of 2.

There is one number we cannot use as a divisor. That number is zero. Since division is the inverse of multiplication, we can say that $0 \div$ (*any number*) $= 0$ because (*any number*) $\times\ 0 = 0$. This means that zero divided by any number (except zero) is equal to zero. However, if 0 is a divisor as in (*number*) $\div\ 0$, we cannot determine the quotient which will equal the *number* when multiplied by zero. This is so because any quotient that may be tried gives a product of zero when multiplied by zero and does not give the *number*. In the case of $0 \div 0$, any positive or negative number can be a quotient, because $0 \times$ (*any number*) $= 0$. This result is said to be **indeterminate.**

Since division by zero yields either no solution or an indeterminate solution,

division by zero is excluded at all times.

We must remember, however, that all other operations with zero are perfectly valid. Thus, when division by zero is indicated, the result is *not defined.*

Example D

$\frac{0}{4} = 0$ since $0 \times 4 = 0$

$\frac{-5}{0}$ is excluded since (*no number*) $\times\ 0 = -5$

$3 + 0 = 3$ and $0 - (-3) = 0 + 3 = 3$

Again *division by zero* is the excluded operation.

Problems involving signed numbers may combine several of the basic operations. In such problems, *all products, quotients, and powers should be evaluated before proceeding with the additions and subtractions.*

Example E Evaluate

$$\frac{+8}{-2}-(-5)(+2)-(-2)^2$$

First we perform the indicated multiplications and divisions. This leads to

$$(-4)-(-10)-(+4)$$

Using the principles for adding and subtracting signed numbers, we have

$$-4+10-4=+2$$

Example F Evaluate

$$\frac{(-3)^2}{-3}+(-5)(-2)-(-1)^3$$

We first evaluate the indicated multiplications and divisions and then use the principles for the addition and subtraction of signed numbers. This leads to the following:

$$\begin{aligned}\frac{(-3)^2}{-3}+(-5)(-2)-(-1)^3 &= \frac{+9}{-3}+(+10)-(-1)\\ &= -3+10+1\\ &= 8\end{aligned}$$

We note that in the first term it is necessary to evaluate $(-3)^2$ before performing the indicated division.

We note here that we did not mention commutativity in connection with division. *Division is not commutative*, since the order in which numbers are divided *does* matter. It can also be shown that *division is not associative.*

Example G Considering the numbers $+2$ and $+4$, we know that $(+2)\div(+4)=+\frac{1}{2}$ and $(+4)\div(+2)=+2$. Therefore, $(+2)\div(+4)\neq(+4)\div(+2)$, which shows that division is not commutative.

Now, consider that $[(+12)\div(+2)]\div(+2)=(+6)\div(+2)=+3$, but that $(+12)\div[(+2)\div(+2)]=(+12)\div(+1)=+12$. Therefore, $[(+12)\div(+2)]\div(+2)\neq(+12)\div[(+2)\div(+2)]$, which shows that division is not associative.

Exercises 1-5

In Exercises 1 through 36, evaluate the given expressions by performing the indicated operations.

1. $(+24) \div (+6)$ **2.** $(+32) \div (+4)$ **3.** $(+27) \div (-3)$ **4.** $(-48) \div (+16)$

5. $\dfrac{-22}{+2}$ **6.** $\dfrac{+20}{-4}$ **7.** $\dfrac{-60}{-4}$ **8.** $\dfrac{-52}{-13}$

9. $0 \div (-6)$ **10.** $0 \div (+3)$ **11.** $\dfrac{-16}{0}$ **12.** $\dfrac{+14}{0}$

13. $\dfrac{(+8)(-4)}{+2}$ **14.** $\dfrac{(-6)(+3)}{-9}$ **15.** $\dfrac{+42}{(-3)(+7)}$ **16.** $\dfrac{-54}{(+2)(-9)}$

17. $\dfrac{(-5)(+15)}{(+25)(-1)}$ **18.** $\dfrac{(-36)(+12)}{(-4)(-27)}$ **19.** $\dfrac{(+27)(-32)}{(-16)(-9)}$ **20.** $\dfrac{(+5)(-64)}{(-8)(+10)}$

21. $\dfrac{(-8)^2}{-4}$ **22.** $\dfrac{(-9)^2}{+3}$ **23.** $\dfrac{-6^2}{+3}$ **24.** $\dfrac{-10^4}{(-2)^4}$

25. $\dfrac{(+10)(-6)(-14)}{(-21)(+8)}$ **26.** $\dfrac{(-22)(+8)(+18)}{(+6)(-44)}$

27. $\dfrac{(+72)(+21)}{(-3)(+8)(-7)}$ **28.** $\dfrac{(-45)(-48)}{(+12)(-6)(+10)}$

29. $(+6)(-3) - \dfrac{-12}{+4} + \dfrac{0}{5}$ **30.** $\dfrac{(-6)(+2)}{-3} - \dfrac{(-2)^3}{+2} - (-4)$

31. $\dfrac{-16}{(-2)(-4)} - \dfrac{-5}{-1} - (-1)^2$ **32.** $\dfrac{(-2)^3}{-4} - (-7)(-2) - (-3)^2$

33. $\dfrac{24}{-3} - 4^2 + \dfrac{(+6)(-1)}{-2}$ **34.** $\dfrac{(+8)(0)}{-2} + 2^3 - \dfrac{-15}{+5}$

35. $5^2 - \dfrac{+56}{-8} - (-2)^2 + \dfrac{-12}{(-1)(+3)}$ **36.** $\dfrac{-36}{+2} - 3^2 - (-1)^2 - \dfrac{+8}{(-2)(-1)}$

In Exercises 37 through 40, set up the given problems in terms of the division of signed numbers. Then perform the operations.

37. A parachutist descended 1000 ft at a constant rate for 4 minutes. What was his change in altitude each minute?

38. The change in altitude of an object falling under gravity is determined by finding the product of -32 and the square of the time of fall and then dividing by 2. Determine the change in altitude (in feet) of an object that has fallen for 4 seconds.

39. As expected, a manufacturing company lost \$20,000 in its first year of operation. It lost \$5000 during the second half of that year. If the losses for each of the first 6 months of that year were equal, express the change in assets per month during this period numerically.

40. A woman borrowed \$5000 in an emergency. If she repaid that amount in 10 payments, express her change in assets per payment numerically.

In Exercises 41 and 42, perform the directed operations.

41. Using the numbers $+8$ and -2, show that division is not commutative. (See Example G.)

42. Calculate the values of $8 \div (4 \div 2)$ and $(8 \div 4) \div 2$, and show that division is not associative. *Calculations within parentheses should be performed first.*

1-6 THE REAL NUMBER SYSTEM; SETS

We have already discussed the basic operations on positive and negative numbers. In this section, we shall discuss various kinds of numbers that constitute what is known as the **real number system.** Also, we shall introduce some terminology that will be useful to this discussion and in certain later topics.

The most familiar group of numbers are those used for the purpose of counting objects. These are the numbers we designate by 1, 2, 3, 4, 5, and so on. They are known as the **counting numbers,** or the **natural numbers.**

For purposes of designating a particular group of numbers, we may consider them as constituting a **set.** *Formally, a set is a collection of objects, and the individual objects are* **elements,** *or* **members,** *of the set.* A set is denoted by placing braces around the elements and commas between the elements. (Although we can form sets of many kinds of objects, we shall deal only with sets of numbers.)

Example A The set of the first five natural numbers is $\{1, 2, 3, 4, 5\}$. The set of all natural numbers can be shown as $\{1, 2, 3, 4, 5, \ldots\}$, where *the three dots indicate that the numbers continue without termination.*

For convenience, so that we can refer to an entire set without always writing down all the elements, we let capital letters represent sets. Thus, if we let P be the set of all natural numbers, we may write $P = \{1, 2, 3, 4, 5, \ldots\}$.

The set of natural numbers is also known as the set of **positive integers.** *The opposites of the positive integers are the* **negative integers,** *which also form a set.* Therefore, if N is the set of negative integers, we may write $N = \{\ldots, -5, -4, -3, -2, -1\}$.

The set that contains all of the members (any given member being listed only once) of two other sets is known as the **union** *of these two sets.* The symbol $\cup$ is used to denote the union of two sets.

Example B If $A = \{1, 2, 3, 4, 5\}$ and $B = \{4, 5, 6, 9\}$, the union of sets A and B is $\{1, 2, 3, 4, 5, 6, 9\}$, since this is the set of all elements contained in either A or B. Thus, we may write $A \cup B = \{1, 2, 3, 4, 5, 6, 9\}$.

The set of all **integers,** *I is therefore now defined as* $I = \{\ldots, -5, -4, -3, -2, -1, 0, 1, 2, 3, 4, 5, \ldots\}$. Thus, we see that I is the union of sets P, N, and O, where $O = \{0\}$. From this arrangement, we also see that all of the elements of P are also elements of I. *This means that P is a* **subset** *of I, which is denoted by* $P \subseteq I$. Following the same reasoning, N and O are also subsets of I, which we can show as $N \subseteq I$ and $O \subseteq I$.

Example C If $A = \{1, 2, 3, 4, 5\}$, if $B = \{4, 5, 6, 9\}$, and if $C = \{1, 2, 3, 4, 5, 6, 7, 8, 9\}$, we see that all of the elements of A are also elements of C. Thus, $A \subseteq C$. Also, all of the elements of B are elements of C. Thus, $B \subseteq C$. However, the elements 6 and 9 in set B are not elements of A. Therefore, B is not a subset of A, or $B \nsubseteq A$.

A **rational number** *is defined as a number that may be expressed as one integer divided by another (not zero).* We see that the definition requires only that it is *possible* to write a given number in this form, even though it may not appear as such originally. Thus, since any integer can be expressed with a denominator of 1, the set of integers I is a subset of the rational numbers Q.

Example D Examples of rational numbers are

$$\frac{-3}{7} \quad \frac{6}{5} \quad \frac{0}{2} \quad \text{and} \quad \frac{112}{4137}$$

The integer 3 is a rational number since $3 = \frac{3}{1}$. Also, -5 is rational since $-5 = \frac{-5}{1}$.

There are many numbers which cannot be expressed as one integer divided by another. They are called **irrational** *numbers.* Some examples of irrational numbers are $\sqrt{2}$ (the square root of 2), $\sqrt{3}$, $\sqrt{7}$, and π (the number equaling the circumference of a circle divided by its diameter). *The union of the set of rational numbers Q and the set of irrational numbers Z is the set of the* **real** *numbers R.* Thus, $Q \cup Z = R$, $Q \subseteq R$, and $Z \subseteq R$.

The set containing all of the elements that are contained in both of two other sets is called the **intersection** *of these other sets.* The intersection of two sets is denoted by the symbol $\cap$ and is illustrated in Example E.

Example E If $A = \{1, 2, 3, 4, 5\}$ and $B = \{4, 5, 6, 9\}$, the intersection of sets A and B is $\{4, 5\}$ because 4 and 5 are the elements that appear in both set A and set B. Thus, $A \cap B = \{4, 5\}$.

Because of their definitions, the set of rational numbers Q and the set of irrational numbers Z have no elements in common. Therefore, their intersection has no elements. *A set with no elements is called the* **empty set** (*or* **null set**). It is denoted by $\{\quad\}$ or ϕ. Thus, we can state that $Q \cap Z = \phi$.

Example F Since there are no positive integers less than $+1$, we can state that the set of positive integers less than $+1$ is the empty set. In the same way, we may state that the set of integers between 3 and 4 is $\{\quad\}$, or ϕ.

We must be careful to note that the empty set has *no* elements. The set $\{0\}$ is not the empty set because it contains the element 0.

As we showed in Section 1-2, we can represent positive and negative numbers on a number scale (recall Fig. 1-1). Now we see that the units of the line are marked off by the set of integers. The other rational numbers are found between the integers. Although we cannot prove it here, the irrational numbers account for all other positions on the line. Thus, every point on the line represents some real number. A few real numbers, rational and irrational, are shown in Fig. 1-2.

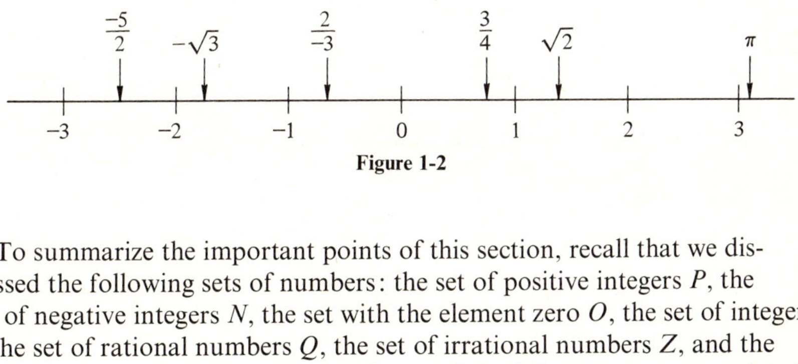

Figure 1-2

To summarize the important points of this section, recall that we discussed the following sets of numbers: the set of positive integers P, the set of negative integers N, the set with the element zero O, the set of integers I, the set of rational numbers Q, the set of irrational numbers Z, and the set of the real numbers R. We also discussed the following concepts related to sets: the union of two sets, denoted by $\cup$; the intersection of two sets, denoted by $\cap$; a subset of a set, denoted by $\subseteq$; and the empty set, denoted by $\{\quad\}$ or by ϕ. Some of the basic relationships among these sets are: $Q \cup Z = R$; $Q \cap Z = \{\quad\}$; $I \subseteq Q$; $P \subseteq I$, $N \subseteq I$, and $O \subseteq I$.

Exercises 1-6

In the following exercises, the sets P, N, O, I, Q, Z, and R are the sets we have discussed in this section.
In Exercises 1 through 8, determine the indicated sets.

1. The set of positive integers less than 8.
2. The set of positive integers between 3 and 7.
3. The set of negative integers between -5 and -1.
4. The set of negative integers greater than -3.
5. The set of integers greater than 10.
6. The set of integers less than -3.
7. The subset of the rational numbers of the set $\left\{-2, -\frac{1}{2}, \sqrt{2}, \frac{\pi}{2}, \frac{11}{3}\right\}$.
8. The subset of the rational numbers of the set $\left\{-\frac{\sqrt{3}}{2}, 0, \frac{1}{3}, \frac{9}{\pi}, 8\right\}$.

In Exercises 9 through 16, use sets $A = \{-2, 0, 1, 3, 5, 6, 7\}$, $B = \{-1, 3, 6, 9, 10\}$, $C = \{0, 1, 5, 6\}$, *and sets I, P, and* ϕ.

9. Is $B \subseteq A$? $C \subseteq A$?
10. Is $B \subseteq I$? $C \subseteq P$?
11. Find $A \cup B$.
12. Find $B \cup C$.
13. Find $A \cap B$.
14. Find $B \cap C$.
15. Find $\phi \cup C$.
16. Find $\phi \cap C$.

In Exercises 17 through 24, determine the indicated sets.

17. If $E = \{2, 4, 6, \ldots\}$, what is the relationship between sets E and P?
18. If $F = \{-6, -4, -2, 0, 2, 4, 6\}$, what is the relationship between sets F and I?
19. A **whole number** can be defined as a nonnegative integer. Express the set of whole numbers W in terms of the union of two of the sets discussed in this section.
20. Referring to the set of whole numbers as defined in Exercise 19, what set is $W \cap P$? What set is $W \cup N$?
21. From the discussion in this section, we know that $Z \subseteq R$. What set is $Z \cup R$?
22. From the discussion in this section, we know that $I \subseteq Q$. What set is $I \cap Q$?
23. Determine all of the subsets of the set $\{0, 1, 2\}$. (*Hint:* The set itself and $\{\ \ \}$ are subsets.)
24. Determine all of the subsets of the set $\{-1, 5\}$. (See Exercise 23.)

1-7 NUMBER BASES

In your past work in mathematics, you have probably not considered the properties of numbers that are basic to writing them and that lead to the methods used in performing the basic operations of addition, subtraction, multiplication, and division. In this section, we shall briefly describe these features in order to develop some understanding of the nature of numbers.

One very important feature used in writing numbers is that of **positional notation,** *which means that the actual position of any given numeral in a number is significant. The other basic feature is that of the* **base** *of the number system.* The base of the number system used in ordinary arithmetical work is ten. Example A illustrates the significance of positional notation and the base ten.

Example A We know that the number 3582 is read as "three thousand five hundred eighty (meaning eight tens)-two." We could think of it as

$$3582 = 3000 + 500 + 80 + 2$$
$$= 3(1000) + 5(100) + 8(10) + 2$$
$$= 3(10)^3 + 5(10)^2 + 8(10) + 2$$

The significance of both position and base can now be seen. The right numeral (2) indicates the number of units; the next numeral (8) indicates the number of tens; the next numeral (5) indicates the number of hundreds (or the number of times we count 10^2); the next numeral (3) represents the number of thousands (10^3). Thus, the number of times a given power of ten (the base) is counted is determined by the value and position of a given numeral. This property can be readily seen by rearranging or omitting any of the numerals. For example.

$$8235 = 8(10)^3 + 2(10)^2 + 3(10) + 5 \qquad \text{and} \qquad 358 = 3(10)^2 + 5(10) + 8$$

It is also noted here that in writing numbers in base ten, we use ten different symbols: 0, 1, 2, 3, 4, 5, 6, 7, 8, and 9. This is a fundamental characteristic of a base.

The probable reason for ten being the base of our number system is that we have ten fingers, and many ancient peoples used their fingers for counting purposes. However, many other bases have been used throughout history. The Babylonians had a number system with a base of sixty. The Pueblo Indians used a system with a base of five, and the Mayan Indians used a base of twenty. Many people, including Napoleon, have contended that twelve would be a more convenient and better base than ten. Today, the base two (the binary system) is used extensively in computer systems.

Example B illustrates several numbers written (with the meaning according to the above expression) in bases two, five, and ten.

Example B

Number	Base Two	Base Five	Base Ten
One	1	1	1
Two	10 $(2 + 0)$	2	2
Three	11 $(2 + 1)$	3	3
Four	100 $(2^2 + 0 + 0)$	4	4
Five	101 $(2^2 + 0 + 1)$	10 $(5 + 0)$	5
Six	110 $(2^2 + 2 + 0)$	11 $(5 + 1)$	6
Seven	111 $(2^2 + 2 + 1)$	12 $(5 + 2)$	7
Eight	1000 $(2^3 + 0 + 0 + 0)$	13 $(5 + 3)$	8
Nine	1001 $(2^3 + 0 + 0 + 1)$	14 $(5 + 4)$	9
Ten	1010 $(2^3 + 0 + 2 + 0)$	20 $(2 \cdot 5 + 0)$	10 $(10 + 0)$
Eleven	1011 $(2^3 + 0 + 2 + 1)$	21 $(2 \cdot 5 + 1)$	11 $(10 + 1)$
Twelve	1100 $(2^3 + 2^2 + 0 + 0)$	22 $(2 \cdot 5 + 2)$	12 $(10 + 2)$

Other numbers in bases two and five are given below.

fifteen, base two: 1111 $(2^3 + 2^2 + 2 + 1)$
fifteen, base five: 30 $(3 \cdot 5 + 0)$
thirty, base two: 11110 $(2^4 + 2^3 + 2^2 + 2 + 0)$
thirty, base five: 110 $(5^2 + 5 + 0)$
sixty, base two: 111100 $(2^5 + 2^4 + 2^3 + 2^2 + 0 + 0)$
sixty, base five: 220 $(2 \cdot 5^2 + 2 \cdot 5 + 0)$
one hundred twenty-three,
base two: 1111011 $(2^6 + 2^5 + 2^4 + 2^3 + 0 + 2 + 1)$
one hundred twenty-three,
base five: 443 $(4 \cdot 5^2 + 4 \cdot 5 + 3)$

It should be noted that the base comes out written as 10. However, this 10 means either two, five, or ten, depending upon which base is being used.

From this point, we shall denote the base of the number by writing the base, as a word, to the right and slightly below the number. For example, the number 201 in base five will be written as 201_{five}. If no base is shown, we will consider the base to be ten unless it is expressly understood to be otherwise.

We shall now show an example of how a number in base two is written in base ten. Then we shall indicate how a number in base ten may be written in base five.

Example C Change 110010_{two} into a number in base ten.

We know that 110010 means $2^5 + 2^4 + 2$, or

$$32 + 16 + 2 = 50$$

Thus, 110010_{two} equals 50_{ten}.

Example D Change 179_{ten} into a number in base five.

We look for the largest power of five which will divide into 179. Since $5^3 = 125$, we know that 125 will divide into 179 once, with a remainder of 54. Therefore, we now know that $179 = 5^3 + 54$. Next we find the next smallest power of five which will divide into 54. Since $5^2 = 25$, we know that 25 will divide into 54 twice with a remainder of 4. Thus, we now know that $179 = 5^3 + 2 \cdot 5^2 + 4$. Since five will not divide into 4 evenly or with a remainder, we have

$$179 = 1 \cdot 5^3 + 2 \cdot 5^2 + 0 \cdot 5 + 4$$

or 179_{ten} equals 1204_{five}. The general method of changing a number in base ten to a number in another base can be seen from this example.

Example E We now show additional numbers in base ten changed into base two.

	2^{10}	2^9	2^8	2^7	2^6	2^5	2^4	2^3	2^2	2^1	Units 1
Base Ten	**1024**	**512**	**256**	**128**	**64**	**32**	**16**	**8**	**4**	**2**	**1**
25							1	1	0	0	1
34						1	0	0	0	1	0
47						1	0	1	1	1	1
67					1	0	0	0	0	1	1
72					1	0	0	1	0	0	0
129				1	0	0	0	0	0	0	1

Therefore, $25_{ten} = 11001_{two}$, $34_{ten} = 100010_{two}$, and so on.

We shall conclude this section by illustrating how addition is performed in base five.

Example F Given the two numbers 324_{five} and 412_{five}, find their sum in base five.

The addition is set up in the ordinary way, that is,

$$\begin{array}{r} 324 \\ \underline{412} \end{array}$$

One must keep in mind the basic method of addition that is used in ordinary arithmetic in base ten. We proceed by adding 4 and 2, which results in six. However, in base five, six is 11, or one five and one unit. Thus, a 1 is placed below the 4 and 2, and a 1 is carried, to be added in the fives column. Here, that 1 which is carried is added to the 2 and 1 already present, giving 4. This means we have 4 fives; the 4 is placed below the 2 and 1. Moving to the 5^2 column, we add 3 and 4 to get seven. Since seven in base five is 12, we place the two under the 3 and 4 and the 1 to its left, indicating one 5^3. Thus, the result is

$$\begin{array}{r} 324 \\ \underline{412} \\ 1241 \end{array}$$

This can be checked by writing all numbers in base ten. Since 324_{five} equals 89_{ten} and 412_{five} equals 107_{ten}, the sum of 89 and 107 is 196, which is 1241_{five}. Thus, the answer checks.

Throughout this section, it can be seen that the numeral 0 is used to place the other numerals properly when a given power of the base is not present in the number. The use of 0 enables us to distinguish between numbers such as 54 and 504 and 50,004.

Writing numbers in other number systems and doing simple computations in these systems should provide a better understanding of the decimal system. Also, it should help in your understanding of similar features of ordinary arithmetic. Any difficulties you may encounter are due to your unfamiliarity with the system, not to the reasoning underlying it.

Exercises 1-7

In Exercises 1 through 4, write the given numbers in base ten in terms of powers of ten, as in Example A.

1. 352 **2.** 9402 **3.** 8617 **4.** 70,576

In Exercises 5 through 8, write the given base ten numbers
(a) as numbers in base two, and
(b) as numbers in base five.

5. 14 **6.** 20 **7.** 69 **8.** 112

In Exercises 9 through 12, write the given base two numbers as numbers in base ten.

9. 1110_{two} **10.** 10001_{two} **11.** 10101_{two} **12.** 1101101_{two}

In Exercises 13 through 16, write the given base five numbers as numbers in base ten.

13. 32_{five} **14.** 401_{five} **15.** 2114_{five} **16.** 4303_{five}

In Exercises 17 and 18, add the given numbers in base two. Check by changing the numbers to base ten.

17. 101_{two} and 111_{two} **18.** 1101_{two} and 11011_{two}

In Exercises 19 and 20, add the given numbers in base five. Check by changing the numbers to base ten.

19. 24_{five} and 213_{five} **20.** 4014_{five} and 3241_{five}

In Exercises 21 through 24, use the symbols 0, 1, 2, *and* 3 *for base four numbers.*

21. Write the base ten numbers from 1 through 20 in base four.

22. Represent the number 2103_{four} as a number in base ten.

23. Represent the number 78_{ten} as a number in base four.

24. Add the numbers 223_{four} and 132_{four} in base four. Check by changing the numbers to base ten.

In Exercises 25 through 28, use the symbols 0, 1, 2, 3, 4, 5, 6, 7, 8, 9, t, *and* e *for base twelve numbers. Note that it is necessary to introduce additional symbols in order to use a base larger than ten.*

25. Write the base ten numbers from 1 through 25 in base twelve.

26. Represent the number $2\text{e}8_{\text{twelve}}$ as a number in base ten.

27. Represent the number 298_{ten} as a number in base twelve.

28. Express the number $14\text{t}_{\text{twelve}}$ as a number in base ten.

In Exercises 29 through 32, make the indicated changes by first changing the given number to a base ten number.

29. Express the number 100101_{two} as a number in base five.

30. Express the number 324_{five} as a base two number.

31. Express the number 503_{six} as a number in base four.

32. Express the number 2131_{four} as a number in base six.

In Exercises 33 through 40, use the following information. Another method of changing a number from base ten to another base is by successive divisions of the number by the new base, the number in the new base being the sequence of remainders written in reverse order. For example, to change 134 *to base six, we have*

Divisor	Quotient	Remainder
6	134	
6	22	2
6	3	4
	0	3

Thus, 134_{ten} *is* 342_{six}.

33. Check the above result by changing 342_{six} to base ten.

34. Given that 876_{ten} is 2361_{seven}, and 321_{ten} is 1253_{six}. Thus,

7	876		6	321	
7	125	1	6	53	3
7	17	6	6	48	5
	2	3	6	8	2
				1	

Check the above results by changing 2361_{seven} and 1253_{six} to base ten numbers.

35. Change 238_{ten} to a number in base eight.

36. Change 305_{ten} to a number in base twelve.

37. Change 182_{ten} to a number in base four.

38. Change 16_{ten} to a number in base two.

39. Change 51_{ten} to a number in base eight.

40. Change 51_{ten} to a number in base nine.

1-8 SUMMARY

In this chapter, we introduced positive and negative numbers and the basic operations on them. Also, the rational number system, sets, and number bases were discussed.

We noted that the two uses of plus and minus signs should be carefully distinguished: One is for the representation of a signed number, and the other is to indicate addition or subtraction.

We established the following principles for the basic operations on positive and negative numbers: (1) We add numbers of like signs by adding the absolute values of the numbers and giving the sum the common sign. To add two numbers with unlike signs, we find the difference of their absolute values and give the difference the sign of the number of larger absolute value. (2) Subtracting a positive number is equivalent to adding a negative number of the same absolute value. Subtracting a negative number is equivalent to adding a positive number of the same absolute value. (3) The product, or quotient, of two numbers of like signs is positive, and the product, or quotient, of two numbers of unlike signs is negative.

The commutative and associative laws of addition and multiplication were illustrated. We also determined that division by zero is excluded.

In our discussion of the real number system, we noted that the set of real numbers is the union of the set of rational numbers and the set of irrational numbers. The set of integers is a subset of the set of rational numbers. A rational number is one which can be expressed as one integer divided by another integer (not zero). We also discussed sets, the union of two sets, the intersection of two sets, subsets, and the empty set.

Finally, the section on number bases explained positional value and base.

Listed below are the important terms introduced in the chapter. These should now be familiar to you. If any of them are not, the appropriate material in the chapter should be reviewed.

Absolute value
Associative laws
Base
Braces
Commutative laws
Cube (exponent)
Denominate numbers
Denominator
Difference
Dividend
Divisor
Element of set
Empty set
Exponent
Factor
Greater than
Indeterminate result
Integers
Intersection of sets
Inverse operations
Irrational numbers
Less than
Member of set
Natural numbers
Negative numbers
Null set
Numerator
Number scale
Origin
Positional notation
Positive numbers
Power
Product
Quotient
Rational numbers
Real numbers
Set
Signed numbers
Square (exponent)
Subset
Sum
Union of sets
Unsigned numbers

Exercises which should be useful in reviewing the material of the chapter are given below.

Exercises 1-8

In Exercises 1 through 8, state whether the commutative law of addition, the associative law of addition, the commutative law of multiplication, or the associative law of multiplication is illustrated.

1. $5 + (7 + 9) = (5 + 7) + 9$
2. $5 + 16 = 16 + 5$
3. $(3)(14) = (14)(3)$
4. $3 \times (4 \times 9) = (3 \times 4) \times 9$
5. $3 + (9 + 15) = (9 + 15) + 3$
6. $(4 + 12) + 6 = 4 + (12 + 6)$
7. $(5 \times 10) \times 9 = 5 \times (10 \times 9)$
8. $6 \times (2 \times 15) = (2 \times 15) \times 6$

In Exercises 9 through 12, insert the proper sign, $>$ or $<$, between the given numbers.

9. $-3 \quad +1$
10. $-2 \quad -9$
11. $|-2| \quad -3$
12. $-|-4| \quad 3$

In Exercises 13 and 14, locate the approximate positions of the given numbers on a number scale.

13. $+2.5; -\frac{5}{4}$
14. $\frac{3}{7}; -\frac{2}{9}$

In Exercises 15 and 16, find the absolute value of each of the given numbers.

15. $\frac{2}{3}; -\frac{1}{4}$
16. $-15; \frac{7}{4}$

In Exercise 17 through 52, evaluate the given expressions by performing the indicated operations.

17. $(+4) + (-6)$
18. $(-6) + (-2)$
19. $(-5) - (+8)$
20. $(-3) - (-7)$
21. $(-9)(+7)$
22. $(-8)(-12)$
23. $(-63) \div (-7)$
24. $(-72) \div (+9)$
25. $(-3) - (-9) + (+4)$
26. $(+10) + (-4) - (-7)$
27. $(+9) - (+2) - (-12)$
28. $(-6) - (+5) + (-8)$
29. $(-2)(-6)(+8)$
30. $(-1)(+4)(+7)$
31. $(+5)(+3)(-2)(+2)$
32. $(-10)(+6)(-2)(-3)$
33. $\dfrac{(-5)(-6)}{+3}$
34. $\dfrac{(-8)(+7)}{-14}$
35. $\dfrac{(-18)(+4)}{(+6)(+2)}$
36. $\dfrac{(+8)(-9)}{(-4)(-3)}$
37. $(-4)^4; -4^4$
38. $(-7)^3; -7^3$
39. $-5^3; (-5)^3$
40. $(-2)^8; -2^8$
41. $\dfrac{(-1)(-5)(+45)}{(+9)(+5)(+1)}$
42. $\dfrac{(-8)(-16)(+3)}{(-2)(+6)(-4)}$
43. $\dfrac{-6^2}{-3}$
44. $\dfrac{(-1)(+10)}{-1^4}$
45. $(-2)(+4) - \dfrac{(-6)}{(+2)} - (-5)$

46. $-(-1)(-5)+\dfrac{(-9)}{(+3)}-\dfrac{(-16)}{(+4)}$

47. $-\dfrac{(-2)(-3)}{-6}+(-4)(+2)-\dfrac{-8}{+4}$

48. $(-5)-(-2)(-5)+\dfrac{+4}{-2}$

49. $-(-7)-(-6)(+2)-(-3)^2$

50. $(-4)-\dfrac{-8}{+2}-(-2)^4$

51. $\dfrac{(-3)(-14)}{-7}+(-5)(-3)-4^2$

52. $(-11)+\dfrac{-16}{-4}-3^3-(-1)^3$

In Exercises 53 through 56, use sets $A=\{-3, 0, \frac{1}{2}, \sqrt{2}, \frac{11}{3}\}$, $B=\{-9, 0, \frac{2}{3}, 2, \frac{11}{3}\}$, *and those identified in Section 1-6.*

53. Find $A \cap B$.

54. Find $A \cup B$.

55. Is $A \subseteq Q$? Is $A \subseteq R$?

56. Is $B \subseteq Q$? Is $B \subseteq I$?

In Exercises 57 through 60, the symbols used earlier for base two and base five numbers are to be used.

57. Express 1011101_{two} as a number in base ten.
58. Express 4032_{five} as a number in base ten.
59. Express 182_{ten} as a number in base five.
60. Express 100_{ten} as a number in base two.

In Exercises 61 through 64, use the set of symbols $\{0, 1, 2, 3, 4, 5, 6, 7\}$ *for writing base eight numbers.*

61. Express 362_{eight} as a number in base ten.
62. Express 362_{ten} as a number in base eight.
63. Express 25_{eight} as a number in base two.
64. Add the numbers 657_{eight} and 241_{eight} in base eight. Check the results by changing the numbers to base ten.

In Exercises 65 through 90, solve the given problems and answer the given questions. Any problem involving operations on signed numbers should first be properly set up in terms of these operations.

65. In a chemistry book, look up the word "valence." You will see that signed numbers are used to indicate that an atom has extra (in a special sense) electrons or that it lacks electrons. What is the valence of sodium? Of chlorine?
66. If the loss on the sale of an item was negative, what conclusion can be drawn?
67. If 3 s after the launching of a spacecraft is shown as $+3$, how should the instant 10 s *before* blastoff be shown? (When watching this type of event on television, note the use of signed numbers.)
68. If 20 gal of water is pumped out of a tank, what is the net change of the volume of water in the tank?
69. If 17 min before the hour is denoted by -17, how would you indicate 18 min past the hour?
70. If driving forward by 50 m is denoted by $+50$ m, how would driving backward by 20 m be shown?

71. In a golf tournament, a golfer was 2 strokes over par in the first round, 1 stroke under par in the second round, 3 strokes under par in the third round, and 2 strokes under par for the fourth round. What was his net score relative to par for the tournament? (If you watch these events on television, you will see signed numbers used in calculating scores.)

72. In three successive months in a certain area, the cost of living decreased 0.1 percent, increased 0.3 percent, and decreased 0.4 percent. What was the net change in the cost of living for these three months?

73. The boiling point of mercury is about 357°C, and its melting point is about −39°C. What is the difference in the boiling and melting points of mercury?

74. After the first half of a basketball game, one team was 5 points behind. They eventually won the game by 12 points. By how many points did they outscore their opponents in the second half?

75. After reaching the bottom of a steep incline, a skier slows down at a rate of 2 ft/s each second. What is the net change in her speed 10 s later?

76. A jet climbs at the rate of 200 m/min for 10 min. It then descends for 6 min at the rate of 500 m/min. What is its net change in altitude?

77. Death Valley is 276 ft below sea level and Mt. Whitney is 14,502 ft above sea level. What is the vertical distance from the bottom of Death Valley to the top of Mt. Whitney?

78. Mr. Rich has $350 in the bank. If he withdraws $25 each week for the next 6 weeks, what will his balance be? (Set up and solve using a withdrawal as a negative number and future time as positive.)

79. Mrs. Mountain has $95 in the bank. If she has withdrawn $5 each week for the past 6 weeks, what was her balance before the first withdrawal? (See note after Exercise 78.)

80. A rocket was traveling at 10,000 km/h. Its speed then increased 1000 km/h for each of the following 7 h. Then its speed decreased 500 km/h for each of the next 5 h. What was its speed after these changes? (Use an increase in velocity as a positive number and solve by using signed numbers.)

81. The temperature in a certain city is −5°C. For the past 4 h, the temperature decreased 2°C each hour. For the 3 h before that, it rose 3°C each hour. What was the temperature 7 h earlier? (Use an increase in temperature as positive and past time as negative. Solve by using signed numbers.)

82. On a combination lock, assume that clockwise movement is positive and counterclockwise movement is negative. A certain such lock has 50 notches, and the first two parts of the combination are R12, L26. How many notches are moved between the settings?

83. Determine the set of all integers less than 4.

84. Determine the set of all even integers.

85. What is the set of rational numbers that can be formed from the elements of the set $\{1, 2, 3\}$?

86. What is the set of rational numbers that can be formed from the elements of the set $\{1, 2, 4\}$?

87. Look up the words "number" and "numeral" in a dictionary. How are they distinguished?

88. What is the basic inconvenience of using a number system with a large base, say sixty, with positional notation?

89. Is $|(-4)+(+3)| = |(-4)| + |(+3)|$?
Is $|(-4)+(-3)| = |(-4)| + |(-3)|$?

90. Is $|(-1)+(-2)-(-5)| = |(-1)| + |(-2)| - |(-5)|$?

2

INTRODUCTION TO ALGEBRA

2-1 LITERAL NUMBERS AND FORMULAS

Now that we have established the basic operations of signed numbers, we can start the development of algebra. In this chapter, we shall explore the meaning of algebra and some of the important terminology associated with it. We shall also examine the basic operations, although they will be more fully developed in later chapters.

Essentially algebra is a generalization of arithmetic which uses letters to represent numbers. The various other symbols, such as the plus sign and the equals sign, are used to represent the various operations and relations among them. We shall now see why and how letters are used in algebra.

If we wished to find the cost of wall-to-wall carpeting for a given room, we would multiply the area, say in square yards, by the cost of carpeting per square yard. If the room is rectangular, we find the area by multiplying the length by the width. Therefore, it is possible to say that the cost equals the length (in yards) times the width (in yards) times the cost of carpeting (per square yard).

Rather than writing out such statements as the one just given, we can write

$$C = l \cdot w \cdot c$$

where it is specified and understood that C is the total cost, l is the length of the room, w is the width of the room, and c is the cost per unit area of the carpeting.

The above **formula** has been determined by writing an algebraic expression for a given statement. In this way, we may look upon algebra as a *language. It is a language in which statements involving numbers can be written in a simplified form and with a general meaning.*

If we wish to find the cost of carpeting for a particular room, we need only to **substitute** *the proper numbers into the formula.* Although the formula remains the same, the values obtained by using it may vary, depending on the values used for the various letters.

Example A If a given room is 18 ft (6 yd) long and 12 ft (4 yd) wide, and the cost of a certain type of carpeting is \$10 per square yard, we have

$$l = 6 \text{ yd}, \qquad w = 4 \text{ yd}, \quad \text{and} \quad c = \$10 \text{ per square yard}$$

Thus,

$$C = l \cdot w \cdot c$$

becomes

$$\begin{aligned} C &= 6 \cdot 4 \cdot 10 \\ &= \$240 \end{aligned}$$

It would cost \$240 to use that carpeting in the room.

In developing the language of algebra for the operations on **literal numbers** (*the letters which represent numbers*), we use certain procedures and terms in writing algebraic expressions. These are adopted because they make statements and notation concise and convenient. Many of these procedures and terms will be introduced as various topics are developed. Since they are used throughout the development of mathematics, it is important that the precise meaning of each be understood as it is introduced.

At this point, we shall discuss basic notation and terminology involving numbers being multiplied. As we stated in our discussion of the operations on signed numbers, *numbers to be multiplied together to form a product are known as* **factors** *of the product*. The product of algebraic factors is expressed without using the signs of multiplication. The literal numbers are placed adjacent to each other, as we placed signed numbers to indicate multiplication. *One of these may be an explicitly stated number, in which case it should be written first; it is called the* **numerical coefficient** *of the expression.*

Example B The expression for the cost of the carpeting given earlier would be written as

$$C = lwc$$

If every case of soda in a given shipment contains 12 bottles, then the total number of bottles b is

$$b = 12c$$

where c is the number of cases. Here, the 12 is the numerical coefficient of c.

If a given number appears in a product more than once, we use exponents to denote the number of times it appears as a factor. Exponents are used here in just the same manner as with signed numbers.

Example C Instead of writing aa, meaning a times a, we write a^2. This is the algebraic expression for "a squared."

In the same way, b^3 ("b cubed") means bbb, and x^5 ("the fifth power of x") means $xxxxx$.

The expression $5x^2y^3$ means five times the square of x times the cube of y.

If we wish to group letters and raise the entire group to a power, we use parentheses: $(axy)^2$ means axy times axy.

In a given product, the numbers multiplying any given factor constitute the **coefficient** *of that factor. This is illustrated in Example D.*

Example D In the expression lwc, each of l, w, and c are factors. The coefficient of c is lw and the coefficient of lw is c.

In the expression $3ab^2$, the 3, a, and b^2 are factors. The coefficient of b^2 is $3a$ and the coefficient of ab^2 is 3.

Wherever mathematics may be used, we may choose symbols to represent the quantities involved. A verbal statement must be translated into an algebraic expression in order that it may be appropriately used in algebra. The following example illustrates several verbal statements and their equivalent algebraic formulas. Note the use of literal numbers and the conciseness of the formulas. You should be familiar with most of these.

Example E

Verbal Statement	Formula	Meaning of Literal Symbols
1. The area of a circle is π times the square of the radius of the circle.	$A = \pi r^2$	A is the area. $\pi = 3.14$ (approx.). r is the radius.
2. The volume of a cube equals the cube of the length of one of the edges of the cube.	$V = e^3$	V is the volume. e is the length of an edge.
3. The number of tons of air pollutants emitted is 0.02 times the number of tons of a particular fuel burned.	$p = 0.02n$	p is the number of tons emitted. n is the number of tons burned.
4. The distance an object travels equals the average speed times the time traveled.	$d = rt$	d is the distance. r is the speed. t is the elapsed time
5. The simple interest earned by a principal equals the principal times the rate of interest times the time the money is invested.	$I = Prt$	I is the interest. P is the principal. r is the rate of interest. t is the time
6. The voltage across a resistance in an electric circuit equals the current in the circuit times the resistance.	$V = IR$	V is the voltage. I is the current. R is the resistance.

These formulas are valid under the conditions specified. As mentioned before, *if we wish to determine the result of using any of these formulas for specific values, these values are* **substituted** *for the appropriate letters into the formula, and the result is then calculated.* The following example illustrates this in the case of some formulas.

Example F The volume of a cube 3 in. on an edge is found by substituting 3 for e in the formula $V = e^3$. This gives

$$V = 3^3 = 27 \text{ in.}^3$$

Therefore, if the cube had been 3 in. on an edge, the volume would have been 27 in.^{3} Note the use of exponents with the units to denote cubic inches. Again, refer to Appendix B on the use of units and the symbols used to represent various units.

If the current in a certain electric circuit is 3 A (amperes), the voltage across a 6-Ω (ohm) resistor in the circuit is found by substituting 3 for I and 6 for R in the formula $V = IR$. This gives

$$V = 3 \cdot 6 = 18 \text{ V (volts)}$$

You should again note that the answer is labeled with the proper units. Also, note that the letter V has a different meaning in each of these cases.

Exercises 2-1

In Exercises 1 through 8, state the products of the listed factors.

1. a, b **2.** x, y, z **3.** x, x

4. a, a, a, a **5.** π, r, r **6.** $3, a, a, c$

7. a, a, a, b, b **8.** $3, 2, a, a, b, c, c, c$

In Exercises 9 through 16, identify the indivdual factors, without exponents, of the given algebraic expressions. (List them in the same way as shown for Exercises 1–8.)

9. bc **10.** $2ax$ **11.** $7pqr$ **12.** $3xyz$

13. i^2R **14.** $17a^2b$ **15.** abc^3 **16.** πr^2h

In Exercises 17 through 24, identify the coefficient of the factor that is listed second by examining the expression that is listed first.

17. $6x, x$ **18.** $3s^2, s^2$ **19.** $2\pi r, r$ **20.** $8a^2b, b$

21. $4\pi emr, mr$ **22.** $qBLD, D$ **23.** mr^2w^2, r^2 **24.** $36a^2cd, a^2d$

In Exercises 25 through 36, using the literal symbols indicated after each statement, translate the given statements into algebraic formulas.

25. A first number is twice a second number. (x, y)

26. A first number equals the square of a second number. (x, y)

27. In a given length, the number of inches equals 12 times the number of feet. (i, f)

28. The total surface area of a cube equals 6 times the square of the length of one of its edges. (A, e)

29. The total increase in voter registration in a certain city is 1000 times the number of months after the first of the year. (i, m)

30. The weekly sales of a certain record was 800 times the square of the number of weeks after it was released (for the first ten weeks). (S, w)

31. The volume, in gallons, of a rectangular container equals 7.5 times the length times the width times the depth (these dimensions are measured in feet). (V, l, w, d)

32. The cost of a piece of sheet metal equals the area of the piece times the cost per unit area of the metal. (C, A, c)

33. The number of bacteria in a certain culture equals 8 times the product of the number of hours after the culture was started and the cube of the original number of bacteria. (N, t, n)
34. The heat developed in a resistor in an electric circuit equals the product of the resistance and the square of the current in the circuit. (H, R, i)
35. The distance an object falls due to gravity equals $\frac{1}{2}$ times the acceleration due to gravity times the square of the time of fall. (s, g, t)
36. The rate of emission of energy per unit area of the filament of an electric light bulb equals about 0.00002 times the fourth power of the thermodynamic temperature of the filament. (R, T)

In Exercises 37 through 40, determine the required formulas.

37. The area of a square of side s.
38. The number of square feet in a rectangular area x yd by y yd.
39. The cost of filling a cubical box, each edge of which is s ft, with a liquid costing c cents/ft^3.
40. The number of students there are in a school if there are g grades in the school, n classrooms for each grade, and an average of 25 students in each classroom.

In Exercises 41 through 52, evaluate the required formula for the given values. See Appendix B with reference to the units and the symbols used to represent them.

41. Evaluate $3a^2b$ for $a = 3$ and $b = -2$.
42. Evaluate $2x^3y^2$ for $x = 4$ and $y = -3$.
43. Find the distance traveled by a rocket if its average speed is 25,000 km/h and it travels for 3 days. (See Example E.)
44. Find the simple interest on \$500 at an interest rate of 4% for 3 years. (See Example E.)
45. Find the voltage in a circuit in which the current is 0.075 A and the resistance is 20 Ω. (See Example E.)
46. Find the volume of a cube for which the edge is 4.3 cm. (See Example E.)
47. Find the number of inches in a metal rod which is 16 ft long. (See Exercise 27.)
48. What was the increase in voter registration at the end of October in the city referred to in Exercise 29?
49. The gasoline tank of a certain truck is 1 ft by 2 ft by 6 ft. Find its capacity in gallons. (See Exercise 31.)
50. Find the number of bacteria in the culture described in Exercise 33 after 2 hours, if the original number of bacteria was 1000.
51. Find the heat (in joules) developed in a 12-Ω resistor if the current in the circuit is 2.0 A. (See Exercise 34.)
52. In Exercise 35, find s if $g = 32.2$ ft/s^2 and $t = 6.50$ s.

2-2 BASIC ALGEBRAIC EXPRESSIONS

In the preceding section, we introduced the ideas of using literal numbers, and we used them in certain simple formulas. However, we mentioned only the multiplication of literal numbers. Of course, it is often necessary to add, subtract, multiply, divide, and perform other operations on literal numbers and expressions containing them. An introduction to these operations is presented in the following section. Here we shall establish the need for such operations and introduce additional terminology which will be used in discussions involving algebraic expressions.

A great many algebraic expressions arise through the addition or subtraction of quantities. Example A gives some elementary expressions using addition and subtraction.

Example A The perimeter (distance around) of a triangle with sides a, b, and c is given by the formula

$$p = a + b + c$$

The net profit P from a sale of articles from which the income is I and which originally cost C is

$$P = I - C$$

The total surface area A of a rectangular box of dimensions l, w, and h is

$$A = 2lw + 2lh + 2wh$$

If an algebraic expression consists of several parts connected by plus and minus signs, each part is called a **term** *of the expression. If the literal part of one term is the same as the literal part of another term, then the terms are called* **like terms,** *or* **similar terms.** Example B illustrates like terms.

Example B The algebraic expression $x + 3xy - 2x$ contains three terms: x, $3xy$, and $-2x$. The terms x and $-2x$ are like terms because the literal portions are the same. The $3xy$ is not included since the factor y makes the literal portion different.

The expression $3y^2 - 2xy + y^2 + xy$ has four terms: $3y^2$, $-2xy$, y^2, and xy. The terms $3y^2$ and y^2 are like terms, and the terms $-2xy$ and xy are like terms.

In the basic operations on algebraic expressions, we shall be dealing with various numbers of terms. Specific names have been given to some of these expressions so that one can conveniently refer to them.

A **monomial** *is an expression with one term. A* **binomial** *is an expression with two terms. A* **trinomial** *is an expression with three terms. A* **multinomial** *is an expression with more than one term.* Thus, binomials and trinomials can also be considered as multinomials.

Example C The following table gives several examples of monomials, binomials, trinomials, and multinomials.

	Multinomials		
Monomials	Binomials	Trinomials	Other multinomials
a^2x	$a^2x + y$	$a^2x + y + x$	$a^2x + y + x + 3a$
$3abc$	$3abc + bc$	$3abc + bc - ab$	$3abc + bc - ab + a - 2b$
a	$a + c^2$	$a + c^2 + a^2$	$a + c^2 + a^2 - 2b + a^3 - c$

Turning our attention to multiplication, we recall that, in the preceding section, we discussed products such as lwc and $3x^2y$. However, if we want to multiply two algebraic expressions, one of which is itself a sum of terms or the difference of terms, we need to be able to indicate that the sum or difference is to be considered as one of the factors. For this purpose, we use **parentheses** to *group* the terms together. Example D illustrates this use of parentheses.

Example D A television dealer charges p dollars for a given set. If she paid c dollars for the set, the profit she has on the sale of the set is $p - c$ dollars. If she sells n sets, her total profit is

$$P = n(p - c)$$

We do *not* write this as $np - c$, for such an arrangement would indicate that n and p are to be multiplied and c is to be subtracted from the product np.

To indicate that the sum of two numbers a and b is to be multiplied by $c - d$, we write

$$(a + b)(c - d)$$

This expression is a *one-term* expression, although each of the two factors contains two terms.

In algebra, division of one number by another is most commonly designated as a fraction. This is similar to the first chapter's work with signed numbers. Therefore, $a \div b$ would be written as $\frac{a}{b}$ or occasionally as a/b. In Example E, use of the division of algebraic expressions is illustrated.

Example E An individual's I.Q. (intelligence quotient) is defined as 100 times his mental age (M) divided by his chronological age (C). This can be written as

$$\text{I.Q.} = \frac{100M}{C}$$

In a particular type of electric circuit, the combined resistance C of two resistors r and R is found to be

$$C = \frac{rR}{r + R}$$

Here we note that the numerator is the product of r and R, and the denominator is the sum of r and R.

Another important operation on a number is that of finding its **square root.** *The square root of a given number is that number which, when squared, equals the given number.* The square root of a number N is designated by $\sqrt{N}$. *The symbol* $\sqrt{\ }$ *is called the* **radical sign.** The square root can be shown as

$$\boxed{\sqrt{N} = a \quad \text{if} \quad a^2 = N} \qquad (2\text{-}1)$$

At this time, we are considering only positive numbers for N and a. Methods of actually calculating square roots will be considered later, although a table of square roots appears in Appendix C.

Example F $\sqrt{4} = 2$, since $2^2 = 4$. $\sqrt{100} = 10$, since $10^2 = 100$. $\sqrt{676} = 26$, since $26^2 = 676$. $\sqrt{50} = 7.07$ (approximately).

Example G One application of square roots is the determination of the time it takes an object to fall. Thus, the approximate time (in seconds) an object falls under the influence of gravity equals the square root of the quotient of the distance (in feet) fallen divided by 16. A formula for this is

$$t = \sqrt{\frac{d}{16}}$$

As we pointed out in the preceding section, an algebraic expression can be evaluated for a given set of values by substituting the proper values into the expression. Example H illustrates the evaluation of some of the expressions given in earlier examples.

Example H In Example A, if $a = 5$ cm, $b = 7$ cm, and $c = 11$ cm, the perimeter

$$p = a + b + c$$

of the triangle becomes

$$p = 5 + 7 + 11 = 23 \text{ cm}$$

In Example D, if $p = \$210$, $c = \$120$, and $n = 6$, the profit on these sales,

$$P = n(p - c)$$

becomes

$$P = 6(210 - 120) = 6(90) = \$540$$

In Example G, if an object falls 144 ft, the time of fall,

$$t = \sqrt{\frac{d}{16}}$$

becomes

$$t = \sqrt{\frac{144}{16}} = \sqrt{9} = 3 \text{ s}$$

Exercises 2-2

In Exercises 1 through 4, identify the terms of the given algebraic expressions.

1. $a^2 - 3ab + 7b$

2. $x^2 + 2xy - \dfrac{y}{x}$

3. $3 - 6ab + 7a - \dfrac{8}{b}$

4. $3(x + y) - 6a + 3x$

In Exercises 5 through 12, identify the like terms in the given expressions.

5. $x - 3y + 2xy + 2x$

6. $3a - 6b - 5a + 2b + ab$

7. $3x^3 - 5x^3 + 7x^2 - x^3 + \dfrac{1}{x}$

8. $\dfrac{m}{n} - \dfrac{2m}{n} + mn - m^2$

9. $2m^2 - 2mn + m^2n - mn + \dfrac{m}{n}$

10. $3dx - 4d^2x - dx + dx^2$

11. $2k^2x - ky^2 + 3ky + 2ky^2$

12. $3(a - b) - 5x - 3y + (a - b)$

In Exercises 13 through 16, indicate the multiplication of the given factors.

13. $6, a, a - x$ **14.** $3, x^2, a - x^2$

15. $x^2, a - x, a + x$ **16.** $3, a^2, a^2 + x^2, a^2 - x^2$

In Exercises 17 through 20, express the indicated divisions as fractions.

17. $2 \div 5$ **18.** $x^2 \div a$

19. $6 \div (a - b)$ **20.** $(3x^2 - 2x + 5) \div (x + 2)$

In Exercises 21 through 24, find the indicated square roots.

21. $\sqrt{36}$ **22.** $\sqrt{121}$ **23.** $\sqrt{576}$ **24.** $\sqrt{900}$

In Exercises 25 through 36, write the required expression or formula.

25. Express the perimeter p of a rectangle in terms of its length l and width w.

26. The total amount of a bank account receiving simple interest is the sum of the principal and interest. Express the total amount A in terms of the principal p, the rate of interest r, and the time t the account has been in effect.

27. The weight, in milligrams, of a bacterial culture was found to be 100 plus 5 times the square of the time in hours after the culture was started. Express the weight w in terms of the time t.

28. If a piece x ft long is cut from a board 20 ft long, express the length of the remaining piece in terms of x.

29. The total surface area A of a right circular cylinder equals the product of $2\pi r$ and the sum of r and h, where r is the radius and h the height of the cylinder. Express this statement as a formula.

30. An analysis of research data showed that the percent of persons who responded to a particular stimulus was 10 times the difference of the stimulus strength less 1. Express the percent p of persons responding in terms of the stimulus strength s.

31. The arithmetic average A of n numbers is the sum of these numbers divided by n. Express a formula for the arithmetic average of the numbers a, b, c, d, and e.

32. The value V of a car depreciates such that its value after t years is its original value p divided by the sum of t and 1. Express this statement as a formula.

33. The time T for one complete oscillation of a pendulum equals 2π times the square root of the quotient of the length of the pendulum and the acceleration g due to gravity. Find the resulting formula.

34. The length c of the hypotenuse (side opposite the right angle) of a right triangle equals the square root of the sum of the squares of the other two sides a and b. This is known as the Pythagorean theorem. Express the formula for the Pythagorean theorem.

35. The Celsius temperature C equals $\frac{5}{9}$ times the difference of the Fahrenheit temperature F and 32 (subtract 32 from degrees F). Find the formula for the Celsius temperature in terms of the Fahrenheit temperature.

36. The efficiency E of an engine is defined as the difference of the heat input I and heat output P (subtract P from I) divided by the heat input. Express this statement as a formula.

In Exercises 37 through 46, evaluate the required formula for the given values.

37. Find the perimeter of a rectangle of length 6 in. and width 4 in. (See Exercise 25.)

38. Find the weight of the bacterial culture described in Exercise 27 after 3 h.

39. Find the total area of the right circular cylinder whose radius is 2 m and height is 4 m. Use $\pi = 3.14$. (See Exercise 29.)

40. A student took five tests in a certain course. The grades were 87, 73, 90, 84, and 71. Find the average of these grades. (See Exercise 31.)

41. Find the total amount of the bank account of an individual who invested $500 for 4 years at the rate of 6 percent per year. (See Exercise 26.)

42. If a piece 6 ft long is cut from a board 4 yd long, what is the length of the remaining piece? (See Exercises 28.)

43. Find the average of these grades: 88, 91, 65, 78, 70, 82. (See Exercise 31.)

44. What is the value of a $4000 car after it has depreciated for 3 years? (Use the information from Exercise 32.)

45. Find c in Exercise 34 if $a = 3$ ft and $b = 4$ ft.

46. Find c if $a = 12$ mm and $b = 5$ mm in Exercise 34.

2-3 AN INTRODUCTION TO ALGEBRAIC OPERATIONS

In this section, we shall establish certain basic algebraic operations. The addition and subtraction of algebraic terms will be shown. Then certain operations involving multiplication and division will be demonstrated. We shall find these operations necessary in the development of the sections which follow. However, a more complete and detailed discussion of algebraic operations will be found in Chapter 4.

If a multinomial contains like as well as unlike terms, it may be **simplified** *by combining the coefficients of the like terms.* The result is expressed as a sum of unlike terms. Example A illustrates the simplification of a multinomial.

Example A In the multinomial

$$3a^2 - 2ab + a^2 - b + 4ab$$

the $3a^2$ and a^2 are like terms, and the $-2ab$ and $4ab$ are like terms. Adding $3a^2$ and a^2, we obtain $4a^2$ (the a^2 term has a coefficient of 1, which is not written). Using the principles for the addition and subtraction of signed numbers, we combine $-2ab$ and $4ab$ to obtain $2ab$. Thus, the given multinomial simplifies to

$$4a^2 + 2ab - b$$

Since there are no like terms in this result, we cannot simplify the answer further. Thus, we have

$$3a^2 - 2ab + a^2 - b + 4ab = 4a^2 + 2ab - b$$

In doing the above problem, we made use of the commutative and associative laws of addition, which were introduced in the first chapter. In order to combine like terms, we should actually first rearrange them in order, thus making use of the commutative law. When combining terms in specific groupings, we are using the associative law. Using algebraic notation, *we can now state the* **commutative law of addition** *and the* **associative law of addition,** *respectively, as*

$$\boxed{a + b = b + a} \tag{2-2}$$

and

$$\boxed{a + (b + c) = (a + b) + c} \tag{2-3}$$

Example B illustrates how these laws are actually used in simplifying a multinomial.

Example B The multinomial

$$x^2 + 5x - 3 + 4x + 5$$

is equivalent to

$$x^2 + 5x + 4x - 3 + 5$$

if we interchange the third and fourth terms according to the commutative law. Now grouping the second and third terms together and then grouping the fourth and fifth terms together, according to the associative law, we have

$$x^2 + (5x + 4x) + (-3 + 5)$$

Combining like terms, we obtain

$$x^2 + 9x + 2$$

In general, one does not have to rewrite such expressions in order to add. However, the fact that we can add like terms without concern is due to these laws. Thus,

$$x^2 + 5x - 3 + 4x + 5 = x^2 + 9x + 2$$

Turning our attention to multiplication, we restrict our consideration at this time to the multiplying of certain multinomials by monomials. In order to perform this operation, *we define the* **distributive law,** *which is*

$$\boxed{a(b + c) = ab + ac} \tag{2-4}$$

That this law is valid in arithmetic can be seen from examples such as

$$3(5 + 7) = 3 \cdot 5 + 3 \cdot 7 = 15 + 21 = 36$$

Since $3(5 + 7)$ also equals $3(12) = 36$, we verify the validity of the distributive law. Illustrations of the use of the distributive law are shown in Example C.

Example C

$$3(x + y) = 3x + 3y$$

$$x(3 + 2y) = 3x + 2xy$$

$$2x(x - y) = 2x(x) + 2x(-y) = 2x^2 - 2xy$$

$$-3ab(2a - 5bc) = (-3ab)(2a) + (-3ab)(-5bc) = -6a^2b + 15ab^2c$$

Note that the principles established for the multiplication of signed numbers are followed in using the distributive law.

It should also be noted that the commutative and associative laws of multiplication are used in rearranging and grouping factors. *Algebraically, the* **commutative law of multiplication** *and the* **associative law of multiplication,** *respectively, can be stated as*

$$\boxed{ab = ba} \qquad (2\text{-}5)$$

and

$$\boxed{a(bc) = (ab)c} \qquad (2\text{-}6)$$

Following is an example of the simplification of an expression involving the distributive law and the combining of like terms.

Example D Simplify: $3s + 2(s + 2t) - s(7 - 2t)$.

First removing the parentheses by use of the distributive law and then combining like terms, we have the following solution:

$$\begin{aligned} 3s + 2(s + 2t) - s(7 - 2t) &= 3s + 2s + 4t - 7s + 2st \\ &= -2s + 4t + 2st \end{aligned}$$

The like terms we combined were $3s$, $2s$, and $-7s$. The $4t$ term and the $2st$ term are not similar to the other terms.

Another result of the use of the commutative law of multiplication is that terms which may not appear to be like terms are in fact like terms. Consider the following example.

Example E In simplifying $3xy + 2y(1 - x)$, we obtain $3xy + 2y - 2yx$. Now we must recognize that $3xy$ and $-2yx$ are actually like terms since $xy = yx$. Thus,

$$\begin{aligned} 3xy + 2y(1 - x) &= 3xy + 2y - 2yx \\ &= 3xy + 2y - 2xy \\ &= xy + 2y \end{aligned}$$

In the division of algebraic expressions, we shall restrict our attention at this time to simple cases of division by a monomial. The principle to be followed here is the same as in the simplification of an arithmetic fraction.

Factors common to both numerator and denominator can be divided out.

We followed this principle in simplifying fractions involving positive and negative numbers in Chapter 1. In Example F, division involving algebraic factors is illustrated.

Example F In simplifying the algebraic fraction

$$\frac{3xy}{ax}$$

we note that both the numerator and denominator have a factor of x. Thus, dividing out this common factor of x, we have

$$\frac{3xy}{ax} = \frac{3y}{a}$$

In dividing out the factor of x, we are stating that $x/x = 1$, and we know that multiplying the remaining factors by 1 does not change the resulting product.

In the following illustration, the resulting denominator is 1, and in this case it need not be shown.

$$\frac{3xy}{x} = \frac{3y}{1} = 3y$$

However, if the resulting *numerator* is 1, it must be retained, as in the following illustration:

$$\frac{x}{ax} = \frac{1}{a}$$

In Example G, additional illustrations of dividing out factors common to both the numerator and denominator of a fraction are shown.

Example G

$$\frac{a}{10abc} = \frac{1}{10bc} \qquad \text{(common factor of } a\text{)}$$

$$\frac{4x}{6xyz} = \frac{2}{3yz} \qquad \text{(common factor of } 2x\text{)}$$

$$\frac{3x^2y}{x} = \frac{3xxy}{x} = \frac{3xy}{1} = 3xy \qquad \text{(common factor of } x\text{)}$$

$$\frac{5a^2bc}{10ab} = \frac{5aabc}{2(5)ab} = \frac{ac}{2} \qquad \text{(common factor of } 5ab\text{)}$$

We shall conclude this section with an example that combines the various operations presented in the section.

Example H Evaluate $3a + \dfrac{2ab}{b} + 5a(3 - 2b)$ for $a = -5$ and $b = 7$.

It would be possible to substitute immediately in order to evaluate the expression. However, it almost always makes the evaluation simpler if

the basic algebraic operations are performed first.

Thus,

$$\begin{aligned} 3a + \frac{2ab}{b} + 5a(3 - 2b) &= 3a + 2a + 15a - 10ab \\ &= 20a - 10ab \end{aligned}$$

Evaluating for $a = -5$ and $b = 7$, we have

$$\begin{aligned} 20a - 10ab &= 20(-5) - 10(-5)(7) \\ &= -100 + 350 \\ &= 250 \end{aligned}$$

Exercises 2-3

In Exercises 1 through 28, simplify the given expressions. In other words, perform any indicated multiplications and divisions and then combine like terms.

1. $a - b + 2a$ **2.** $3x + 5xy - x$ **3.** $3a - 5b^2 + 4a + b^2$

4. $2x - y + 5x - y$ **5.** $2s + 3t - s + 4s$ **6.** $-5m + 2n - m + 8m$

7. $5(x + y)$ **8.** $6(s + t)$ **9.** $3(2a - b)$

10. $7(x - 3y)$ **11.** $\dfrac{7xy}{2y}$ **12.** $\dfrac{4abc}{5b}$

13. $\dfrac{x}{2x^2y}$ **14.** $\dfrac{3a}{6ab^2}$ **15.** $3(a - 5b) + 4a$

16. $4(x + 2y) - 6x$ **17.** $2(3x - y) + 3(2x + y)$ **18.** $4a(3 - b) - b(a - 4)$

19. $\dfrac{ax}{a} + 3x$ **20.** $\dfrac{2y}{2} - 4y$ **21.** $\dfrac{2rs^2}{s} - 3r(5 - s)$

22. $\dfrac{3a^2b}{ab} - 2a(3 - b)$ **23.** $\dfrac{6abc^2}{2c} + c(ab - c) + b(ac - 1)$

24. $\dfrac{12mx^3}{3x} - \dfrac{2km^2x^2}{km} - 3mx(2 - x)$ **25.** $2(ax - 2) + \dfrac{9a^2x}{3a} + \dfrac{x}{x}$

26. $5(3a^2 - 2) - \dfrac{15a^3}{3a} + \dfrac{a^2}{a^2}$ **27.** $\dfrac{16a(x - y)}{4a} - \dfrac{3x^2y}{3xy} + 2(x - 3y)$

28. $\dfrac{24x^2(b + c)}{6x} + \dfrac{2b^2x}{2b} - \dfrac{50c^3x^2}{25c^2x}$

In Exercises 29 through 36, express the appropriate expressions in simplest form.

29. Two newspapers print $5x + 1000$ and $4x - 2000$ papers, respectively, each day. What is the sum of their daily printings?

30. One car goes 30 mi/h for t hours and a second car goes 40 mi/h for $t + 2$ hours. Find the expression for the sum of the distances traveled by the two cars.

31. Find an expression for the value, in cents, of $2a + 1$ nickels and $a + b$ dimes.

32. A carpenter cuts $2x$ ft from each of six 20-ft boards and x ft from each of four y-ft boards. Find the expression for the total length of the remaining pieces.

33. One train goes 45 km/h for t hours, and a second train goes 55 km/h for $t + 2$ hours. Find the expression for the sum of the distances traveled by the two trains.

34. Two magazines print $5(x + 246)$ and $4x - 2145$ copies, respectively, each month. What is the sum of their monthly printings?

35. One rectangular area has dimensions x and y, another $x + 2$ and y, and a third 3 and $y + 5$. What is the expression for the sum of the areas?

36. A boat travels $9axy$ mi, and its average fuel consumption is $3x$ mi/gal. How many gallons are used?

In Exercises 37 through 44, evaluate the required expressions for the given values. In each case, evaluate the simplified expression.

37. $3a + 6a - 2b + a + b$ for $a = 9$ and $b = 4$

38. $5x^2 - y - 3x^2 + 8y$ for $x = 5$ and $y = -3$

39. $6a(x + 2y) + \dfrac{4a^2x}{a} - 7ay$ for $a = -2$, $x = 8$, and $y = 9$

40. $\dfrac{15a^2bc}{3a} + a(bc + a) - \dfrac{3ab^2c}{b}$ for $a = 2$, $b = 3$, and $c = 18$

41. If $a = 8$ and $b = 5$, what is the total value of the coins of Exercise 31?

42. If $x = 3$ and $y = 12$, what was the total length of the remaining pieces of board in Exercise 32?

43. If $t = 2.75$ h, find the total distance traveled by the two trains in Exercise 33.

44. If $x = 10$ cm and $y = 15$ cm, find the sum of the areas of the rectangles in Exercise 35.

In Exercises 45 through 48, evaluate the given expressions by (a) substituting directly and by (b) substituting in the simplified expression.

45. $3a(b + 2) + 7b(a - 2) - 6(a + b)$ for $a = 37$ and $b = 19$

46. $\dfrac{8x^2y}{2xy} + 13x(2 - 5a) - 35ax$ for $a = 23$, $x = 57$, and $y = 17$

47. $\dfrac{3a^2(b + 2)}{a} + 7a(b + 5) - 9ab$ for $a = 18$ and $b = 6$

48. $\dfrac{x^2y(a + y)}{xy} + 7x(y - 2a)$ for $a = 8$, $x = 13$, and $y = 29$

2-4 SUMMARY

In this chapter, we introduced some basic terminology and the use of literal numbers in mathematics. We also presented some of the elementary operations of algebra. Since the terminology and operations discussed are used extensively in the remaining portions of the text, it is imperative that you become familiar with them here.

Listed below are the important terms introduced in this chapter.

Binomial
Coefficient
Distributive law
Formula
Like terms
Literal numbers
Monomial
Multinomial
Parentheses
Radical sign
Similar terms
Simplify
Square root
Substitute
Term
Trinomial

If there is any word or phrase in the above list that you do not fully understand, review the section where its meaning is explained. In this way, you should avoid future difficulties due to uncertainty about terminology.

In addition to basic terminology, you should be able to perform the following basic algebraic operations:

Addition and subtraction of like terms.
Dividing out factors common to the numerator and the denominator of a fraction.
Multiplying factors together and the use of exponents.
Multiplying by use of the distributive law.
Simplification of algebraic expressions (combining above operations).
Substitution of values.

Practice these operations until you are sure that you have mastered them. The following exercises should be helpful in reviewing the topics of this chapter.

Exercises 2-4

In Exercises 1 through 28, simplify the given expressions.

1. $x - 3y + 5y + 2x$
2. $5a^2b + ab - 3ab - a^2b$
3. $2a(3a - x)$
4. $-5y(6 - 7y)$
5. $3(x + y) - 2y$
6. $x(2x + 6y) - x^2$
7. $\frac{a^2c}{a} - 2ac + c$
8. $5xy - \frac{xy^2}{y} - y^2$
9. $5(a + b) - 3(a + 2b)$
10. $a(a - x) + a(3a + 2x)$
11. $3(a + 2b) + 2b + b(3 - a)$
12. $x(2x - y) + y(2x + 1)$
13. $\frac{ax^2}{x} + a(x + 2)$
14. $\frac{by}{b} - (7 - y)$
15. $\frac{abx}{b} - \frac{a^2x}{a} + a(2 - x)$
16. $\frac{a^2c}{c} + \frac{a^3c}{ac} + 2a(a + c)$
17. $xy - 2x(y - 1) + 3x$
18. $4ac - a(2c - 3) - 5a$
19. $ab + 2(a - b) + a(2b - 1)$
20. $x(y + 1) + 6(x - y) - y(x - 2)$
21. $\frac{a}{a} + 2(a + 1) + a(3 + x)$
22. $\frac{2ax}{a} + \frac{x^2}{x} + \frac{3b^2x}{b^2} - 8x$
23. $\frac{6x(a + b)}{x} + 3(2a - b) + \frac{ab}{b}$
24. $\frac{c^2x^2}{cx} + c(2x - 3) + \frac{2cx}{x}$
25. $\frac{36a(a + b)}{12a} + \frac{2a(x + y)}{x + y} + \frac{b^2}{b}$
26. $\frac{abc^2}{b} + c(ac - b) + a(3c^2 + b)$
27. $\frac{9x(a - y)^2}{3(a - y)} - 2y(x - a) + ax$
28. $m(2a - n) + \frac{mn^3}{n^2} - a(2m - a)$

In Exercises 29 through 32, identify the coefficient of the factor that is shown in parentheses.

29. $6a^2x$, (x) **30.** $3abc^2$, (c^2) **31.** $12abx$, (ab) **32.** $15mn^2p$, (mn^2)

In Exercises 33 through 44, determine the appropriate formulas.

33. The length of a badminton court is 24 ft more than its width. Express the length l in terms of the width w.

34. What is the formula for the sum S of the areas of a square of side x and a rectangle of sides x and y?

35. In the theory of relativity, Einstein found that the equivalent energy E of a mass m equaled the product of the mass and the square of the speed of light c. Express this statement as a formula.

36. The increase I in population of a certain city was 1000 times the square of the number of years t after a census was taken. Express this statement as a formula.

37. Research indicated that the percentage p of persons responding to a particular stimulus under the influence of a drug was the product of the stimulus strength s and the time t after the drug was given, divided by one more than the square of the time. Find the formula for p in terms of s and t.

38. When a certain wire is heated from temperature t to temperature T, its length increases by an amount equal to the difference of the temperatures divided by 1000. Write a formula for the increase L in the length.

39. A person has n nickels, $n + 2$ dimes, and $n + 5$ quarters. Find a formula for the total value V (in cents) of these coins.

40. A firm pays 20 employees w dollars per week and 10 employees $w - 10$ dollars per week. Find an expression for the total wages T paid to these employees each week.

41. One car travels a mi/h for t hours. A second car travels $2a$ mi/h for $t + 1$ hours. A third car travels $3a$ mi/h for $t + 2$ hours. Find a formula for the total distance d traveled by the three cars.

42. An integer has the numeral u in the units position and a t in the tens position. Express the number N in terms of u and t.

43. If a number N equals the sum of x and y divided by t, express the number N in terms of x, y, and t.

44. If a number N is divided by k, and the sum of the quotient and 1 is g, express the number g in terms of k and N.

In Exercises 45 through 48, use the sets $A = \{a, b, 2a, b + 1\}$ *and* $B = \{2a, 2b, a + 1, b\}$.

45. What is the sum of the elements of $A \cap B$?

46. What is the sum of the elements of $A \cup B$?

47. What is the result when the sum of the elements of A is added to the sum of the elements of B?

48. What is the result when the product of the elements of A is added to the product of the elements of B?

In Exercises 49 through 60, evaluate the required expressions for the given values. In each case, use the simplified expression.

49. $3x^2 - 5y + 2x^2 + 8y$ for $x = -4$ and $y = 6$

50. $18xy + 6x^2 + x(2y + x)$ for $x = 5$ and $y = 11$

51. $15(x + 2a) + 5a(2 - x) + 15x(a - 1)$ for $a = 7$ and $x = 9$

52. $\dfrac{a^2bc^2}{a} + c(9abc - 5) + 5c$ for $a = 6$, $b = 7$, and $c = 12$

53. What percentage of persons responded to the stimulus referred to in Exercise 37 if $s = 40$ units and $t = 3$ min?

54. If $t = 2$ hours and $a = 20$ mi/h, find the sum of the distances traveled by the three cars in Exercise 41.

55. One piece of wire is $2(y + 1)$ ft long and another is $3y + 2$ yd long. What is the combined length, in feet, of these pieces if $y = 17$?

56. One machine produces a articles per hour for t hours. A second machine produces $a + 7$ articles per hour for t hours. A third machine produces a articles per hour for $t - 2$ hours. Find the total number N of articles produced by the machines if $a = 8$ and $t = 5$ hours.

57. A person has p pennies, $p + 3$ nickels, $p + 4$ dimes, and $p + 5$ quarters. Find a formula for the total value V (in cents) of these coins, and then evaluate V for $p = 16$.

58. In Exercise 40, if $w = \$84$, find the total wages T paid to these employees for a period of 6 weeks.

59. One dimension of a rectangle is $2(7x - 2)$ ft and the other dimension is $8x - 3$ yd. What are the length and width of this rectangle if $x = \frac{1}{2}$?

60. In Exercise 56, if the production of the second machine lessened by a articles for t hours and the production of the third machine increased by a articles for $t - 2$ hours, find the total number N of articles produced by the machines if $a = 8$ and $t = 5$ hours.

3

SIMPLE EQUATIONS AND INEQUALITIES

3-1 SOLVING A SIMPLE EQUATION

An **equation** *is a mathematical statement that two algebraic expressions are equal.* Many algebraic forms are found in equations, but we shall concern ourselves in this chapter with equations that contain one literal number, referred to as the **unknown**, for which we are to solve. *By* **solving** *an equation, we mean finding a value of the unknown which, when substituted into the equation, makes the two sides of the equation identically equal.*

Example A $3x - 4 = 2x - 1$ is an equation. In this equation, it is understood that x is the unknown. If we substitute $x = 3$ into each side of the equation, we obtain $5 = 5$. Thus, $x = 3$ is the solution of the equation. This is often states as "3 *satisfies* the equation." If we try any other value of x, we will find that the two sides of the equation are not equal.

We shall use set notation to indicate the solution to an equation. Although the equations we shall consider in this section will have only one number as a solution, situations will arise later where the solution will consist of more than one number. Thus, *the* **solution set** *of an equation has as its elements any solutions of the equation.* In Example A, the solution set of the of the equation is $\{3\}$.

It is the purpose of this chapter to show methods by which we can solve equations such as the one in Example A. At this time, we will treat only equations with the unknown to the first power; the methods of solution are based upon the basic operations discussed in Chapter 2. Later chapters will show how other types of equations are solved.

A significant part of algebra is devoted to the problem of finding solutions to various kinds of equations. Equations are found in many fields of learning, including the physical sciences, the biological sciences, business, and the social sciences. Technical and scientific work, in particular, give rise to equations of all kinds, and their solutions are essential to the attainment or confirmation of scientific information. Example B illustrates some statements about numbers that lead to equations. Other equations arising from various fields of study are considered in the following section.

Example B In each case, we will let x be the unknown number.

Statement	Equation
1. A number decreased by seven equals twelve.	1. $x - 7 = 12$
2. Nine added to a number equals three.	2. $x + 9 = 3$
3. A number divided by two equals seven.	3. $\frac{x}{2} = 7$
4. Five times a number equals twenty.	4. $5x = 20$
5. Twice a number less three equals one more than the number.	5. $2x - 3 = x + 1$
6. Sixteen less three times a number equals two added to four times the number.	6. $16 - 3x = 4x + 2$

Keeping in mind the meaning of an equation—the two sides are equal for the appropriate value of x—we may see how a solution is obtained. Since the two sides are equal,

if the same number is added to each side, the two sides are still equal. Also, if the same number is subtracted from each side, they remain equal to each other. If the two sides are multiplied by the same number, or divided by the same number **(provided it is not zero),** *they remain equal.*

When one of these operations is performed on each side of the equation, the resulting values on each side are still equal, although these resulting values probably differ from the original values of each side of the equation.

The goal in performing these operations is to isolate the unknown on one side of the equation and the other quantities on the other side. The following examples illustrate the method used to solve the equations of Example B.

Example C

1. Solve $x - 7 = 12$.

Note that, when we add 7 to the left side, only x will remain. Thus, by adding 7 to both sides, we obtain

$$x - 7 + 7 = 12 + 7$$
$$x = 19$$

because $-7 + 7 = 0$ and $12 + 7 = 19$. Thus, $x = 19$ is the solution, which in turn means that the solution set is $\{19\}$. The solution is checked (always a good policy) by substituting $x = 19$ into the original equation. This leads to $12 = 12$, which means that the solution is correct.

2. Solve $x + 9 = 3$.

When we subtract 9 from each side, only x will remain on the left side. Performing this operation, we have

$$x + 9 - 9 = 3 - 9$$
$$x = -6$$

Therefore, the solution set is $\{-6\}$. Substitution into the original equation gives $3 = 3$, which means that the solution checks. Such substitutions are the means to determine whether or not a solution is correct.

Example D

1. Solve $\dfrac{x}{2} = 7$.

If the left side is multiplied by 2, it becomes $2x/2$, or x. Thus, by multiplying both sides by 2, we obtain

$$2\left(\frac{x}{2}\right) = 2(7)$$
$$x = 14$$

which means that the solution set is $\{14\}$. Substitution into the original equation gives $7 = 7$, which means the solution checks.

2. Solve $5x = 20$.
If both sides are divided by 5, the left side then becomes x, and we have

$$\frac{5x}{5} = \frac{20}{5}$$

$$x = 4$$

The solution set is $\{4\}$. Substitution into the original equation gives $20 = 20$.

The operations performed in the preceding examples should call to mind a balance scale. Whatever is added to or subtracted from one side of an equation must be added to or subtracted from the other side in order to maintain equality or balance. The same is true when multiplication and division are performed.

Each of the equations in Examples C and D was solved by use of one of the basic operations. However, for most equations, we must perform several operations to obtain the solution. The following examples illustrate the solution of such equations (see Example B). In these examples, explanations of the basic operations performed in proceeding from one step to the next are given, although these operations are not actually shown.

Example E Solve $2x - 3 = 1 + x$.

By combining terms containing x on one side and the other terms on the other side, we may isolate x and thereby find the solution. The solution proceeds as follows.

$$2x - 3 = 1 + x \qquad \text{(original equation)}$$

$$2x = 4 + x \qquad \text{(add 3 to each side)}$$

$$x = 4 \qquad \text{(subtract } x \text{ from each side)}$$

Since the last equation directly gives the desired value of x, the solution is complete. This means that the solution set is $\{4\}$. Check: Substituting into the original equation, we obtain $5 = 5$.

Example F Solve $16 - 3x = 2 + 4x$.

The solution proceeds as follows.

$$16 - 3x = 2 + 4x \qquad \text{(original equation)}$$

$$14 - 3x = 4x \qquad \text{(subtract 2 from each side)}$$

$$14 = 7x \qquad \text{(add } 3x \text{ to each side)}$$

$$2 = x, \quad \text{or} \quad x = 2 \qquad \text{(divide each side by 7)}$$

Thus, the solution set is $\{2\}$. Substituting into the original equation gives $10 = 10$.

The question often arises as to how one knows the order of proceeding in the solution of an equation. No specific order is required, but the form of the equation suggests the procedure. With sufficient experience, solving an equation should become essentially automatic. In the following example, the equation of Example F is solved with a variation in the order of steps.

Example G

$16 - 3x = 2 + 4x$ (original equation)

$16 - 7x = 2$ (subtract $4x$ from each side)

$-7x = -14$ (subtract 16 from each side)

$x = 2$ (divide each side by -7)

We see that the solution agrees with that found in Example F, which means that the solution set is $\{2\}$.

Example H illustrates the solution of an equation in which the distributive law is used and for which the solution is negative.

Example H Solve $3(2x + 5) = 2x + 3$.

We proceed as follows.

$3(2x + 5) = 2x + 3$ (original equation)

$6x + 15 = 2x + 3$ (use the distributive law)

$6x = 2x - 12$ (subtract 15 from each side)

$4x = -12$ (subtract $2x$ from each side)

$x = -3$ (divide each side by 4)

Therefore, the solution set is $\{-3\}$. Checking in the original equation, we obtain $3(-6 + 5) = -6 + 3$, or $-3 = -3$.

Two final points about equations will now be made. First, *some equations are valid for all values of x. Such equations are referred to as* **identities.** *The equations which we have been solving, those which are true only for specific values of the unknown, are often called* **conditional equations.** Second, some equations are not valid for any values of x, since no values of the unknown will satisfy them. Example I illustrates an identity and an equation which has no solution.

Example I

1. $2x + 2 = 2(x + 1)$ is true for all values of x, and is therefore an identity. Proceeding with the solution, we would arrive at $0 = 0$, which is true regardless of the value of x. The solution set may be shown as $\{\text{all } x\}$.

2. $x + 1 = x + 2$ is an equation, although in solving we obtain $1 = 2$, which is not true for any value of x. The solution set is $\{\ \}$.

Exercises 3-1

In Exercises 1 through 28, find the solution set for each of the equations. Check by substituting the value found in the original equation.

1. $x - 2 = 3$ **2.** $x - 4 = 18$ **3.** $x + 4 = 7$

4. $x + 5 = 13$ **5.** $2x = 16$ **6.** $3x = 21$

7. $\frac{x}{7} = 5$ **8.** $\frac{x}{3} = 8$ **9.** $x + 6 = 2$

10. $x - 1 = -7$ **11.** $2x - 5 = 13$ **12.** $3x + 20 = 5$

13. $4x + 11 = 3$ **14.** $3x - 2 = 16$ **15.** $3 + 6x = 24 - x$

16. $5 - x = 8x - 13$ **17.** $3x - 6 = x - 18$ **18.** $14 - 2x = 5x + 7$

19. $2(x - 1) = x - 3$ **20.** $3(x + 2) = 2x + 11$

21. $\frac{x}{2} = x - 4$ **22.** $\frac{x}{3} = 12 - x$

23. $4(x - 2) = 7(x + 4)$ **24.** $6(2x + 1) = 3(x + 8)$

25. $3(x - 1) + x = 2(x - 1)$ **26.** $3(2 - 3x) = 7(x + 1) - 33$

27. $6x - 1 = 3(x - 2) + 6$ **28.** $4(2x + 1) = 3(x - 3) - 7$

In Exercises 29 through 36, set up an equation from the given statement, and then determine the solution set.

29. A number less six equals seven.

30. Five more than a number is eight.

31. Three times a number is thirty-three.

32. A number divided by four is seventeen.

33. Eight more than five times a number equals three times the number.

34. Three less than five times a number is nine more than the number.

35. Four times the sum of a number and three equals twenty more than twice the number.

36. Three times the sum of a number and two is eight less than the number.

In Exercises 37 and 38, answer the given questions involving identities and equations having no solutions.

37. By solving for the solution sets, distinguish between $x + 1 = x - 5$ and $x + 1 = 5 - x$.

38. Which of the following are identities?

(a) $2(x + 1) + 1 = 3 + 2x$

(b) $3(x + 1) = 1 + 3x$

(c) $5x = 4(x + 1) - x$

3-2 SIMPLE FORMULAS AND LITERAL EQUATIONS

In introducing the solution of equations, we considered only equations containing the unknown x and other specific numbers. All the formulas illustrated in Chapter 2 are actually equations, since they express equality of algebraic expressions. These formulas contained more than one literal number. In this section, we shall show how the methods of solving equations are applied to formulas and other equations containing more than one literal number. Consider Example A.

Example A The formula for the average A of two numbers a and b is

$$A = \frac{a + b}{2}$$

There may be times when it is advantageous, or even necessary, to have an expression for one of the numbers, say b, rather than for the average. Such an expression can be found by solving the equation for b. Of course, our answer, the resulting formula for b, will be in terms of A and a, rather than some explicit number, as in the preceding section.

The solution proceeds as follows.

$$A = \frac{a + b}{2} \qquad \text{(original formula)}$$

$$2A = a + b \qquad \text{(multiply each side by 2)}$$

$$2A - a = b \qquad \text{(subtract } a \text{ from each side)}$$

Therefore, our solution is $b = 2A - a$.

Actually, any equation containing literal numbers other than the unknown, or required, literal number is solved in the same general manner as the equations of the preceding section. In each case, many of the basic operations are performed with literal numbers; the result will contain the other literal numbers and not just an explicit number. Consider the following examples.

Example B In the equation $ay + b = 2c$, solve for b.

Since the object is to isolate b, we subtract ay from each side of the equation. This gives

$$b = 2c - ay$$

which is the required solution.

Example C In the study of genetics, the equation $q(u + v) = v$ is used. Solve for u.

$$q(u + v) = v \qquad \text{(original equation)}$$

$$qu + qv = v \qquad \text{(use the distributive law)}$$

$$qu = v - qv \qquad \text{(subtract } qv \text{ from each side)}$$

$$u = \frac{v - qv}{q} \qquad \text{(divide each side by } q\text{)}$$

This last equation gives the required solution. Note that *q cannot be divided out of the numerator and denominator since it is not a factor of the numerator.*

We can illustrate the concept which leads to the fact that *q cannot be divided out* (in Example C) by using an illustration from arithmetic. Consider the following expression.

$$\frac{15 - (3)(4)}{3}$$

We repeat the 15 and also multiply 3 by 4. Thus,

$$\frac{15 - (3)(4)}{3} = \frac{15 - 12}{3}$$

We now subtract 12 from 15, or

$$\frac{15 - 12}{3} = \frac{3}{3}$$

which of course is 1. If we had divided out (or "canceled") the 3 which appears in both numerator and denominator, we would be "cheating" the 15 of its denominator! Also, we see that the apparent result of 11 is incorrect.

$$\frac{15 - (\cancel{3})(4)}{\cancel{3}} \qquad \text{(an incorrect procedure)}$$

In a fraction we do not divide out terms or factors of terms which are not factors of the whole numerator and whole denominator.

Example D In the equation $2a(f + x) - 3b = 4(b + x)$, solve for f.
The solution proceeds as follows.

$$2a(f + x) - 3b = 4(b + x) \qquad \text{(original equation)}$$
$$2af + 2ax - 3b = 4b + 4x \qquad \text{(remove parentheses)}$$
$$2af + 2ax = 7b + 4x \qquad \text{(add } 3b \text{ to each side)}$$
$$2af = 7b + 4x - 2ax \qquad \text{(subtract } 2ax \text{ from each side)}$$
$$f = \frac{7b + 4x - 2ax}{2a} \qquad \text{(divide each side by } 2a\text{)}$$

Thus, we have the solution. Again, note that we cannot divide the $2a$ which appears in both numerator and denominator since it is not a factor of the whole numerator.

In many problems, it is necessary to refer to two or more values of a given quantity. For example, we may wish to denote the prices of several different items, representing the price by the letter p. Instead of choosing a different letter for each of the prices, we may use **subscripts** on the letter p for this purpose. Thus, p_1 could be the price of the first item, p_2 the price of the second item, and so forth. Remember that p_1 and p_2 are different literal numbers. Example E shows the solution of a literal equation involving subscripts.

Example E Solve the equation $as_1 + cs_2 = 3a(s_1 + a)$ for s_1.
The solution proceeds as follows.

$$as_1 + cs_2 = 3a(s_1 + a) \qquad \text{(original equation)}$$
$$as_1 + cs_2 = 3as_1 + 3a^2 \qquad \text{(remove parentheses)}$$
$$cs_2 = 2as_1 + 3a^2 \qquad \text{(subtract } as_1 \text{ from each side)}$$
$$cs_2 - 3a^2 = 2as_1 \qquad \text{(subtract } 3a^2 \text{ from each side)}$$
$$\frac{cs_2 - 3a^2}{2a} = s_1 \qquad \text{(divide each side by } 2a\text{)}$$

Thus,

$$s_1 = \frac{cs_2 - 3a^2}{2a}$$

Among the most important formulas are those related to some of the basic geometric figures. Following are some of these basic formulas for perimeters (distance around), areas, and volumes. In these formulas, p represents perimeter, A represents area, and V represents volume.

triangle with sides a, b, and c:	$p = a + b + c$	(3-1)
rectangle of length l and width w:	$p = 2l + 2w$	(3-2)
	$A = lw$	(3-3)
square of sides s:	$p = 4s$	(3-4)
	$A = s^2$	(3-5)
rectangular solid of length l, width w, and height h:	$V = lwh$	(3-6)
cube of edge e:	$V = e^3$	(3-7)

It is assumed that the reader is familiar with the geometric figures mentioned. These formulas will be of use in some of the examples and exercises in the text.

Example F Solve the formula for the perimeter of a rectangle for the width w.

$$p = 2l + 2w \quad \text{(original formula)}$$

$$p - 2l = 2w \quad \text{(subtract } 2l \text{ from each side)}$$

$$w = \frac{p - 2l}{2} \quad \text{(divide each side by 2, and switch sides)}$$

This final equation is the required result.

An illustration of a statement which leads to a formula, which in turn must be solved for one of its literal numbers, is demonstrated in Example G.

Example G If one car goes v_1 mi/h for 3 h and another car goes v_2 mi/h for $3 + t$ h, the total distance they travel is d. Solve the resulting formula for t.

Since the distance the first car goes is $3v_1$ and the distance the second car goes is $v_2(3 + t)$, we have the total distance d as

$$d = 3v_1 + v_2(3 + t)$$

Solving this formula for t, we have the following.

$$d = 3v_1 + 3v_2 + v_2t \quad \text{(remove parentheses)}$$

$$d - 3v_1 - 3v_2 = v_2t \quad \text{(subtract } 3v_1 \text{ and } 3v_2 \text{ from each side)}$$

$$t = \frac{d - 3v_1 - 3v_2}{v_2} \quad \text{(divide each side by } v_2\text{, and then switch sides)}$$

The last formula is the desired result.

Exercises 3-2

Each of the formulas in Exercises 1 through 20 arises in the area of study listed at the right. Solve for the indicated letter.

1. $R = \dfrac{100M}{C}$, for M (psychology: intelligence rating)
2. $I = Prt$, for t (business: interest)
3. $v_2 = v_1 + at$, for v_1 (physics: motion)
4. $C = a + bx$, for a (economics: cost analysis)
5. $PV = RT$, for T (chemistry: gas law)
6. $A = 4\pi r^2$, for r^2 (geometry: area of sphere)
7. $V = \dfrac{kN}{t}$, for N (biology: growth rate)
8. $r = \dfrac{ms_1}{s_2}$, for s_2 (mathematics: statistics)
9. $E = \dfrac{mv^2}{2}$, for m (physics: energy)
10. $A = 180 - (B + C)$, for B (geometry: angles of triangle)
11. $F = \dfrac{9C + 160}{5}$, for C (science: temperature)
12. $I = \dfrac{S_2 - S_1}{S_1}$, for S_2 (psychology: stimuli)
13. $A = P(1 + rt)$, for r (business: interest)
14. $L = L_0(1 + at)$, for a (physics: temperature expansion)
15. $a = V(k - PV)$, for k (biology: muscle contractions)
16. $T_2w = q(T_2 - T_1)$, for T_1 (chemistry: energy)
17. $R_2 = R_1 - k(P - a)$, for P (psychology: learning)
18. $F = A_2 - A_1 + P(V_2 - V_1)$, for V_2 (chemistry: energy)
19. $R = \dfrac{wL}{H(w + L)}$, for H (interior design: lighting)
20. $A = \dfrac{n_1p_1 + n_2p_2}{n_1 + n_2}$, for p_1 (economics: average price)

In Exercises 21 through 36, solve the given equations for the indicated letter.

21. $a = bc + d$, for d
22. $x - 2y = 3t$, for x
23. $ax + 3y = f$, for x
24. $3ay + b = 7q$, for y
25. $2a(x + y) = 3y$, for x
26. $3x(a - b) = 2x$, for a
27. $\dfrac{a}{2} = b + 2$, for a
28. $\dfrac{y}{x} = 2 - a$, for y
29. $x_1 = x_2 + a(3 + b)$, for b
30. $s_1 + s_2 = 2(a - b)$, for a

31. $R_3 = \dfrac{R_1 + R_2}{2}$, for R_2

32. $m + n = \dfrac{2n + 1}{3}$, for n

33. $7a(y + z) = 3(y + 2)$, for z

34. $3x(x + y) = 2(3 - x)$, for y

35. $3(x + a) + a(x + y) = 4x$, for y

36. $2a(a - x) = 3a(a - b) + 2ax$, for b

In Exercises 37 through 42, set up the required formula and solve for the indicated letter.

37. On a certain vote, the number of "yes" votes y was 100 more than the number of "no" votes n. Solve for n.

38. A carpenter cuts x ft from each of three boards, obtaining a total of y ft. Solve for x.

39. The average A of three numbers a, b, and c equals their sum divided by 3. Solve for c.

40. An integer N has the numeral t in its tens position and a u in its units position. Solve for t.

41. A man has x dimes, $x + 2$ quarters, and $x - 1$ half-dollars. These coins give him a total of y cents. Solve for x.

42. A machine produces a articles per hour for t h. A second machine produces $a + 5$ articles per hour for t h. Together, they produce a total of b articles. Solve for a.

3-3 SIMPLE INEQUALITIES

In the first section of this chapter, we defined an equation to be a mathematical statement that two algebraic expressions are equal. In the same manner, *an* **inequality** *is a mathematical statement that one algebraic expression is greater than, or less than, another algebraic expression.*

As has been pointed out, equation solving does play a very important role in mathematics and its applications. However, there are also occasions that require the solutions of inequalities. Some of these will be shown through the examples and exercises of this section.

In Chapter 1, we introduced the signs of inequality, $>$ and $<$. *These signs define the* **sense** *of the inequality. The two sides of the inequality are known as* **members** *of the inequality.*

Example A The inequalities $x + 4 > 1$ and $3 - x < 2$ have opposite senses. The left and right members of the first inequality are $x + 4$ and 1, respectively. The left and right members of the second inequality are $3 - x$ and 2, respectively.

The **solution** *of an inequality is the set of those values of the variable that satisfy the inequality.* That is, when any of the values of the solution set is substituted into the inequality, it becomes a correct statement mathematically. For most of the inequalities that we shall consider, the solution set will consist of an unlimited number of values, bounded by a specific real number.

Example B If we substitute any number greater than -3 for x in the inequality $x + 4 > 1$, we will find that the inequality is satisfied. That is, if we substitute -2, 0, 4, π, $-\sqrt{2}$, and so on for x, the inequality is satisfied. Thus, the solution set consists of *all* real numbers, rational and irrational, which are greater than -3. No value that is -3 or less satisfies the inequality.

In Example B, we see that the solution set of an inequality consists of all real numbers in a specified range. In stating the solution set, it is convenient to use a particular notation. For example, we would state the solution set of the inequality of Example B as $\{x|x > -3\}$. This is read as "the set of all x such that x is greater than -3." Read the vertical bar as "such that."

Example C The solution set of the inequality $3 - x < 2$ is $\{x|x > 1\}$, which means "the set of all x such that x is greater than 1." Thus, any value of x greater than 1 satisfies this inequality.

When finding the solution set of an inequality, we perform on each member operations that are similar to those used in solving equations. That is,

we may add the same number to each member, subtract the same number from each member, or multiply or divide each member by the same positive number and still maintain the sense of the inequality.

There is one operation which must be handled with specific care.

If each member of the inequality is multiplied or divided by a negative number, the sense of the inequality is reversed.

These operations are verified in the following examples.

Example D We know that $2 < 6$. If 3 is added to each member, we obtain $2 + 3 < 6 + 3$, or $5 < 9$, which is still a correct inequality.

If 3 is subtracted from each member of the inequality $2 < 6$, we obtain $2 - 3 < 6 - 3$, or $-1 < 3$, which is a correct inequality.

If each member of the inequality $2 < 6$ is multiplied by 3, we obtain $2 \times 3 < 6 \times 3$, or $6 < 18$, which is a correct inequality.

If each member of the inequality $2 < 6$ is divided by 3, we obtain $\frac{2}{3} < \frac{6}{3}$, or $\frac{2}{3} < 2$, which is a correct inequality.

Example E If we multiply each member of the inequality $2 < 6$ by -3, we obtain -6 for the left member and -18 for the right member. However, we know that $-6 > -18$. Thus, we obtain $-6 > -18$; the sense of the inequality has been reversed.

If each member of the inequality $2 < 6$ is divided by -2, we obtain $-1 > -3$; and again the sense of the inequality is reversed.

The goal in performing these operations is to isolate the unknown on one side of the inequality and the other numbers on the other side. In this way, we can solve the inequality and determine its solution set. This is the same as in the case of solving an equation. The following examples illustrate solving inequalities.

Example F

1. Solve the inequality $x + 6 < 9$.

By subtracting 6 from each member, we obtain $x < 3$, which means that values of x less than 3 satisfy the inequality. Thus, the solution set is $\{x|x < 3\}$.

2. Solve the inequality $x - 6 < 9$.

By adding 6 to each member, we obtain $x < 15$. Thus, the solution set is $\{x|x < 15\}$.

Example G

1. Solve the inequality $\dfrac{x}{2} > 5$.

By multiplying each member by 2, we obtain $x > 10$. Thus, the solution set is $\{x|x > 10\}$.

2. Solve the inequality $2x > 16$.

By dividing each member by 2, we obtain $x > 8$. Thus, the solution set is $\{x|x > 8\}$.

Example H Solve the inequality $6 - 2x < 20$.

By first subtracting 6 from each member, we obtain $-2x < 14$. Then if each member is divided by -2, we obtain $x > -7$, noting that the sense of the inequality has been reversed. Thus, the solution set is $\{x|x > -7\}$.

Example I illustrates the solution of an inequality that requires several basic operations. The operations performed are indicated beside the steps.

Example I Solve the inequality $6(x - 5) > 10 + x$.

$6(x - 5) > 10 + x$	(original inequality)
$6x - 30 > 10 + x$	(use distributive law)
$6x > 40 + x$	(add 30 to each member)
$5x > 40$	(subtract x from each member)
$x > 8$	(divide each member by 5)

Thus, the solution set is $\{x|x > 8\}$.

Example J illustrates the use of an inequality in an applied situation.

Example J A man earns \$4 per hour, and his wife earns \$100 each week. If we wish to determine how many hours he must work so that their combined weekly income is over \$240, we solve the inequality $4x + 100 > 240$, where x is the number of hours he must work during the week.

$$4x + 100 > 240 \quad \text{(original inequality)}$$
$$4x > 140 \quad \text{(subtract 100 from each member)}$$
$$x > 35 \quad \text{(divide each member by 4)}$$

Thus, he must work over 35 hours in a week for their combined income to exceed \$240.

Exercises 3-3

In Exercises 1 through 8, perform the indicated operations on each member of the given inequalities.

1. (a) $3 > 1$, add 5 (b) $-2 < 4$, subtract 3

2. (a) $-2 < 8$, add 4 (b) $7 > 1$, subtract 4

3. (a) $1 > -2$, multiply by 3 (b) $5 < 10$, divide by 5

4. (a) $3 < 10$, multiply by 4 (b) $-8 > -16$, divide by 8

5. (a) $-1 < 0$, multiply by -2 (b) $6 < 9$, divide by -3

6. (a) $2 < 9$, multiply by -3 (b) $4 > -4$, divide by -1

7. $-2 < 8$, add 4, and then divide by -2.

8. $11 > -3$, subtract 4, and then divide by -7.

In Exercises 9 through 28, solve the given inequalities.

9. $x - 7 > 5$ **10.** $x + 8 < 4$ **11.** $x + 2 < 7$

12. $x - 2 > 9$ **13.** $\frac{x}{3} > 6$ **14.** $\frac{x}{4} < 5$

15. $4x < -24$ **16.** $3x > 6$ **17.** $-2x > -8$

18. $-x < 5$ **19.** $2x + 7 < 5$ **20.** $5x - 3 > -13$

21. $2 - x > 6$ **22.** $3 - x < 1$ **23.** $4x - 1 < x - 7$

24. $2x + 1 > 10 - x$ **25.** $3(x - 2) > -9$ **26.** $6(x + 5) < x + 10$

27. $2(2x + 1) < 5x - 4$ **28.** $3(1 - 2x) > 1 - 4x$

In Exercises 29 through 34, set up inequalities from the given statements and then solve.

29. If any of the numbers in a given set is multiplied by 2, the product is greater than two more than the number. Find the set of numbers.

30. If five is subtracted from any of the numbers of a given set, the difference is less than three times the number. Find the set of numbers.

31. A company determines that they must spend less than \$100,000 on new equipment. They decide to spend \$12,500 for a particular piece of new equipment. How much can they spend on other new pieces?

32. In a particular housing development, a family is ineligible for housing if their total annual income is more than \$12,000. If one of two wage-earners in a household earns \$6500 annually, how much can the other earn in order for them to be eligible for housing?

33. What average speeds will allow a traveler to go more than 150 mi in 3 h?

34. A student has grades of 80, 92, 86, and 78 on the four tests in a certain course. The final exam in the course counts as two tests. What exam grades would cause his average to fall below 80 for the course? (See Exercise 31 of Exercises 2-2.)

3-4 MORE ON INEQUALITIES

In the previous section, we introduced simple inequalities and methods of solving them. There are also occasions when we want to combine an inequality with an equality. In this section, we shall consider this type of situation, and we shall also show how solution sets of inequalities may be represented on a number line.

When we combine an equality with an inequality, we use the symbols $\leq$ *and* $\geq$, *which are read as "less than or equal to" and "greater than or equal to," respectively*. Algebraic statements which use these symbols are called "inequalities" even though part of their meaning is that of an equality. *This type of inequality is solved in the same manner as the inequalities of the previous section*. Consider the following examples.

Example A Solve the inequality $x + 3 \geq 11$.

If we subtract 3 from each member, we obtain $x \geq 8$. This means that the solution set is $\{x | x \geq 8\}$. This means that any number *greater than or equal to* 8 satisfies the inequality. If $x = 8$, the expression $x + 3 = 11$, and the equality part is satisfied. If $x > 8$, $x + 3$ is greater than 11, and the inequality part is satisfied.

Example B Solve the inequality $2x + 6 \le 9$.

In finding the solution to this inequality, we are to find those values of x for which $2x + 6$ is *less than or equal to* 9.

The solution proceeds as follows:

$$2x + 6 \le 9 \qquad \text{(original inequality)}$$

$$2x \le 3 \qquad \text{(subtract 6 from both members)}$$

$$x \le \frac{3}{2} \qquad \text{(divide both members by 2)}$$

This means that the solution set includes not only values of x less than $\frac{3}{2}$ but also the value $\frac{3}{2}$ itself. Thus, the solution set is $\{x \mid x \le \frac{3}{2}\}$.

Example C Solve the inequality $5(2x - 3) \ge 6 + 3x$.

$$5(2x - 3) \ge 6 + 3x \qquad \text{(original inequality)}$$

$$10x - 15 \ge 6 + 3x \qquad \text{(use distributive law)}$$

$$10x \ge 21 + 3x \qquad \text{(add 15 to both members)}$$

$$7x \ge 21 \qquad \text{(subtract } 3x \text{ from both members)}$$

$$x \ge 3 \qquad \text{(divide both members by 7)}$$

Here the solution set includes values of x greater than 3, and 3 itself. Thus, the solution set is $\{x \mid x \ge 3\}$.

To help visualize the solution set of an inequality, it is often helpful to show the solution set on a number line. The method of doing this is demonstrated in the following example.

Example D For the inequality $x + 6 < 9$, the solution set is $\{x \mid x < 3\}$ (see Example F of the previous section). This solution set may be represented on a number line by placing a right parenthesis at $x = 3$, and then drawing an arrow from the parenthesis toward the left. See Figure 3-1.

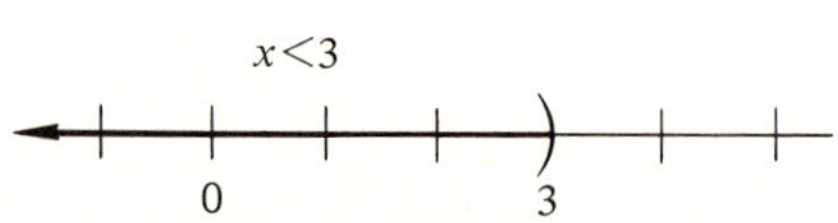

Figure 3-1

The parenthesis represents the bounding value, and the arrow indicates that all values to the left are included. If we had the solution set $\{x|x > 3\}$, we would have used a left parenthesis, and the arrow would have pointed to the right.

For the inequality in Example C, the solution set is $\{x|x \geq 3\}$. Since the value 3 is included as part of the solution, we use a bracket at the boundary value. The solution set is shown in Fig. 3-2.

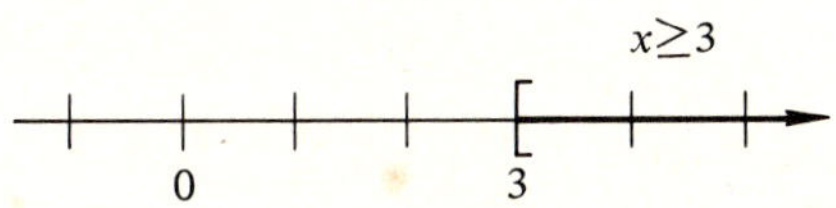

Figure 3-2

Example E Solve the inequality $3(2 - 3x) > x - 14$ and show its solution set on the number line.

The solution proceeds as follows:

$3(2 - 3x) > x - 14$	(original inequality)
$6 - 9x > x - 14$	(use distributive law)
$-9x > x - 20$	(subtract 6 from each member)
$-10x > -20$	(subtract x from each member)
$x < 2$	(divide each member by -10)

This means that the solution set is $\{x|x < 2\}$. Since the solution set does not include the value 2, we use a right parenthesis when showing the solution set on the number line. See Fig. 3-3.

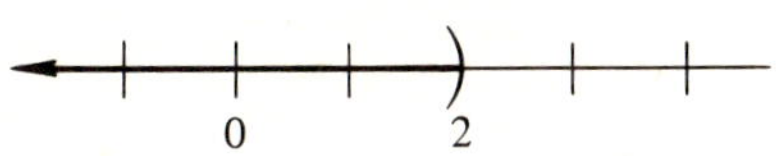

Figure 3-3

Example F Solve the inequality $1 - 3x \leq 4 - (6 + x)$, and show its solution set on the number line.

The solution proceeds as follows:

$$1 - 3x \leq 4 - (6 + x) \quad \text{(original inequality)}$$
$$1 - 3x \leq 4 - 6 - x \quad \text{(use distributive law)}$$
$$-3x \leq -3 - x \quad \text{(subtract 1 from each member)}$$
$$-2x \leq -3 \quad \text{(add } x \text{ to each member)}$$
$$x \geq \frac{3}{2} \quad \text{(divide each member by } -2\text{)}$$

Therefore, the solution set is $\{x \mid x \geq \frac{3}{2}\}$. Since the value $\frac{3}{2}$ is in the solution set, a left bracket is used as shown in Fig. 3-4.

Figure 3-4

Exercises 3-4

In Exercises 1 through 16, solve the given inequalities.

1. $x - 3 \leq 5$
2. $x + 2 \geq 3$
3. $3x \geq 15$
4. $4x \leq -28$
5. $\frac{x}{2} \leq 4$
6. $\frac{x}{5} \geq -1$
7. $2x - 1 \geq 11$
8. $6x + 5 \leq -7$
9. $8 - x \leq 5$
10. $1 - 2x \geq 7$
11. $2x - 9 \geq -10$
12. $3 + 5x \leq 20$
13. $-\frac{x}{5} + 2 \leq 3$
14. $\frac{x}{3} - 3 \geq 5$
15. $2(2x + 3) \geq 9 - x$
16. $3(4x - 7) \leq 10x - 8$

In Exercises 17 through 20, show the given solution sets on the number line.

17. $\{x \mid x > -1\}$
18. $\{x \mid x < -2\}$
19. $\{x \mid x \leq \frac{5}{2}\}$
20. $\{x \mid x \geq \frac{4}{3}\}$

In Exercises 21 through 28, solve the given inequalities and show the solution sets on the number line.

21. $x - 7 > -9$
22. $x + 10 \leq 12$
23. $3x + 2 \geq 8$
24. $4x - 1 < 15$
25. $2(1 - x) \leq x + 5$
26. $5(2 - x) > 6 - 4x$
27. $x + 3(x - 2) < 7 - x$
28. $3(4x - 1) \leq 2 + x$

In Exercises 29 through 32, set up inequalities from the given statements and then solve.

29. If any of the numbers in a given set is divided by 3, the quotient is less than or equal to four more than the number. Find the set of numbers.

30. If four is added to any of the numbers of a given set, the sum is equal to or greater than two times the number. Find the set of numbers.

31. A woman has $150 in a bank account. How much must she deposit in the account such that she has at least $225 in the account?

32. A copier service charges 8¢ per copy made of a given document. How many copies can be made if the cost is $4 or less?

3-5 RATIO AND PROPORTION

One might wonder how the formulas that are used in the various fields of study are formulated. Some of the formulas that are used result from basic definitions. Other formulas are derived through observation and experimental evidence. In this and in the following section, we will show how a number of formulas are established in these ways.

The **ratio** *of one number to another is the first number divided by the second number.* That is, the ratio of a to b is a/b. (Another notation is $a:b$.) By this definition we see that a ratio is a comparison of numbers by division.

Example A

1. The ratio of 12 to 8 is $12 \div 8$, or $\frac{12}{8}$. Simplifying the fraction, we obtain $\frac{3}{2}$, which means that the ratio of 12 to 8 is the same as the ratio of 3 to 2. It also means that 12 is $\frac{3}{2}$ as large as 8.
2. The ratio of 2 to 9 is $\frac{2}{9}$.
3. The ratio of 10 to 2 is $\frac{10}{2}$, or $\frac{5}{1}$.

Every measurement is a ratio of the measured magnitude to an accepted unit of measurement. When we say that an object is 4 m long, we are saying that the length of the object is four times as long as the unit of length, the meter. Other examples of ratios are scales of measurements on maps, density (mass/volume), and pressure (force/area). As shown by these examples, ratios may compare quantities of the same kind or they may express divisions of magnitudes of different quantities.

Ratios are often used to compare like quantities. When they are used for this purpose, we should express each of the quantities in the same units. The following example illustrates this use of ratios.

Example B

1. The length of a certain room is 18 ft, and the width of the room is 12 ft. Therefore, the ratio of the length of the room to the width is 18 ft/12 ft, or $\frac{3}{2}$.

2. If the width of the room is expressed as 4 yd, we should not express the ratio as 18 ft/4 yd, or 9 ft/2 yd. It is preferable and more meaningful to change one of the measurements to the same units as the other. Changing the width from 4 yd to 12 ft, we express the ratio as $\frac{3}{2}$, as above. From this ratio we easily see that the length of the room is $\frac{3}{2}$ as long as the width.

3. It is essential to observe the order of the quantities being compared in a ratio. Comparison of the width of the above room to its length would be 12 ft/18 ft, or $\frac{2}{3}$. The room is $\frac{2}{3}$ as wide as it is long.

As we have noted, ratios are often used to compare like quantities, but they may also be used to compare unlike quantities. When we do compare unlike quantities, we must attach the proper units to the ratio. This is illustrated in the following example.

Example C If a car travels 80 mi in 2 h and consumes 4 gal of gasoline, the ratio of distance to time is 80 mi/2 h, or 40 mi/h. Also, the ratio of distance to fuel consumed is 80 mi/4 gal, or 20 mi/gal. In each case, we must note the units of the ratio.

A statement of equality of two ratios is a **proportion.** By this definition,

$$\boxed{\frac{a}{b} = \frac{c}{d}} \tag{3-8}$$

is a proportion, equating the ratios a/b and c/d. (Another way of denoting a proportion is $a:b = c:d$, which is read as "a is to b as c is to d.") Thus, we see that a proportion is an equation.

Example D

1. The ratio $\frac{16}{6}$ equals the ratio $\frac{8}{3}$. Thus, we may state this equality by writing the proportion $\frac{16}{6} = \frac{8}{3}$.

2. $\frac{3}{4} = \frac{12}{16}$ is a proportion since multiplying each of 3 and 4 by 4 results in 12 and 16, respectively.

3. $\dfrac{2}{6x} = \dfrac{1}{3x}$ is a proportion since dividing each of 2 and $6x$ by 2 results in 1 and $3x$, respectively.

4. $\frac{18}{12} = \frac{3}{2}$ and $\frac{80}{4} = \frac{20}{1}$ are proportions.

5. 80 km/2 h = 40 km/1 h = 40 km/h is a proportion.

In a given proportion, if one ratio is known and one part of the other ratio is also known, then the unknown part of the second ratio can be determined. This is done by letting an appropriate literal symbol represent the unknown part and then solving the resulting equation. This is illustrated in the following examples.

Example E The ratio of a given number to 3 is the same as the ratio of 16 to 6. Find the number.

First, let $x =$ the number. We then set up the indicated proportion as

$$\frac{x}{3} = \frac{16}{6}$$

Solving for x, we multiply each side by 3:

$$\frac{3x}{3} = \frac{16(3)}{6}$$

which is simplified to

$$x = 8$$

Checking, we see that

$$\frac{8}{3} = \frac{16}{6}$$

Example F On a certain blueprint, a measurement of 25 ft is represented by 2 in. What is the actual distance between two points if they are 5 in. apart on the blueprint?

Let $x =$ the required distance and then note that 25 is to 2 as x is to 5. Thus,

$$\frac{25}{2} = \frac{x}{5}$$

By multiplying each side by 5, we solve the equation. This leads to

$$\frac{25(5)}{2} = \frac{5x}{5}$$

or

$$x = 62.5 \text{ ft}$$

Example G A certain alloy is 5 parts tin and 2 parts lead. How many kilograms of each are there in 35 kg of the alloy?

First, we let $x =$ the number of kilograms of tin in the given amount of alloy. Next, we note that there are 7 total parts of alloy, of which 5 are tin. Thus, 5 is to 7 as x is to 35. This gives us the equation

$$\frac{5}{7} = \frac{x}{35}$$

Multiplying each side by 35, we get

$$\frac{35(5)}{7} = \frac{35x}{35}$$

or

$$x = 25 \text{ kg}$$

Thus, there are 25 kg of tin and 10 kg of lead in the given amount of alloy.

We have shown that a proportion is a statement of equality. Now, to explain a special property associated with a proportion, we restate Eq. (3-8).

$$\frac{a}{b} = \frac{c}{d} \tag{3-8}$$

The **means** *of this proportion are the quantities b and c, and the* **extremes** *of this proportion are the quantities a and d.* If we multiply b by c, we get bc, and if we multiply a by d, we get ad. Now if we multiply each side of Eq. (3-8) by bd, we find that

$$\frac{abd}{b} = \frac{cbd}{d} \quad \text{or} \quad ad = bc$$

Therefore, in a proportion *the product of the means is equal to the product of the extremes.*

Example H Find the value of x in $\frac{x}{6} = \frac{4}{3}$.

$\frac{x}{6} = \frac{4}{3}$ (the original proportion)

$3x = 24$ (multiply means and extremes)

$x = 8$ (divide both members by 3)

Checking, we see that $\frac{8}{6} = \frac{4}{3}$.

Example I Write the proportion that is equivalent to the equation $3x = 4y$.

Since the product of the means equals the product of the extremes, and using the left side of the equation as the product of the means, we identify 3 and x to be the means and 4 and y to be the extremes. Thus, the proportion is

$$\frac{4}{3} = \frac{x}{y}$$

We could just as well have used the left member of the equation as the product of the extremes. This would have yielded the proportion

$$\frac{3}{4} = \frac{y}{x}$$

which is equivalent to the proportion obtained with the order reversed.

Sometimes the means of a proportion happen to be equal. When that occurs, we have

$$\frac{a}{b} = \frac{b}{c}$$

Here b is called *the* **mean proportional** *between a and c*. Now, using the method of the products of means and extremes, we obtain $b^2 = ac$. We shall explain how to solve such equations in Chapter 8.

Exercises 3-5

In Exercises 1 through 4, set up the given ratios as fractions and simplify when possible.

1. 12 to 5; 3 to 23 **2.** 2 to 11; 13 to 7

3. 21 to 3; 2 to 12 **4.** 5 to 45; 60 to 15

In Exercises 5 through 16, find the indicated ratios.

5. 30 A to 8 A **6.** 4 mi to 22 mi **7.** 9 cm to 30 cm

8. 50 W to 16 W **9.** 8 in. to 4 ft **10.** 5 cm to 40 mm

11. 80 s to 3 min **12.** 500 lb to 3 t **13.** 12 m to 6 s

14. 2 in. to 100°C **15.** 8 lb to 36 ft^3 **16.** 25 L to 35 h

In Exercises 17 through 24, solve the given proportions for x.

17. $\frac{x}{2} = \frac{5}{8}$ **18.** $\frac{x}{3} = \frac{7}{12}$ **19.** $\frac{3}{14} = \frac{x}{4}$ **20.** $\frac{8}{15} = \frac{x}{9}$

21. $\frac{3}{x} = \frac{7}{5}$ **22.** $\frac{18}{5} = \frac{9}{x}$ **23.** $\frac{1}{x} = \frac{8}{3}$ **24.** $\frac{1}{x} = \frac{8}{9}$

In Exercises 25 through 28, write the proportions for the given equations. Consider the left side of each equation to be the product of the means.

25. $2x = 8y$ **26.** $7t = 14u$ **27.** $3m = 27n$ **28.** $5a = 25b$

In Exercises 29 through 42, solve the given problems by first setting up the proper proportion.

29. The ratio of a number to 6 is the same as the ratio of 70 to 30. Find the number.

30. The ratio of a number to 15 is the same as the ratio of 17 to 60. Find the number.

31. The ratio of a number to 40 is the same as the ratio of 7 to 16. Find the number.

32. The ratio of a number to 44 is the same as the ratio of 8 to 33. Find the number.

33. 908 g = 2 lb; what weight in grams is 10 lb?

34. 60 mi/h = 88 ft/s; what speed in miles per hour is 66 ft/s?

35. A rectangular picture 20 in. by 15 in. is to be enlarged so that the width of the enlargement is 25 in. What should be the length of the enlargement?

36. If 40 lb of sodium hydroxide will neutralize 49 lb of sulfuric acid, how many pounds of sodium hydroxide are needed to neutralize 21 lb of sulfuric acid?

37. A board 15 m long is cut into two pieces, the lengths of which are in the ratio of 2 to 3. What is the length of each piece?

38. The ratio of a person's federal income tax to state income tax is 10 to 3. If the person pays a total income tax of $2600, how much was paid to the federal government and how much to the state government?

39. A certain gasoline company sells regular gas and lead-free gas in the ratio of 5 to 2. If, in a month, the company sells a total of 14,000,000 gal, how many gallons of each type were sold?

40. A tablet of medication contains two substances in the ratio of 9 to 5. If a tablet contains 280 mg of medication, how much of each type is in the tablet?

41. A metal bar which is 35 in. long is divided into two parts. If their lengths are in a ratio of 3 to 4, find the lengths of the parts.

42. A woman has $3600 to distribute to her two favorite charities. She distributes the amount in the ratio of 2 to 3. Find the two amounts.

3-6 VARIATION

Variation can be explained through the use of applied situations. This is so because in many such situations the ratio of one quantity to another remains the same. This fact is the basis for determining many important formulas. Consider the illustrations in the following example.

Example A

1. In a given length, the ratio of the length in inches to the length in feet is 12. This may be written as

$$\frac{i}{f} = 12$$

2. Experimentation shows that the ratio of the electric resistance of a wire to the length of the wire is always the same. This may be written as

$$\frac{R}{l} = k$$

3. It can be shown that for a given mass of gas, if the pressure remains constant, the ratio of the volume of the gas to the thermodynamic temperature remains constant. This may be written as

$$\frac{V}{T} = k$$

This leads us to the concept of a **variable**. *This is a quantity that may take on different values in a given discussion.* On the other hand, *a* **constant** *takes on only one value during a given discussion.* We may express the fact that a ratio of two variables remains constant by the equation

$$\boxed{\frac{y}{x} = k} \tag{3-9}$$

where x and y are the variables and k is the constant.

Example B

1. In the first illustration of Example A, the number of inches i and the number of feet f are variables. Each may take on essentially any value so long as their ratio remains 12. This means that when we consider many different distances, and therefore many different values of i and f, the ratio is always 12.

2. In the second illustration of Example A, the resistance R and the length l are the variables and k is the constant. The value of k may differ for different wires, but *for a given wire* k takes on a specific value, regardless of the length of the wire.

3. In the third illustration of Example A, the volume V and temperature T are the variables and k is the constant. Again, k may differ from one body of gas to another, but for a given body of gas k takes on a specific value, whereas the volume and temperature of that body of gas may change.

By convention, letters near the end of the alphabet, such as x, y, and z, are used to denote variables. Letters near the beginning of the alphabet, such as a, b, and c, are used as constants. The meaning of other letters, such as k in this case, are specified in the problem.

If we solve the equation $y/x = k$ for y, we get

$$\boxed{y = kx} \tag{3-10}$$

This equation is read as "y **is proportional to** *x" or "y* **varies directly as** *x." This type of relationship is known as* **direct variation.**

Given a set of values for x and y, we can determine the value of the **constant of proportionality** k. Then we can substitute this value for k to obtain the relationship between x and y. Then we can find y for any other value of, say, x and conversely. Consider the following example.

Example C If y varies directly as x, and $x = 6$ when $y = 18$, find the value of y when $x = 5$.

First, we write

$$y = kx$$

to denote that y varies directly as x. Next, we substitute $x = 6$ and $y = 18$ into the equation. This leads to

$$18 = 6k$$

or $k = 3$. Thus, for the present discussion the constant of proportionality is 3, and this may be substituted into $y = kx$, giving

$$y = 3x$$

as the equation between y and x. Now for any given value of x we may find the value of y by substitution. For $x = 5$, we get

$$y = 3(5) = 15$$

In many problems one variable will vary directly as some power other than the first of another variable. The following example illustrates this in the case of an applied problem.

Example D The distance d that an object falls under the influence of gravity is proportional to the square of the time t of fall. If an object falls 64 ft in 2 s, how far does it fall in 6 s?

To express the fact that d varies directly as the square of t, we write

$$d = kt^2$$

Then, using the fact that $d = 64$ ft when $t = 2$ s, we get

$$64 = k(2^2)$$

which gives us $k = 16$ ft/s^2. Thus, in general,

$$d = 16t^2$$

We now substitute $t = 6$ s, which gives

$$d = 16(6^2) = 16(36) = 576 \text{ ft}$$

This means that an object falling under the influence of gravity will fall 576 ft in 6 s.

Note that the constant of proportionality in Example D has a set of units associated with it. This will be the case unless the quantities related by k have precisely the same units. We can determine the units for k by solving the equation for k and noting the units on the other side of the equation.

Example E In Example D, when we solve for k, we can find its units as well as its value if we include the units in the calculation. In this case, we have

$$\begin{aligned} d &= kt^2 \\ 64 \text{ ft} &= k(2 \text{ s})^2 \\ k &= \frac{64 \text{ ft}}{4 \text{ s}^2} = 16 \text{ ft/s}^2 \end{aligned}$$

In the first illustration of Example A, the 12 has units of inches per foot. In the second illustration, if R is measured in ohms and l in feet, the units for k are ohms per foot. In the third illustration, if V is measured in liters and T in kelvins, the units for k are liters per kelvin. The units actually used in a given problem will determine the units in which k will be measured in that problem.

Another important type of variation is **inverse variation**, *which is expressed by the equation*

$$y = \frac{k}{x} \tag{3-11}$$

Here y **varies inversely as** *x, and k is the constant of proportionality.*

Example F A business firm found that the volume V of sales of a certain item was inversely proportional to the price p.

This statement is expressed as

$$V = \frac{k}{p}$$

If we know that $V = 1000$ sales/week for $p = \$5$, we get

$$1000 = \frac{k}{5}$$

or $k = 5000$ dollar-sales/week. This means that the equation relating V and p is

$$V = \frac{5000}{p}$$

For any value of p, we may find the corresponding value of V. For example, if $p = \$4$, we get

$$V = \frac{5000}{4} = 1250 \text{ sales/week}$$

Finally, it is possible to relate one variable to more than one other variable by means of **combined variation**. Let us consider the following example.

Example G

1. The equation expressing the fact that the force F between two electrically charged particles, with charges q_1 and q_2, varies directly as the product q_1 and q_2 is

$$F = kq_1q_2$$

2. The equation expressing the fact that y varies directly as x and inversely as z is

$$y = \frac{kx}{z} \tag{3-12}$$

Note that the word *and* appears in the statement, but the formula contains only products and quotients. The *and* is used only to note that y varies in more than one way; it does not imply addition.

3. The equation expressing the fact that s varies directly as the square of t and inversely as the cube of v is

$$s = \frac{kt^2}{v^3}$$

Note again that only a product and quotient appear in the formula.

As before, in each case we can determine k by knowing one set of values of the variables. Then we can evaluate one of the variables for a given set of values of the others.

Exercises 3-6

In Exercises 1 through 12, express the given statements as equations.

1. y varies directly as t.

2. x varies directly as s.

3. y varies directly as the square of s.

4. s varies directly as the cube of t.

5. t varies inversely as y.

6. y varies inversely as the square of x.

7. y varies directly as the product st.

8. s varies directly as the product xyz.

9. y varies directly as s and inversely as t.

10. y varies directly as s and inversely as the square of t.

11. x varies directly as the product yz and inversely as the square of t.

12. v varies directly as the cube of s and inversely as t.

In Exercises 13 through 22, give the equation relating the variables after evaluating the constant of proportionality for the given set of values.

13. y varies directly as s, and $y = 25$ when $s = 5$.

14. y varies inversely as t, and $y = 2$ when $t = 7$.

15. s is proportional to the cube of t, and $s = 16$ when $t = 2$.

16. v is proportional to the product st, and $v = 18$ when $s = 2$ and $t = 3$.

17. u is inversely proportional to the square of d, and $u = 17$ when $d = 4$.

18. q is inversely proportional to the square root of p, and $q = 5$ when $p = 9$.

19. y is directly proportional to x and inversely proportional to t, and $y = 6$ when $x = 2$ and $t = 3$.

20. t is directly proportional to n and inversely proportional to p, and $t = 21$ when $n = 3$ and $p = 5$.

21. d is directly proportional to s and inversely proportional to the square root of v, and $d = 20$ when $s = 6$ and $v = 4$.

22. w is inversely proportional to v and directly proportional to the square root of t, and $w = 12$ when $v = 8$ and $t = 25$.

In Exercises 23 through 32, find the required value by setting up the general equation and then evaluating.

23. Find s when $t = 4$ if s varies directly as t and $s = 20$ when $t = 5$.

24. Find y when $x = 5$ if y varies directly as x and $y = 36$ when $x = 2$.

25. Find q when $p = 5$ if q varies inversely as p and $q = 8$ when $p = 4$.

26. Find y when $x = 6$ if y varies inversely as x and $y = 15$ when $x = 3$.

27. Find z when $x = 3$ if z is directly proportional to the square of x and $z = 64$ when $x = 4$.

28. Find v when $t = 7$ if v varies inversely as the square of t and $v = 1$ when $t = 6$.

29. Find y when $x = 4$ and $z = 6$ if y varies directly as the product xz and $y = 9$ when $x = 3$ and $z = 5$.

30. Find s when $p = 75$ and $q = 5$ if s varies directly as p and inversely as the square of q and $s = 100$ when $p = 4$ and $q = 6$.

31. Find d when $s = 12$ and $v = 25$ if d varies inversely as s and directly as the square root of v and $d = 5$ when $s = 8$ and $v = 81$.

32. Find w when $u = 8$ and $t = 4$ if w varies directly as u and inversely as the sum $u + t$ and $w = 6$ when $u = 4$ and $t = 3$.

Exercises 33 through 46 are applied problems. To solve each one, set up the solution in terms of the indicated variation. Then proceed as required.

33. If the velocity v of an object falling under the influence of gravity is proportional to the time t of fall, find the equation relating v and t if $v = 64$ ft/s when $t = 2$ s.

34. An apartment owner charges rent R proportional to the floor area A of the apartment. Find the equation she uses relating R and A if an apartment of 100 m^2 rents for \$200 per month.

35. In biology, it is found that under certain conditions the rate of increase v of bacteria is proportional to the number N of bacteria present. If $v = 50$ bacteria/h when $N = 1000$ bacteria, what is v when $N = 3500$ bacteria?

36. It was determined that a particular type of automobile engine produces p cubic feet of carbon monoxide proportional to the time t that it idles. If $p = 3$ ft^3 for $t = 5$ min, find p for $t = 8$ min.

37. If the ratio of the circumference to the diameter of a circle is always π, find the radius of the circle whose circumference is 62.8 in. (Use $\pi = 3.14$.)

38. If the ratio of the area to the radius squared of a circle is always π, find the area of the circle whose radius is 5.00 cm. (Use $\pi = 3.14$.)

39. In economics, it is often found that the demand D for a product varies inversely as the price P. If the demand is 500 units per week at a cost of \$8 per unit, what would the demand be if the price was \$10 each?

40. A psychologist determined that the number of words W remembered from a particular list by a person was inversely proportional to the time t that had passed since the person last studied the list. If 10 words are remembered after 4 min, how many words are remembered after 8 min?

41. The number of gallons V required to fill a cubical container varies directly as the cube of the edge e of the container. If 60 gal fills a container of edge 2 ft, find V if $e = 4$ ft.

42. The time t necessary for an elevator to rise varies directly as the product of the weight w being lifted and the distance s. If 10 s is required to lift 400 lb through 60 ft, how long will it take the elevator to lift 500 lb through 80 ft?

43. The illuminance I of a light source varies inversely as the square of the distance d from the source. Determine the equation between I and d if $I = 25$ units when $d = 4$ m.

44. A person on a diet found that the number of pounds P lost in a week was inversely proportional to the square root of the number N of the week on the diet. If he lost 2 lb during the fourth week, how many pounds did he lose in the sixteenth week?

45. In chemistry, the general gas law states that the pressure P of a gas varies directly as the absolute temperature T and inversely as the volume V. Express this statement as a formula. Use R as the constant of proportionality.

46. The electric resistance R of a wire varies directly as the length l and inversely as the square of the diameter d of the wire. Express this as an equation, and then solve for l.

3-7 SUMMARY

In this chapter, we introduced the basic methods of solving equations and inequalities. These methods are used to isolate the required literal number on one side of the equation or inequality by placing all other quantities on the other side. The methods used for solving equations are:

1. Adding the same quantity to each side of the equation.
2. Subtracting the same quantity from each side of the equation.
3. Multiplying each side of the equation by the same number (not zero).
4. Dividing each side of the equation by the same number (not zero).

These same methods are used when solving inequalities, with one special situation: when each member of the inequality is multiplied or divided by a negative number, the sense of the inequality is reversed.

We also presented an introduction to ratio, proportion, and variation. In this presentation, we showed that a proportion is a statement of equality of two ratios. Also, the method of developing formulas through variation was shown.

In discussing these topics, we introduced additional important terminology. The reader who has completed this chapter should be familiar with the following list of mathematical terms.

Conditional equation
Constant
Constant of proportionality
Extremes
Equation
Identity
Inequality
Mean proportional
Means
Member of inequality
Proportion
Proportional to
Ratio
Sense of inequality
Solution set
Solving an equation
Subscript
Unknown
Variable
Variation
  combined
  direct
  inverse

The following exercises will help the reader review the topics of the chapter.

Exercises 3-7

In Exercises 1 through 12, find the solution set for each of the equations. Check by substituting the value found in the original equation.

1. $x - 3 = 5$

2. $x + 14 = 11$

3. $3y = 27$

4. $26q = 13$

5. $4x + 21 = 5$

6. $2n - 1 = 17$

7. $3(x + 1) = x + 11$

8. $2(y - 1) = 19 - 5y$

9. $2(1 - 2t) = 11 - t$

10. $19 - 3x = 5(1 - 2x)$

11. $7(s - 1) + 2(s + 2) = 3s$

12. $8(t + 2) = 3(2t + 13)$

In Exercises 13 through 24, solve for the indicated letter in terms of the others. Fields of study in which each formula was developed are in parentheses.

13. $S = C + E + P$, for P (business: sales and profits)

14. $cE = 1 - g$, for g (psychology: learning)

15. $F = \frac{wa}{g}$, for g (physics: force)

16. $A = \frac{ab}{2}$, for a (mathematics: geometry)

17. $M_1V_1 + M_2V_2 = PT$, for M_1 (economics: finance)

18. $Q_1 = P(Q_2 - Q_1)$ for Q_2 (refrigeration)

19. $T = \dfrac{L(D-d)}{2L_1}$, for D (machine design)

20. $Q = \dfrac{kAT(t_2 - t_1)}{d}$, for t_2 (physics: heat conduction)

21. $a(2+3x) = 3y + 2ax$, for y

22. $c(ax+c) = b(x+c)$, for a

23. $a(x+b) = b(x+c)$, for c

24. $3a(a+2x) + a^2 = a(2+a)$, for x

In Exercises 25 through 40, solve the given inequalities.

25. $x - 2 < 7$

26. $x + 4 \leq 2$

27. $2x \geq 10$

28. $6x > 18$

29. $3x + 10 < 1$

30. $5x - 1 < 19$

31. $6 - x \geq 10$

32. $5 - x > 4$

33. $5(x+1) < x - 3$

34. $4(x+2) \leq x - 4$

35. $12 - 2x > 3(x-1)$

36. $9 - x > 3(x-5)$

37. $6(x-2) \quad x + 3$

38. $4x + 3 \geq 2(x+7)$

39. $-2(3+x) \leq 2(-2x+1)$

40. $3(x-5) \leq -4(-x+2)$

In Exercises 41 through 52, express the given statements as formulas.

41. y varies directly as the square of z.

42. y varies inversely as the cube of q.

43. t is directly proportional to s and inversely proportional to the fourth power of x.

44. s is directly proportional to the square root of a and inversely proportional to p.

45. w is directly proportional to the cube of the product mn.

46. x is directly proportional to the cube root of the product st.

47. y is inversely proportional to the square of the product pq.

48. p is inversely proportional to the square root of the product rt.

49. x varies directly as the square of the product mn.

50. y varies inversely as the cube of the product pq.

51. w is inversely proportional to s and directly proportional to the square root of the product rt.

52. v is directly proportional to s and inversely proportional to the square root of the product pq.

In Exercises 53 through 60, solve the given proportions for x. In Exercises 57 through 60, compare the solutions of parts (a) and (b) in each exercise.

53. $\dfrac{x}{4} = \dfrac{9}{10}$

54. $\dfrac{x}{5} = \dfrac{16}{9}$

55. $\dfrac{3}{10} = \dfrac{x}{15}$

56. $\dfrac{6}{5} = \dfrac{x}{10}$

57. (a) $\dfrac{2}{9} = \dfrac{x}{3}$ (b) $\dfrac{9}{2} = \dfrac{3}{x}$

58. (a) $\dfrac{1}{12} = \dfrac{7}{x}$ (b) $\dfrac{12}{1} = \dfrac{x}{7}$

59. (a) $\dfrac{21}{x} = \dfrac{2}{5}$ (b) $\dfrac{x}{21} = \dfrac{5}{2}$

60. (a) $\dfrac{2}{13} = \dfrac{11}{x}$ (b) $\dfrac{13}{2} = \dfrac{x}{11}$

In Exercises 61 through 64, solve the given problems using variation.

61. y varies directly as x and inversely as z; $y = 6$ when $x = 2$ and $z = 3$. Find the resulting equation relating y, x, and z; and solve for x.

62. s varies directly as the square of t and inversely as v; $s = 18$ when $t = 3$ and $v = 4$. Find the resulting equation relating s, v, and t; and solve for v.

63. y varies directly as the product $s\sqrt{t}$; $y = 12$ when $s = 3$ and $t = 4$. Find y for $s = 5$ and $t = 9$.

64. m varies directly as n and inversely as the square of p; $m = 6$ when $n = 3$ and $p = 2$. Find n for $m = 16$ and $p = 3$.

In Exercises 65 through 76, set up the indicated equations, inequalities, ratios, and proportions from the given statements and then determine the solution set.

65. Five more than twice a number equals eleven. Find the number.

66. Two more than a number equals three times the number. Find the number.

67. Twice the sum of a number and four equals the number. Find the number.

68. One more than a number is three times the sum of the number and five. Find the number.

69. If any of the numbers in a given set is doubled, the product is less than three less than the number. Find the set of numbers.

70. If seven is added to any of the numbers of a given set, the sum is greater than six less than twice the number. Find the set of numbers.

71. If three is subtracted from any of the numbers in a given set, the result is less than or equal to three more than three times the number. Find the set of numbers.

72. If any of the numbers of a given set of numbers is doubled, the result is greater than or equal to seven more than the number. Find the set of numbers.

73. The ratio of a number to 5 is the same as the ratio of 7 to 15.

74. The ratio of a number to 7 is the same as the ratio of 25 to 28.

75. The ratio of a number to 12 is the same as the ratio of 7 to 8.

76. The ratio of a number to 25 is the same as the ratio of 13 to 15.

In Exercises 77 through 88, solve the given applied problems.

77. The United States House of Representatives has 435 members. If there are D Democrats and R Republicans and all members are in one of these two parties, set up the equation with R and D and solve for R.

78. The total cost T of producing x items is F dollars of fixed costs plus \$5 per item produced. Set up the equation with T, F, and x and solve for x.

79. A person determines that she must spend less than \$30 for two items. If one of them costs \$8, what possible prices may the other item have?

80. The area of a rectangle is its length times its width. What values may the length l of a rectangle be if its width is 5 m and its area is greater than 45 m^2?

81. A manufacturer determines that the number r of aluminum cans that can be made by recycling n used cans is proportional to n. How many recycled cans can be made from 50,000 used cans if $r = 115$ cans for $n = 125$ cans?

82. Due to inflation, the increase in cost of a certain item is proportional to the time since 1960. If the price increased \$5 from 1960 to 1970, what should be the increase in price from 1960 to 1980?

83. The weight w on the end of a spring varies directly as the length x that the spring stretches. If a weight of 6 lb stretches a spring 2 in., what weight will stretch it 3 in.?

84. A relation which has been used in psychology is that the strength S of a sensation is proportional to the nth power of the stimulus intensity I. Express this statement as an equation.

85. The period of a pendulum varies directly as the square root of its length. If a pendulum 4 ft long has a period of 2.22 s, what is the period of a pendulum 9 ft long?

86. The cost C of using an electric appliance varies as the product of the wattage w of the appliance, the time t it is used, and the electric rate r. If it costs 12¢ to operate a 400-W color television set for 6 h at a rate of 5¢/kW·h, how much does it cost to operate a 75-W electric light bulb for 10 h at 6¢/kW·h?

87. The amount of heat H which passes through a wall t ft thick is proportional to the product of the area A of the wall and $T_2 - T_1$, the difference in temperature of the surfaces of the wall, and is inversely proportional to t. Express this as an equation, and solve for T_2.

88. Let s indicate the outcomes favorable to the occurrence of an event E, and let f indicate the outcomes unfavorable to the occurrence of that event E. Then the probability P of the occurrence of E is equal to the ratio of the favorable outcomes to the total outcomes. Express this as an equation.

4

BASIC ALGEBRAIC OPERATIONS

4-1 ADDITION AND SUBTRACTION

In Section 2-3, we introduced the basic operations in algebra. These operations were used in solving the equations and inequalities in Chapter 3. However, it is necessary to develop these basic operations further for our later work. Therefore, in this chapter, we shall reconsider these operations, developing them in greater detail.

We recall from Section 2-3 that, *in the addition of algebraic expressions, we may combine only like terms; we can only indicate the sum of unlike terms.* Also, we showed the use of parentheses for grouping terms. Example A illustrates algebraic addition.

Example A Add $4xy + 3a - x$ and $2xy - 5a + 2x$.

This addition may be written as

$$\begin{array}{l} 4xy + 3a - x \\ \underline{2xy - 5a + 2x} \\ 6xy - 2a + x \end{array}$$

where the sum is $6xy - 2a + x$. We were able to combine only the like terms. We also note that the sum of $3a$ and $-5a$ is $-2a$. A more common method of expressing addition in algebra is by means of parentheses. Thus, for the sum given above, we write

$$(4xy + 3a - x) + (2xy - 5a + 2x)$$

It is not possible to proceed with the addition until we write an equivalent expression without parentheses. Only then are we able to combine the like terms. Removing the parentheses from this expression, we have

$$4xy + 3a - x + 2xy - 5a + 2x$$

Combining like terms, we have $6xy - 2a + x$, which is the above result. Thus, we may state that

$$\begin{aligned}(4xy + 3a - x) + (2xy - 5a + 2x) &= 4xy + 3a - x + 2xy - 5a + 2x\\ &= 6xy - 2a + x\end{aligned}$$

Generalizing on the results of this example, we see that

when parentheses are preceded by a plus sign, they may be removed if the sign of each term within the parentheses is retained.

From Section 1-3, recall that, when subtracting a signed number, we may change the sign of the number and proceed as in addition. This principle is important for the subtraction of one algebraic expression from another. Example B illustrates the use of this principle in the subtraction of algebraic expressions.

Example B Subtract $2xy - 5a + 2x$ from $4xy + 3a - x$.

This subtraction may be written as

$$\begin{array}{r}4xy + 3a - x\\ \underline{2xy - 5a + 2x}\\ 2xy + 8a - 3x\end{array}$$

where the difference is $2xy + 8a - 3x$. In this subtraction, we may consider three separate subtractions. The first is $4xy - 2xy$, which results in $2xy$. The second is $(+3a) - (-5a)$, which, by the principle of subtracting signed numbers, becomes $(+3a) + (+5a) = 3a + 5a = 8a$. The third subtraction is $(-x) - (+2x)$, which becomes

$$(-x) + (-2x) = -x - 2x = -3x$$

The combination of these three differences gives the result shown above.

Using parentheses to indicate the subtraction, we have

$$(4xy + 3a - x) - (2xy - 5a + 2x)$$

which is a more common algebraic type of expression. Since in each individual subtraction shown above it is necessary to change the sign of the number being subtracted, we remove the second parentheses here and also change the sign of each term within them. This leads to

$$4xy + 3a - x - 2xy + 5a - 2x = 2xy + 8a - 3x$$

which agrees with the above result.

Generalizing on the results of this example, we see that

when parentheses are preceded by a minus sign, they may be removed if the sign of each term within the parentheses is changed.

Example C

$$\begin{aligned}4a - (a - 3x + 2y) + (-4x + y) &= 4a - a + 3x - 2y - 4x + y \\ &= 3a - x - y\end{aligned}$$

When we remove the first set of parentheses, the sign of a (understood to be positive) is changed to negative, the sign of $-3x$ is changed to positive, and the sign of $2y$ is changed to negative. When we remove the second set of parentheses, the signs of $-4x$ and y are retained. In removing the parentheses, we also remove the signs preceding the parentheses.

Brackets [ ] *and* **braces** { } *are also used to group terms, particularly when a group of terms is contained within another group.* (Another symbol of grouping, the **bar** or **vinculum,** is used, especially for radicals and fractions. The expression $\sqrt{a + b}$ means the square root of the *quantity* $a + b$, and

$$\frac{c}{a + b}$$

means c divided by the *quantity* $a + b$.) In simplifying expressions containing more than one type of grouping symbols, we may remove the symbols one at a time. In general, it is better to remove the innermost symbols first. Consider the illustrations of Example D and Example E.

Example D Simplify the expression $5a - [2a - (3a + 6)]$.

We shall start by removing the parentheses, and then we will remove the brackets. This leads to

$$\begin{aligned}5a - [2a - (3a + 6)] &= 5a - [2a - 3a - 6] \\ &= 5a - 2a + 3a + 6 \\ &= 6a + 6\end{aligned}$$

Example E In simplifying the expression $18 - \{[-(a-1)+3a] - 4a\}$, we shall remove the parentheses, brackets, and braces, in that order:

$$\begin{aligned} 18 - \{[-(a-1)+3a] - 4a\} &= 18 - \{[-a+1+3a] - 4a\} \\ &= 18 - \{-a+1+3a-4a\} \\ &= 18 + a - 1 - 3a + 4a \\ &= 17 + 2a \end{aligned}$$

Algebraic addition and subtraction and the associated use of symbols of grouping are often encountered in equations. Consider Example F.

Example F Solve the equation $2 - (3a - 4x) = 2x + (7 - a)$ for x.

We shall first remove the parentheses, then collect terms with x on the left and the other terms on the right. This leads to

$$\begin{aligned} 2 - (3a - 4x) &= 2x + (7 - a) \\ 2 - 3a + 4x &= 2x + 7 - a \\ 4x - 2x &= 7 - a - 2 + 3a \\ 2x &= 5 + 2a \\ x &= \frac{5 + 2a}{2} \end{aligned}$$

Exercises 4-1

In Exercises 1 through 4, add the given expressions.

1. $\begin{array}{r} 2s - 3xy + 5a \\ s + 2xy - 6a \\ \hline \end{array}$

2. $\begin{array}{r} 3t^2 - 4as - p \\ -9t^2 - 5as + 2h \\ \hline \end{array}$

3. $\begin{array}{r} 5y - 2x + 4a - 8xy \\ 4y - x - 8a - xy \\ \hline \end{array}$

4. $\begin{array}{r} 8u - rs + 2y - 3s \\ -2u - 3rs - 7y - 6s \\ \hline \end{array}$

In Exercises 5 through 8, subtract the lower expression from the upper expression.

5. $\begin{array}{r} 5x - 6xy \\ 2x - 8xy \\ \hline \end{array}$

6. $\begin{array}{r} 4as - 9py - s \\ 6as + 2py + 4s \\ \hline \end{array}$

7. $\begin{array}{r} 4x^2 - 6xy - 4s \\ 7x^2 - 9xy - 2s \\ \hline \end{array}$

8. $\begin{array}{r} -7y + 2w - 5u + 7uy \\ 6y - 3w - 4u - 8uy \\ \hline \end{array}$

In Exercises 9 through 24, remove, the symbols of grouping and simplify the given expressions.

9. $4a + (3 - 2a)$

10. $6x + (-5x + 4)$

11. $3 - (4x + 7)$

12. $9y - (4y - 8)$

13. $(4s - 9) + (8 - 2s)$

14. $(5x - 2 + 3y) + (4 - 3x)$

15. $(t - 7 + 3y) - (2 - y + t)$

16. $-(6b^3 - 3as + 4x^2) - (2s - 3b^3 - 6x^2)$

17. $4 + [6x - (3 - 4x)]$

18. $3 + [5n - (2 - n)]$

19. $3s - [2 - (4 - s)]$

20. $7 - [2 - (4x^2 - 2)]$

21. $-(7 - x) - [(2x + 3) - (7x - 2)]$

22. $-(5 - x) - [(5x - 7) - (2 - 3x + b)]$

23. $(t - 5x) - \{[(6p^2 - x) - 9] - (6t + p^2 - x)\}$

24. $8 - \{6xy - [7 - (2xy - 5)] - [6 - (xy - 8)]\}$

In Exercises 25 through 32, solve the given equations for x.

25. $3x = 7 - (4 - 2x)$

26. $5 - (x - 3) = x$

27. $5 - 7x = 2 - (5x - 1)$

28. $6 - (3 - x) = (5 + 4x) - (x - 1)$

29. $-(x - 8) = 2x - (5 - 7x)$

30. $4a + (2x - a) = 5 - (a - x)$

31. $(t - x) - (x + t) = (2x - 1) - (5 - x)$

32. $-(2a - x) - (4x - 5) = -[5x - (3 - x)]$

In Exercises 33 through 36, solve the given problems.

33. Twice a number equals the sum of the number and six subtracted from three. Find the number.

34. In a reforesting program, $2x + 1000$ trees were planted in one area and $x - 2000$ were planted in a second area. How many more trees were planted in the first area?

35. In physics, when studying momentum, one encounters the equation

$$I = -(mv - mv_0)$$

Solve for v.

36. The net loss L that a firm had on the sale of an item can be expressed as

$$L = C - (P_2 - P_1)$$

where C is the overhead cost, P_2 is the selling price, and P_1 is the cost to the firm. Solve for P_2.

4-2 MULTIPLICATION

In Chapter 2, we introduced the multiplication of algebraic expressions. We also noted that the use of exponents is directly associated with multiplication. We shall now show a general expression for multiplication of powers of a number.

The expression a^n is read as "a to the nth power," where n is the **exponent** *and a is called the* **base.** The following example leads us to a conclusion regarding the multiplication of terms with the same base.

Example A

$$(a^3)(a^4) = (a \cdot a \cdot a)(a \cdot a \cdot a \cdot a) = a^7$$

However, $a^{3+4} = a^7$. We see that we can multiply a^3 by a^4 as follows:

$$(a^3)(a^4) = a^{3+4} = a^7$$

Generalizing on the results of Example A, we see that if we multiply a^m by a^n, a^m has m factors of a, a^n has n factors of a, and therefore

$$\boxed{(a^m)(a^n) = a^{m+n}} \qquad (4\text{-}1)$$

Another result that is important to our work with exponents is encountered when we raise a power of a number to a power. Consider Example B.

Example B

$$(a^3)^4 = (a^3)(a^3)(a^3)(a^3) = (a \cdot a \cdot a)(a \cdot a \cdot a)(a \cdot a \cdot a)(a \cdot a \cdot a) = a^{12}$$

However, $a^{3 \times 4} = a^{12}$. We see that we can raise a^3 to the fourth power by writing

$$(a^3)^4 = a^{3 \times 4} = a^{12}$$

Generalizing on the results of Example B, we see that, if we raise a^m to the nth power, we have

$$\boxed{(a^m)^n = a^{mn}} \qquad (4\text{-}2)$$

The distinction between the meaning of Eqs. (4-1) and (4-2) should be carefully noted and understood. These equations will be of considerable importance in our future work, since they will be of direct use on numerous occasions.

To multiply terms containing more than one base, the exponents of like bases are combined, respectively. Example C demonstrates this.

Example C

$$(a^3b^2)(a^2b^4) = a^3b^2a^2b^4 = a^3a^2b^2b^4 = a^{3+2}b^{2+4} = a^5b^6$$

We can only indicate the multiplication of a^5 and b^6 as shown. It is not possible to simplify further, just as the addition of algebraic expressions is complete when like terms have been combined.

The power of a product leads to another important result in operating with exponents. The following examples will enable us to develop a general expression for this case.

Example D

$$(xy)^4 = (xy)(xy)(xy)(xy) = x^4y^4$$

$$(3c^2d^5)^3 = (3c^2d^5)(3c^2d^5)(3c^2d^5) = 27c^6d^{15}$$

However, $3^3(c^2)^3(d^5)^3 = 27c^6d^{15}$.

We see from Example D that *if a product of factors is raised to a power n, the result is the product of each factor raised to the power n.* Generalizing this result, we have

$$(ab)^n = a^nb^n \qquad (4\text{-}3)$$

When multiplying monomials, we use the notation of exponents and the rules for multiplying signed numbers. We first multiply the numerical coefficients and then the literal factors. The product of the new numerical coefficient and the literal factors is the required product.

Example E

$$2ab^3(3a^2bc) = (2 \cdot 3)(a \cdot a^2 \cdot b^3 \cdot b \cdot c) = 6a^3b^4c$$

$$-2x^2y^3(4ax^2y) = -8ax^4y^4$$

$$(-p^2q)(3pq^2r)(-2qrs) = 6p^3q^4r^2s$$

To multiply a monomial and a multinomial, we make use of the distributive law (see Section 2-3), which states that we must *multiply each term of the multinomial by the monomial.* Also, we must be careful to assign the correct sign to each term of the result in accordance with the rules for multiplication of signed numbers.

Example F

$$\begin{aligned} 5ab^2(-3ac - 2b^3x) &= 5ab^2(-3ac) + 5ab^2(-2b^3x) \\ &= -15a^2b^2c - 10ab^5x \end{aligned}$$

$$\begin{aligned} 2xy^2(3x + y - 4x^2y) &= 2xy^2(3x) + 2xy^2(y) + 2xy^2(-4x^2y) \\ &= 6x^2y^2 + 2xy^3 - 8x^3y^3 \end{aligned}$$

To multiply one multinomial by another, we multiply each term of one by each term of the other, again using the rules outlined above. This is a result of the use of the distributive law.

Example G

$$(x-2)(x+7) = x(x) + x(7) + (-2)(x) + (-2)(7)$$
$$= x^2 + 7x - 2x - 14$$
$$= x^2 + 5x - 14$$

$$(x^2 - 2xy)(ab + a^2) = x^2(ab) + x^2(a^2) + (-2xy)(ab) + (-2xy)(a^2)$$
$$= abx^2 + a^2x^2 - 2abxy - 2a^2xy$$

Example H

$$(2a+4)(2a^2 + 3a - b) = (2a)(2a^2) + (2a)(3a) + (2a)(-b) + (4)(2a^2)$$
$$+ (4)(3a) + (4)(-b)$$
$$= 4a^3 + 6a^2 - 2ab + 8a^2 + 12a - 4b$$
$$= 4a^3 + 14a^2 - 2ab + 12a - 4b$$

Example I The expression $(a + 2b)^2$ is equivalent to $(a + 2b)(a + 2b)$. Therefore, we square $(a + 2b)$ as follows:

$$(a + 2b)^2 = (a + 2b)(a + 2b)$$
$$= a(a) + a(2b) + 2b(a) + 2b(2b)$$
$$= a^2 + 2ab + 2ab + 4b^2$$
$$= a^2 + 4ab + 4b^2$$

A common error which is made in squaring (or raising to any power) an expression is simply to square the individual terms. Here we see that $(a + 2b)^2$ is not $a^2 + 4b^2$. The result must include the $4ab$ term.

Exercises 4-2

In Exercises 1 through 56, perform the indicated multiplications.

1. x^3x^7 **2.** n^2n^6 **3.** y^5y^2y **4.** $p^2p^3p^5$

5. $(x^3)^7$ **6.** $(n^2)^6$ **7.** $(y^2y)^5$ **8.** $(p^2p^3)^5$

9. $(3ax)(-2a^2xy)$ **10.** $(-6yrs^2)(-3xy^2r^3)$ **11.** $(-5act^2)(9at^5)$

12. $(8dr^3s^2)(-4drs^4)$ **13.** $(-4rs)(-3st^3)(-7rt)$ **14.** $(-2axy^3)(7ay^4)(-ax^4y)$

15. $(-2st^3x)^3$ **16.** $(-3axt^7)^4$ **17.** $2x(x^2 + 4a)$

18. $3a(a^3 - 6)$ **19.** $3a^2(2x - a^2)$ **20.** $4b^3(3 + 2b^2)$

21. $(-2st)(sx - t^2y)$ **22.** $(-8y)(8y + t^2)$ **23.** $(-3xy)(x^2y - 3axy^6)$

24. $(-5y^6)(-uy^7 - hpy)$ **25.** $(x + 3)(x - 1)$ **26.** $(x - 5)(x + 2)$

27. $(s-2)(s+3)$
28. $(x-6)(x-4)$
29. $(2x+3)(x-2)$
30. $(3x-4)(2x+1)$
31. $(x+1)(2x-1)$
32. $(3a-2)(5a+4)$
33. $(5v+1)(2v+3)$
34. $(3v-2)(4v-3)$
35. $(a-x)(a-2x)$
36. $(x+2y)(x-y)$
37. $(2a-c)(3a+2c)$
38. $(3s+4t)(5s+2t)$
39. $(2x-5t)(x+9)$
40. $(4x-9uy)(2+3uy)$
41. $(2a-9py)(2a+9py)$
42. $(s-3xu^2)(-xu^2-8s)$
43. $(2a+1)(a^2-3a-5)$
44. $(3x-2)(2x^2+x-3)$
45. $(a-x)(a+2xy-3x)$
46. $(2-x)(5+x-x^2)$
47. $(x-2)(x+3)(x-4)$
48. $(2x-3)(x+1)(3x-4)$
49. $(x-2)^2$
50. $(a+5)^2$
51. $(x+2y)^2$
52. $(2a-3b)^2$
53. $(x+1)^3$
54. $(x+2b)^3$
55. $(y-4c)^3$
56. $(2a-x)^3$

In Exercises 57 through 60, solve the given problems involving multiplication.

57. The length of a rectangle is $x+2$ ft, and the width of the rectangle is $x-1$ ft. Find an expression for the area of the rectangle.
58. In determining the focal length of a lens, one encounters the expression

$$(u-1)(r_2-r_1)$$

Carry out the multiplication.

59. Show by substitution that $(2x-y)(2x+y)=4x^2-y^2$ for the values $x=3$ and $y=-4$.
60. Show by substitution that $(a-b)(a^2+ab+b^2)=a^3-b^3$ for the values $a=-2$ and $b=-3$.

4-3 DIVISION: THE USE OF EXPONENTS

As in multiplication, the use of exponents plays an important role in division. Example A leads us to the basic formula for the use of exponents in division.

Example A Divide a^5 by a^2.

We may express this division as a fraction as stated in Section 2-2. This gives us

$$\frac{a^5}{a^2}$$

Now we use the principle stated in that section, namely, that if a fraction has a factor common to the numerator and denominator, this factor may be divided out. We note the common factor of a^2 in the numerator and the denominator, or

$$\frac{a^5}{a^2}=\frac{a^2a^3}{a^2}$$

Dividing out this factor of a^2, we have

$$\frac{a^5}{a^2} = a^3$$

However, $a^{5-2} = a^3$. We see that we can divide a^5 by a^2 as follows:

$$\frac{a^5}{a^2} = a^{5-2} = a^3$$

From Example A, we see that, if we divide a^m by a^n (a not zero) where $m > n$, we have

$$\boxed{\frac{a^m}{a^n} = a^{m-n}} \qquad (4\text{-}4)$$

Let us now consider the case of dividing a^m by a^n where $m < n$. This division is illustrated in Example B.

Example B Divide a^2 by a^5.

We express this division as a fraction and then divide out the common factor. This procedure leads to

$$\frac{a^2}{a^5} = \frac{a^2}{a^2a^3} = \frac{1}{a^3}$$

Here we see that

$$\frac{a^2}{a^5} = \frac{1}{a^{5-2}} = \frac{1}{a^3}$$

where we have subtracted the smaller exponent from the larger one.

Applying Eq. (4-4) to the fraction, we have

$$\frac{a^2}{a^5} = a^{2-5} = a^{-3}$$

If these two results are to be consistent, it must be that

$$\frac{1}{a^3} = a^{-3}$$

From Example B, we see that, if the exponent in the numerator is less than that in the denominator, we may express the result with a positive exponent and the resulting factor in the denominator. Also, we may express the result with a negative exponent, if we define

$$\boxed{a^{-n} = \frac{1}{a^n}} \tag{4-5}$$

This shows us that we may use negative integers as exponents.

Example C

$$\frac{x^3}{x^4} = \frac{1}{x^{4-3}} = \frac{1}{x} \quad \text{or} \quad \frac{x^3}{x^4} = x^{3-4} = x^{-1} = \frac{1}{x}$$

$$\frac{c^2}{3c^6} = \frac{1}{3c^{6-2}} = \frac{1}{3c^4} \quad \text{or} \quad \frac{c^2}{3c^6} = \frac{c^{2-6}}{3} = \frac{c^{-4}}{3} = \frac{1}{3c^4}$$

We now see that Eq. (4-5) shows us that *we may move a* **factor** *from the numerator to the denominator or from the denominator to the numerator by changing the sign of its exponent*. Consider the illustrations of Example D.

Example D

$$b^{-3} = \frac{1}{b^3} \qquad \frac{1}{b^{-3}} = b^3$$

$$a^2x = \frac{x}{a^{-2}} \qquad xa^{-2} = \frac{x}{a^2}$$

$$\frac{x}{2a^{-2}} = \frac{xa^2}{2} \qquad \frac{2a^{-2}}{x} = \frac{2}{a^2x}$$

Note that, in the last four illustrations, only the factor a is raised to the power 2 or -2.

Example E

1. $4x^{-2} = \dfrac{4}{x^2}$ since only x has the exponent of -2.

2. $(4x)^{-2} = \dfrac{1}{(4x)^2} = \dfrac{1}{16x^2}$ since the quantity $4x$ has the exponent of -2, as indicated by the parentheses.

3. $$\frac{1}{3a^{-1}} = \frac{1}{3\left(\frac{1}{a}\right)} = \frac{1}{\frac{3}{a}} = 1 \cdot \frac{a}{3} = \frac{a}{3}$$

In general, with certain specific exceptions, negative exponents are not used in the expression of a final result. They are, however, used in intermediate steps in many operations.

To divide one monomial by another, we use the basic principle of dividing out any factor common to the numerator and the denominator. As in multiplication, *we can combine only the exponents of the same base.*

Example F

$$-16a^2b \div 4a = \frac{-16a^2b}{4a} = -4ab$$

$$36a^2b^3 \div (-12ab) = \frac{36a^2b^3}{-12ab} = -3ab^2$$

$$-18x^3y \div (-12xy^4) = \frac{-18x^3y}{-12xy^4} = \frac{3x^2}{2y^3}$$

$$-8ab^2x^5 \div (-14a^2b^2x) = \frac{-8ab^2x^5}{-14a^2b^2x} = \frac{4x^4}{7a}$$

In each of the preceding illustrations, we have subtracted the smaller exponent from the larger exponent of any given base. In the first two illustrations, the exponents in the numerator are larger than those in the denominator. In the third illustration, the exponent of x is larger in the numerator while the exponent of y is larger in the denominator. In the fourth illustration, the exponent of a is larger in the denominator; the exponent of x is larger in the numerator. Also, b^2 is part of the common factor; and therefore, b does not appear in the final result.

Example G In reducing the fraction

$$\frac{x^2yz^3}{xy^4z^2}$$

to its simplest form, we may place all factors in the numerator by the use of negative exponents and combine exponents of the same base in order to determine the result. This leads to

$$\frac{x^2yz^3}{xy^4z^2} = x^2x^{-1}yy^{-4}z^3z^{-2} = x^{2-1}y^{1-4}z^{3-2} = xy^{-3}z = \frac{xz}{y^3}$$

As we stated previously, negative exponents are generally not used in the expression of a final result. They are, however, often used in intermediate steps.

Finally, let us turn to the division of a^m by a^n where $m = n$. In such a case, we are dividing a quantity by itself; and the result must be 1. That is, $a^m/a^m = 1$. Using Eq. (4-4), we have $a^m/a^m = a^{m-m} = a^0$. Since a has not been specified,

$$\boxed{a^0 = 1} \tag{4-6}$$

for any a not zero. This equation states that *any algebraic expression which is not zero and which is raised to the zero power is* 1.

Example H

$$5^0 = 1 \qquad x^0 = 1$$

$$(rs)^0 = 1 \qquad (ax - p)^0 = 1$$

$$(9x^2 - 8x - 3)^0 = 1 \qquad \left(\frac{ax - b}{cx + d}\right)^0 = 1$$

$$5x^0 = 5(1) = 5 \qquad (5x)^0 = 1$$

Note that in the expression $5x^0$, only x has the exponent zero.

Zero and negative exponents may be used in the other laws of exponents just as any positive exponent. For reference, we now list the laws of exponents we have encountered.

$$(a^m)(a^n) = a^{m+n} \tag{4-1}$$

$$(a^m)^n = a^{mn} \tag{4-2}$$

$$(ab)^n = a^n b^n \tag{4-3}$$

$$\frac{a^m}{a^n} = a^{m-n} \quad (a \text{ not zero}) \tag{4-4}$$

$$a^{-n} = \frac{1}{a^n} \quad (a \text{ not zero}) \tag{4-5}$$

$$a^0 = 1 \quad (a \text{ not zero}) \tag{4-6}$$

Example I

1. Using Eq. (4-1): $(a^5)(a^0) = a^{5+0} = a^5$

2. Using Eqs. (4-1) and (4-5): $a^{-2}b^{-1}a^5 = a^{5-2}b^{-1} = \dfrac{a^3}{b}$

3. Using Eqs. (4-2) and (4-5): $(a^5)^{-1} = a^{-5} = \dfrac{1}{a^5}$

4. Using Eqs. (4-3) and (4-5): $(x^{-2}y^3)^{-2} = x^{(-2)(-2)}y^{3(-2)} = x^4y^{-6}$

$$= \frac{x^4}{y^6}$$

5. Using Eq. (4-4): $\frac{x^{-2}}{x^{-7}} = x^{-2-(-7)} = x^{-2+7} = x^5$

6. Using Eqs. (4-5), (4-6), and (4-1): $\frac{r^{-3}s^2t^0}{rs^{-1}t^{-3}} = \frac{(s^2)(s)(1)(t^3)}{(r^3)(r)} = \frac{s^{2+1}t^3}{r^{3+1}} = \frac{s^3t^3}{r^4}$

Exercises 4-3

In Exercises 1 through 60, write each of the given expressions in the simplest form that contains only positive exponents.

1. $x^7 \div x^4$
2. $x^8 \div x^2$
3. $p^3 \div p^{13}$
4. $n \div n^7$
5. $\frac{a^5}{a^4}$
6. $\frac{y^9}{y^3}$
7. $\frac{a^2}{a^{11}}$
8. $\frac{s}{s^6}$
9. b^{-4}
10. c^{-5}
11. $\frac{1}{a^{-4}}$
12. $\frac{1}{b^{-5}}$
13. 8^0
14. $(10)^0$
15. $(a - 5b)^0$
16. $\frac{1}{(4x - 8)^0}$
17. $\frac{1^0}{1}$
18. $\frac{1}{1^0}$
19. $2ab^0$
20. $5x^0y^2$
21. $\frac{(4x - 2)^0}{2x^0}$
22. $\frac{(3y - 9)^0}{3^0y}$
23. $2^0(x - y^0)$
24. $x^0(x^0 - y)$
25. $\frac{8n^4}{2n}$
26. $\frac{9m}{6m^8}$
27. $\frac{a^2x}{-a}$
28. $\frac{-x^4y^3}{x^2y}$
29. $(-4a^2ct^4) \div (-2ac^3)$
30. $(9c^3p^4y) \div (-12cpy^6)$
31. $(-5rx^2y) \div (20r^3xy)$
32. $(-rst^4) \div (-rs^4t^2)$
33. $2b^{-1}$
34. $(2b)^{-1}$
35. $\frac{1}{2b^{-1}}$
36. $\frac{1}{(2b)^{-1}}$
37. $3a^{-1}b$
38. $3ab^{-1}$
39. $(3b)^{-2}$
40. $3b^{-2}$
41. $\frac{-7t^3uy}{-6ty}$
42. $\frac{9abc^4ds}{15bc^4d^4}$
43. $\frac{8x^2y^3z}{4xy^5z^4}$

44. $\dfrac{27ab^4c^6}{3a^3b^4c^7}$

45. $a^2x^{-1}a^{-3}$

46. $4b^{-2}c^{-3}b^5$

47. $2(c^{-2})^4$

48. $6(x^{-1})^{-1}$

49. $\dfrac{6x^{-1}}{x}$

50. $\dfrac{6(x^{-1})^2}{x^{-2}}$

51. $\dfrac{6(x^{-2})^{-2}}{x^{-4}}$

52. $\dfrac{(xy^{-1})^{-2}}{x^2y^{-3}}$

53. $\dfrac{y^{-2}a^0b^{-4}}{a^2b^0y^{-2}}$

54. $\dfrac{6^{-1}a^{-2}b}{b^{-2}a^{-2}}$

55. $\dfrac{3^{-1}a}{3a^{-1}}$

56. $\dfrac{(m^{-2}n)^0}{m^3n^{-4}}$

57. $\dfrac{x^{-5}y^{-1}}{xy^{-3}}$

58. $\dfrac{2^{-2}x^{-2}y^{-4}}{2x^{-6}y^0}$

59. $\dfrac{(axy^{-1})^{-2}}{(a^{-2}x^{-1}y)^{-3}}$

60. $\dfrac{(3a^{-2}bc^{-1})^{-1}}{6a^{-3}(bc)^{-1}}$

In Exercises 61 through 66, solve the given problems involving the use of exponents.

61. The area of a rectangle is $4a^2xy$. If the length of the rectangle is $2ax$, what is its width?

62. The product of the cube of a first number x and the square of a second number y is divided by the product of the numbers. Find the quotient.

63. Under certain economic circumstances, the price p for each of x commodities is given by the formula $p = 10(3^{-x})$. Express this formula without the use of negative exponents.

64. When using logarithms (see Chapter 11), you will find that negative and zero powers of 10 are important. Since $100 = 10^2$, $\dfrac{1}{100} = \dfrac{1}{10^2} = 10^{-2}$. Similarly, express $\dfrac{1}{10}$ and $\dfrac{1}{1000}$ as negative powers of 10.

65. When working with numbers of very small magnitude, you will often use negative powers of 10. The weight of an electron is about $9(10^{-28})$ g. Express this number in ordinary notation.

66. In the theory dealing with electricity, we encounter the expression $e^{-i(wt-ax)}$. Rewrite this expression so that no minus signs appear in the exponents.

4-4 DIVISION OF ALGEBRAIC EXPRESSIONS

In the preceding section, we saw the use of exponents when dividing one algebraic expression by another. Also, the case of the division of one monomial by another was considered. In this section, we shall show how multinomials are divided by monomials and by other multinomials.

We may draw from the arithmetic of fractions to show how a multinomial is to be divided by a monomial. When adding fractions, say $\frac{2}{7}$ and $\frac{3}{7}$, we have

$$\frac{2}{7}+\frac{3}{7}=\frac{2+3}{7}$$

Considering the right-hand side of this equation, we have the sum of two numbers divided by one number; and this equals, reading right to left, the sum obtained by dividing each of the numbers of the numerator separately by the number of the denominator. Since algebraic expressions represent numbers, we have the following method of dividing a multinomial by a monomial.

Divide each term of the multinomial by the monomial, and add the resulting terms algebraically to obtain the quotient.

Example A

$$(4x^2y + 2xy^3) \div 2xy = \frac{4x^2y + 2xy^3}{2xy}$$

Dividing each term of the binomial by the monomial is performed by first expressing the division as two separate fractions. Thus,

$$\frac{4x^2y + 2xy^3}{2xy} = \frac{4x^2y}{2xy} + \frac{2xy^3}{2xy}$$
$$= 2x + y^2$$

In the following two illustrations the same procedure is used.

$$(28a^3b^3 + 35a^2b^2) \div (-7ab^2) = \frac{28a^3b^3 + 35a^2b^2}{-7ab^2}$$
$$= \frac{28a^3b^3}{-7ab^2} + \frac{35a^2b^2}{-7ab^2} = -4a^2b - 5a$$

$$(axy^2 + ax^3 - 4a^3x^2) \div (-ax) = \frac{axy^2 + ax^3 - 4a^3x^2}{-ax}$$
$$= \frac{axy^2}{-ax} + \frac{ax^3}{-ax} + \frac{-4a^3x^2}{-ax}$$
$$= -y^2 - x^2 + 4a^2x$$

In dividing algebraic expressions of more than one term by another such expression, we shall deal exclusively with **polynomials.** *A polynomial in a variable is a term or sum of terms, each of which is either a constant or the product of a constant and a whole-number power of the variable.* In a polynomial the variable may not appear in the denominator or under a radical. We may also have polynomials in two or more variables. The **degree** of a polynomial is the highest power of the variable which appears in any term.

Example B The expression $2x^2 + \frac{1}{x}$ is not a polynomial in x due to the x in the denominator of the second term. The expression $2x^2 + x$ is a polynomial. The degree of this polynomial is 2, since the highest power of x which appears is 2.

The expression $4x^3 - \sqrt{x} - 7$ is not a polynomial due to the $\sqrt{x}$ term. The expression $4x^3 - x - 7$ is a polynomial of degree 3.

The expression $x^3y - 4xy^2 - 6$ is a polynomial in x and y. The degree of the polynomial in x is 3 and the degree of the polynomial in y is 2.

To divide one polynomial by another, we arrange both the dividend and the divisor in descending powers of the same literal factor. Then we divide the first term of the dividend by the first term of the divisor. The result of this division is the first term of the quotient. We then multiply the entire divisor by the first term of the quotient and subtract this product from the dividend. Now we repeat these operations by dividing the first term of the divisor into the first term of the difference just obtained. The result of this division is the second term of the quotient. We multiply the entire divisor by this second term and subtract the result from the first difference. We repeat this process until the difference is either zero or a quantity whose degree is less than that of the divisor (that is, the exponent of the literal factor must be less than the highest exponent of the literal factor in the divisor).

Example C Divide $2x^2 + 5x - 3$ by $2x - 1$.

Since both the dividend and divisor are in descending powers of x, no rearrangement is necessary. Then we set up the division as follows:

$$2x - 1 \,\overline{)\, 2x^2 + 5x - 3}$$

We now determine that $2x^2$ divided by $2x$ is x. Thus, x becomes the first term of the quotient.

$$\begin{array}{r l} & \quad x \\ 2x - 1 & \overline{)\, 2x^2 + 5x - 3} \end{array}$$

We now multiply $2x - 1$ by x and place the product below the dividend.

$$\begin{array}{r l} & \quad x \\ 2x - 1 & \overline{)\, 2x^2 + 5x - 3} \\ & \ \ \underline{2x^2 - \ x} \end{array}$$

The $2x^2 - x$ is now *subtracted* from the dividend.

$$\begin{array}{r|l} & x \\ \hline 2x-1 & 2x^2+5x-3 \\ & \underline{2x^2-x} \\ & 6x-3 \end{array}$$

We now determine that $2x$, the first term of the divisor, divided into $6x$, the first term of the remainder, is 3. Thus, 3 becomes the second term of the quotient.

$$\begin{array}{r|l} & x\ +3 \\ \hline 2x-1 & 2x^2+5x-3 \\ & \underline{2x^2-x} \\ & 6x-3 \end{array}$$

The divisor is now multiplied by 3, and the product is placed below the remainder and subtracted.

$$\begin{array}{r|l} & x\ +3 \\ \hline 2x-1 & 2x^2+5x-3 \\ & \underline{2x^2-x} \\ & 6x-3 \\ & \underline{6x-3} \\ & 0 \end{array}$$

Since the remainder is zero and all terms of the dividend have been used, the division is complete. Thus,

$$(2x^2+5x-3) \div (2x-1) = x+3$$

Example D Divide $3x^2 - 4x + x^3 - 12$ by $x + 2$.

The dividend is arranged in descending powers of x, and the division proceeds as shown.

$$\begin{array}{lr|rl} & & x^2 + x - 6 & \text{(quotient)} \\ \hline \text{(divisor)} & x+2 & x^3 + 3x^2 - 4x - 12 & \text{(dividend)} \\ & & x^3 + 2x^2 & \text{(subtract)} \\ \hline & & x^2 - 4x - 12 & \\ & & x^2 + 2x & \text{(subtract)} \\ \hline & & -6x - 12 & \\ & & -6x - 12 & \text{(subtract)} \\ \hline & & 0 & \text{(remainder)} \end{array}$$

The zero remainder shows that the division is exact.

The division may be checked by multiplying the quotient by the divisor. In this case,

$$\begin{aligned}(x+2)(x^2+x-6) &= x^3 + x^2 - 6x + 2x^2 + 2x - 12 \\ &= x^3 + 3x^2 - 4x - 12\end{aligned}$$

Example E Divide $4y^3 + 6y^2 + 1$ by $2y - 1$.

Since no first-power term in y appears in the dividend, we shall insert one with a zero coefficient to simplify the division.

$$\begin{array}{rll}
 & 2y^2 + 4y \;\; + 2 & \\
2y - 1 \,) & 4y^3 + 6y^2 + 0(y) + 1 & \\
 & 4y^3 - 2y^2 & \text{(subtract)} \\
 & \quad\;\; 8y^2 + 0(y) + 1 & \\
 & \quad\;\; 8y^2 - 4y & \text{(subtract)} \\
 & \qquad\qquad\; 4y + 1 & \\
 & \qquad\qquad\; 4y - 2 & \text{(subtract)} \\
 & \qquad\qquad\quad\; + 3 & \text{(remainder)}
\end{array}$$

The quotient in this case is

$$2y^2 + 4y + 2 + \frac{3}{2y - 1}$$

Note how the remainder is expressed as part of the quotient.

The solution of equations uses a combination of the basic operations. Example F illustrates such a solution.

Example F Solve the equation $2(3 - x) = 8 - 4(x - a)$ for x.

The solution proceeds as follows:

$$\begin{aligned}
2(3 - x) &= 8 - 4(x - a) \\
6 - 2x &= 8 - 4x + 4a \\
4x - 2x &= 8 + 4a - 6 \\
2x &= 2 + 4a \\
x &= \frac{2 + 4a}{2} = \frac{2}{2} + \frac{4a}{2} \\
&= 1 + 2a
\end{aligned}$$

Exercises 4-4

In Exercises 1 through 36, perform the indicated divisions.

1. $(6ab + 5a) \div a$

2. $(2x^2 - 3x) \div x$

3. $(9m^2 - 3m) \div 3m$

4. $(8s^3 + 2s^2) \div 2s$

5. $(a^3x^4 - a^2x^3) \div (-ax^2)$

6. $(-2xy^6 - 4x^2y^5) \div (2xy^3)$

7. $(3xy^4 - 6x^2y^5) \div (3xy^2)$

8. $(-14p^3q^5 + 49p^4q^2) \div (-7p^2q^2)$

9. $\dfrac{a^2b^2c^3 - a^3b^4c^6 - 2a^3bc^2}{a^2bc^2}$

10. $\dfrac{8rs^2t^5 - 18r^3st^4 - 16rst^3}{-2rst^2}$

11. $\dfrac{a^2b^3 - 2a^3b^4 - ab - ab^2}{-ab}$

12. $\dfrac{5m^2n^2y - 30mn^2 + 35m^2n^8}{-5mn^2}$

13. $(x^2 - 2x - 3) \div (x + 1)$

14. $(x^2 + 3x - 10) \div (x - 2)$

15. $(2x^2 - 5x - 3) \div (x - 3)$

16. $(6x^2 + 7x - 3) \div (3x - 1)$

17. $(3x^2 + 6x - 1) \div (x + 4)$

18. $(3x^2 + 4x - 5) \div (x + 2)$

19. $\dfrac{3x^2 + 8x - 3}{3x - 1}$

20. $\dfrac{4x^2 + 5x - 6}{4x - 3}$

21. $\dfrac{12y^2 - 11y + 2}{3y - 2}$

22. $\dfrac{20y^2 - 23y + 6}{5y - 2}$

23. $(9x - 2 + 56x^2) \div (7x + 2)$

24. $(6 - 10v + 6v^2) \div (3v + 1)$

25. $(72x^2 - 19x - 3) \div (8x - 3)$

26. $(8x^2 + 6x - 7) \div (2x + 3)$

27. $(x^3 - 3x^2 - x + 6) \div (x - 2)$

28. $(2x^3 + 5x^2 - 11x + 4) \div (x + 4)$

29. $(5x - 5x^2 + 2x^3 - 6) \div (x - 2)$

30. $(1 - x^2 + 6x^3) \div (1 + 2x)$

31. $\dfrac{x^3 - 5x^2y + 2xy^2 + 2y^3}{x - y}$

32. $\dfrac{2s^3 - 9s^2t + 6st^2 - t^3}{2s - t}$

33. $\dfrac{6x^4 - 5x^3 + 7x^2 + x - 3}{3x - 1}$

34. $\dfrac{2x^4 + 11x^3 - 5x^2 - 12x - 5}{2x + 5}$

35. $\dfrac{a^2 - 12b^2 + ab}{a + 4b}$

36. $\dfrac{xy + 2x^2 - 6y^2}{x + 2y}$

In Exercises 37 through 44, solve the given equations for x.

37. $2(2 - x) = 7 + x - 3a$

38. $2(x - 3) = -4(6 - b)$

39. $5x + 8b = 4a + x - 12c$

40. $6(2 - x) = 9 - 3(x - 2a)$

41. $(a - 1)x = a^2 + 4a - 5$

42. $(3b - 1)x = 3b^3 - b^2 + 3b - 1$

43. $(2c + 3)x = 2c^2 - 13c - 24$

44. $(a + 4)x = a^3 + 4a^2 - 5a - 20$

In Exercises 45 through 48, solve the given problems involving division.

45. A total of $2x$ books weigh $4x^2 + 6x$ lb. How much does each book weigh?

46. The revenue R obtained from selling x units at a price of p is related to x and p by the formula $R = px$. Under certain conditions, $R = p_0x - kx^2$, where p_0 is the highest price and k is a constant depending upon the economic conditions. Substitute this expression for R, and solve the resulting equation for p.

47. Under certain conditions when dealing with the electronics of coils, one may encounter the expression

$$\frac{60r^2}{6r + 1.2}$$

Perform the indicated division.

48. The sum of four times the square of a number, six times the number, and two is divided by one more than twice the number. Find the expression for the quotient if x represents the number.

4-5 SUMMARY

This chapter has been devoted to a further development of the algebraic operations first introduced in Chapter 2.

We discussed the addition and subtraction of algebraic expressions in some detail, with particular emphasis upon the proper removal of symbols of grouping. Recall that, as the symbols of grouping are removed, the signs of all terms of a group are changed when the group is preceded by a minus sign. Also, when more than one symbol of grouping is involved, it is generally best to remove the innermost symbol first.

In discussing multiplication and division of algebraic expressions, we illustrated some important operations with exponents. These are shown in Eqs. (4-1) to (4-6). Here we showed that the addition of exponents is associated with multiplication and the subtraction of exponents is associated with division. We also showed the meanings of negative and zero exponents.

In the multiplication and division of algebraic expressions, we showed the basic operations with monomials and polynomials. We again saw the importance of the distributive law in multiplication. When a multinomial is divided by a monomial, each term of the multinomial is divided by the monomial. We also illustrated the basic procedure for dividing one polynomial by another.

Along with the operations which were developed in the chapter, we introduced new terminology. A list of the important terms follows. You should understand each of these well.

Bar (grouping symbol)
Base
Braces (grouping symbol)
Brackets (grouping symbol)
Degree of polynomial
Negative exponent
Polynomial
Zero exponent

Exercises which should be helpful in reviewing the material of the chapter are given on the following pages.

Exercises 4-5

In Exercises 1 through 48, simplify the given expression by performing the indicated operations. Express all results with positive exponents.

1. $a - (x - 2a)$

2. $x - (3s - 2x)$

3. $2(x - 5y) - (3y - 7x)$

4. $-(y + 2s) - (5s - y)$

5. $(-3a^2b)(2ab^5)$

6. $(-2s^3t^2)(-4st^4)$

7. $(-2xy^2z)^3(-7xy^3z^5)$

8. $(-2x^2y)^2(-3xy^5)$

9. $(-8a^2px^4) \div (2apx)$

10. $(-18rs^3t^5) \div (-24r^4st^6)$

11. $\dfrac{2x^3y^{-1}}{3x^{-2}y^2}$

12. $\dfrac{(2a)^0(b^{-1}c)}{4a^2bc^{-3}}$

13. $\dfrac{(-a)^{-2}y^0}{a^{-1}y^2}$

14. $\dfrac{3^{-1}ab^{-1}}{(6a)^{-1}}$

15. $\dfrac{4^{-2}(ab)^{-2}}{(4a)^{-1}}$

16. $\dfrac{(-b)^{-1}(-a)^{-2}(ab)^0}{(ab)^{-2}(ab)}$

17. $\dfrac{(2x + y)^0(xy)^{-1}}{(2x + y)xy^{-1}}$

18. $\dfrac{px^{-2}y^{-1}y^{-2}}{p^{-2}xy^{-2}y^2}$

19. $4y + [(2y - b) - (b + 2y)]$

20. $-(xyz - x) - 4[xyz + y - (xyz - y)]$

21. $2x + [(2x - a) - (a - x)]$

22. $-(3xy - y) - 2[2x - y - (y - xy)]$

23. $-\{2b - [b - (4 - 5b)] + (6 - b)\}$

24. $2x - y - 2\{3x - [y - (3y - 4x)]\}$

25. $-2a(a^2x - at)$

26. $-3a^2j(3j^4 - 4aj)$

27. $(x - 3)(2x + 7)$

28. $(3x - 2)(x + 5)$

29. $(2a - 5b)(3a + 2b)$

30. $(2x - y)(y - 5x)$

31. $(2b - 2a)(3b - 3a)$

32. $(4x - 4y)(5x - 5y)$

33. $n(2b + a)(7b - a)$

34. $k(3x + y)(9x - y)$

35. $2(2 - x)(x^2 - x - 4)$

36. $-3(xy - q)(x - 2y + 3q)$

37. $(2x^3y^5 - 3x^6y^2) \div (-x^3y^2)$

38. $(-6a^2b^3 - 9a^3b^4 + 12ab^5) \div (-3ab^2)$

39. $\dfrac{h^2j^4 - 3hj^6 - 6h^4j^7}{hj^4}$

40. $\dfrac{-18f^3g^2k^2 + 24f^2gk^6 - 36fgk^4}{-6fgk^2}$

41. $(x^2 + x - 12) \div (x - 3)$

42. $(6x^2 - 7x - 5) \div (2x + 1)$

43. $\dfrac{2x^3 + x^2 - x - 8}{2x + 3}$

44. $\dfrac{3x^2 + 5xy - 2y^2}{3x - y}$

45. $\dfrac{6x^4 + x^3 + 5x^2 + 2}{2x^2 - x + 1}$

46. $\dfrac{-a^2b - 2b^2 + 3a^4}{a^2 - b}$

47. $\dfrac{2x^4 - x^3 - 2x - 8}{x^2 + 2}$

48. $\dfrac{6x^4 + x^3 + 5x^2 + 2}{2x^2 - x + 1}$

In Exercises 49 through 52, solve the given equations for x.

49. $ax - 3 - a(2 - x) = 3(4a - 1)$

50. $b(x - 3b) - (b^2 - x) = x - b(1 - b)$

51. $(a^2 - 1)x = a^4 - 1$

52. $(c + 2)x = 2c^2 + c - 6$

In Exercises 53 through 68, solve the given problems involving the basic operations.

53. One factory produced $x + 12$ units in a given day and a second factory produced $2x + 1$ units in the same day. What was the total production of the two factories?

54. How many more units were produced by the second factory in Exercise 53 than by the first factory?

55. The total revenue R obtained from selling x units of a certain item is given by $R = x(36 - 4x)$. By performing the indicated multiplication, find an equivalent expression for R.

56. When finding the center of a certain geometric figure, one encounters the expression

$$(2x - x^2) - (3x^2 - 6x)$$

Simplify this expression.

57. When one is dealing with the electron, one finds the expression $m(v_2 - v_1)(v_2 + v_1)$. Multiply out this expression.

58. In optics, the expression

$$\frac{ID}{d}\left(n + \frac{1}{2}\right) - \frac{ID}{d}\left(n - \frac{1}{2}\right)$$

is found. Simplify this expression.

59. When dealing with the expansion of a surface which has been heated, one finds the expression

$$lh(1 + at)^2$$

Perform the indicated multiplication.

60. When one is studying the interference of light from a double source, the expression

$$(2x + d)^2 - (2x - d)^2$$

arises. Simplify this expression.

61. In the biological study of muscles, the expression

$$(p + a)(V + b)$$

is found. Perform the indicated multiplication.

62. Under certain conditions, the distance s (in feet) that an object is above the ground is given by the formula $s = 16t(5 - 2t)$, where t is the time in seconds. Perform the indicated multiplication.

63. The psychological theory of learning utilizes the expression $(1 - p)(1 + p - r)$. Perform the indicated multiplication.

64. A rectangular box is to be made from a piece of cardboard 8 cm by 10 cm by cutting out a square of side x from each corner and turning up the sides. Show that the volume of the box is $V = x(8 - 2x)(10 - 2x)$ and then perform the indicated multiplication.

65. In the theory dealing with the efficiency of an engine, there is the expression

$$\frac{Q_1 - Q_2}{Q_1}$$

By performing the indicated division, find an equivalent expression.

66. The area of a certain rectangle is $2a^2 + 11a + 14$, and one side is $a + 2$. Find an expression for the length of the other side.

67. In the mathematics dealing with geometric progressions, the equation

$$S(1 - r) = a(1 + r + r^2 + r^3) - a(r + r^2 + r^3 + r^4)$$

may be used. Solve for S.

68. A car traveled a distance of $6x^2 - x - 2$ miles in $2x + 1$ hours, traveling at an average rate of 43 mi/h. Determine the actual distance traveled.

5

FACTORING

5-1 THE PRINCIPLE OF FACTORING

In the preceding chapter, we discussed the basic operations of algebra. Among these essential operations was that of multiplication, which is performed when we wish to determine the product of two or more algebraic expressions. Many times, however, we will find that we have an algebraic expression and wish to determine what expressions, when multiplied together, equal the given expression. We know that, when an algebraic expression is the product of two or more quantities, each of these quantities is called a factor of the expression. Therefore, *determining these factors is known as* **factoring.** We shall find this process important to the work on fractions in the next chapter, as well as to the work in later sections.

Example A Since $x^2 - 4 = (x - 2)(x + 2)$, the quantities $x - 2$ and $x + 2$ are factors of $x^2 - 4$. Since $2ax - a^2x = ax(2 - a)$, the quantities ax and $2 - a$ are factors of $2ax - a^2x$. In fact, ax itself has factors of a and x, which means that we could state that the factors of $2ax - a^2x$ are a, x, and $2 - a$.

As Example A shows, there are times when the factors of a given expression may be stated in different ways. Factors of the integers can also show this point. For example, $12 = 4 \cdot 3$ or $12 = 6 \cdot 2$, where the factors are 4 and 3 or 6 and 2. Note that $+1$ or -1 may always be stated as a factor of a given expression. Consider the various possibilities shown in Example B.

Example B

$$24 = (+1)(+24) \qquad 24 = (+3)(-1)(-8)$$
$$= (-1)(-24) \qquad = (-3)(-1)(+8)$$
$$= (+2)(+12) \qquad = (+4)(+6)(+1)$$
$$= (+2)(-1)(-12) \qquad = (-1)(-2)(-2)(-6)$$
$$= (+2)(+2)(+2)(+3)$$

Innumerable other possibilities also exist.

In our work on factoring, we shall consider only the factoring of polynomials (which include the integers), all terms of which will have integers as coefficients. Also, all factors will have integral coefficients and will contain no radicals such as $\sqrt{2}$, for example. *Such a polynomial may be called* **prime** *if it contains no factors other than* $+1$ *or* -1, *or plus or minus itself. We then say that an expression is* **factored completely** *if it is expressed as a product of its prime factors.*

Example C $12 = (2)(2)(3)$. Here the factors of 12 are seen to be 2, 2, and 3 (any given factor may appear more than once). Since neither 2 nor 3 may be expressed as the product of any other pair of integers not including 1, they are prime. If we factor 12 as $12 = (6)(2)$, we have not factored it completely, since $6 = (2)(3)$.

Similarly, $24 = (2)(2)(2)(3)$ is the complete factorization of 24. Again, $24 = (6)(4)$ is not the complete factorization of 24, since $6 = (2)(3)$ and $4 = (2)(2)$.

Now, $x^2 - 1 = (x - 1)(x + 1)$ represents the complete factorization of the expression $x^2 - 1$. However, even though $x - 1 = (\sqrt{x} - 1)(\sqrt{x} + 1)$, we do not include this type of factoring, since it includes radicals in the factors. In the same sense, $x^2 - 3$ is not factorable even though $(x - \sqrt{3})(x + \sqrt{3}) = x^2 - 3$.

In Example C we noted that $12 = (2)(2)(3)$. We also noted that 2 and 3 are prime numbers. Here 12 is a number whose factors are three integers other than $+1$, -1, $+12$ or -12. Again, $24 = (2)(2)(2)(3)$, and 24 is a number whose factors are four integers other than $+1$, -1, $+24$, or -24. These two cases are illustrations of numbers which are not prime. *Numbers such as these which are not prime are known as* **composite numbers.** *Therefore, a composite number is one which is the product of two or more integers other than* $+1$ *or* -1, *or plus or minus itself.*

We shall not attempt to develop any specific method for determining whether or not a factor is prime. For the restricted factoring we shall encounter, a knowledge of basic algebraic multiplication is sufficient, for any set of factors may be checked by multiplication.

Example D The expressions $x - 1$, $x^2 + 1$, $x^2 - 5$, and $a^2 - ax + x$ are prime. The expressions $2x - 2$, $x^2 - 4$, and $a^2 - 2ax + x^2$ are not prime, since

$$2x - 2 = 2(x - 1) \qquad x^2 - 4 = (x - 2)(x + 2)$$

and

$$a^2 - 2ax + x^2 = (a - x)^2$$

Methods of determining factors such as those in Example D will be discussed in the sections which follow, although we must be able to recognize prime factors when they occur.

Exercises 5-1

In Exercises 1 through 4, identify the prime integers.

1. 2, 3, 5, 7, 9
2. 4, 11, 17, 21, 23
3. 15, 19, 26, 27, 29
4. 13, 18, 31, 33, 37

In Exercises 5 through 8, determine the prime integers between the given integers.

5. Between 40 and 50
6. Between 50 and 60
7. Between 60 and 70
8. Between 70 and 80

In Exercises 9 through 20, factor the given integers into a product of prime integers.

9. 10
10. 18
11. 28
12. 36
13. 50
14. 72
15. 88
16. 98
17. 140
18. 117
19. 210
20. 315

In Exercises 21 through 24, state the prime factors of the indicated products.

21. $2(x - 1)(x + 3)$
22. $2a(x - 8)$
23. $9c^2(ax - b)(cx + d)$
24. $abc(c - a)(b + d)$

In Exercises 25 and 26, determine whether or not the given expressions have been factored completely.

25. $5x - 25x^2 = 5x(1 - 5x)$
26. $2x^2 - 7x - 4 = (2x + 1)(x - 4)$

In Exercises 27 and 28, determine whether or not the given expressions have been correctly factored. Check by multiplication.

27. $x^2 + 4 = (x + 2)^2$

28. $2x^2 - x = 2x(x - 1)$

In Exercises 29 through 32, determine the quantity that must be used to multiply the first expression so that the product is the second expression given.

29. $x - 5, 3x - 15$

30. $x - b, 2bx - 2b^2$

31. $x - 3, x^2 - 9$

32. $x + 1, x^2 + 2x + 1$

In Exercises 33 through 38, answer the given questions on prime integers and expressions.

33. Are the following expressions prime: $x + 1, x^2 + 2, x^2 + ax + 1$?

34. Are the following expressions prime: $3x - 8, 4x - 8, 3x - 9$?

35. Let P be the set of all prime integers and I the set of all integers. What is the relation of set P to set I?

36. Let A be the set of all prime integers greater than 2, and let B be the set of all odd integers greater than 2. What is the relation of set A to set B?

37. Are there any prime integers greater than 5 which end in 5?

38. What are the first 25 prime integers?

5-2 COMMON MONOMIAL FACTORS

Often an algebraic expression contains a monomial that is common to each term of the expression. *The first step in factoring any expression is to determine whether there is such a* **common monomial factor.** If there is, then the expression is written as the product of the common monomial factor and the quotient, obtained by dividing the expression by the monomial factor.

Example A Factor $ax - 5a$.

We note that each of the terms of the expression $ax - 5a$ contains a factor of a. If $ax - 5a$ is divided by a, the quotient is $x - 5$. Thus,

$$ax - 5a = a(x - 5)$$

Here, a is the common monomial factor and $a(x - 5)$ is the required factored form of $ax - 5a$.

Example B Factor $3ax^2 + 3x^3y - 6x^2z$.

The numerical factor 3 and the literal factor x^2 are common to each term of the polynomial and they are common monomial factors. Normally, monomial factors are not reduced to prime factors; and therefore we say that $3x^2$ is the common monomial factor. The quotient obtained by dividing the above polynomial by $3x^2$ is $a + xy - 2z$. Therefore, the factoring of $3ax^2 + 3x^3y - 6x^2z$ is expressed as

$$3ax^2 + 3x^3y - 6x^2z = 3x^2(a + xy - 2z)$$

There is a systematic method of finding common monomial factors. *When each of the terms of a polynomial is expressed as the product of its prime factors, the highest common factor is the product of the factors common to all terms.* Example C illustrates this technique.

Example C Factor $4x^3 + 8x^2y - 24xy^2$.

Expressing each of the terms as the product of prime factors, we have

$$4x^3 = 2 \cdot 2 \cdot x \cdot x \cdot x \qquad 8x^2y = 2 \cdot 2 \cdot 2 \cdot x \cdot x \cdot y$$

$$-24xy^2 = -1 \cdot 2 \cdot 2 \cdot 2 \cdot 3 \cdot x \cdot y \cdot y$$

We now note that 2 appears twice in each of the products and x appears once in each. Even though 2 appears three times in each of the second and third terms, it appears only twice in the first term; hence, we may use only two factors of 2 in the highest common factor. Similar arguments may be given for x and y. Therefore, the highest common factor is $2 \cdot 2 \cdot x = 4x$. The remaining factors of the first term give the product x^2; those of the second term give $2xy$, and those of the third term give $-6y^2$. Therefore,

$$4x^3 + 8x^2y - 24xy^2 = 4x(x^2 + 2xy - 6y^2)$$

In practice, one determines the common monomial factor of a given polynomial by inspection. The method of finding the highest common factor as in Example C is a general technique that can be followed regardless of the nature of the terms. However, it is rarely necessary actually to write out the factors of the individual terms. The factors of each term can be seen by inspection, and the common monomial factor is thereby found. Example D gives several illustrations of factoring problems done by inspection.

Example D

$$2x - 2y = 2(x - y) \qquad 2x^2 + 4y = 2(x^2 + 2y)$$

$$3ax + 6a^2 = 3a(x + 2a) \qquad 6p^2q - 18pq^2 = 6pq(p - 3q)$$

$$4x^2 - 8xy - 20xy^2 = 4x(x - 2y - 5y^2)$$

$$6a^2b^3c - 9a^3bc^2 + 3a^2bc = 3a^2bc(2b^2 - 3ac + 1)$$

Note the presence of the 1 in the last illustration. Since each of the factors of this term is common to all other terms, we must include the factor of 1. We can also recognize that this inclusion is necessary by multiplying the right-hand side as a check. Also, we find the polynomial in parentheses by dividing the polynomial on the left by $3a^2bc$, and the division of the last term gives 1.

Exercises 5-2

In Exercises 1 through 32, factor the given expressions by determining any common monomial factors that may exist.

1. $5x + 5y$
2. $3x^2 - 3y$
3. $7a^2 - 14bc$
4. $3s - 12t$
5. $a^2 + 2a$
6. $3p + pq$
7. $2x^2 - 4x$
8. $5h + 10h^2$
9. $3ab - 3a$
10. $4x^2 - 4x$
11. $4p - 6pq$
12. $6s^3 - 15st$
13. $3a^2b + 9ab$
14. $6xy^3 - 9xy^2$
15. $a^2bc^2f - 4acf$
16. $2rs^2t - 8r^2st^2$
17. $2x + 2y - 2z$
18. $3r - 3s + 3t$
19. $5x^2 + 15xy - 20y^3$
20. $4rs - 14s^2 - 16r^2$
21. $6x^2 + 4xy - 8x$
22. $5s^3 + 10s^2 - 15s$
23. $12pq^2 - 8pq - 28pq^3$
24. $18x^2y^2 - 24x^2y^3 + 54x^3y$
25. $35a^3b^4c^2 + 14a^2b^5c^3 - 21a^3b^2$
26. $15x^2yz^3 - 45x^3y^2z^2 + 16x^2y^2z$
27. $6a^2b - 3a + 9ab^2 - 12a^2b^2$
28. $4r^2s - 8r^3s^2 + 16r^4s - 4r^2s^3$
29. $24kp^2r - 2k^2pr^2 + 8pr + 32p^2r$
30. $9hnt + 3hn^2p + 27nt + 81h^3n^3t^3$
31. $28xyz - 56x^2y^2z^2 + 64x^2z^3 + 36y^2z$
32. $3x^3 - 4x^5 + 5yx - 7xz$

In Exercises 33 through 36, solve the given problems, which involve factoring.

33. The total surface area A of a rectangular solid is given by the formula

$$S = 2lw + 2lh + 2hw$$

where l, w, and h are the length, width, and height of the solid. Factor the right-hand side of this formula.

34. Under certain circumstances, the distance s which an object is above the surface of the earth is given by the formula $s = 128 + 64t - 32t^2$, where t is the time. Factor the right-hand side of this formula.

35. In computing the value of a 6-month loan on which payments of R dollars per month are made, we use the equation $P = Rv + Rv^2 + Rv^3 + Rv^4 + Rv^5 + Rv^6$, where v is a factor involving the interest which is paid. Factor the right-hand side of this equation.

36. Under certain circumstances, in discussing magnetic fields, the equation

$$H = 2iMAC + 2iMBC - 10iMDC$$

could be used. Factor the right-hand side of this equation.

5-3 THE DIFFERENCE OF TWO SQUARES

If we multiply $(a + b)$ by $(a - b)$, the product is $a^2 - b^2$. This product contains two perfect squares, a^2 and b^2. In general, *a* **perfect square** *is any quantity which is an exact square of a rational quantity.*

Example A 4 is a perfect square because $4 = 2^2$. The quantity $9y^2$ is a perfect square because $9y^2 = (3y)^2$. The quantity $25a^4$ is a perfect square because $25a^4 = (5a^2)^2$. Other perfect squares are 16, $49x^2$, x^2y^2, b^4, p^2y^4, and $36x^8y^2$.

The number 7 is not a perfect square because there is no rational number which, when squared, equals 7. The quantity x^3 is not a perfect square because there is no integral power of x which, when squared, equals x^3.

Having noted that

$$\boxed{(a + b)(a - b) = a^2 - b^2} \tag{5-1}$$

we see that the multiplication results in the **difference of the squares** of the two terms that appear in the binomials. Therefore, we can easily recognize the factors of the binomial $a^2 - b^2$, which are $(a + b)$ and $(a - b)$. We must recognize the perfect squares represented by a^2 and b^2 when we factor expressions of this type. We can now apply the result of Eq. (5-1) to the problem of factoring the difference of any two squares.

Example B

$$x^2 - 9 = x^2 - 3^2 = (x + 3)(x - 3)$$

$$9 - 4y^2 = 3^2 - (2y)^2 = (3 + 2y)(3 - 2y)$$

$$36p^2 - 49q^2 = (6p)^2 - (7q)^2 = (6p + 7q)(6p - 7q)$$

$$4s^4 - 25t^4 = (2s^2)^2 - (5t^2)^2 = (2s^2 + 5t^2)(2s^2 - 5t^2)$$

$$16y^4x^2 - p^6 = (4y^2x)^2 - (p^3)^2 = (4y^2x + p^3)(4y^2x - p^3)$$

In actually writing down the result of the factoring, we do not usually need to write the middle step shown in this example. Thus, the first illustration would be written simply as $x^2 - 9 = (x + 3)(x - 3)$. The middle steps are given here to indicate the perfect squares involved.

Equation (5-1) can be used to perform certain multiplications very rapidly. Example C demonstrates how this is done.

Example C To multiply 49 by 51, we may express the product as

$$(49)(51) = (50 - 1)(50 + 1) = (50)^2 - 1^2 = 2500 - 1 = 2499$$

In Eq. (5-1), we are letting $x = 50$ and $y = 1$.

Other illustrations of this method of multiplication are:

$$(32)(28) = (30 + 2)(30 - 2) = (30)^2 - 2^2 = 900 - 4 = 896$$

$$(104)(96) = (100 + 4)(100 - 4) = (100)^2 - 4^2 = 10{,}000 - 16 = 9984$$

$$(43)(57) = (50 - 7)(50 + 7) = (50)^2 - 7^2 = 2500 - 49 = 2451$$

The major value of performing multiplication in this way is to attain facility in working with the difference of squares. It is not too often that one comes across the type of product which may be obtained in this manner, for it must be of a special form.

As indicated previously, if it is possible to factor out a common monomial factor, this factoring should be done first. If the resulting factor is the difference of squares, the factoring is not complete until this difference is also factored. Thus, **complete factoring** *often requires more than one step*. Also, be sure to include *all* factors in writing the result in complete factoring.

Example D In factoring $3x^2 - 12$, we note that there is a common factor of 3 in each of the terms. Therefore, $3x^2 - 12 = 3(x^2 - 4)$. However, the factor $x^2 - 4$ is itself the difference of two perfect squares. Therefore, $3x^2 - 12$ is completely factored as

$$3x^2 - 12 = 3(x^2 - 4) = 3(x + 2)(x - 2)$$

Example E In factoring $x^4 - 16$, we note that we have the difference of squares. Therefore, $x^4 - 16 = (x^2 + 4)(x^2 - 4)$. However, the factor $x^2 - 4$ is again the difference of squares. Therefore,

$$x^4 - 16 = (x^2 + 4)(x^2 - 4) = (x^2 + 4)(x - 2)(x + 2)$$

The factor $x^2 + 4$ is prime. We note that this is the *sum* of squares. Whereas the *difference* of squares is always factorable, the *sum* of squares is factorable *only* to the extent of removing a common monomial factor, if one exists.

This product and its factors, Eq. (5-1), should become very familiar, for the product of the sum and difference of two terms appears frequently in algebra.

Exercises 5-3

In Exercises 1 through 4, determine which of the given quantities are perfect squares.

1. $9, 25, 4x^2, 9x$
2. $18, 121, 8x^2, x^8$
3. $a^2b^4, x^2yz^4, 16xy^2$
4. $9r^6, s^2t^5x^2, 36d^4y^{10}, 81r^{12}s$

In Exercises 5 through 12, determine the products by direct multiplication.

5. $(x + y)(x - y)$
6. $(a - 2)(a + 2)$
7. $(2a - b)(2a + b)$
8. $(3xy + 1)(3xy - 1)$
9. $(7ax^2 + p^3)(7ax^2 - p^3)$
10. $(5x^2 - 6y^5)(5x^2 + 6y^5)$
11. $(x + 2)(x + 2)$
12. $(a + b)(a + b)$

In Exercises 13 through 16, perform the indicated multiplications by means of Eq. (5-1). Check by direct multiplication.

13. (21)(19) **14.** (78)(82) **15.** (210)(190) **16.** (144)(156)

In Exercises 17 through 40, factor the given expressions completely.

17. $a^2 - h^2$ **18.** $x^2 - 4$ **19.** $16 - x^2$
20. $25 - y^2$ **21.** $4x^4 - y^2$ **22.** $64 - x^2y^2$
23. $9y^2 - v^2$ **24.** $81 - y^2v^2$ **25.** $25w^2 - u^2$
26. $49 - w^2u^2$ **27.** $a^2b^2 - 9y^2$ **28.** $4q^2 - (rs)^2$
29. $81x^4 - 4y^6$ **30.** $49b^6 - 25c^4$ **31.** $9v^4 - 81w^6$
32. $36q^4 - 9p^2$ **33.** $100x^6 - 25z^4$ **34.** $256x^6 - 64y^4$
35. $4x^2 - 100y^2$ **36.** $24x^2 - 54a^2$ **37.** $x^4 - 1$
38. $a^4 - 81b^4$ **39.** $4x^2 + 36y^2$ **40.** $75a^2x^2 + 27b^4y^4$

In Exercises 41 through 46, solve the given problems, which involve factoring.

41. In discussing the energy of a moving object, one encounters the expression $mv_1^2 - mv_2^2$. Factor this expression.

42. In economics, one sometimes uses the following formula relating a quantity supplied, x, and the price p per unit quantity:

$$x = C^2 - k^2p^2$$

where C and k are constants depending on the economic conditions. Factor the right-hand side of this formula.

43. Find the area of a rectangle whose sides are 37 in. and 43 in. by performing the multiplication by the method of this section.

44. Find the area of a rectangle whose sides are 55 cm and 65 cm. Use the multiplication method of this section.

45. Set up and then factor the expression for the difference in volumes of a cube of side x and a rectangular solid of sides x, y, and y.

46. The difference of the energy radiated by an electric light filament at temperature T_2 and that radiated by a filament at temperature T_1 is given by the formula

$$R = kT_2^4 - kT_1^4$$

Factor the right-hand side of the formula.

5-4 FACTORING TRINOMIALS

In this section, we shall discuss the factoring of trinomials of the form $ax^2 + bx + c$. As we stated in Section 5-1, we shall consider only those cases where a, b, and c are integers. Such a trinomial may or may not be factorable into two binomial factors, the coefficients of which are integers.

We shall first discuss the trinomial $x^2 + bx + c$; that is, the case in which $a = 1$. *The binomial factors of this trinomial (if they exist) will be of the form*

$(x + p)$ *and* $(x + q)$. That is, we shall determine whether factors exist such that

$$\boxed{x^2 + bx + c = (x + p)(x + q)} \tag{5-2}$$

Multiplying out the right-hand side of this equation, we obtain

$$(x + p)(x + q) = x^2 + px + qx + pq = x^2 + (p + q)x + pq$$

We now see that $b = p + q$ and $c = pq$. It is therefore necessary to find two integers p and q whose sum is b and whose product is c. The quantities p and q are readily found by inspection.

Example A In order to factor the trinomial $x^2 + 5x + 4$, we must find two integers such that their product is $+4$ and their sum is $+5$. The possible factors of $+4$ are: $+2$ and $+2$, -2 and -2, $+4$ and $+1$, and -4 and -1. Of these possibilities, only $+4$ and $+1$ have a sum of $+5$. Hence, the required factors are $(x + 4)$ and $(x + 1)$. Therefore,

$$x^2 + 5x + 4 = (x + 4)(x + 1)$$

When we factor $x^2 - 5x + 4$, an examination of the possible factors of $+4$, which are indicated above, reveals that the only factors whose sum is -5 are -4 and -1. Therefore, the factors are $(x - 4)$ and $(x - 1)$. Hence,

$$x^2 - 5x + 4 = (x - 4)(x - 1)$$

When we factor $x^2 + 3x + 4$, an examination of the possible factors of $+4$, which are indicated in the first part of this example, reveals that none of the combinations has a sum of $+3$. Therefore, $x^2 + 3x + 4$ is prime.

Example B In order to factor $x^2 + x - 6$, we must find two integers whose product is -6 and whose sum is $+1$. The possible factors of -6 are -1 and $+6$, $+1$ and -6, -2 and $+3$, and $+2$ and -3. Of these factors, only -2 and $+3$ have a sum of $+1$. Therefore,

$$x^2 + x - 6 = (x - 2)(x + 3)$$

When factoring $x^2 - 5x - 6$, we see that the necessary factors of -6 are $+1$ and -6. Hence,

$$x^2 - 5x - 6 = (x + 1)(x - 6)$$

When factoring $x^2 + 4x - 6$, we see that this expression is prime, since none of the pairs of factors of -6 has a sum of $+4$.

We now turn our attention to factoring the trinomial $ax^2 + bx + c$, where a does not equal 1. The binomial factors of this trinomial will be of the form $(rx + p)$ and $(sx + q)$. That is, we shall determine whether factors exist such that

$$\boxed{ax^2 + bx + c = (rx + p)(sx + q)} \tag{5-3}$$

Multiplying out the right-hand side of this equation, we obtain

$$(rx + p)(sx + q) = rsx^2 + rqx + spx + pq = rsx^2 + (rq + sp)x + pq$$

We now see that $a = rs$, $b = rq + sp$, and $c = pq$. Therefore, we must find four such integers r, s, p, and q.

You probably realize that finding these integers is relatively easy so far as the ax^2 term and the c term are concerned. We simply need to find integers r and s whose product is a and integers p and q whose product is c. However, we must find the right integers so that the bx term is correct. This is usually the most troublesome part of factoring trinomials. We must remember that *the bx term will result from the sum of two terms* when the factors are multiplied together.

It is sometimes helpful to indicate the two terms which are combined to form the bx term as in this diagram:

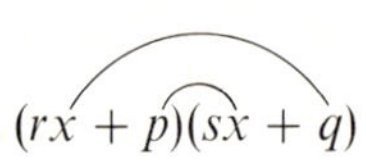

$$(rx + p)(sx + q)$$

We see that the "outer" product rqx and the "inner" product spx are the two terms of the product which together form the bx term.

The problem therefore is to determine factors of a and c, the sum of whose **cross products** *is equal to b.* We may determine the cross products $rq + sp$ in this manner: Arrange the factors r, s, p, and q at the corners of an imaginary square, as

We find the product rq by multiplying the factors on the diagonal r—q, while we find sp by multiplying the factors on the diagonal s—p. We shall *always* take r and s to be positive if a is positive.

Example C When factoring $2x^2 - 11x + 5$, we take the factors of 2 to be $+2$ and $+1$ (we shall take only the positive factors of a since it is $+2$). The factors of $+5$ are $+1$ and $+5$, and -1 and -5. Since the sum of the cross products must equal -11, only negative factors can possibly be used (both p and q are of the same sign, and are to be multiplied by positive numbers). We now set up the factors to determine the cross product:

$$\begin{matrix} +2 & -5 \\ +1 & -1 \end{matrix}$$

The sum of the cross products is $-2 - 5 = -7$. Since this arrangement of factors did not give us the sum of -11, we now try another arrangement:

$$\begin{matrix} +1 & -5 \\ +2 & -1 \end{matrix}$$

This sum is $-1 - 10 = -11$, the required sum. The numbers in this arrangement are the required coefficients of the binomial factors. Thus,

$$2x^2 - 11x + 5 = (x - 5)(2x - 1)$$

In checking, we note that the "outer" product is $-x$ and the "inner" product is $-10x$. Their sum is $-11x$, which is the middle term of $2x^2 - 11x + 5$.

Example D Factor $4y^2 + 7y + 3$.

The positive factors of $+4$ are $+1$ and $+4$, and $+2$ and $+2$. Since b is $+7$ and c is $+3$, both cross products must be formed from positive numbers, which means we are limited to the positive factors of $+3$. The positive factors of $+3$ are $+1$ and $+3$. We now set up our factors to determine the cross products:

$$\begin{matrix} +1 & +1 \\ +4 & +3 \end{matrix}$$

The sum of the cross products is $+3 + 4 = +7$, the required value. Therefore,

$$4y^2 + 7y + 3 = (y + 1)(4y + 3)$$

It is unnecessary to try the other pair of factors, $+2$ and $+2$.

Example E Factor $6x^2 + 5x - 4$.

The positive factors of $+6$ are $+6$ and $+1$, and $+3$ and $+2$. The factors of -4 are $+4$ and -1, -4 and $+1$, and $+2$ and -2. Since b is $+5$, we know that the cross product of larger absolute value will be the product of positive numbers. This eliminates several of the possible combinations, such as

$$\begin{array}{ll} +6 & +1 \\ +1 & -4 \end{array}$$

We find that the factors $+6$ and $+1$ do not work with any of the factors of -4. Trying the factors $+3$ and $+2$, we find that the combination

$$\begin{array}{ll} +3 & +4 \\ +2 & -1 \end{array}$$

has a sum of cross products of $-3 + 8 = +5$, the required value. Therefore,

$$6x^2 + 5x - 4 = (3x + 4)(2x - 1)$$

As we mentioned in the previous sections, *we must always be on the alert for a common monomial factor*. A check should be made for such a factor first. Consider Example F.

Example F Factor $2x^2 + 6x - 8$.

At first, it might appear that we are to find the factors of $+2$ and -8. However, it should be noted that there is a common factor of 2 in each term. Therefore, we can write

$$2x^2 + 6x - 8 = 2(x^2 + 3x - 4)$$

It is now necessary only to factor $x^2 + 3x - 4$. The factors of this trinomial are $x + 4$ and $x - 1$. Therefore,

$$2x^2 + 6x - 8 = 2(x + 4)(x - 1)$$

Many trinomials which are factorable by the methods of this section will contain a literal factor in each term of the binomial factors. Example G illustrates this type of problem.

Example G Factor $4x^2 + 22xy - 12y^2$.

We first note that there is a common factor of 2 in each term. This means $4x^2 + 22xy - 12y^2 = 2(2x^2 + 11xy - 6y^2)$. The problem now is to factor $2x^2 + 11xy - 6y^2$. Since the product of

$$(rx + py)(sx + qy) = rsx^2 + (rq + sp)xy + pqy^2$$

we see that this is essentially the same as the problems previously encountered. The only difference is the presence of the xy in the middle term and the y^2 in the last term. However, the process of finding the values of r, s, p, and q is the same. We find that the arrangement

$$\begin{matrix} +2 & -1 \\ +1 & +6 \end{matrix}$$

works for this combination. Therefore,

$$4x^2 + 22xy - 12y^2 = 2(2x - y)(x + 6y)$$

Much of the work done in factoring the trinomials in these examples can be done mentally. With experience, you will be able to write the factors of such expressions by inspection.

Exercises 5-4

In Exercises 1 through 40, factor the given trinomials, when possible.

1. $x^2 + 3x + 2$ **2.** $x^2 - x - 2$ **3.** $x^2 + x - 12$

4. $s^2 + 7s + 12$ **5.** $y^2 - 4y - 5$ **6.** $x^2 + 6x - 7$

7. $x^2 + 10x + 25$ **8.** $t^2 + 3t - 10$ **9.** $2q^2 + 11q + 5$

10. $2a^2 - a - 3$ **11.** $3x^2 - 8x - 3$ **12.** $2x^2 - 7x + 5$

13. $5c^2 + 34c - 7$ **14.** $3x^2 + 4x - 7$ **15.** $3x^2 - x - 5$

16. $7n^2 - 16n + 2$ **17.** $2s^2 - 13st + 15t^2$ **18.** $3x^2 - 14x + 8$

19. $5x^2 + 17x + 6$ **20.** $3x^2 - 17x - 6$ **21.** $4x^2 - 8x + 3$

22. $6x^2 + 19x - 7$ **23.** $12q^2 + 20q + 3$ **24.** $8y^2 + 5y - 3$

25. $6t^2 + 7tu - 10u^2$ **26.** $4x^2 + 33xy + 8y^2$ **27.** $8x^2 + 6x - 9$

28. $12x^2 - 7xy - 12y^2$ **29.** $4x^2 + 21x - 18$ **30.** $6s^2 - 7s - 20$

31. $8n^2 + 2n - 15$ **32.** $9y^2 + 28y - 32$ **33.** $2x^2 - 22x + 48$

34. $3x^2 - 12x - 15$ **35.** $4x^2 + 2xz - 12z^2$ **36.** $5x^2 + 15x + 25$

37. $2x^3 + 6x^2 + 4x$ **38.** $2x^4 + x^3 - 10x^2$

39. $10ax^2 + 23axy - 5ay^2$ **40.** $54a^2 - 45ab - 156b^2$

In Exercises 41 through 48, factor the indicated expressions.

41. To find the side of a rectangle of a given area, it is necessary to factor the expression $x^2 - 6x - 40$. Factor this expression.

42. To find the side of another rectangle of a given area, it is necessary to factor the expression $x^2 - 8x - 128$. Factor this expression.

43. To determine the speed of a train under certain circumstances, it is necessary to factor the expression $3s^2 + 30s - 9000$. Factor this expression.

44. Under certain conditions, in order to find the total profit of an article selling for p dollars, it is necessary to factor the expression $2p^2 - 108p + 400$. Factor this expression.

45. To find the time which a projectile has been in the air, it is necessary to factor the expression $16t^2 - 128t - 320$. Factor this expression.

46. A medical experiment which measured the concentration C (in parts per million) of a particular drug in a person's blood found that $C = 0.2 + 0.1t - 0.1t^2$, where t is in hours. Factor the right-hand side of this equation.

47. An open box is made from a sheet of cardboard, 12 in. square, by cutting equal squares of side x from the corners and then bending up the sides. The volume of the box is given by the formula $V = 4x^3 - 48x^2 + 144x$. Factor the right-hand side of this formula.

48. A certain problem reads: A rectangular field bounded on one side by a river is to be fenced in. No fencing is required along the river, and the area to be fenced in is 2500 ft^2. What is the length of the side of the field that is perpendicular to the river if 225 ft of fencing is used? The solution to this problem involves the expression $2x^2 - 225x + 2500$. Factor this expression.

5-5 FACTORING OTHER TYPES OF EXPRESSIONS

The basic techniques of factoring we have discussed to this point allow us to factor many important types of expressions. There are, however, many additional types of expressions which require additional methods in order that they may be factored. In this section, we discuss certain of these additional methods.

First, we consider binomial expressions which are the **difference of cubes** and the **sum of cubes.** *Both types have* **perfect cubes** *for each term, and both types are factorable.* (Recall that the difference of squares is factorable, but the sum of squares is prime.) Equations (5-4) and (5-5) show the form of each along with its factored form.

$$a^3 - b^3 = (a - b)(a^2 + ab + b^2) \tag{5-4}$$

$$a^3 + b^3 = (a + b)(a^2 - ab + b^2) \tag{5-5}$$

The validity of these equations can be verified by multiplying out the right-hand sides.

Example A Factor $x^3 - 8y^3$.

Since x^3 is a perfect cube, and $8y^3 = (2y)^3$ is also a perfect cube, this expression fits the form of the left side of Eq. (5-4), with the a and b equal to x and $2y$, respectively. Thus, $(a - b)$ is equivalent to $(x - 2y)$, and $(a^2 + ab + b^2)$ is equivalent to $[x^2 + x(2y) + (2y)^2]$. Therefore,

$$x^3 - 8y^3 = (x - 2y)[x^2 + x(2y) + (2y)^2]$$
$$= (x - 2y)(x^2 + 2xy + 4y^2)$$

Example B Factor $27x^3 + 64y^3$.

Since $27x^3 = (3x)^3$, and $64y^3 = (4y)^3$, then $27x^3 + 64y^3 = (3x)^3 + (4y)^3$. This means that $3x$ and $4y$ are the a and b of Eq. (5-5). We may now write the factors by noting that $(a + b)$ is equivalent to $(3x + 4y)$, and $(a^2 - ab + b^2)$ is equivalent to $[(3x)^2 - (3x)(4y) + (4y)^2]$. The factorization of $27x^3 + 64y^3$ is, therefore,

$$27x^3 + 64y^3 = (3x + 4y)[(3x)^2 - (3x)(4y) + (4y)^2]$$
$$= (3x + 4y)(9x^2 - 12xy + 16y^2)$$

Grouping of terms of an expression sometimes allows us to find a factor which is common to each of the groups. Following such a procedure is known as **factoring by grouping.** The following examples illustrate the method.

Example C Factor $a^2 - a + ab - b$ by grouping.

If we group the first two terms together, and group the last two terms together we have

$$(a^2 - a) + (ab - b)$$

We now note that the first grouping has a common factor of a, and the second grouping has a common factor of b. Factoring each group, therefore, leads to

$$a(a - 1) + b(a - 1)$$

We now see that each group has a factor of $(a - 1)$. Therefore, factoring out this common factor, we have

$$(a - 1)(a + b)$$

Thus,

$$a^2 - a + ab - b = (a^2 - a) + (ab - b)$$
$$= a(a - 1) + b(a - 1)$$
$$= (a - 1)(a + b)$$

Example D Factor $cw^2 - cv^2 - dw^2 + dv^2$ by grouping.

Grouping the terms in pairs, we have

$$(cw^2 - cv^2) - (dw^2 - dv^2)$$

Note that enclosing the last two terms in parentheses caused a change in the signs of the terms because of the minus sign preceding the term dw^2. Then we factor the common monomials of the pairs, respectively:

$$c(w^2 - v^2) - d(w^2 - v^2)$$

The next step consists of factoring out $(w^2 - v^2)$ from each term:

$$(c - d)(w^2 - v^2)$$

To factor completely, we must consider the difference of squares, which can be factored:

$$(c - d)(w + v)(w - v)$$

Thus,

$$cw^2 - cv^2 - dw^2 + dv^2 = (c - d)(w + v)(w - v)$$

Example E Factor $a^2 + 2ab + b^2 - 1$.

By grouping the first three terms together, we note that $a^2 + 2ab + b^2 = (a + b)^2$, which means that the first three terms together form a perfect square. Since 1 is also a perfect square, we have the expanded form of the difference of squares. Thus,

$$\begin{aligned} a^2 + 2ab + b^2 - 1 &= (a^2 + 2ab + b^2) - 1 \\ &= (a + b)^2 - 1^2 \\ &= [(a + b) + 1][(a + b) - 1] \\ &= (a + b + 1)(a + b - 1) \end{aligned}$$

Example F Factor $x^2 + 6xy + 9y^2 - 4$.

The first three terms consist of the perfect trinomial square $(x + 3y)^2$. Since 4 is also a perfect square, we have an expanded form of the difference of squares. Therefore,

$$\begin{aligned} x^2 + 6xy + 9y^2 - 4 &= (x + 3y)^2 - 2^2 \\ &= [(x + 3y) + 2][(x + 3y) - 2] \\ &= (x + 3y + 2)(x + 3y - 2) \end{aligned}$$

Exercises 5-5

In Exercises 1 through 40, factor the given exercises completely, if possible.

1. $x^3 + y^3$
2. $k^3 - z^3$
3. $(2x)^3 - y^3$
4. $(3x)^3 + (2y)^3$
5. $64x^3 - 125$
6. $27v^3 + 125u^3$
7. $a^3b^3 + 1$
8. $x^3y^3 - 1$
9. $8x^3 - 27y^3$
10. $125w^3 - 64v^3$
11. $16x^3 + 54y^3$
12. $216n^3 + 343m^3$
13. $4x^3 - 500$
14. $3y^3 - 192$
15. $16x^3 + 81y^3$
16. $8x^3 + 54y^3$
17. $4x + xy + 4y + y^2$
18. $7b + bd + 7d + d^2$
19. $4bx + 7x - 16bv - 28v$
20. $6kt - 7t - 36kd + 42d$
21. $2c + d^2 + cd + 2d$
22. $2m^2 + 3n + 3m + 2mn$
23. $3a - b^2 + ab - 3b$
24. $2s^2 - tu - 2su + st$
25. $x^3 + 2x^2 - 4x - 8$
26. $y^3 + 4y^2 - 16y - 64$
27. $kn^2 + pq^2 - kq^2 - pn^2$
28. $pq^2 + nm^2 - pm^2 - nq^2$
29. $y^3 - 8y^2 - 6y + 48$
30. $z^3 - 9z^2 - 7z + 63$
31. $2c - d^2 + cd - 2d$
32. $a^3 - 2a^2 + 3a - 6$
33. $a^2 - 2ab + b^2 - 1$
34. $x^2 + 2xy + y^2 - 4$
35. $x^2 + 6xy + 9y^2 - 4$
36. $4z^2 + 12zv + 9v^2 - y^2$
37. $y^2 - x^2 + 6x - 9$
38. $16y^2 - 4x^2 + 12x - 9$
39. $4m^2 - 4n^2 - 4n - 1$
40. $9a^2 - b^2 + 6b - 9$

In Exercises 41 through 44, solve the given problems.

41. Verify Eq. (5-4) by expanding the right-hand side.
42. Verify Eq. (5-5) by expanding the right-hand side.
43. If the side of one cube is x and the side of another cube is $5y$, express the difference of the volumes in factored form.
44. If the area of one figure is $y^2 - 2y + 1$, and the side of a square is $4x$, express the difference of the areas of these figures in factored form.

5-6 SUMMARY

This chapter was devoted to the specific but important topic of factoring. Factoring is the process of determining which algebraic expressions, when multiplied together, equal a given expression.

Numerous methods can be developed for factoring various types of algebraic expressions. We have restricted our attention to basic types of factoring. They are: (a) finding common monomial factors, (b) factoring the difference of two perfect squares, (c) factoring trinomials, and other types of factoring, including (d) the difference (and sum) of perfect cubes and (e) factoring by grouping.

After a given expression has been factored, each resulting factor should be inspected to see whether it in turn can be factored. Only when each of the factors finally determined is prime is the expression factored completely.

Important new terms were introduced in this chapter. The reader should be certain he or she knows these terms as well as those introduced in earlier chapters.

Common monomial factor
Composite number
Cross product
Difference of cubes
Difference of squares
Factored completely
Factoring
Factoring by grouping
Perfect cube
Perfect square
Prime factor
Sum of cubes

Continued practice in factoring is the best way of learning the techniques introduced in this chapter. The following exercises should help develop facility in this algebraic operation.

Exercises 5-6

In Exercises 1 through 76, factor the given expressions completely, if possible.

1. $5a - 5c$ **2.** $4r + 8s$ **3.** $3a^2 + 6a$

4. $6t^3 - 8t^2$ **5.** $12a^2b + 4ab$ **6.** $15t^3y - 10ty^2$

7. $3xyz + 18xy^2z$ **8.** $5pqr^2 - pqr$ **9.** $4x^2 - y^2$

10. $p^2 - 9u^2v^2$ **11.** $16y^4 - x^2$ **12.** $r^2s^2t^2 - 4x^2$

13. $x^2 + 2x + 1$ **14.** $x^2 - 2x - 3$ **15.** $x^2 - 7x + 6$

16. $x^2 + 2x - 63$ **17.** $x^2 - 12x + 20$ **18.** $x^2 - 2x - 35$

19. $ax^2 + 3a^2x - a^3$ **20.** $18r^2t - 9r^3t^2 - 6r^2t^3$

21. $2nm^3 - 4n^2m^2 + 6n^3m$ **22.** $8y^2 + 24y^3z - 32y^2z^4$

23. $4p^3t^2 - 12t^4 - 4t^2 + 4at^2$ **24.** $22r^2s^2t^2 - 121rst^2 - 22r^3st^2 + 33r^4s^2t^2$

25. $(4rs)^2 - 9y^2$ **26.** $49r^4t^4 - y^6$ **27.** $36w^2x^2 + y^4$

28. $(a + b)^2 - c^2$ **29.** $2x^2 + 9x + 7$ **30.** $3y^2 + y - 10$

31. $14t^2 - 19t - 3$ **32.** $5x^2 + x - 4$ **33.** $4x^2 - 16x - 9$

34. $12x^2 - 5x - 2$ **35.** $9x^2 + 6x + 1$ **36.** $8r^2 + 2r - 15$

37. $x^2 + 3xy + 2y^2$ **38.** $6a^2 - 17ab + 12b^2$ **39.** $10c^2 + 23cd - 5d^2$

40. $4p^2 - 12pq + 9q^2$ **41.** $88x^2 - 19x - 84$ **42.** $16y^2 + 56y + 49$

43. $48y^3 - 64y^4 + 16y^5$ **44.** $18a^2u^2 + 23a^2u - 6a^2$ **45.** $5x^4 - 125$

46. $4a^8 - 64$ **47.** $2x^2 - 18y^2$ **48.** $4r^2t^2 - 36p^2q^2$

49. $4x^2 + 36$ **50.** $9a^2 + 81b^2$ **51.** $8y^4x^6 - 32y^2x^4$

52. $3m^5n - 27mn^3$ **53.** $3ax^2 + 3ax - 36a$ **54.** $36c^2x - 34cx - 30x$

55. $54r^3 - 45r^2s - 156rs^2$ **56.** $8c^2x^2 + 52c^2x + 72c^2$ **57.** $16x^4 - 1$

58. $x^8 - 1$ **59.** $36x^4 - 4$ **60.** $2x^{12} - 32$

61. $x^3y^3 + p^3q^3$ **62.** $w^3v^3 - r^3s^3$ **63.** $8x^3 - y^3z^3$

64. $125a^3 + 27b^3$ **65.** $8a^6 + b^3$ **66.** $64x^3 - y^6$

67. $2a^2 - 3a + 2ab - 3b$ **68.** $2mn + 2m + 3n + 3$

69. $2t^2 - s + st - 2t$ **70.** $y - 3x - y^2 + 3xy$

71. $8x + 2xy + 2y^2 + 8y$ **72.** $21b + 3bd + 3d^2 + 21d$

73. $4a^2 - 8ab + 4b^2 - 36$ **74.** $2x^2 + 12xy + 18y^2 - 32$

75. $3 - 3x^2 + 6xy - 3y^2$ **76.** $16 - 16a^2 + 32ab - 16b^2$

In Exercises 77 through 88, solve the given problems, which involve factoring.

77. When determining the number of teeth on a circular gear, one encounters the expression $PN + 2P$. Factor this expression.

78. In biology, when studying muscle contractions, the expression $Pv + av$ is encountered. Factor the expression.

79. In psychology, in the theory related to learning, the expression $2at - at^2$ is found. Factor this expression.

80. For a certain projectile fired from ground level, its distance s above the ground after t seconds is given by the formula $s = 256t - 16t^2$. Factor the right-hand side of this formula.

81. Under specified conditions, the price p per item for x items is given by the formula $p = x^2 - 8x + 15$. Factor the right-hand side of this formula.

82. A company determined that its profit P from selling c calculators is given by $P = c^2 + 6c - 72$. Factor the right-hand side of this equation.

83. In determining the side of a rectangle whose area is A, one encounters, under specified conditions, the equation $2x^2 + 5x - A = 0$. If $A = 52$ in.^{2}, factor the left-hand side of this equation.

84. If the side of one cube is $x + 2$ and the side of another cube is x, the difference of their volumes is $(x + 2)^3 - x^3$. Expand $(x + 2)^3$, simplify the resulting expression, and then factor.

85. At an altitude of h ft above sea level the boiling point of water is lower by a certain number of degrees T than the boiling point at sea-level, which is 212°F. The difference can be approximated by solving equation $T^2 + 520T - h = 0$. Assuming that $h =$ 5300 ft, factor the left-hand side of the equation.

86. When dealing with the theory of light reflection, we use the expression

$$\frac{u^2 - 2u + 1}{u^2 + 2u + 1}$$

Factor both the numerator and the denominator of this expression.

87. When studying radioactive tracers in biology and medicine, the equation $TT_b + TT_p = T_bT_p$ arises. By first factoring the left-hand side of this equation, solve for T.

88. Of three squares, one has a side 1 cm longer than the smallest and the third square has a side 2 cm longer than the smallest. Set up the expression for the sum of the areas of the two larger squares less the area of the smallest square, in terms of the side of the smallest square. Factor this expression.

6

FRACTIONS

6-1 EQUIVALENT FRACTIONS

As we explained in Chapter 2, a fraction indicates the division of one quantity by another. Using the terminology introduced earlier, the numerator is the dividend and the denominator is the divisor. The purpose of this chapter is to develop the basic operations of addition, subtraction, multiplication, and division of fractions consisting of algebraic expressions. Since algebraic expressions are general representations of numbers, the basic operations on fractions from arithmetic will form the basis of the algebraic operations.

One very important principle underlies the operations with fractions. It is the **fundamental principle of fractions,** which states that

> ***if the numerator and denominator of any fraction are multiplied or divided by the same quantity, not zero, the value of the fraction will remain unchanged.***

The use of the fundamental principle of fractions on any given fraction will yield an **equivalent fraction.** Consider Example A.

Example A If we multiply the numerator and denominator of the fraction $\frac{3}{4}$ by 2, the result is $\frac{6}{8}$, which is an equivalent fraction. If we divide the numerator and denominator of the fraction $\frac{18}{24}$ by 6, we obtain the fraction $\frac{3}{4}$, which is an equivalent fraction. Therefore, $\frac{3}{4}$, $\frac{6}{8}$, and $\frac{18}{24}$ are equivalent.

There are many other fractions equivalent to these: $\frac{21}{28}$ and $\frac{57}{76}$, for example. Thus, we may write

$$\frac{3}{4} = \frac{6}{8} = \frac{18}{24} = \frac{21}{28} = \frac{57}{76}$$

Example B If we multiply the numerator and denominator of the fraction

$$\frac{3x^2}{5x^3}$$

by x^2, we obtain the equivalent fraction

$$\frac{3x^4}{5x^5}$$

This can be seen by

$$\frac{(3x^2)(x^2)}{(5x^3)(x^2)} = \frac{3x^4}{5x^5}$$

If we divide the numerator and denominator of the fraction

$$\frac{3x^2}{5x^3}$$

by x^2, we obtain the equivalent fraction

$$\frac{3}{5x}$$

This can be seen by

$$\frac{\frac{3x^2}{x^2}}{\frac{5x^3}{x^2}} = \frac{3}{5x}$$

Therefore,

$$\frac{3x^2}{5x^3} \qquad \frac{3x^4}{5x^5} \quad \text{and} \quad \frac{3}{5x}$$

are equivalent, and we may write

$$\frac{3x^2}{5x^3} = \frac{3x^4}{5x^5} = \frac{3}{5x}$$

With an algebraic fraction, as with a fraction in arithmetic, *one of the most important operations is* **reducing** *it to its* **lowest terms,** *or* **simplest form.** As in the case of an arithmetic fraction, *we remove the factors which are common to both numerator and denominator by dividing both the numerator and the denominator by the common factors.* By finding the highest common factor of the numerator and denominator, and dividing each by this factor, we obtain the simplest form.

Example C In simplifying the fraction $\frac{21}{28}$, we note that both numerator and denominator have a factor of 7. Therefore, dividing each by 7, we obtain $\frac{3}{4}$, which is the simplest form of the fraction $\frac{21}{28}$.

In order to simplify the fraction

$$\frac{15x^2y^3z}{20xy^4z}$$

we note that the highest common factor of the numerator and denominator is $5xy^3z$. Dividing both numerator and denominator by this highest common factor, we have

$$\frac{(15x^2y^3z) \div (5xy^3z)}{(20xy^4z) \div (5xy^3z)} = \frac{3x}{4y}$$

Example D When simplifying the fraction

$$\frac{x^2 - x - 2}{x^2 + 3x + 2}$$

we must first factor both numerator and denominator in order to determine whether there is a common factor. We obtain

$$\frac{x^2 - x - 2}{x^2 + 3x + 2} = \frac{(x - 2)(x + 1)}{(x + 2)(x + 1)} = \frac{x - 2}{x + 2}$$

Here it is seen that the highest common factor is $(x + 1)$. This factor is divided into numerator and denominator.

Special note: Remember that in simplifying fractions, both numerator and denominator must be divided by the common factor. This process is called **cancellation.** However, many students remove *terms* that appear in both numerator and denominator. This is an *incorrect* application of cancellation. Example E illustrates common errors in the simplification of fractions.

Example E When simplifying the expression

$$\frac{x^2 - x - 2}{x^2 + 3x + 2}$$

many students would "cancel" the terms x^2 and 2, since they appear in both numerator and denominator. This is *incorrect*, for removing them in this way is equivalent to *subtracting* terms from the numerator and from the denominator, which violates the fundamental principles of fractions.

When simplifying the expression

$$\frac{2x + 3}{2(x - 1)(x + 3)}$$

many students tend to "cancel" the 3's and 2's. Again, this is *incorrect*. Actually, this expression cannot be further simplified, since there is no common *factor* in the numerator and the denominator. The 2 in the numerator is *not* a factor of the entire numerator. Thus, we see that in this process "cancel" means divide; it does not mean delete.

Example F illustrates proper cancellation in the reduction of fractions to their lowest terms.

Example F When simplifying the fraction

$$\frac{5a^2b^2c^4}{30ab^6c}$$

we note that the factor $5ab^2c$ is common to both numerator and denominator. Thus, dividing each by this factor, we obtain

$$\frac{5a^2b^2c^4}{30ab^6c} = \frac{ac^3}{6b^4}$$

When simplifying the fraction

$$\frac{2x^2 + 7x - 4}{4x^2 - 1}$$

we first factor the numerator and the denominator. Thus, we obtain

$$\frac{2x^2 + 7x - 4}{4x^2 - 1} = \frac{(2x - 1)(x + 4)}{(2x - 1)(2x + 1)}$$

Now we note the common factor of $2x - 1$, which is to be divided out. Therefore,

$$\frac{2x^2 + 7x - 4}{4x^2 - 1} = \frac{(2x - 1)(x + 4)}{(2x - 1)(2x + 1)} = \frac{x + 4}{2x + 1}$$

There are cases, especially in the addition and subtraction of fractions, when a fraction in its simplest form must be changed to an *equivalent*, not the *simplest*, form. This is accomplished by the *multiplication* of both numerator and denominator by the same quantity.

When simplifying fractions, we shall often find that the numerator and denominator have factors which differ only in *sign*. We can show that

$$\boxed{(a - b) = -(b - a)} \tag{6-1}$$

since $-(b - a) = -b + a = a - b$. *This equation shows us that we may change all the signs of the terms of a factor and not change the sign of the expression so long as we also introduce a factor of* -1. The following examples illustrate the simplification of fractions where a change of signs is necessary.

Example G We may simplify the fraction

$$\frac{x - 3}{3 - x}$$

by expressing the denominator as $-(x - 3)$. This leads to

$$\frac{x - 3}{3 - x} = \frac{x - 3}{-(x - 3)} = \frac{1}{-1} = -1$$

After rewriting the denominator, we divide out the common factor of $x - 3$.

Example H The fraction

$$\frac{4 - x^2}{x^2 - 6x + 8} = \frac{(2 - x)(2 + x)}{(x - 2)(x - 4)}$$

is seen to have factors in the numerator and denominator which differ only in sign. Changing $(2 - x)$ to $-(x - 2)$, we have

$$\frac{(2 - x)(2 + x)}{(x - 2)(x - 4)} = \frac{-(x - 2)(2 + x)}{(x - 2)(x - 4)} = \frac{-(2 + x)}{x - 4}$$

This result is one form of the simplest form of the original fraction. We can also express it as

$$\frac{2 + x}{4 - x}$$

by replacing $x - 4$ by $-(4 - x)$. Therefore, acceptable forms of the simplest form are

$$\frac{-(2+x)}{x-4} = -\frac{2+x}{x-4} = \frac{2+x}{-(x-4)} = \frac{2+x}{4-x}$$

Changing signs of factors of fractions often causes students difficulty. This need not be the case if Eq. (6-1) is understood and used correctly.

Exercises 6-1

In Exercises 1 through 8, multiply the numerator and denominator of each of the given fractions by each of the two given factors and obtain the two equivalent fractions.

1. $\frac{2}{3}$; 2, 5 **2.** $\frac{7}{5}$; 3, 6 **3.** $\frac{3ax}{b}$; $2a^2, ab$ **4.** $\frac{5b}{3x}$; $2x, 7b^2x$

5. $\frac{x-y}{x+y}$; $x-y, x+y$ **6.** $\frac{2(x-2y)}{2x-y}$; $x+2y, 2x+y$

7. $\frac{6-3b}{b-2}$; $b-2, b+2$ **8.** $\frac{3(8-b)}{b-2}$; $b-2, 8+b$

In Exercises 9 through 16, divide the numerator and denominator of each of the given fractions by the given factor and obtain an equivalent fraction.

9. $\frac{16}{28}$; 4 **10.** $\frac{27}{39}$; 3 **11.** $\frac{6a^2x}{9a^3x}$; $3ax$ **12.** $\frac{14r^2st}{28rst^2}$; $7rt$

13. $\frac{2x^2+5x-3}{x^2+4x+3}$; $x+3$ **14.** $\frac{6x^2-11x+3}{2x^2-3x}$; $2x-3$

15. $\frac{6a^2+a-2}{6a+4}$; $3a+2$ **16.** $\frac{12y^2-17y-7}{9y+3}$; $3y+1$

In Exercises 17 through 44, determine the simplest form of each of the given fractions.

17. $\frac{8}{36}$ **18.** $\frac{42}{63}$ **19.** $\frac{48}{180}$ **20.** $\frac{125}{150}$

21. $\frac{2x}{6x}$ **22.** $\frac{5a}{30a}$ **23.** $\frac{12b}{36b}$ **24.** $\frac{7k}{49k}$

25. $\frac{5a^2b}{20a}$ **26.** $\frac{12xyz^2}{21xyz^3}$ **27.** $\frac{18r^3st^2}{63r^5s^2t}$ **28.** $\frac{91a^3b^2c}{26ab^2c^5}$

29. $\frac{(x-4)(3x-2)}{(3x-2)(x+4)}$ **30.** $\frac{(2x-5)(2x-3)}{(2x+3)(2x-5)}$

31. $\frac{(x-1)(x+1)(x+2)}{2(x-1)(x+3)}$ **32.** $\frac{3(x+5)(2x-1)(3x-2)}{9(5x+1)(3x-2)(2x-1)}$

33. $\frac{x^2-1}{x^2-2x+1}$ **34.** $\frac{x^2+5x+4}{x^2+3x-4}$

35. $\dfrac{3x^2 - x}{3x^2 + 5x - 2}$

36. $\dfrac{4x^2 + 9x - 9}{4x^2 + 12x}$

37. $\dfrac{6x^2 - 19x + 10}{8x^2 - 14x - 15}$

38. $\dfrac{4x^2 - 2x - 12}{12x^2 + 8x - 4}$

39. $\dfrac{x^2 - 9y^2}{3xy - 9y^2}$

40. $\dfrac{5a^2 + 39ab - 8b^2}{5a^2 + 4ab - b^2}$

41. $\dfrac{(3x - 1)(5 - x)}{(x - 5)(3x + 1)}$

42. $\dfrac{(x - 2)(x - 3)(4 - x)}{(3 - x)(x - 4)(2x - 1)}$

43. $\dfrac{20 - 9x + x^2}{8 + 2x - x^2}$

44. $\dfrac{4ab - b^2}{b^2 - 16a^2}$

In Exercises 45 through 48, solve the problems which involve the simplification of fractions.

45. Evaluate the fraction

$$\frac{4a - 8}{4(a + 2)}$$

for the value $a = 5$ before and after reducing it to simplest form.

46. Evaluate the fraction

$$\frac{2x^2 + 5xy - 3y^2}{2x^2 + 3xy - 2y^2}$$

for the values $x = 2$ and $y = 3$ before and after reducing it to simplest form.

47. Evaluate the fraction

$$\frac{x - 2}{2 - x}$$

for the value $x = 7$, and compare it with the value obtained when it is reduced to lowest terms.

48. Which of the following fractions are in simplest form?

(a) $\dfrac{x - 2}{x + 2}$ (b) $\dfrac{2x}{2x + 3}$ (c) $\dfrac{3x + 1}{3x + 5}$ (d) $\dfrac{x^2 - x + 2}{x^2 + 2}$ (e) $\dfrac{x^2 + x + 4}{x^2(x + 4)}$

6-2 MULTIPLICATION AND DIVISION OF FRACTIONS

From arithmetic, we recall that *the product of two fractions is a fraction whose numerator is the product of the numerators and whose denominator is the product of the denominators of the given fractions.* Since algebra is a generalization of arithmetic, the same definition holds in algebra. Symbolically, this is written as

$$\frac{a}{b} \cdot \frac{c}{d} = \frac{ac}{bd} \quad (b \text{ or } d \text{ not zero}) \tag{6-2}$$

Example A

$$\frac{5}{6}\cdot\frac{7}{9}=\frac{(5)(7)}{(6)(9)}=\frac{35}{54}$$

$$\frac{3a}{b}\cdot\frac{x^2}{5b}=\frac{(3a)(x^2)}{(b)(5b)}=\frac{3ax^2}{5b^2}$$

Closely associated with the multiplication of fractions is the power of a fraction. To find $(a/b)^n$, we would have n factors of (a/b), which would result in a numerator of a^n and a denominator of b^n. Therefore,

$$\left(\frac{a}{b}\right)^n=\frac{a^n}{b^n} \quad (b \text{ not zero}) \tag{6-3}$$

Example B

$$\left(\frac{2}{3}\right)^3=\frac{2^3}{3^3}=\frac{8}{27}$$

$$\left(\frac{2a^2}{x}\right)^4=\frac{(2a^2)^4}{x^4}=\frac{(2^4)(a^2)^4}{x^4}=\frac{16a^8}{x^4}$$

Another basic arithmetical operation is the division of one fraction by another. To do this, *we invert the divisor* (*the fraction in the denominator*) *and multiply the dividend* (*the fraction in the numerator*) *by the inverted fraction.* This procedure can be explained on the basis of the basic principle of fractions. If a/b is divided by c/d, we multiply the numerator and the denominator by d/c. We may write

$$\frac{\frac{a}{b}}{\frac{c}{d}}=\frac{\frac{a}{b}\cdot\frac{d}{c}}{\frac{c}{d}\cdot\frac{d}{c}}=\frac{\frac{ad}{bc}}{\frac{cd}{dc}}=\frac{\frac{ad}{bc}}{1}=\frac{ad}{bc}$$

Therefore, showing the division of fractions symbolically, we have

$$\frac{\frac{a}{b}}{\frac{c}{d}}=\frac{ad}{bc} \quad (b, c, \text{ and } d \text{ not zero}) \tag{6-4}$$

Example C

$$\frac{2}{5} \div \frac{3}{7} = \frac{\frac{2}{5}}{\frac{3}{7}} = \frac{2}{5} \cdot \frac{7}{3} = \frac{(2)(7)}{(5)(3)} = \frac{14}{15}$$

$$\frac{\frac{4xy}{3b}}{\frac{7b}{xyz}} = \frac{4xy}{3b} \cdot \frac{xyz}{7b} = \frac{(4xy)(xyz)}{(3b)(7b)} = \frac{4x^2y^2z}{21b^2}$$

Another mathematical term that is defined in terms of division is **reciprocal.** *The reciprocal of a number is* 1 *divided by that number*. That is, the reciprocal of a (a not zero) is $1/a$. Since

$$\frac{1}{\frac{a}{b}} = 1 \cdot \frac{b}{a} = \frac{b}{a}$$

we see that the reciprocal of the fraction a/b is the inverted fraction b/a.

Example D The reciprocal of $2a$ is $\frac{1}{2a}$.

The reciprocal of $\frac{1}{b}$ is $\frac{b}{1}$, or b.

The reciprocal of $-\frac{xy}{a}$ is $-\frac{a}{xy}$. When we are finding the reciprocal of a fraction, the sign of the fraction is unaltered.

One immediate use of the term reciprocal is in an alternative definition of the division of one fraction by another. We may state that, *in dividing one fraction by another, we multiply the dividend by the reciprocal of the divisor.*

As we brought out in Section 6-1, results which are fractions are usually expressed in simplified form. Due to the process of multiplying fractions, it is usually a relatively simple matter to simplify the product or quotient of two fractions. Remember that, in simplifying a fraction, we must divide the numerator and denominator by factors which are common to both. Therefore, if *we factor the numerators and denominators of the fractions being multiplied or divided and only* **indicate** *the appropriate product*, we can readily determine any common factors. If this factoring is not done, it is very possible that the result obtained by multiplying out the fractions will be a fraction containing expressions that are extremely difficult to factor and hence a fraction that is difficult to simplify. The following examples illustrate the multiplication and division of algebraic fractions.

Example E Multiply $\dfrac{8x^2y^3}{21abc}$ by $\dfrac{15ax^2}{16b^2y^2}$.

$$\frac{8x^2y^3}{21abc} \cdot \frac{15ax^2}{16b^2y^2} = \frac{(8)(15)ax^4y^3}{(21)(16)ab^3cy^2}$$

Now we see that the highest common factor of the numerator and denominator is $(8)(3)ay^2$. Dividing numerator and denominator by this factor, we have $5x^4y/14b^3c$. Therefore,

$$\frac{8x^2y^3}{21abc} \cdot \frac{15ax^2}{16b^2y^2} = \frac{5x^4y}{14b^3c}$$

Example F Perform the multiplication

$$\frac{x^2 - y^2}{x + 2y} \cdot \frac{3x + 6y}{x - y}$$

If we perform the multiplication directly by multiplying numerators together and denominators together, we obtain

$$\frac{3x^3 + 6x^2y - 3xy^2 - 6y^3}{x^2 + xy - 2y^2}$$

Although the numerator is factorable by grouping, it is not necessary to perform the multiplication. If we only *indicate* the multiplication as

$$\frac{x^2 - y^2}{x + 2y} \cdot \frac{3x + 6y}{x - y} = \frac{(x^2 - y^2)(3x + 6y)}{(x + 2y)(x - y)}$$

the numerator is already partially factored and the denominator is completely factored. The additional work now required is considerably less than that required after direct multiplication. Completing the factoring of the numerator, since $x^2 - y^2 = (x + y)(x - y)$ and $3x + 6y = 3(x + 2y)$, we have

$$\begin{aligned}\frac{x^2 - y^2}{x + 2y} \cdot \frac{3x + 6y}{x - y} &= \frac{(x^2 - y^2)(3x + 6y)}{(x + 2y)(x - y)}\\ &= \frac{3(x - y)(x + y)(x + 2y)}{(x + 2y)(x - y)}\\ &= 3(x + y)\end{aligned}$$

Here it is seen that the highest common factor of the numerator and denominator is $(x - y)(x + 2y)$.

Example G Perform the division

$$\frac{b^2 - 2b + 1}{9b^2 - 1} \div \frac{5b - 5}{12b - 4}$$

We will first indicate the product of the dividend and the reciprocal of the divisor. We then factor both the numerator and denominator. Next, the highest common factor is divided out. These steps are as follows:

$$\frac{b^2 - 2b + 1}{9b^2 - 1} \div \frac{5b - 5}{12b - 4} = \frac{(b^2 - 2b + 1)(12b - 4)}{(9b^2 - 1)(5b - 5)}$$

$$= \frac{(b - 1)(b - 1)(4)(3b - 1)}{(3b + 1)(3b - 1)(5)(b - 1)}$$

$$= \frac{4(b - 1)}{5(3b + 1)}$$

The highest common factor that was divided out was $(b - 1)(3b - 1)$. The result has been left in factored form. Although it would be perfectly correct to multiply out both the numerator and denominator, the factored form often is more convenient. Therefore, it is definitely permissible to leave the result in factored form.

Exercises 6-2

In Exercises 1 through 4, find the reciprocal of each expression.

1. $8n; \frac{1}{13s}$ **2.** $a^2b; \frac{a^2}{b}$ **3.** $\frac{a}{3b}; \frac{3b}{a}$ **4.** $-\frac{2x^2}{y}; -\frac{5cd}{3ax}$

In Exercises 5 through 46, perform the indicated multiplications and divisions, expressing all answers in simplest form with positive exponents.

5. $\frac{2}{9} \cdot \frac{3}{5}$ **6.** $\frac{6}{13} \cdot \frac{7}{2}$ **7.** $\frac{17rs}{12t} \cdot \frac{3t^2}{51s}$ **8.** $\frac{24xy^2}{5z} \cdot \frac{125z^2}{8y^3}$

9. $\left(\frac{3}{4}\right)^4$ **10.** $\left(\frac{7}{5}\right)^3$ **11.** $\left(\frac{a^2}{2x}\right)^5$ **12.** $\left(\frac{3s}{4t^2}\right)^4$

13. $\left(\frac{4x}{8y^{-2}}\right)^5$ **14.** $\left(\frac{18b^{-1}}{36b^2}\right)^3$ **15.** $\frac{6}{17} \div \frac{7}{68}$ **16.** $\frac{15}{8} \div \frac{25}{18}$

17. $\dfrac{17x}{22} \div \dfrac{68m}{11}$

18. $\dfrac{15}{8z} \div \dfrac{25}{18y}$

19. $\dfrac{3x}{25y} \div \dfrac{27x^2}{5y^2}$

20. $\dfrac{3a^2x}{4ay^2} \div \dfrac{6ax}{5a^3x^2}$

21. $\dfrac{9a^2b^2}{10ab^3} \div \dfrac{72a^{-3}b^4}{40}$

22. $\dfrac{4x}{3x^2y^{-2}} \div \dfrac{20xy}{9y^{-3}}$

23. $\dfrac{a-5b}{a+b} \cdot \dfrac{a+3b}{a-5b}$

24. $\dfrac{5n+10}{3n-9} \cdot \dfrac{n-3}{15}$

25. $\dfrac{x^2-y^2}{14x} \cdot \dfrac{35x^2}{3x+3y}$

26. $\dfrac{a+b}{4a} \cdot \dfrac{3a^2}{(a+b)^2}$

27. $\dfrac{2b+3}{5} \div \dfrac{4b+6}{5b-2}$

28. $\dfrac{6}{3x+4} \div \dfrac{2x+2}{9x^2+12x}$

29. $\dfrac{x^2+2x-3}{x^2-4} \cdot \dfrac{x^2-x-6}{x^2-5x+4}$

30. $\dfrac{3x^2+10x-8}{36x^2-16} \cdot \dfrac{9x^2+15x+6}{x^2+3x-4}$

31. $\dfrac{x-3}{2x+1} \div \dfrac{x-3}{2x+3}$

32. $\dfrac{2-t}{3s-t} \div \dfrac{2t}{3s-t}$

33. $\dfrac{2x-6}{2x-8} \cdot \dfrac{3x-12}{x-3}$

34. $\dfrac{(x-4)^2}{x^2-16} \cdot \dfrac{3x-12}{4x-16}$

35. $\dfrac{3a^2-3b^2}{a^2-4b^2} \div \dfrac{a^2+2ab+b^2}{a+2b}$

36. $\dfrac{20x}{x^2-8x+15} \div \dfrac{10x}{x^2-2x-15}$

37. $\dfrac{s^2-5s-14}{s^2-9s-36} \div \dfrac{s^2+4s-77}{s^2+10s+21}$

38. $\dfrac{p^2+5pq+6q^2}{2p^2+pq-q^2} \div \dfrac{p^2-9q^2}{(p+q)^2}$

39. $\dfrac{4x^2-4}{3x^2-13x-10} \div \dfrac{4x+4}{x^2-6x+5}$

40. $\dfrac{y^2-y-2}{2y^2+y-10} \div \dfrac{6y^2-y-7}{12y^2+16y-35}$

41. $\dfrac{3a-3b}{6a} \cdot \dfrac{(a+b)^2}{a^2-b^2} \cdot \dfrac{4a^2}{2a^2-ab-b^2}$

42. $\left(\dfrac{2x^2-13x+21}{x^2-6x+9} \cdot \dfrac{x^2-9}{4x^2-28x+49}\right) \div \dfrac{3x+9}{4x-28}$

43. $\left(\dfrac{2x^2+x-15}{4x^2-5x-21} \cdot \dfrac{x^2-6x+9}{x^2-9}\right) \div \dfrac{4x^2-25}{4x+7}$

44. $\left[\dfrac{(x-y)^{-3}}{a^2} \cdot \dfrac{3x-3y}{a^{-3}}\right] \div \dfrac{3^{-3}}{4x^2-9y^2}$

45. $\left[\dfrac{(x+y)^{-1}}{b^2} \cdot \dfrac{2x+2y}{b^{-1}}\right] \div \dfrac{4^{-4}}{9x^2-16y^2}$

46. $[(3x+7)(8x^2-10x-25)^{-1}] \cdot [(14-x-3x^2)(x^2-4)^{-1}]$

In Exercises 47 and 48, solve the given problems involving the multiplication and division of fractions.

47. Under certain conditions, the product of the price and the demand of a given commodity is constant. If the price of each of x units is $\frac{2x+6}{3}$ and the demand for these is $\frac{6000}{x^2+3x}$, determine the expression for the product of the price and demand.

48. The current in a simple electric circuit is the voltage in the circuit divided by the resistance. If the voltage changes with time according to the formula

$$V = \frac{3t}{t+2}$$

and the resistance changes with time according to the formula

$$R = \frac{t^2+4}{t^2+4t+4}$$

what is the formula for the current in terms of t?

6-3 THE LOWEST COMMON DENOMINATOR

From basic arithmetic, we recall that *the sum of a set of fractions all having the same denominator is the sum of the numerators divided by the denominator.* The reason for this fact is reviewed in Example A.

Example A When determining the sum $\frac{2}{7} + \frac{4}{7}$, we are essentially finding the total number of one-sevenths. The first fraction represents two and the second fraction four of the quantity one-seventh. Thus, the total is six one-sevenths. This is represented as

$$\frac{2}{7} + \frac{4}{7} = \frac{2+4}{7} = \frac{6}{7}$$

Therefore, finding the sum of a set of fractions having the same denominator is a relatively easy task. However, if the fractions do not all have the same denominator, we must first change each to an equivalent fraction so that each member of the resulting set of fractions has the same denominator. Since this step is of utmost importance in the addition of fractions, this section is devoted entirely to finding the common denominator of a given set of fractions. The following section will then discuss the complete method for the addition and subtraction of fractions.

The most convenient and useful denominator of a set of fractions is the **lowest common denominator.** *This is the denominator that contains the smallest number of prime factors of the given denominators and that is exactly divisible by each denominator. That is, it is the smallest number or simplest expression into which all given denominators will divide evenly.* The basic method for determining this lowest common denominator will be outlined after Example B.

Example B When computing the sum $\frac{3}{8} + \frac{5}{16}$, we note that both 8 and 16 will divide evenly into 16. Therefore, the sum may be set up as $\frac{6}{16} + \frac{5}{16}$, from which the result, $\frac{11}{16}$, is easily found. It is possible to use 32 as a common denominator. This would result in our having to change both fractions into equivalent fractions with 32 as the denominator, and the resulting fraction would then have to be simplified to $\frac{11}{16}$. This is shown below:

$$\frac{3}{8} + \frac{5}{16} = \frac{12}{32} + \frac{10}{32} = \frac{12 + 10}{32} = \frac{22}{32} = \frac{11}{16}$$

Now this procedure was not particularly difficult to follow. However, the product of 8 and 16 would also serve as a common denominator. Using 128, we would have

$$\frac{3}{8} + \frac{5}{16} = \frac{48}{128} + \frac{40}{128} = \frac{48 + 40}{128} = \frac{88}{128}$$

We can simplify $\frac{88}{128}$ to $\frac{11}{16}$ once we note that both 88 and 128 are divisible by 8. However, we did much more work than we did to add $\frac{6}{16}$ and $\frac{5}{16}$. Also, when working with algebraic fractions, we find that simplifying the resulting fraction is made much more difficult if the simplest possible denominator is not used. For if the *lowest* common denominator is not used, it is always possible to reduce the resulting fraction even though it may not seem to be reducible.

To determine the lowest common denominator of a set of fractions, *we first find all the prime factors of each denominator and then form the product of these prime factors, giving each factor the largest exponent it has in any of the given denominators.*

Example C

1. Find the lowest common denominator of $\frac{3}{8}$ and $\frac{5}{16}$.
Factoring 8 and 16 into their prime factors, we have

$$8 = 2 \cdot 2 \cdot 2 = 2^3 \quad \text{and} \quad 16 = 2 \cdot 2 \cdot 2 \cdot 2 = 2^4$$

The only prime factor that appears is 2, and it is to the fourth power for 16. Thus, $2^4 = 16$ is the lowest denominator, for 2^3 will divide evenly into 2^4; that is, $2^4/2^3 = 2$.

2. Find the lowest common denominator of $\frac{5}{12}$ and $\frac{7}{18}$.
Factoring 12 and 18 into their prime factors gives

$$12 = 2 \cdot 2 \cdot 3 = 2^2 \cdot 3 \quad \text{and} \quad 18 = 2 \cdot 3 \cdot 3 = 2 \cdot 3^2$$

The prime factors that must be considered are 2 and 3. The largest exponent of each is 2. Therefore, the lowest common denominator is $2^2 \cdot 3^2 = 36$. No number smaller than 36 is divisible evenly by both 12 and 18.

3. When we are finding the lowest common denominator of

$$\frac{1}{4x^2y} \quad \text{and} \quad \frac{7a}{6xy^4}$$

we factor the denominators into their prime factors. This gives

$$4x^2y = 2^2x^2y \quad \text{and} \quad 6xy^4 = 2 \cdot 3 \cdot xy^4$$

The prime factors to be considered are 2, 3, x, and y. The largest exponent of 2 is 2; the largest exponent of 3 is 1; the largest exponent of x is 2; and the largest exponent of y is 4. Therefore, the lowest common denominator is $2^2 \cdot 3 \cdot x^2y^4 = 12x^2y^4$. This is the simplest expression into which $4x^2y$ and $6xy^4$ both divide evenly.

We shall now consider the problem of finding the lowest common denominator of a set of algebraic fractions. The basic procedure is the same as for the arithmetic fractions we have considered to this point. However, *in forming the product that is the lowest common denominator, we only* **indicate** *the multiplication of the necessary factors; we do not actually multiply out the product.* This procedure prevents us from losing the identity of the several factors and therefore makes the process of finding the appropriate equivalent fractions much easier. (Our reasoning will be shown in the following section, when we actually add and subtract algebraic fractions.) The following examples illustrate the procedure for finding the lowest common denominator for given sets of algebraic fractions.

Example D Find the lowest common denominator of the fractions

$$\frac{x-9}{x^2+x-6} \quad \text{and} \quad \frac{x+2}{x^2-5x+6}$$

Factoring the denominators, we have

$$x^2+x-6=(x+3)(x-2) \quad \text{and} \quad x^2-5x+6=(x-2)(x-3)$$

The necessary factors are $(x-2)$, $(x-3)$, and $(x+3)$. Since the highest power to which each appears is 1, the lowest common denominator is

$$(x-2)(x-3)(x+3)$$

The product of these three factors is $x^3-2x^2-9x+18$. It is readily apparent that we would lose the identity of the factors in the multiplication. In forming the equivalent fractions, we would actually have to divide this expression by each of the individual denominators. This procedure would create a great deal of extra work because, for the factored form, this division can be done by inspection. Therefore, the form $(x-2)(x-3)(x+3)$ is preferable.

Example E Find the lowest common denominator of the fractions

$$\frac{2}{5x^2} \qquad \frac{3x}{10x^2-10} \qquad \frac{5}{x^2-x}$$

Factoring the denominators, we have

$$5x^2 = 5x^2$$

$$10x^2-10=10(x^2-1)=2\cdot 5(x-1)(x+1)$$

$$x^2-x=x(x-1)$$

The prime factors of the lowest common denominator are 2, 5, x, $x-1$, and $x+1$. Each appears to the first power except x, which appears to the second power. Therefore, the lowest common denominator is

$$2\cdot 5x^2(x-1)(x+1)=10x^2(x-1)(x+1)$$

(Here we have multiplied the numerical factors in the coefficient. However, the others have been left in factored form.)

Example F Find the lowest common denominator of the fractions

$$\frac{3x}{x^2+4x+4} \qquad \frac{5}{x^2-4} \qquad \frac{x}{6x+12}$$

Factoring the denominators, we have

$$x^2+4x+4 = (x+2)(x+2) = (x+2)^2$$
$$x^2-4 = (x-2)(x+2)$$
$$6x+12 = 6(x+2) = 2 \cdot 3(x+2)$$

The prime factors of the lowest common denominator are 2, 3, $x+2$, and $x-2$. All appear to the first power except $x+2$, which appears to the second power. Therefore, the lowest common denominator is

$$2 \cdot 3(x+2)^2(x-2) = 6(x+2)^2(x-2)$$

The lowest common denominator of many simpler fractions, especially arithmetic fractions, can be determined by inspection. However, whenever the *lowest* common denominator is not obvious, the method outlined above should be followed.

Exercises 6-3

Find the lowest common denominator of each of the following sets of fractions.

1. $\frac{1}{4}, \frac{1}{6}$ **2.** $\frac{7}{15}, \frac{8}{25}$ **3.** $\frac{5}{18}, \frac{11}{45}$ **4.** $\frac{7}{40}, \frac{7}{72}$

5. $\frac{9}{60}, \frac{1}{105}, \frac{9}{28}$ **6.** $\frac{3}{8}, \frac{1}{24}, \frac{5}{27}$ **7.** $\frac{36}{125}, \frac{7}{15}, \frac{5}{6}$ **8.** $\frac{9}{16}, \frac{1}{12}, \frac{37}{72}$

9. $\frac{2}{5}, \frac{3}{5a}$ **10.** $\frac{4}{3x}, \frac{1}{x}$ **11.** $\frac{1}{5a^2}, \frac{7}{3a^2}$ **12.** $\frac{2}{7t}, \frac{1}{4t^4}$

13. $\frac{3}{5s^2}, \frac{4}{15s}$ **14.** $\frac{5}{4x^5}, \frac{9}{8x^2}$ **15.** $\frac{2}{7k^2}, \frac{8}{21k}, \frac{3}{4k}$

16. $\frac{7}{4x^2y^2}, \frac{10}{3y^3}, \frac{6}{5x^3y}$ **17.** $\frac{15}{32ab^3}, \frac{11}{12a^3b^2}$ **18.** $\frac{8}{27abc^5}, \frac{7x}{3a^2bc^4}$

19. $\frac{27}{4acx^3}, \frac{5}{12a^2cx}, \frac{13b}{20acx}$ **20.** $\frac{4}{25p^2q}, \frac{8}{15q^2r}, \frac{16}{27prs^2}$ **21.** $\frac{1}{x(x+1)}, \frac{2}{x^2}$

22. $\dfrac{1}{4y(y+5)}, \dfrac{1}{2y^2}$

23. $\dfrac{3}{a^2(a-3)}, \dfrac{5}{2a(a-3)}$

24. $\dfrac{6}{x^3(2x+1)}, \dfrac{3}{(2x+1)^2}$

25. $\dfrac{5}{4x-4}, \dfrac{3}{8x}$

26. $\dfrac{7}{6y^3-12y^2}, \dfrac{3y}{4y-8}$

27. $\dfrac{7}{3x-3}, \dfrac{4}{9x}, \dfrac{7}{(x-1)^2}$

28. $\dfrac{3}{8y^3-16y^2}, \dfrac{21y}{(1+2y)^2}, \dfrac{5}{12y+24y^2}$

29. $\dfrac{5}{3a^2x-9ax}, \dfrac{7x}{6a^2-18a}$

30. $\dfrac{9}{8a^2x^3+2a^3x^2}, \dfrac{26}{12x+3a}$

31. $\dfrac{2}{3x(x-1)}, \dfrac{4}{x^2(x+1)}, \dfrac{1}{(x-1)(x+1)}$

32. $\dfrac{5}{2x^2(x-2)}, \dfrac{7}{(x-2)^2}, \dfrac{3}{4x^2}$

33. $\dfrac{3x}{2x-2y}, \dfrac{5}{x^2-xy}, \dfrac{7x}{6x^2-6y^2}$

34. $\dfrac{1}{x^4-x^3}, \dfrac{3}{4x^2-4}, \dfrac{a}{x^3+x^2}$

35. $\dfrac{a+3b}{a^2-ab-2b^2}, \dfrac{a+b}{a^2-4b^2}$

36. $\dfrac{x-5}{x^2-3x+2}, \dfrac{x}{2x^2-4x+2}$

37. $\dfrac{x-1}{4x^2-36}, \dfrac{x+7}{3x^2+18x+27}$

38. $\dfrac{7}{2t^2-5t-12}, \dfrac{5t}{2t^2+10t+6}$

39. $\dfrac{x-7y}{3x^2-7xy-6y^2}, \dfrac{x+y}{x^2-9y^2}, \dfrac{7}{2x-6y}$

40. $\dfrac{5}{6x^4-6y^4}, \dfrac{x+y}{4x^2+4y^2}, \dfrac{2x-5y}{3x^2-6xy+3y^2}$

6-4 ADDITION AND SUBTRACTION OF FRACTIONS

Now that we have discussed the basic method of finding the lowest common denominator of a set of fractions, we are in a position to perform the operations of addition and subtraction of algebraic fractions. As we have pointed out, if we want to add or subtract fractions whose denominators are the same, we place the sum of numerators over the denominator. *If the denominators differ, we must first change the fractions to equivalent ones with a common denominator equal to the lowest common denominator. We then combine the numerators of the equivalent fractions algebraically, placing this result over the common denominator.* The following examples illustrate the method.

Example A Combine

$$\frac{7}{5ax} + \frac{3b}{5ax} - \frac{4b}{5ax}$$

Since the denominators are the same for each fraction, we need only to combine the numerators over the common denominator. We therefore have

$$\frac{7}{5ax} + \frac{3b}{5ax} - \frac{4b}{5ax} = \frac{7 + 3b - 4b}{5ax} = \frac{7 - b}{5ax}$$

Note that the signs of the terms of the numerator are the same as those of the fractions from which they were obtained.

Example B Combine

$$\frac{3}{ax} + \frac{x}{2a^2} - \frac{7}{2x}$$

By observation of the denominators, we see that the factors necessary in the lowest common denominator are 2, a, and x. Both 2 and x appear only to the first power, and a appears squared. Thus, the lowest common denominator is $2a^2x$. We now wish to write each of the fractions in an equivalent form with $2a^2x$ as denominator. Since the denominator of the first fraction contains factors of a and x, it is necessary to introduce additional factors of 2 and a; or in other words, we must multiply the numerator and denominator of the first fraction by $2a$. For similar reasons, we must multiply the numerator and denominator of the second fraction by x and the numerator and denominator of the third fraction by a^2. This leads to

$$\begin{aligned}\frac{3}{ax} + \frac{x}{2a^2} - \frac{7}{2x} &= \frac{3(2a)}{(ax)(2a)} + \frac{x(x)}{(2a^2)(x)} - \frac{7(a^2)}{(2x)(a^2)} \\ &= \frac{6a}{2a^2x} + \frac{x^2}{2a^2x} - \frac{7a^2}{2a^2x} \\ &= \frac{6a + x^2 - 7a^2}{2a^2x}\end{aligned}$$

Note that the signs of the terms of the numerator are the same as those of the fractions from which they were obtained.

Example C Combine

$$\frac{4}{x + 3} + \frac{3}{x + 2}$$

Since there is only one factor in each of the denominators and the factors are different, the lowest common denominator of these fractions is the product $(x + 3)(x + 2)$. Thus, we must multiply the numerator and denominator of the first fraction by $x + 2$ and the numerator and denominator of the second fraction by $x + 3$ in order to make equivalent fractions that have $(x + 3)(x + 2)$ as the denominator. Therefore, performing the addition, we have

$$\frac{4}{x+3} + \frac{3}{x+2} = \frac{4(x+2)}{(x+3)(x+2)} + \frac{3(x+3)}{(x+2)(x+3)}$$

$$= \frac{4(x+2) + 3(x+3)}{(x+3)(x+2)}$$

$$= \frac{4x + 8 + 3x + 9}{(x+3)(x+2)}$$

$$= \frac{7x + 17}{(x+3)(x+2)}$$

Example D Combine

$$\frac{x}{x+y} - \frac{2y^2}{x^2 - y^2}$$

Factoring the denominator of the second fraction, we have

$$x^2 - y^2 = (x + y)(x - y)$$

Since the factor $x + y$ appears in the first fraction and the only other factor present is $x - y$, the lowest common denominator is $(x + y)(x - y)$. Therefore, we must multiply the numerator and denominator of the first fraction by $x - y$, whereas the second fraction remains the same. This leads to

$$\frac{x}{x+y} - \frac{2y^2}{x^2-y^2} = \frac{x}{x+y} - \frac{2y^2}{(x+y)(x-y)}$$

$$= \frac{x(x-y)}{(x+y)(x-y)} - \frac{2y^2}{(x+y)(x-y)}$$

$$= \frac{x^2 - xy - 2y^2}{(x+y)(x-y)}$$

$$= \frac{(x+y)(x-2y)}{(x+y)(x-y)}$$

$$= \frac{x - 2y}{x - y}$$

We note here that, when the numerators were combined, the result was factorable. One of the factors also appeared in the denominator, which means it was possible to reduce the resulting fraction. Remember, *we must express the result in simplest form.*

Example E Combine

$$\frac{5}{a^2 - a - 6} + \frac{1}{a^2 - 5a + 6} - \frac{2}{a^2 - 4a + 4}$$

Factoring the denominators, we have

$$\begin{aligned} a^2 - a - 6 &= (a - 3)(a + 2) \\ a^2 - 5a + 6 &= (a - 3)(a - 2) \\ a^2 - 4a + 4 &= (a - 2)(a - 2) = (a - 2)^2 \end{aligned}$$

The prime factors of the lowest common denominator are $a - 3$, $a + 2$, and $a - 2$, where $a - 2$ appears to the second power. The lowest common denominator is therefore

$$(a - 3)(a + 2)(a - 2)^2$$

To have equivalent fractions with the lowest common denominator, it is necessary to multiply the numerator and the denominator of the first fraction by $(a - 2)^2$, those of the second fraction by $(a - 2)(a + 2)$, and those of the third fraction by $(a + 2)(a - 3)$. This leads to

$$\begin{aligned} &\frac{5}{a^2 - a - 6} + \frac{1}{a^2 - 5a + 6} - \frac{2}{a^2 - 4a + 4} \\ &\quad = \frac{5}{(a - 3)(a + 2)} + \frac{1}{(a - 3)(a - 2)} - \frac{2}{(a - 2)^2} \\ &\quad = \frac{5(a - 2)^2}{(a - 3)(a + 2)(a - 2)^2} + \frac{1(a - 2)(a + 2)}{(a - 3)(a - 2)(a - 2)(a + 2)} \\ &\qquad - \frac{2(a + 2)(a - 3)}{(a - 2)^2(a + 2)(a - 3)} \\ &\quad = \frac{5(a - 2)^2 + (a - 2)(a + 2) - 2(a + 2)(a - 3)}{(a - 3)(a + 2)(a - 2)^2} \\ &\quad = \frac{5(a^2 - 4a + 4) + a^2 - 4 - 2(a^2 - a - 6)}{(a - 3)(a + 2)(a - 2)^2} \\ &\quad = \frac{5a^2 - 20a + 20 + a^2 - 4 - 2a^2 + 2a + 12}{(a - 3)(a + 2)(a - 2)^2} \\ &\quad = \frac{4a^2 - 18a + 28}{(a - 3)(a + 2)(a - 2)^2} = \frac{2(2a^2 - 9a + 14)}{(a - 3)(a + 2)(a - 2)^2} \end{aligned}$$

Since $2a^2 - 9a + 14$ is not factorable, it is not possible to simplify the form of the result any further.

Exercises 6-4

In Exercises 1 through 8, change the indicated sum of fractions to an indicated sum of equivalent fractions with the proper lowest common denominator. Do not combine.

1. $\frac{5}{9}-\frac{7}{12}$

2. $\frac{1}{40}+\frac{5}{84}$

3. $\frac{5}{ax}+\frac{1}{bx}-\frac{4}{a}$

4. $\frac{6}{a^3b}-\frac{5}{4ab}-\frac{1}{6a^2}$

5. $\frac{4}{x^2-x}-\frac{3}{2x^3-2x^2}$

6. $\frac{5b}{3ab-6ac}-\frac{7c}{3b^2-12c^2}$

7. $\frac{x}{2x-4}+\frac{5}{x^2-4}-\frac{3x}{x^2+4x+4}$

8. $\frac{a+1}{3a-3}-\frac{8}{a^2-5a+4}-\frac{1}{9}$

In Exercises 9 through 44 combine the given fractions, expressing all results in simplest form.

9. $\frac{1}{2}-\frac{5}{8}+\frac{3}{20}$

10. $\frac{11}{12}+\frac{13}{40}-\frac{4}{15}$

11. $\frac{2}{3}-\frac{1}{2x}+\frac{1}{10}$

12. $\frac{9}{10}+\frac{11}{14}-\frac{6}{5a}$

13. $\frac{2}{5x}+\frac{3}{2x}$

14. $\frac{2}{3a}-\frac{5}{3b}$

15. $\frac{8}{by}-\frac{1}{y^2}$

16. $\frac{5}{p^2q}-\frac{1}{6pq}$

17. $\frac{2}{x}+\frac{5}{y}-\frac{3}{x^2y}$

18. $\frac{2}{a^2}-\frac{5}{6b}+\frac{9}{8ab}$

19. $\frac{7}{2x}-\frac{5}{4y}+\frac{1}{6z}$

20. $\frac{1}{6pq}+\frac{7}{3p}-\frac{9}{2p}$

21. $\frac{3}{x(2x+1)}+\frac{5}{x^2}$

22. $\frac{3}{2(t+3)}-\frac{t}{4}$

23. $\frac{2}{y(3x+2)}+\frac{3}{y^2}$

24. $\frac{7}{3(a+2)}-\frac{a}{6}$

25. $\frac{y}{3y-9}-\frac{4}{5y+15}$

26. $\frac{p}{q-pq}+\frac{3}{q+pq}$

27. $\frac{x}{2x-6}-\frac{3}{4x+12}$

28. $\frac{b}{a^2-ab}+\frac{2}{a^3+a^2b}$

29. $\frac{y-1}{3y+9}+\frac{2}{y-3}-\frac{8}{3y-9}$

30. $\frac{3}{4x-16}-\frac{x}{4-x}-\frac{1}{4x}$

31. $\frac{x-1}{4x+6}+\frac{2x}{6x+9}-\frac{5}{6}$

32. $\frac{c+d}{5c-2d}-\frac{c-d}{5c+2d}-\frac{4c}{15c-6d}$

33. $\frac{4}{4-9x^2}-\frac{x-5}{2+3x}$

34. $\frac{q-3}{q^2-q-12}+\frac{q+1}{q^2-4q}$

35. $\frac{3x}{x^2-4}+\frac{2}{x-2}-\frac{5x}{x+2}$

36. $\frac{4}{x}-\frac{1}{3x-1}+\frac{x+1}{3x+1}$

37. $\dfrac{p+2q}{2p^2-20pq+50q^2}+\dfrac{7p}{4p-20q}-\dfrac{7}{8}$

38. $\dfrac{2}{x^2+2x-15}+\dfrac{1}{x^2-9}-\dfrac{7}{x^2+8x+15}$

39. $\dfrac{3p+2q}{18p^2+24pq+8q^2}+\dfrac{2q}{6p-10q}-\dfrac{4}{5p}$

40. $\dfrac{3p-2q}{2q^2-9q-5}+\dfrac{3p+2q}{9p^2-4q^2}-\dfrac{q-5}{2q+1}$

41. $\left[\dfrac{7}{t-2}+\dfrac{t+3}{t^2+t-6}\right]\div\left[\dfrac{t^4-16}{7t-14}\right]$

42. $\left[\dfrac{2z}{z+1}+\dfrac{2z^2}{z^2-1}\right]\div\left[\dfrac{z^4}{z-1}\right]$

43. $\left[\dfrac{1}{x}+\dfrac{2}{y}\right]\div\left[\dfrac{1}{x}-\dfrac{2}{y}\right]$

44. $\left[\dfrac{3b}{a}+\dfrac{3a}{b}\right]\div\left[\dfrac{3}{a^2}-\dfrac{3}{b^2}\right]$

In Exercises 45 through 48, solve the given problems by combining the appropriate fractions.

45. The total electric resistance of two resistances under particular conditions is given by the formula

$$R=\frac{R_1R_2}{R_1+R_2}$$

If $R_1=\frac{5}{6}$ ohm and $R_2=\frac{7}{8}$ ohm, find R.

46. The total cost Q of producing x units of a commodity is given by the formula

$$Q=\frac{19}{6}+\frac{x}{8}+\frac{x^2}{18}$$

How many dollars does it cost to produce 50 units?

47. A psychologist studying the nature of the learning process may have use of the expression

$$\frac{1}{M}+\frac{M-1}{M(M+1)}$$

Combine these fractions.

48. In determining the characteristics of a specific optical lens, one uses the expression

$$\frac{2n^2-n-4}{2n^2+2n-4}+\frac{1}{n-1}$$

Combine and simplify.

6-5 EQUATIONS WITH FRACTIONS

Many important equations that arise in various areas of study have fractions in them. Although the solution of these equations will still involve the use of the basic operations as stated in Section 3-1, an additional procedure can be used to eliminate the fractions and thereby help in the solution. The method is to

multiply each term of the equation by the lowest common denominator.

The resulting equation will not involve fractions and can be solved by methods previously discussed. The following examples illustrate how to solve equations with fractions.

Example A Solve for x: $\frac{x+1}{6} + \frac{3}{2} = x.$

We first note that the lowest common denominator of the terms of the equation is 6. Therefore, we multiply each term by 6. This gives

$$\frac{6(x+1)}{6} + \frac{6(3)}{2} = 6x$$

We reduce each term to its lowest terms, and solve the resulting equation.

$$\begin{aligned} x + 1 + 9 &= 6x \\ x + 10 &= 6x \\ 10 &= 5x \\ 2 &= x \text{ or } x = 2 \end{aligned}$$

Thus, the solution set is $\{2\}$. When we check this solution *in the original equation*, we obtain 2 on each side of the equals sign. Therefore, the solution checks.

Example B Solve for x: $\frac{x}{a} - \frac{3}{4a} = \frac{1}{2}.$

We first determine that the lowest common denominator of the terms of the equation is $4a$. We then multiply each term by $4a$ and continue with the solution.

$$\frac{4a(x)}{a} - \frac{4a(3)}{4a} = \frac{4a(1)}{2}$$

$$\begin{aligned} 4x - 3 &= 2a \\ 4x &= 2a + 3 \\ x &= \frac{2a+3}{4} \end{aligned}$$

Checking shows that the left side of the equation is $\frac{1}{2}$, which means that the solution checks.

Example C Solve for x: $\dfrac{x}{a} - \dfrac{1}{2b} = \dfrac{3x}{ab}$.

The lowest common denominator of the terms of the equation is $2ab$. We therefore multiply each term of the equation by $2ab$ and proceed with the solution:

$$\frac{(2ab)x}{a} - \frac{2ab(1)}{2b} = \frac{2ab(3x)}{ab}$$

$$2bx - a = 6x$$

$$2bx - 6x = a$$

To isolate x, we must factor x from each term on the left. This leads to

$$x(2b - 6) = a$$

$$x = \frac{a}{2b - 6}$$

Checking will lead to the expression

$$\frac{3}{2b(b - 3)}$$

on each side of the original equation.

Example D Solve for t: $\dfrac{2}{t} = \dfrac{3}{t + 2}$.

After determining that the lowest common denominator of the terms of the equation is $t(t + 2)$, we multiply each term by this expression and proceed with the solution.

$$\frac{t(t + 2)(2)}{t} = \frac{t(t + 2)(3)}{t + 2}$$

$$2(t + 2) = 3t$$

$$2t + 4 = 3t$$

$$4 = t \quad \text{or} \quad t = 4$$

Thus, the solution set is $\{4\}$. Checking shows that each side of the original equation is $\frac{1}{2}$ for $t = 4$.

Example E When developing the equations which describe the motion of the planets, the equation

$$\frac{1}{2}v^2 - \frac{GM}{r} = -\frac{GM}{2a}$$

is used. Solve for M.

We first determine that the lowest common denominator of the terms of the equation is $2ar$. Multiplying each term by $2ar$ and proceeding with the solution, we have

$$\frac{2ar(v^2)}{2} - \frac{2ar(GM)}{r} = -\frac{2ar(GM)}{2a}$$

$$arv^2 - 2aGM = -rGM$$

$$rGM - 2aGM = -arv^2$$

$$M(rG - 2aG) = -arv^2$$

$$M = -\frac{arv^2}{rG - 2aG} \quad \text{or} \quad \frac{arv^2}{2aG - rG}$$

The second form of the result is obtained by using Eq. (6-1). Again, note the use of factoring to arrive at the final result.

Example F Solve for x: $\dfrac{2}{x+1} - \dfrac{1}{x} = -\dfrac{2}{x^2+x}$.

Multiplying each term by the lowest common denominator $x(x+1)$ and continuing with the solution, we have

$$\frac{2(x)(x+1)}{x+1} - \frac{x(x+1)}{x} = -\frac{2x(x+1)}{x(x+1)}$$

$$2x - (x+1) = -2$$

$$2x - x - 1 = -2$$

$$x = -1$$

Checking this solution *in the original equation*, we see that we have zero in the denominators of the first and third terms of the equation. Since division by zero is undefined (see Section 1-5), $x = -1$ cannot be a solution. Thus, *there is no solution to this equation.* In this case, the solution set is $\{\ \}$. This example points out clearly why it is necessary to check solutions in the original equation. It also shows that, *whenever we multiply each term by a lowest common denominator which contains the unknown, it is possible to obtain a solution that is not a solution of the original equation. Such a solution is termed* **extraneous.** Only certain equations will lead to extraneous solutions, but we must be careful to identify them when they occur.

Exercises 6-5

In Exercises 1 through 24, solve for s, t, v, x or y.

1. $\frac{x}{8} = \frac{1}{2}$

2. $\frac{y}{6} = \frac{1}{3}$

3. $\frac{2s}{3} - 1 = s$

4. $\frac{x}{4} + 6 = x$

5. $\frac{2x}{5} + 3 = \frac{7x}{10}$

6. $\frac{3v}{8} - 2 = \frac{v}{2}$

7. $\frac{t+1}{2} + \frac{1}{4} = \frac{3}{8}$

8. $\frac{2x-1}{9} - \frac{1}{3} = 2x$

9. $\frac{x}{b} + 3 = \frac{1}{b}$

10. $\frac{y}{c} - 2 = \frac{3}{c}$

11. $\frac{y}{2a} - \frac{1}{a} = 2$

12. $\frac{x}{b} + \frac{1}{3b} = 1$

13. $\frac{1}{2a} - \frac{3s}{4} = \frac{1}{a^2}$

14. $\frac{2}{5c} + \frac{2t}{5} = \frac{1}{c^3}$

15. $\frac{2x}{b} + \frac{4}{b^2} = \frac{x}{4b}$

16. $\frac{3s}{4a} - \frac{s}{a^2} = \frac{1}{8}$

17. $\frac{1}{x+2} = \frac{1}{2x}$

18. $\frac{1}{2v-1} = \frac{1}{3}$

19. $\frac{2}{3s+1} = \frac{1}{4}$

20. $\frac{3}{3x+2} = \frac{2}{x}$

21. $\frac{5}{2x+4} + \frac{3}{x+2} = 2$

22. $\frac{2}{s-3} - \frac{3}{2s-6} = 4$

23. $\frac{1}{y^2-y} - \frac{1}{y} = \frac{1}{y-1}$

24. $\frac{2}{x^2-1} - \frac{2}{x+1} = \frac{1}{x-1}$

In Exercises 25 through 32, solve for the indicated letter.

25. $3 + \frac{1}{a} = \frac{1}{b}$; solve for a

26. $\frac{3}{a} - \frac{2}{c} = \frac{7}{ac}$; solve for c

27. $\frac{x}{n} + \frac{3}{4n} = \frac{x}{2}$; solve for x

28. $\frac{7}{p^2} = t - \frac{3}{2p} + \frac{t}{p}$; solve for t

29. An equation obtained in analyzing a certain electric circuit is

$$\frac{V-6}{5} + \frac{V-8}{15} + \frac{V}{10} = 0$$

Solve for V.

30. A psychologist studying adaption levels may use the equation

$$M = A - \frac{A}{N}$$

Solve for A.

31. In economics, when determining discounts, the equation

$$r = \frac{A}{Pt} - \frac{1}{t}$$

arises. Solve for A.

32. In biology, an equation used when analyzing muscle reactions is

$$P + \frac{a}{V^2} = \frac{k}{V}$$

Solve for a.

6-6 SUMMARY

The primary concern of this chapter was the development of the basic operations with fractions. *Basic to all of these operations is the* **fundamental principle of fractions,** which was stated in Section 6-1 as follows.

If the numerator and denominator of any fraction are multiplied or divided by the same quantity, not zero, the value of the fraction will remain unchanged.

The importance of using this principle cannot be overstated.

We reviewed the basic operations of addition, subtraction, multiplication, and division of fractions, noting that these operations are the same as the equivalent operations with arithmetic fractions. Also, we noted that the final result of an operation with fractions should be put in simplest form if it is also a fraction. We placed special emphasis on the proper understanding of the operation of cancellation, because many students do not perform this operation correctly. Remember, cancellation is a process of division.

The key step in the addition and subtraction of fractions is that of finding the proper lowest common denominator of the set of fractions. For this reason, we devoted an entire section to determining lowest common denominators.

In addition to developing the basic operations with algebraic fractions, we also showed how to solve equations with fractions. Here, the initial key step of multiplying each term by the lowest common denominator was emphasized.

In this chapter, several new terms were introduced. The following list contains the important terminology which was discussed in the chapter. If any of these terms are not clear, they should be reviewed.

Cancellation
Equivalent fractions
Extraneous solution
Fundamental principle of fractions
Highest common factor
Lowest common denominator
Lowest terms
Reciprocal
Simplest form

The following set of exercises should help in the review of the material of the chapter.

Exercises 6-6

In Exercises 1 through 16, reduce the given fractions to simplest form, expressing all results with positive exponents.

1. $\frac{60}{126}$
2. $\frac{63}{105}$
3. $\frac{3xy}{18x^2}$
4. $\frac{7a^2b}{14b}$
5. $\frac{2a^2bc}{6ab^2c^3}$
6. $\frac{76x^3yz^3}{19xz^5}$
7. $\frac{(2a-b)(a+3b)}{(a+3b)(a-2b)}$
8. $\frac{(x-4y)(2x+y)}{(x-2y)(x-4y)}$
9. $\frac{a^2x-a^2y}{ax^2-ay^2}$
10. $\frac{p^2+pq}{3p+2p^3}$
11. $\frac{4a-12ab}{5b-15b^2}$
12. $\frac{2a^2+2ab-2ac}{4ab+4b^2-4bc}$
13. $\frac{p^2-3p-4}{p^2-p-12}$
14. $\frac{6x^2-7xy-3y^2}{4x^2-8xy+3y^2}$
15. $\frac{2y^2-14y+20}{7y-2y^2-6}$
16. $\frac{x^4-16}{3x^2+12}$

In Exercises 17 through 48, perform the indicated operations.

17. $\frac{12}{35}\cdot\frac{55}{64}$
18. $\frac{20}{77}\div\frac{8}{231}$
19. $\frac{31}{9}\div\frac{32}{27}$
20. $\frac{35}{16}\cdot\frac{12}{3}$
21. $\frac{2}{3}-\frac{5}{6}+\frac{3}{10}$
22. $3-\frac{4}{9}+\frac{7}{15}$
23. $\frac{3x^2}{4y^3}\cdot\frac{5y^4}{x^3}$
24. $\frac{7p}{12q^3}\cdot\frac{20q^4}{35p^3}$
25. $\frac{ab^3}{bc}\div a^2c$
26. $\frac{20m^3n^2}{12mn^4}\div\frac{27mn}{8m^2n^3}$
27. $\frac{2}{a^2}-\frac{3}{5ab}$
28. $\frac{3x}{4y}-\frac{5y}{6x}$
29. $\frac{5}{c}-\frac{3}{c^2d}+\frac{1}{2cd}$
30. $\frac{3}{2x^2}+\frac{1}{4x}-\frac{a}{6x^3}$
31. $\frac{2x}{x^2-1}\cdot\frac{x-1}{x^2}$
32. $\frac{5a}{a^2-a-6}\cdot\frac{a+2}{10a^2}$
33. $\frac{x^2-2x-15}{x^2-9}\cdot\frac{x^2-6x+9}{4x-12}$
34. $\frac{2x^2-14x+24}{x^2-4x+3}\cdot\frac{3x^2-27}{4x-20}$
35. $(9x^2-4y^2)\div\frac{3x-2y}{y-2x}$
36. $\frac{6r^2-rs-s^2}{4r^2-16s^2}\div\frac{2r^2+rs-s^2}{r^2+3rs+2s^2}$
37. $\frac{3}{x^2}+\frac{2}{x^2+3x}$
38. $\frac{1}{a^2-2a}-\frac{3}{2a}$
39. $\frac{2x}{x-2}-\frac{x^2-3}{x^2-4x+4}$
40. $\frac{3}{3x+y}-\frac{7}{3x^2-5xy-2y^2}$
41. $\frac{2x-3}{2x^2-x-15}-\frac{3x}{x^2-9}$
42. $\frac{2x}{x^2+2x-8}+\frac{3x}{2x+8}$

43. $\left[\dfrac{2}{a-b}\cdot\dfrac{a+b}{5}\right]\div\left[\dfrac{a^2+2ab+b^2}{a^2-b^2}\right]$

44. $\left[\dfrac{x^4-1}{(x-1)^2}\div(x^2+1)\right]\cdot\left[\dfrac{x-1}{x+1}\right]$

45. $\left[\dfrac{1}{x-2}\div\dfrac{3-x}{x+1}\right]\div\left[\dfrac{2+x}{x^2-x-6}\right]$

46. $\left[\dfrac{2s}{s-1}+\dfrac{s^2}{s^2-1}\right]\div\left[\dfrac{s^3}{s-1}\right]$

47. $\left[\dfrac{2}{x}+\dfrac{5}{y}\right]\div\left[\dfrac{2}{x}-\dfrac{5}{y}\right]$

48. $\left[\dfrac{1}{x}+\dfrac{1}{y}\right]\div\left[\dfrac{1}{x^2}-\dfrac{1}{y^2}\right]$

In Exercises 49 through 60, solve the given equations for x.

49. $\dfrac{x}{8}+\dfrac{3}{2}=1$

50. $\dfrac{x}{6}+\dfrac{10}{3}=x$

51. $\dfrac{x}{5}+5=\dfrac{x+7}{2}$

52. $x-\dfrac{1}{3}=\dfrac{x+8}{6}$

53. $\dfrac{x}{a}-\dfrac{1}{4a}=\dfrac{1}{2}$

54. $\dfrac{2x}{c}-\dfrac{1}{5}=\dfrac{3}{10c}$

55. $\dfrac{bx}{a}-\dfrac{1}{4}=x+\dfrac{3}{2a}$

56. $\dfrac{4(x-y)}{5b}-\dfrac{x}{b^2}=\dfrac{y}{5b}$

57. $\dfrac{6}{7x-2}=\dfrac{1}{2x}$

58. $\dfrac{5}{6x+1}=\dfrac{3}{1-x}$

59. $\dfrac{5}{x+1}-\dfrac{3}{x}=\dfrac{-5}{x^2+x}$

60. $\dfrac{20}{y^2-25}-\dfrac{1}{y+5}=\dfrac{2}{y-5}$

In Exercises 61 through 72, solve the given problems involving the basic operations with fractions.

61. A particular motorboat uses $\frac{7}{8}$ gal of gasoline each hour. How long will $4\frac{1}{4}$ gal last?

62. An ice cube melts in such a way that its edge changes from $\frac{5}{6}$ cm to $\frac{3}{4}$ cm. What is the change in volume of the ice cube?

63. In the theory of electricity, the expression

$$L^2\omega^2-\frac{2L}{C}+\frac{1}{C^2\omega^2}$$

is found. Combine these fractions and factor the resulting numerator for the final answer.

64. If a certain stock always pays a dividend of \$1, and this is always a 5% dividend, the present value P of the stock is found from the formula

$$P=\frac{\dfrac{1}{1+i}}{1-\dfrac{1}{1+i}}$$

where i is the percentage expressed as a decimal. Simplify the right-hand side of this equation.

65. In the theory related to photographic developing, the expression

$$1 - \frac{2g_1}{g_2} + \frac{g_1^2}{g_2^2}$$

arises. Combine the terms of this expression.

66. In biology, when discussing the theory in the use of x-rays, the expression

$$\frac{ds}{L} - \frac{ds^3}{24L^3}$$

arises. Combine the terms of this expression.

67. A sphere of radius r floats in a liquid. When one is determining the height h which the sphere protrudes above the surface, the expression

$$\frac{1}{4r^2} - \frac{h}{12r^3}$$

is found. Combine and simplify.

68. In psychology, when analyzing certain learning curves, the expression

$$\frac{1}{M} + \frac{M-1}{M^2+M}$$

is found. Combine the terms of this expression.

69. A Norman window has the shape of a rectangle surmounted by a semicircle. Its area can be expressed as

$$A = \frac{1}{2}wp - \frac{1}{2}w^2 - \frac{\pi}{8}w^2$$

where p is its perimeter and w is its width. Solve for p.

70. In chemistry, when dealing with ideal gas vapors, the equation

$$\frac{1}{P} = \frac{1}{P_A} + \frac{x}{P_B} - \frac{x}{P_A}$$

arises. Solve for P_B.

71. In the study of the convection of heat, the equation

$$\frac{1}{U} = \frac{x}{k} + \frac{1}{h}$$

is used. Solve for x.

72. In photography, when one is analyzing the focusing of cameras, one uses the equation

$$\frac{q_2 - q_1}{d} = \frac{f + q_1}{D}$$

Solve for q_1.

7

THE TECHNIQUE OF PROBLEM SOLVING

7-1 INTERPRETING VERBAL STATEMENTS

Most people appreciate mathematics because it enables them to solve applied problems. Some of these problems are in direct formula form and therefore lend themselves to direct solution. However, we often encounter problems in the form of verbal statements, and it is necessary to translate such statements into mathematical terms before the solution can be obtained by direct application of the methods we are developing. Already, in a number of situations, we have put verbal statements into mathematical symbols and then proceeded to the required solution. However, these have been very explicit statements. A number of problems which can be solved quite directly by elementary mathematical methods are not as explicitly stated as those we have met so far.

The purpose of this chapter is to indicate how to solve such problems. In this section, we shall concentrate on the interpretation of verbal statements. It is not possible to state certain specific rules that tell you how to interpret verbal statements for mathematical formulation. *Certain conditions are* **implied,** *and these are the ones which we must be careful to recognize properly. A very important step to the solution is a careful reading of the statement to make sure that all terms and phrases used are understood.*

The remainder of this section is devoted to examples. Keep in mind that the goal is to obtain an equation from the statement; this equation is then solved to complete the solution. Our concern here is interpreting the statement; the complete solution will be taken up in the following section.

Example A A problem reads: A rectangle whose length is 2 in. more than its width has a perimeter of 36 in. Find the dimensions.

In proceeding with the interpretation of this statement, we must understand the meaning of all terms used. For example, here we must know what is meant by rectangle, perimeter, and dimensions. Once we are sure of the terms, we must recognize what is required. Here we are told to "find the dimensions." This should mean to us "find the length and the width of the rectangle." Once this is established, we let x (or any appropriate symbol) represent one of these quantities. Thus, we write

let $x =$ the width, in inches, of the rectangle

Next we look to the statement for other pertinent information. The order in which we find it useful may not be the order in which it is presented. For example, here we are told that the "length is 2 in. more than its width." Since we let x be the width, then

$x + 2 =$ the length, in inches, of the rectangle

Thus, we have a representation of both required quantities.

Now that we have identified the unknown quantities, we must look for information by which an equation may be established. We look to the portion of the statement which is as yet unused. This is "A rectangle . . . has a perimeter of 36 in." Recalling that the perimeter of a geometric figure is the distance around it, we know that the perimeter of a rectangle equals twice the length plus twice the width. Thus, we can write the equation by multiplying the width x by 2, adding this to twice the length $x + 2$, and equating this sum to 36, the known value of the perimeter. This leads to $2x + 2(x + 2) = 36$.

Our analysis is complete. Probably a great deal of what was just stated seemed very obvious. However, even more involved problems require a similar technique of reading and interpreting, which is the key to finding the solution.

Example B A sponsor paid a total of \$4300 to run a commercial on two different television stations. One station charged \$700 more than the other. What did each station charge to run the commercial?

The question at the end identifies the required quantities. We may choose the charge of either station as x, so we

let $x =$ charge made by the less expensive station

The sentence "one station charged \$700 more than the other" tells us that

$x + 700 =$ charge made by the more expensive station

The initial statement that the "sponsor paid a total of \$4300" gives us the information that leads to the equation $x + (x + 700) = 4300$. The analysis is now complete.

Example C A person has \$1.40 in nickels and dimes. How many of each has he if there are 17 coins in all?

The phrase "how many of each" tells us that the number of nickels and the number of dimes are to be determined. Thus, choosing one quantity as x, we write:

let $x =$ the number of nickels

Then the phrase "17 coins in all" means that

$17 - x =$ the number of dimes

Since the total value of the coins is \$1.40, by multiplying the number of nickels, x, by 5 and the number of dimes, $17 - x$, by 10 and equating this sum to 140 (the number of cents), we find the equation, which is $5x + 10(17 - x) = 140$.

Example D Find three consecutive integers whose sum is 69.

Here we are asked to "find three . . . integers." Recalling that an integer represents a whole number (see Section 1-6), we write:

let $x =$ the smallest integer

The word "consecutive" means that the other integers follow without a break, or

$x + 1 =$ the second integer

and

$x + 2 =$ the third integer

Since the sum of these integers is 69, we write the equation

$x + (x + 1) + (x + 2) = 69.$

Example E A car traveling 40 mi/h is gone for 2 h when a second car starts along the same route going 60 mi/h. When will the second car overtake the first car?

The one word "when" means "at what time" or "for what value of t." We may let t represent the time either car has been gone; so, choosing one quantity, we write:

let t = the time, in hours, after the first car left

The fact that the first car "is gone for 2 h when a second car starts" means that

$t - 2$ = the time, in hours, after the second car left

The key to setting up the equation is the word "overtake," which implies, although it does not state explicitly, that *the cars will have gone the same distance*. Since, in general, distance equals rate times time, the equation is established by equating the distance the first car travels, $40t$, to the distance the second car travels, $60(t - 2)$. The equation is $40t = 60(t - 2)$.

Exercises 7-1

In each of the exercises that follow, the answers will consist of identifying the unknown quantities and stating the necessary equation that will lead to the solution.

(a) *In each problem, after noting the statement that identifies the required unknown quantities, write*

"*let* $x = \ldots$"

and complete the statement, clearly identifying the unknown represented by x.

(b) *Then, after identifying the relation among the unknowns, clearly specify each of the unknowns other than* x *in terms of* x.

(c) *After identifying the statement which leads to the equation necessary for the solution, state the equation to be solved, but do not solve it.*

Each of the examples of this section illustrated this procedure.

1. A rectangular table top has a perimeter of 140 in. The length is 10 in. more than the width. Find the dimensions of the table top.
2. A rectangular billiard table is twice as long as it is wide. If the perimeter is 30 ft, find the dimensions.
3. The perimeter of an isosceles triangle (two sides are equal) is 29 cm. Find the sides of the triangle if each of the equal sides is 2 cm less than the third side.
4. The side of one square is 3 m more than the side of another square. The perimeter of the larger square is 36 m. What is the perimeter of the smaller square?

5. Divide 87 into two parts such that one part is twice the other.

6. An 8-ft board is cut into two pieces, one of which is 10 in. longer than the other. What are the lengths of the pieces?

7. A man whose total annual income is \$13,200 spends ten times as much as he saves. How much does he save in a year?

8. Three men together weigh 562 lb. One of the men weighs 7 lb more than another and 8 lb less than the third. What is the weight of each?

9. Twenty-one nickels and quarters have a total value of \$2.65. How many of each are there?

10. A drawer contains fifty-two 13-cent and 15-cent stamps, which have a total face value of \$7.38. How many of each are there?

11. A man invested \$2000, part of the amount at 5% and the remainder at 6%. If the total annual income from the investment is \$105, how much was invested at 5%?

12. Thirty coins, all dimes and quarters, have a total value of \$4.20. How many of each are there?

13. The amount needed for a student's education at a certain community college comes from three sources. The state and county each contribute \$100 more than the student's tuition. If the cost per student is \$2300 per year, what is the student's tuition?

14. The House of Representatives has 435 members. On the vote on a particular bill there were 158 more "yes" votes than "no" votes, with 37 members not voting. What was the vote on the bill?

15. A certain newspaper is published each day of the week. It prints 50,000 more copies on Sunday than on a weekday, and it prints a total of 1,100,000 copies in a week. What is the number printed each day of the week?

16. A music student practiced the piano every day except Sunday. She practiced the same length of time each day except Saturday, when she practiced 15 min longer. If she practiced a total of 4 h during the week, how long did she practice each day?

17. Find two consecutive integers whose sum is 31.

18. Find two consecutive odd integers whose sum is 64.

19. Find three consecutive integers whose sum is 156.

20. Find four consecutive even integers whose sum is 140.

21. Two cars, originally 800 mi apart, start at the same time and travel toward each other. Find the speed of each if one travels 10 mi/h faster than the other and they meet in 8 h.

22. A man travels to and from a city along the same route in 12 h. His average speed going was 45 mi/h, and on the return trip it was 55 mi/h. How long did each part of the trip take?

23. Mr. Jones attended a conference 200 km from home. His friend drove him to the airport. He traveled by plane to the conference city; this distance was 40 km less than 15 times the distance he traveled by car. How many kilometers did his friend drive him to the airport?

24. Ms. Jones drove her car for 3 h to a meeting. She returned to her home office because of an emergency. Not wanting to miss the conclusion of the meeting, she drove again to the meeting but at a faster rate: for 2 h at 30 km/h more than the rate of her first trip there. At what rate did she drive the first time?

7-2 SOLVING STATED PROBLEMS

Of the steps required to solve a stated problem, establishing the necessary equation is usually the one which causes the most difficulty. For this reason we spent the entire preceding section interpreting the statements of verbal problems. This interpretation is the key to the proper equation. Of course, the complete solution is the ultimate goal. Therefore, in this section, we shall outline basic guidelines for the solution and then present several examples.

The basic steps to follow in solving a stated problem are:

1. *Read the statement of the problem carefully.*
2. *Clearly identify the unknown quantities and then assign an appropriate letter to represent one of them, stating this choice clearly.*
3. *Specify each of the other unknown quantities in terms of the one specified in step 2.*
4. *Analyze the statement of the problem in order to establish the necessary equation.*
5. *Solve the equation, clearly stating the solution.*
6. *Check the solution obtained* **with the original statement** *of the problem.*

Example A Divide the number 132 into two parts such that one part is 12 more than twice the other part.

Noting from the statement of the problem that we want two numbers whose sum is 132, we write:

let $x =$ the smaller part of 132

and

let $132 - x =$ the larger part of 132

From the statement, we see that

$$132 - x = 12 + 2x$$

The solution of this equation is as follows:

$$132 = 12 + 3x$$
$$120 = 3x$$
$$x = 40$$

Checking, we note that, if 40 is doubled and 12 added, we obtain 92. The sum of 40 and 92 is 132. Thus, 40 checks, and the two required parts of 132 are 40 and 92.

Example B A woman invested \$5000, part of the amount at 5% and the remainder at 9%. If the total annual income from the investments is \$362, how much was invested at 5%?

The last portion of the statement of the problem leads us to write:

let x = the amount, in dollars, invested at 5%

Also,

let $5000 - x$ = the amount, in dollars, invested at 9%

The annual income from each part equals the amount times the percent. Thus,

$$0.05x + 0.09(5000 - x) = 362$$

since we know that the total income is \$362. We now solve the equation.

$$0.05x + 450 - 0.09x = 362$$

$$-0.04x = -88$$

$$x = \$2200$$

Checking, we know that 5% of \$2200 is \$110 and 9% of \$2800 is \$252. The sum of these amounts is \$362. The solution checks. Therefore, the required answer is \$2200.

Example C A basketball team won 24 of their first 30 games. Then they lost several consecutive games such that their "percentage" of wins was 0.600. How many games in a row did they lose?

Let x = the number of consecutive games lost after the thirtieth game. This means that

$x + 30$ = the total number of games played

We know that they had won 24 games, which means that

$$\frac{24}{x + 30} = 0.600$$

is the required equation, since the "percentage" of wins is the number of wins divided by the total number of games played. Thus,

$$24 = 0.600(x + 30)$$

$$24 = 0.600x + 18$$

$$6 = 0.600x$$

$$x = 10 \text{ games}$$

Checking, we find that by losing 10 consecutive games after the thirtieth game, they had won 24 of 40 games, which gave them a "percentage" of 0.600. Therefore, the solution checks.

Example D A traveler drove 9 h to go from city A to city B. In returning along the same route, he drove 7 h while going 10 km/h faster. Find the distance from city A to city B.

Let $d =$ the distance, in kilometers, between city A and city B. The equation is based upon the statement about the rates. Since, in general, $D = RT$, then $R = D/T$. The rate in going was $d/9$ and this is 10 km/h less than the rate coming, $d/7$. Thus,

$$\frac{d}{9} = \frac{d}{7} - 10$$

Solving, we have

$$7d = 9d - 630$$

$$-2d = -630$$

$$d = 315 \text{ km}$$

Checking, we find that the rate going was $\frac{315}{9} = 35$ km/h and the rate coming was $\frac{315}{7} = 45$ km/h, which is 10 km/h more. Thus, the solution checks.

Example E One hundred pounds of a cement-sand mixture is 40% sand. How many pounds of sand must be added such that the resulting mixture will be 60% sand?

Let $n =$ the number of pounds of sand to be added

The equation is established by expressing the number of pounds of sand in the final mixture as a sum of the sand originally present, 40 lb, and that which is added, n lb, and equating this to the amount in the final mixture, which is $0.60(100 + n)$. (The final mixture is 60% sand, and there is a total of $100 + n$ lb.) Thus,

$$40 + n = 0.60(100 + n)$$

$$40 + n = 60 + 0.60n$$

$$0.40n = 20$$

$$n = 50 \text{ lb}$$

Checking this with the original statement, we find that the final mixture will be 150 lb, of which 90 lb will be sand. Since $\frac{90}{150} = 0.60$, the solution checks.

If you would like to check additional worked examples, complete those of Section 7-1. The numerical results for these are:

A. Width = 8 in., and length = 10 in.
B. \$1800 and \$2500.
C. 6 nickels and 11 dimes.
D. 22, 23, and 24.
E. Six hours after the first car left.

Exercises 7-2

Solve each of the following problems. Be certain to do the following.

(a) *Assign an appropriate letter to one of the unknowns of each problem, stating your choice clearly.*
(b) *Specify any other unknowns in terms of the one selected in step (a).*
(c) *Set up and solve an appropriate equation.*
(d) *Check the solution in the original statement of the problem.*

Note: Trial and error is not considered an appropriate method of solution.

1. The sum of two numbers is 93. Three more than 4 times the smaller equals the larger of the numbers. Find the numbers.

2. Divide 99 into three parts such that the smallest part is 30 less than the largest part, and the middle part is 12 less than the largest.

3. Approximately 1,309,000 auto hulks were processed in 1972 in the New England states and in the Middle Atlantic states by scrap and waste materials dealers and merchant wholesalers. There were 255,000 more hulks processed in the Middle Atlantic states than in the New England states. How many were processed in each area?

4. A woman paid \$4800 in state and federal income taxes in a year. Her federal income tax was five times as great as her state income tax. How much did she pay on each of these income taxes?

5. The length of a rectangle is 1 in. less than twice the width. Find the dimensions if the perimeter is 40 in.

6. If each side of a certain square were reduced by 2 cm, the perimeter would be 24 cm. What is the side of the square?

7. A triangular plate in the shape of an equilateral triangle (all sides are equal) expanded because it was heated. Each side became 1 cm longer, and the resulting perimeter was 273 cm. What was the original perimeter of the triangle?

8. The length of a rugby field, which is rectangular, is 10 yd more than twice its width. Find the dimensions if its perimeter is 470 yd.

9. A man earns \$1 less per hour on a part-time job than on his regular job. In a week, he works 40 h on his regular job and 6 h on his part-time job and earns \$155. What is his hourly rate on each job?

10. A student took four tests in a given course. On each test, he scored 10 points more than on the previous test. His average for the tests was 77. What were his grades on the tests?

11. As a result of contract negotiations, each employee of a certain firm received an annual salary increase six times as large as the value of the fringe benefits. If the total additional amount the firm had to pay for salary and fringe benefits for each employee was $840, what was the increase in salary?

12. Tickets to a certain show were $1.50 for adults and $0.50 for children. There were twice as many children's tickets sold as adult tickets, with the total receipts being $275. How many tickets were sold?

13. Find two consecutive integers whose sum is 87.

14. Find two consecutive odd integers such that the larger is 17 less than twice the smaller.

15. Find three consecutive even integers such that the sum of the first two is 66 more than half the third.

16. Find four consecutive integral multiples of 3 such that the sum of the three largest is 120 more than one-sixth of the smallest.

17. Eighteen dimes and quarters have a total value of $4.05. How many dimes are there?

18. A person has 3 times as many 15-cent stamps as 10-cent stamps. The value of the 15-cent stamps is $4.20 more than that of the 10-cent stamps. How many of each has he?

19. A woman invests part of $8400 at $5\frac{1}{2}\%$ and the remainder at $6\frac{1}{2}\%$ such that the income from each portion is the same. How much is invested at $5\frac{1}{2}\%$?

20. A man invests a certain amount of money at 8% and $2000 more than this amount at 12%. The income from the 12% investment exceeds that of the investment at 8% by $340 per year. How much was invested at 12%?

21. Two cars start from a city, traveling in opposite directions. After 4 h they are 320 mi apart, with one car averaging 8 mi/h more than the other. What is the average speed of each?

22. Two cars, 598 mi apart, start toward each other. One travels at the rate of 50 mi/h and the other at 42 mi/h. When will they meet?

23. A man traveled 1140 km, the first part on a bus which averaged 40 km/h, and the last part on a plane which averaged 400 km/h. If he was on the bus one hour longer than on the plane, how long was the trip, assuming he went immediately from one to the other?

24. A plane travels with a speed relative to the air of 560 mi/h. The trip between two given cities takes one-half hour longer when the plane is flying against a 40 mi/h wind than when it is flying with it. How far apart are the cities?

25. How much water must be added to 15 L of alcohol to make a solution which is 40% alcohol?

26. How many pounds of coffee worth $3.30/lb must be mixed with 20 lb of a blend worth $2.40/lb to get a final mixture worth $3.00/lb?

27. Nuts worth $2.00/lb are to be mixed with nuts worth $3.20/lb. How many pounds of each should be used to make 20 lb of a mixture worth $2.72/lb?

28. A pharmacist has 350 mL of a 20% solution of a certain chemical. She wishes to dilute this solution to a 10% solution of the same chemical. How much water should she add?

For additional problems, see the following section, or complete those of Section 7-1. The answers for the completed problems of Section 7-1 are found following those of this section.

7-3 SUMMARY

In this chapter, we have analyzed and interpreted verbal statements with the goal of solving problems mathematically. We then solved stated problems, noting that the key to most solutions was a proper interpretation of the statement of the problem, which enabled us to establish the necessary equation.

The steps that should be followed in the solution of verbal problems are found in the introduction to section 7-2. The following exercises are provided to give the reader more experience in solving stated problems.

Exercises 7-3

In each of the following exercises, follow the instructions given for Exercises 7-2.

1. A golfer had a score of 87 for 18 holes. He took 5 more strokes on the first nine holes than on the last nine holes. What was his score on each nine?
2. The power of one engine is 2 horsepower (hp) more than that of a second engine, which has 6 hp more than a third engine. The three engines have a total of 26 hp. What is the power of each?
3. The area of Idaho is 15,000 mi^2 more than that of Washington. The area of both states together is 6000 mi^2 more than the area of Montana. What is the area of each state if together they cover 300,000 mi^2?
4. A shopper spent one-half of her money in the first store at which she stopped. She spent one-third of the remainder at a second store. After spending an additional \$5, she had \$15 left. How much money did she start with?
5. A student obtained a score of 68 on her first test and a score of 81 on her second test. What score must she get on the third test to have an average of 80 for the three tests?
6. A student had an 82 on the first test, a 73 on the third test, and a 62 on the fourth test, but he could not remember the grade on the second test. His instructor said he had an average of 71 after the fourth test. What was his grade on the second test?
7. The area of a square equals the area of a rectangle whose width is 12 cm. The side of the square is 14 cm. Find the length of the rectangle.
8. The side of one square metal part is 3 mm more than the side of another square part. The perimeter of the larger part is 36 mm. What is the perimeter of the smaller part?
9. If one pair of parallel sides of a square was doubled and the other pair was halved, the perimeter would be 20 in. What is the side of the square?
10. The length of a rectangle is 6 in. more than the side of a square. The width of the rectangle is 4 in. less than the side of the square. What are the dimensions of each figure if the areas are the same?

11. Eighty coins—nickels, dimes, and one quarter—have a total value of \$5.55. How many of each coin are there?

12. Two thousand dollars are invested at a certain percent and \$3000 at 2% less. What is the rate on the \$2000 investment if the income from each is the same?

13. Find two consecutive even integers and the next odd integer whose sum is 77.

14. The sum of three consecutive odd integers is 21. Find the three integers.

15. A piece of wire 60 cm long is to be made into the shape of a rectangle, of which the ratio of the length to the width is $\frac{3}{2}$. Find the dimensions of the rectangle.

16. In a certain fraction, the denominator exceeds the numerator by 6. If 7 is added to both numerator and denominator, the fraction is equivalent to $\frac{2}{3}$. What is the original fraction?

17. In constructing a bridge, a certain number of 50-ft lengths of steel girders are required to go the length of the span. If the girders were 45 ft long, 10 more would be necessary. How long is the bridge?

18. In a biology experiment, the number of white mice in each generation was double that in the previous generation. If there are a total of 98 mice in three generations, how many were in the first generation?

19. The reign of Louis XIV of France lasted one-eighth longer than that of Queen Victoria of England. Together their reigns totaled 136 years. How long did Louis XIV reign?

20. A man left $\frac{1}{2}$ of his estate to his wife, $\frac{1}{5}$ to his daughter, and the remaining \$13,200 to his son. How much was the estate?

21. If the larger of two integers, their sum being 65, is divided by the smaller, the quotient is 7 with a remainder of 1. What are the integers?

22. One integer is 6 more than 3 times the other. If 8 is added to each, their quotient is 2. What are the numbers?

23. An architect designs a rectangular window such that the width is 18 in. less than the height. If the perimeter of the window is 180 in., what are its dimensions?

24. An air sample contains 4 ppm (parts per million) of two harmful pollutants. The concentration of one is four times the other. What is the concentration of each?

25. *A* can do a job in 3 days and *B* can do the same job in 4 days. How long would it take for them to do the job together?

26. Three pipes can individually empty a given pool in 3 h, 5 h, and 6 h, respectively. How long would it take the three pipes together to empty the pool?

27. A square tract of land is enclosed with fencing and then divided in half by additional fencing parallel to two of the sides. If 75 m of fencing is used, what is the length of one side of the tract?

28. A woman travels to and from a city along the same route in 12 h. Her average speed on the trip to the city is 85 km/h, and on the return trip it is 95 km/h. How long did each part of the trip take?

29. Two cars, 570 mi apart, start toward each other, one at 30 mi/h and the other at 42 mi/h. When will they meet?

30. Two cars, 800 mi apart, start toward each other. One is traveling 15 mi/h faster than the other. Where will they meet in 10 h?

31. The current in a river flows at 2 km/h. A motorboat can go 6 km/h in still water. It takes 2 h longer to go upstream than it does to go downstream between the same two points. How far apart are the points?

32. A jet took 6 h to fly against headwinds of 50 mi/h from city *A* to city *B*, and 5 h on the return trip when the winds became tailwinds of 50 mi/h. How fast does the jet travel relative to the air?

33. How many quarts of pure alcohol must be added to 10 qt of a solution which is 50% alcohol in order to make a solution which is 70% alcohol?

34. Eighty grams of solder which is 50% tin is to be melted with solder which is 10% tin. How many grams of the second type of solder must be used if the resulting solder is to be 20% tin?

35. A chemist has 40 cm^3 of an 8% solution of sulfuric acid. How many cubic centimeters of a 50% solution must she add to get a 12% solution?

36. Thirty-three percent of the voters in a certain congressional district with 240,000 voters are registered with party *X* and 37% are registered with party *Y*; the remaining voters are independent. In redistricting considerations, the members of party *X* would like to include part of an area where the registration is 50% in party *X*, 20% in party *Y*, and 30% independent. How many voters in this area would have to be included with the district in order that the registration in the resulting district would be equal for the two parties?

8

ROOTS AND RADICALS

8-1 ROOTS OF A NUMBER

In Chapter 2, we introduced the idea of the square root of a number. However, at that time we limited the discussion to finding the square roots of a few simple numbers. In the work which follows this chapter, we will need a more thorough understanding of square roots and other roots of numbers as well as of certain basic operations with them. Therefore, this chapter is devoted to the further development of these topics.

We defined a **square root** *of a given number to be that number which, when squared, equals the given number.* Considering now our work with signed numbers, we see that there are actually two numbers which could be considered to be the square root of a given positive number. Also, there are no numbers among those we have discussed so far which could be the square root of a negative number.

Example A Since $(+3)^2 = 9$ and $(-3)^2 = 9$, we see that the square root of 9 can be either $+3$ or -3 by the definition we have used up to this point. Also, since the square of $+3$ and the square of -3 are both 9, we have no apparent result for the square root of -9.

Since there are two numbers, one positive and the other negative, whose squares equal a given positive number, *we define the* **principal square root** *of a positive number to be positive*. That is,

$$\boxed{\sqrt{N} = x} \tag{8-1}$$

where $\sqrt{N}$ *is the principal square root of N and equals* (*positive*) x. Also, of course, $x^2 = N$. This avoids the ambiguity of two possible answers for a square root.

Example B

$\sqrt{16} = 4 \qquad \sqrt{25} = 5 \qquad \sqrt{121} = 11$

We do *not* accept answers such as $\sqrt{144} = -12$.

By defining the principal square root of a number in this way, we do not mean that problems involving square roots cannot have negative results. By $-\sqrt{N}$, we mean the negative of the principal value, which means therefore the negative of a positive signed number, and this gives a negative result.

Example C

$-\sqrt{16} = -(+4) = -4 \qquad -\sqrt{25} = -5 \qquad -\sqrt{121} = -11$

We do *not* accept answers such as $-\sqrt{144} = -(-12) = 12$.

Returning to the problem of the square root of a negative number, we find it necessary to define a new kind of a number to provide the required result. Thus, we define

$$\boxed{(bi)^2 = -b^2} \tag{8-2}$$

where bi is called an **imaginary number.** (The name imaginary is simply the name of the number; these numbers are not imaginary in the usual meaning of the word.) By this definition, it can be seen that

$$\boxed{i^2 = -1 \quad \text{or} \quad \sqrt{-1} = i} \tag{8-3}$$

This is the first time a number we have used is not a member of the set of real numbers.

Example D

$$\sqrt{-16} = 4i \qquad \sqrt{-25} = 5i$$

$$-\sqrt{-121} = -(11i) = -11i \qquad \sqrt{-289} = 17i$$

Also,

$$\sqrt{-144} = 12i \quad \text{and } \textit{not} \quad -12i$$

Furthermore,

$$(7i)^2 = -49 \qquad (2i)^2 = -4 \qquad (-3i)^2 = (-3)^2 i^2 = (9)(-1) = -9$$

Actually, *bi is the* **imaginary part** *of what is known as the* **complex number** *$a + bi$, where a is the* **real part** *of the complex number*. Thus, $5 + 6i$ is such a number. What we use as *an imaginary number can be considered as a complex number with the real part equal to zero*. Such a number is $0 + 4i$. Also, *a* **real number** *can be considered as a complex number with the coefficient b of bi equal to zero*. Thus, $5 + (0)(i)$ is a real number.

Most of our work with roots will deal with square roots, but we will have some occasion to consider other roots of a number. In general, *the* **principal *n*th root** *of a number N is designated as*

$$\boxed{\sqrt[n]{N} = x} \tag{8-4}$$

where $x^n = N$. (If $n = 2$, it is not usually written in the radical, as we have already seen.) If N is positive, x is positive. If N is negative, and n is odd, x is negative. We will not consider the case of N being negative and n being even and larger than 2.

Example E

$$\sqrt[3]{64} = 4 \qquad (\text{since } 4^3 = 64)$$

$$\sqrt[3]{-64} = -4 \qquad (\text{since } (-4)^3 = -64)$$

$$\sqrt[4]{81} = 3 \qquad (\text{since } 3^4 = 81)$$

$$\sqrt[5]{32} = 2 \qquad (\text{since } 2^5 = 32)$$

$$\sqrt[5]{-32} = -2 \qquad (\text{since } (-2)^5 = -32)$$

If a negative sign precedes the radical sign, we express the result as the negative of the principal value. Consider Example F.

Example F

$$-\sqrt[3]{64} = -4 \qquad -\sqrt[3]{-64} = -(-4) = 4$$

$$-\sqrt[4]{16} = -2 \qquad -\sqrt[5]{-32} = 2 \qquad -\sqrt[7]{-128} = 2$$

We shall conclude this section with a brief discussion of the type of number which the root of a number represents. In Section 1-6, we introduced the meanings of rational and irrational numbers. *If the root of a number can be found* **exactly** *in decimal form, the root is* **rational.** *Otherwise, it is* **irrational.** By this, we mean that it is possible to express certain roots as the ratio of one integer to another, but that many cannot be expressed in this way. All the rational numbers and irrational numbers constitute the real numbers.

Example G

$\sqrt{4} = 2$, and since 2 is an integer, it is rational.

$\sqrt[3]{0.008} = 0.2 = \frac{1}{5}$, and since $\frac{1}{5}$ is the ratio of one integer to another, it is rational.

$\sqrt{2}$ is approximately 1.414, but it is not possible to find two integers whose ratio is exactly $\sqrt{2}$. Therefore, $\sqrt{2}$ is irrational.

$\sqrt[3]{7}$ and $\sqrt[5]{-39}$ are irrational since we cannot determine an exact decimal form.

$\sqrt{-4} = 2i$, which we have already found to be an imaginary number.

Therefore, we see that in discussing roots of numbers we are considering numbers of which many—in fact, most—are irrational.

Exercises 8-1

In Exercises 1 through 32, find the principal roots or the indicated negatives of these roots.

1. $\sqrt{36}$ **2.** $\sqrt{81}$ **3.** $-\sqrt{169}$ **4.** $-\sqrt{225}$

5. $\sqrt{0.25}$ **6.** $\sqrt{0.01}$ **7.** $-\sqrt{0.04}$ **8.** $-\sqrt{0.36}$

9. $\sqrt{400}$ **10.** $\sqrt{900}$ **11.** $-\sqrt{1600}$ **12.** $-\sqrt{2500}$

13. $\sqrt[3]{8}$ **14.** $\sqrt[3]{27}$ **15.** $\sqrt[3]{-8}$ **16.** $\sqrt[3]{-27}$

17. $-\sqrt[3]{125}$ **18.** $-\sqrt[3]{-125}$ **19.** $\sqrt[3]{0.125}$ **20.** $\sqrt[3]{-0.001}$

21. $\sqrt[4]{16}$ **22.** $-\sqrt[4]{625}$ **23.** $\sqrt[5]{243}$ **24.** $-\sqrt[5]{-243}$

25. $\sqrt[6]{64}$ **26.** $-\sqrt[8]{256}$ **27.** $\sqrt{-4}$ **28.** $\sqrt{-100}$

29. $-\sqrt{-400}$ **30.** $-\sqrt{-900}$ **31.** $\sqrt{-0.49}$ **32.** $-\sqrt{-0.0001}$

In Exercises 33 through 36, perform the indicated operations.

33. $(5i)^2$ **34.** $5i^2$ **35.** $(3i)(4i)$ **36.** $(-4i)^2$

In Exercises 37 through 40, classify each given number as rational, irrational, or imaginary.

37. $\sqrt{4}, \sqrt{3}, \sqrt[3]{8}, \sqrt[3]{-8}, \sqrt{-8}$

38. $\sqrt[3]{27}, \sqrt{15}, \sqrt{50}, \sqrt[5]{-32}, \sqrt{-9}$

39. $\sqrt[3]{100}, \sqrt[4]{0.0001}, \sqrt{-1}, \sqrt[5]{64}, \sqrt{0.16}$

40. $\sqrt{144}, \sqrt[4]{250}, \sqrt[3]{-0.027}, \sqrt{-81}, \sqrt{0.64}$

In Exercises 41 through 44, solve the given problems.

41. The period T, in seconds, of a pendulum of length l, in feet, is given by the equation

$$T = \frac{\pi}{2}\sqrt{\frac{l}{2}}$$

What is the period of a pendulum 8 ft long?

42. The ratio of the rates of diffusion of two gases is given by $\frac{r_1}{r_2} = \frac{\sqrt{m_2}}{\sqrt{m_1}}$, where m_1 and m_2 are the masses of the molecules of the gases. Find the ratio $\frac{r_1}{r_2}$ if $m_1 = 25$ units and $m_2 = 81$ units.

43. A cubical water tank holds 512 m^3. What is the length of an edge of the tank?

44. At the end of a certain biology experiment, it was observed that the number of bacteria in a certain culture equaled 1000 times the fourth root of the number originally present. If 160,000 bacteria were originally present, how many were observed at the end of the experiment?

8-2 SIMPLIFYING RADICALS

A basic property of radicals allows us to change many of them into simpler, more convenient, forms. This property is that *the root of a product equals the product of the roots of its factors.* That is,

$$\sqrt{ab} = \sqrt{a}\sqrt{b} \tag{8-5}$$

To avoid difficulties with imaginary numbers, *we shall assume that all letters under radicals represent positive real numbers.*

Example A We know that $\sqrt{36} = 6$. If we now consider

$$\sqrt{36} = \sqrt{4 \cdot 9} = \sqrt{4}\sqrt{9} = 2 \cdot 3$$

we see that the same result, 6, is obtained.

In Eq. (8-5), *if the root of either a or b can be found exactly, we may simplify the radical so that the remaining indicated root is that of a number smaller than the number that was originally present.* Consider Example B.

Example B By writing $\sqrt{48} = \sqrt{16 \cdot 3}$, we have the following result:

$$\sqrt{48} = \sqrt{16 \cdot 3} = \sqrt{16}\sqrt{3} = 4\sqrt{3}$$

Thus, $\sqrt{48}$ can be expressed in terms of $\sqrt{3}$. Also,

$$\sqrt{175} = \sqrt{25 \cdot 7} = \sqrt{25}\sqrt{7} = 5\sqrt{7}$$

In order to use Eq. (8-5) properly to reduce the number under the radical, it is essential that we are able to factor any given number into two factors, one of which is a perfect square. Until the number under the radical contains no perfect square factors, it is not simplified completely.

Example C In Example B, we could have written $\sqrt{48} = \sqrt{6 \cdot 8}$. However, with these factors, no simplification can be done because neither 6 nor 8 is a perfect square. If we noted that $\sqrt{48} = \sqrt{4 \cdot 12} = \sqrt{4}\sqrt{12} = 2\sqrt{12}$, the radical has been simplified, but not completely. We must then note that $\sqrt{12} = \sqrt{4 \cdot 3} = \sqrt{4}\sqrt{3} = 2\sqrt{3}$, which means that $\sqrt{48} = 2\sqrt{12} = 2(2\sqrt{3}) = 4\sqrt{3}$.

Other illustrations of the use of Eq. (8-5) in simplifying radicals follow:

$$\sqrt{72} = \sqrt{36 \cdot 2} = \sqrt{36}\sqrt{2} = 6\sqrt{2}$$
$$\sqrt{54} = \sqrt{9 \cdot 6} = \sqrt{9}\sqrt{6} = 3\sqrt{6}$$
$$\sqrt{126} = \sqrt{9 \cdot 14} = \sqrt{9}\sqrt{14} = 3\sqrt{14}$$

Table 1 in Appendix C shows the square roots of numbers from 1.0 to 9.9. However, using Eq. (8-5), we can extend the use of the table to find the square root of any number. Of course, if a calculator with a $\sqrt{x}$ key is used, a table may not be necessary. *In using the table for numbers not directly listed, we rewrite the given number as the product of a listed number and a power of* 10 *for which the root is easily determined.* This is shown in the following example.

Example D In finding $\sqrt{5.8}$ from Table 1, we simply look opposite 5.8 in the $\sqrt{N}$ column. Thus, $\sqrt{5.8} = 2.408$.

In order to find $\sqrt{58}$, we note that $\sqrt{58} = \sqrt{(5.8)(10)}$. This means that we find $\sqrt{58}$ opposite 5.8 in the $\sqrt{10N}$ column. Thus, $\sqrt{58} = 7.616$.

The value of $\sqrt{580}$ is not directly in the table. However, $\sqrt{580} = \sqrt{(5.8)(100)} = \sqrt{5.8}\sqrt{100} = (2.408)(10) = 24.08$.

In the same manner,

$$\begin{aligned}
\sqrt{5800} &= \sqrt{(58)(100)} = \sqrt{58}\sqrt{100} \\
&= (7.616)(10) = 76.16 \\
\sqrt{58000} &= \sqrt{(5.8)(10000)} = \sqrt{5.8}\sqrt{10000} \\
&= (2.408)(100) = 240.8 \\
\sqrt{0.58} &= \sqrt{(58)(0.01)} = \sqrt{58}\sqrt{0.01} \\
&= (7.616)(0.1) = 0.7616 \\
\sqrt{0.058} &= \sqrt{(5.8)(0.01)} = \sqrt{5.8}\sqrt{0.01} \\
&= (2.408)(0.1) = 0.2408
\end{aligned}$$

Square roots of products involving literal numbers can also be simplified by use of Eq. (8-5). The simplification is based upon the fact that the square root of the square of a positive real number is that number. That is,

$$\sqrt{a^2} = a \qquad (a > 0) \tag{8-6}$$

By use of Eq. (8-5) an algebraic product can be expressed as the product of square roots. Then, for any of these products for which Eq. (8-6) is applicable, the expression can be simplified. Three illustrations are shown in Example E.

Example E

$$\begin{aligned}
\sqrt{a^2b^2c} &= \sqrt{a^2}\sqrt{b^2}\sqrt{c} = ab\sqrt{c} \\
\sqrt{25a^4} &= \sqrt{25}\sqrt{a^4} = 5\sqrt{(a^2)^2} = 5a^2 \\
\sqrt{6a^2b^6x} &= \sqrt{6}\sqrt{a^2}\sqrt{b^6}\sqrt{x} = (\sqrt{6})(a)(\sqrt{(b^3)^2})(\sqrt{x}) \\
&= (\sqrt{6})(a)(b^3)(\sqrt{x}) = ab^3\sqrt{6}\sqrt{x} = ab^3\sqrt{6x}
\end{aligned}$$

If a radical appears in the denominator of a fraction, the evaluation of the expression usually becomes a more lengthy process. For this reason, another operation in the simplification of radicals is the **rationalization** of denominators. *We perform this step by multiplying both numerator and denominator of the fraction by the proper expression such that no radicals appear in the resulting denominator.* We proceed as follows:

$$\sqrt{\frac{a}{b}} = \sqrt{\frac{a \cdot b}{b \cdot b}} = \frac{\sqrt{ab}}{\sqrt{b^2}} = \frac{\sqrt{ab}}{b} \qquad (b \text{ not zero}) \tag{8-7}$$

The use of Eq. (8-7) is demonstrated in the following examples.

Example F

$$\sqrt{\frac{1}{3}} = \sqrt{\frac{1 \cdot 3}{3 \cdot 3}} = \frac{\sqrt{3}}{\sqrt{3^2}} = \frac{\sqrt{3}}{3}$$

$$\sqrt{\frac{12}{5}} = \sqrt{\frac{4 \cdot 3}{5}} = \sqrt{\frac{4 \cdot 3 \cdot 5}{5 \cdot 5}} = \frac{2\sqrt{3 \cdot 5}}{\sqrt{5^2}} = \frac{2\sqrt{15}}{5}$$

Example G

$$\sqrt{\frac{a^2}{b}} = a\sqrt{\frac{1}{b}} = a\sqrt{\frac{b}{b^2}} = \frac{a\sqrt{b}}{b}$$

$$\sqrt{\frac{a^4b}{c^3}} = \sqrt{\frac{a^4b}{c^2c}} = \frac{a^2}{c}\sqrt{\frac{b}{c}} = \frac{a^2}{c}\sqrt{\frac{bc}{c^2}} = \frac{a^2\sqrt{bc}}{c^2}$$

$$\sqrt{\frac{4a^3}{7b}} = 2a\sqrt{\frac{a}{7b}} = 2a\sqrt{\frac{a \cdot 7b}{(7b)^2}} = \frac{2a\sqrt{7ab}}{7b}$$

We have devoted our attention thus far to the simplification of expressions involving square roots since they are of primary importance in our later work. However, we can carry out similar operations with expressions involving other roots. Example H illustrates this type.

Example H

$$\sqrt[3]{16} = \sqrt[3]{8}\sqrt[3]{2} = 2\sqrt[3]{2}$$

$$\sqrt[4]{3a^7} = \sqrt[4]{3}\sqrt[4]{a^4}\sqrt[4]{a^3} = (\sqrt[4]{3})(a)(\sqrt[4]{a^3}) = a\sqrt[4]{3a^3}$$

$$\sqrt[5]{32x^8} = (\sqrt[5]{32})(\sqrt[5]{x^5})(\sqrt[5]{x^3}) = (2)(x)(\sqrt[5]{x^3}) = 2x\sqrt[5]{x^3}$$

In these three illustrations, we have removed from under the radicals factors which are a perfect cube, a perfect fourth power, and a perfect fifth power, respectively.

In conclusion, *we shall consider a radical as being expressed in simplest form if all perfect powers have been removed and if all denominators are rationalized.*

Exercises 8-2

In Exercises 1 through 32, express the given radicals in simplest form.

1. $\sqrt{8}$ **2.** $\sqrt{18}$ **3.** $\sqrt{27}$ **4.** $\sqrt{28}$

5. $\sqrt{63}$ **6.** $\sqrt{99}$ **7.** $\sqrt{150}$ **8.** $\sqrt{98}$

9. $\sqrt{108}$ **10.** $\sqrt{162}$ **11.** $\sqrt{192}$ **12.** $\sqrt{250}$

13. $\sqrt{\frac{1}{2}}$ **14.** $\sqrt{\frac{2}{3}}$ **15.** $\sqrt{\frac{9}{8}}$ **16.** $\sqrt{\frac{4}{27}}$

17. $\sqrt{ac^2}$ **18.** $\sqrt{3a^4}$ **19.** $\sqrt{a^3b^2}$ **20.** $\sqrt{12a^5}$

21. $\sqrt{\frac{a}{2}}$ **22.** $\sqrt{\frac{4a}{5}}$ **23.** $\sqrt{\frac{27a}{b}}$ **24.** $\sqrt{\frac{8b^3}{x}}$

25. $\sqrt{4a^2bc^3}$ **26.** $\sqrt{9a^3b}$ **27.** $\sqrt{80x^4yz^5}$ **28.** $\sqrt{240xy^7z^6}$

29. $\sqrt{\frac{ab^2}{12}}$ **30.** $\sqrt{\frac{c^3e^5}{44}}$ **31.** $\sqrt{\frac{2x^2y}{5a^8}}$ **32.** $\sqrt{\frac{13xy^5}{40a^3}}$

In Exercises 33 through 44, use Table 1 to evaluate the given square roots.

33. $\sqrt{50}$ **34.** $\sqrt{29}$ **35.** $\sqrt{660}$ **36.** $\sqrt{7700}$

37. $\sqrt{17000}$ **38.** $\sqrt{350000}$ **39.** $\sqrt{0.4}$ **40.** $\sqrt{0.85}$

41. $\sqrt{0.0093}$ **42.** $\sqrt{0.018}$

43. Evaluate $\frac{\sqrt{2}}{\sqrt{3}}$ without rationalizing and then by evaluating $\frac{\sqrt{6}}{3}$.

44. Evaluate $\frac{\sqrt{15}}{\sqrt{8}}$ without rationalizing and then by evaluating $\frac{\sqrt{30}}{4}$.

In Exercises 45 through 52, simplify the given radicals.

45. $\sqrt[3]{54}$ **46.** $\sqrt[3]{24}$ **47.** $\sqrt[3]{8a^4}$ **48.** $\sqrt[3]{6a^{10}}$

49. $\sqrt[4]{16a^9}$ **50.** $\sqrt[4]{243a^{11}}$ **51.** $\sqrt[5]{64x^7}$ **52.** $\sqrt[8]{a^9b^{16}c^5}$

In Exercises 53 through 56, solve the given problems.

53. Find the side of the square whose area is 640 cm^2 by first simplifying the appropriate radical and then using Table 1.

54. A rectangular lot is twice as long as it is wide and has an area of 16,200 ft^2. What are its dimensions?

55. The expression for the resonant frequency of a certain electric circuit is given by

$$f = \frac{1}{2\pi}\sqrt{\frac{1}{LC}}$$

Express this equation in rationalized form.

56. In statistics, the standard deviation of a set of numbers is that number which, when subtracted from and added to the average of the set, gives the values between which about two-thirds of the numbers lie. In the study of the I.Q.'s of 21 children, the average I.Q. was found to be 105 and the standard deviation was $\sqrt{4000/21}$. Find the values of I.Q. between which two-thirds of those of the children lie.

8-3 BASIC OPERATIONS WITH RADICALS

Now that we have seen how radicals may be simplified, we can show how the basic operations of addition, subtraction, multiplication, and division are performed on them. These operations follow the basic algebraic operations, the only difference being that radicals are involved. Also, these radicals must be expressed in simplest form in order that the final result is in simplest form.

In adding and subtracting radicals, we add and subtract **like** *radicals*, just as we add and subtract like terms when we add and subtract algebraic expressions we have already encountered. Consider Example A.

Example A

$$\begin{aligned}\sqrt{3} + 4\sqrt{5} - 3\sqrt{3} + \sqrt{5} &= (\sqrt{3} - 3\sqrt{3}) + (4\sqrt{5} + \sqrt{5}) \\ &= -2\sqrt{3} + 5\sqrt{5}\end{aligned}$$

However, when working with radicals, we must be certain that all radicals are in simplest form. We must make sure because *many radicals that do not appear to be similar in form actually are similar*, as shown in Example B.

Example B $\sqrt{3} + \sqrt{80} - \sqrt{27} + \sqrt{5}$ appears to have four different types of terms, none of them being similar. However, noting that

$$\sqrt{80} = \sqrt{16 \cdot 5} = 4\sqrt{5}$$

and

$$\sqrt{27} = \sqrt{9 \cdot 3} = 3\sqrt{3}$$

we have the equivalent expression $\sqrt{3} + 4\sqrt{5} - 3\sqrt{3} + \sqrt{5}$, which, of course, is the same as that which appears in Example A. Therefore, the third term and the first term are similar, as are the second and fourth terms. Thus,

$$\sqrt{3} + \sqrt{80} - \sqrt{27} + \sqrt{5} = \sqrt{3} + 4\sqrt{5} - 3\sqrt{3} + \sqrt{5} = -2\sqrt{3} + 5\sqrt{5}$$

The following examples further illustrate the method of addition and subtraction of radicals.

Example C

$$\begin{aligned}\sqrt{8}+\sqrt{18}-6\sqrt{2}+3\sqrt{32} &= \sqrt{4\cdot 2}+\sqrt{9\cdot 2}-6\sqrt{2}+3\sqrt{16\cdot 2}\\ &= 2\sqrt{2}+3\sqrt{2}-6\sqrt{2}+12\sqrt{2}\\ &= 11\sqrt{2}\end{aligned}$$

$$\begin{aligned}\sqrt{20}+3\sqrt{5}-7\sqrt{12}+2\sqrt{45} &= \sqrt{4\cdot 5}+3\sqrt{5}-7\sqrt{4\cdot 3}+2\sqrt{9\cdot 5}\\ &= 2\sqrt{5}+3\sqrt{5}-14\sqrt{3}+6\sqrt{5}\\ &= 11\sqrt{5}-14\sqrt{3}\end{aligned}$$

Example D

$$\begin{aligned}\sqrt{a^2b}+\sqrt{9b}-2\sqrt{16c^2b} &= a\sqrt{b}+3\sqrt{b}-2(4c)\sqrt{b}\\ &= (a+3-8c)\sqrt{b}\end{aligned}$$

$$\begin{aligned}\sqrt{3a^3c^2}-\sqrt{12ab^2}+\sqrt{24ac^3} &= \sqrt{(3a)(a^2c^2)}-\sqrt{(3a)(4b^2)}+\sqrt{(6ac)(4c^2)}\\ &= ac\sqrt{3a}-2b\sqrt{3a}+2c\sqrt{6ac}\end{aligned}$$

Example E

$$\begin{aligned}3\sqrt[3]{2}+\sqrt[3]{16}-\sqrt{2} &= 3\sqrt[3]{2}+\sqrt[3]{8\cdot 2}-\sqrt{2}\\ &= 3\sqrt[3]{2}+2\sqrt[3]{2}-\sqrt{2}\\ &= 5\sqrt[3]{2}-\sqrt{2}\end{aligned}$$

$$\begin{aligned}\sqrt[4]{48a}+2\sqrt[3]{ab^3c^4}-\sqrt[4]{3a} &= \sqrt[4]{16\cdot 3a}+2\sqrt[3]{ab^3cc^3}-\sqrt[4]{3a}\\ &= 2\sqrt[4]{3a}+2bc\sqrt[3]{ac}-\sqrt[4]{3a}\\ &= \sqrt[4]{3a}+2bc\sqrt[3]{ac}\end{aligned}$$

When multiplying expressions containing radicals, we proceed just as in any algebraic multiplication. When simplifying the result, we use the relation

$$\sqrt{a}\sqrt{b}=\sqrt{ab}$$

which is actually Eq. (8-5). The following examples illustrate the method.

Example F

$$\begin{aligned}\sqrt{2}(\sqrt{6}+3\sqrt{5}) &= \sqrt{2}\sqrt{6}+3\sqrt{2}\sqrt{5}\\ &= \sqrt{12}+3\sqrt{10}\\ &= 2\sqrt{3}+3\sqrt{10}\end{aligned}$$

$$\begin{aligned}(\sqrt{3}-2\sqrt{2})(3\sqrt{3}+5\sqrt{2}) &= 3\sqrt{3}\sqrt{3}-\sqrt{3}\sqrt{2}-10\sqrt{2}\sqrt{2}\\ &= 3\sqrt{9}-\sqrt{6}-10\sqrt{4}\\ &= 3(3)-\sqrt{6}-10(2)\\ &= 9-\sqrt{6}-20\\ &= -11-\sqrt{6}\end{aligned}$$

Example G

$$
\begin{aligned}
(\sqrt{a} - 3\sqrt{b})(2\sqrt{a} + \sqrt{b}) &= 2\sqrt{a}\sqrt{a} - 5\sqrt{a}\sqrt{b} - 3\sqrt{b}\sqrt{b} \\
&= 2\sqrt{a^2} - 5\sqrt{ab} - 3\sqrt{b^2} \\
&= 2a - 5\sqrt{ab} - 3b \\
(\sqrt{2} - 3\sqrt{x})(\sqrt{2} + \sqrt{3}) &= \sqrt{2}\sqrt{2} + \sqrt{2}\sqrt{3} - 3\sqrt{x}\sqrt{2} - 3\sqrt{x}\sqrt{3} \\
&= 2 + \sqrt{6} - 3\sqrt{2x} - 3\sqrt{3x}
\end{aligned}
$$

In the preceding section, we showed how to rationalize a fraction in which a radical appeared in the denominator. However, we restricted our attention to the simpler cases of rationalization. Example H illustrates rationalizing the denominator of a fraction in which the numerator is a sum of terms.

Example H

$$
\frac{\sqrt{3} + 5}{\sqrt{2}} = \frac{\sqrt{2}(\sqrt{3} + 5)}{\sqrt{2}\sqrt{2}} = \frac{\sqrt{2}\sqrt{3} + 5\sqrt{2}}{\sqrt{4}} = \frac{\sqrt{6} + 5\sqrt{2}}{2}
$$

If the denominator is the sum of two terms, the fraction can be rationalized by multiplying both numerator and denominator by the difference of the same two terms. This is so because

$$
\boxed{(\sqrt{a} + \sqrt{b})(\sqrt{a} - \sqrt{b}) = a - b} \qquad (8\text{-}8)
$$

which results in an expression without radicals. Example I illustrates the method.

Example I

$$
\begin{aligned}
\frac{\sqrt{3}}{\sqrt{2} - \sqrt{5}} &= \frac{\sqrt{3}(\sqrt{2} + \sqrt{5})}{(\sqrt{2} - \sqrt{5})(\sqrt{2} + \sqrt{5})} = \frac{\sqrt{6} + \sqrt{15}}{2 - 5} \\
&= -\frac{\sqrt{6} + \sqrt{15}}{3}
\end{aligned}
$$

$$
\begin{aligned}
\frac{\sqrt{a}}{2\sqrt{a} - 3} &= \frac{\sqrt{a}(2\sqrt{a} + 3)}{(2\sqrt{a} - 3)(2\sqrt{a} + 3)} = \frac{2\sqrt{a}\sqrt{a} + 3\sqrt{a}}{2 \cdot 2 \cdot \sqrt{a}\sqrt{a} - 3 \cdot 3} \\
&= \frac{2a + 3\sqrt{a}}{4a - 9}
\end{aligned}
$$

Exercises 8-3

In Exercises 1 through 50, perform the indicated operations, expressing each answer in simplest form.

1. $\sqrt{3} + 3\sqrt{3} - 2\sqrt{3} + \sqrt{3}$
2. $\sqrt{13} + 7\sqrt{13} - \sqrt{13} + 3\sqrt{13}$
3. $\sqrt{7} - \sqrt{5} + 3\sqrt{7} + 2\sqrt{5}$
4. $\sqrt{11} + 8\sqrt{11} - \sqrt{17} + 3\sqrt{11}$
5. $2\sqrt{15} + \sqrt{14} - 7\sqrt{15} - \sqrt{14}$
6. $3\sqrt{17} - 2\sqrt{10} + \sqrt{17} + 3\sqrt{10}$
7. $2\sqrt{2} - 3\sqrt{3} + 4\sqrt{2} + \sqrt{3}$
8. $3\sqrt{2} - 3\sqrt{5} + \sqrt{2} + 2\sqrt{5}$
9. $2\sqrt{3} + \sqrt{27}$
10. $3\sqrt{5} + \sqrt{75}$
11. $2\sqrt{40} + 5\sqrt{10}$
12. $3\sqrt{44} - 2\sqrt{11}$
13. $\sqrt{8} + \sqrt{18} + \sqrt{32}$
14. $\sqrt{12} + \sqrt{27} + \sqrt{48}$
15. $\sqrt{28} - 2\sqrt{63} + 5\sqrt{7}$
16. $2\sqrt{24} + \sqrt{6} - 2\sqrt{54}$
17. $2\sqrt{8} - 2\sqrt{12} - \sqrt{50}$
18. $2\sqrt{20} - \sqrt{44} + 3\sqrt{11}$
19. $\sqrt[3]{16} + 3\sqrt[3]{2} - \sqrt[3]{24}$
20. $\sqrt[4]{4} - 2\sqrt[4]{64} - \sqrt[3]{4}$
21. $\sqrt[4]{81} - \sqrt[3]{27} + \sqrt[5]{64}$
22. $2\sqrt[3]{32} + 3\sqrt[3]{4} - \sqrt[5]{32}$
23. $\sqrt{a} + \sqrt{9a}$
24. $\sqrt{5x} + \sqrt{20x}$
25. $\sqrt{2a} + \sqrt{8a} + \sqrt{32a^3}$
26. $\sqrt{ac} + \sqrt{4ac} + \sqrt{16a^3c}$
27. $a\sqrt{2} + \sqrt{72a^2} - \sqrt{12a}$
28. $\sqrt{x^2yz} - \sqrt{y^3z} + 2x\sqrt{y^2z}$
29. $\sqrt[4]{a^2b^4c^5} + 3bc\sqrt[4]{a^2c}$
30. $a\sqrt[3]{8xy^3} - ay\sqrt[3]{x}$
31. $\sqrt{3}(\sqrt{7} - 3\sqrt{6})$
32. $\sqrt{5}(\sqrt{20} - 6\sqrt{3})$
33. $\sqrt{2}(\sqrt{8} - \sqrt{32} + 5\sqrt{18})$
34. $\sqrt{7}(\sqrt{14} + 2\sqrt{6} - \sqrt{56})$
35. $\sqrt{a}(\sqrt{ab} + 3\sqrt{ac})$
36. $\sqrt{2a}(\sqrt{8} + \sqrt{6a^3})$
37. $(\sqrt{2} + \sqrt{3})(2\sqrt{2} - \sqrt{3})$
38. $(\sqrt{7} - 2\sqrt{5})(3\sqrt{7} + \sqrt{5})$
39. $(\sqrt{5} - 3\sqrt{3})(2\sqrt{5} + \sqrt{27})$
40. $(2\sqrt{11} - \sqrt{8})(3\sqrt{2} + \sqrt{22})$
41. $(\sqrt{a} - 3\sqrt{c})(2\sqrt{a} + 5\sqrt{c})$
42. $(\sqrt{2b} + 3)(\sqrt{2} - \sqrt{b})$
43. $\dfrac{\sqrt{2} + \sqrt{3}}{\sqrt{2}}$
44. $\dfrac{\sqrt{3} + \sqrt{5}}{\sqrt{5}}$
45. $\dfrac{\sqrt{7}}{\sqrt{7} + \sqrt{3}}$
46. $\dfrac{2\sqrt{5}}{\sqrt{5} - \sqrt{11}}$
47. $\dfrac{\sqrt{3} + \sqrt{6}}{2\sqrt{3} - \sqrt{6}}$
48. $\dfrac{\sqrt{2} - 3\sqrt{7}}{2\sqrt{2} + \sqrt{7}}$
49. $\dfrac{\sqrt{a}}{\sqrt{a} + 2\sqrt{b}}$
50. $\dfrac{\sqrt{x}}{2\sqrt{x} - 3\sqrt{y}}$

In Exercises 51 through 54, solve the given problems.

51. Three square pieces of land are along a straight road, as shown in Fig. 8-1. Find the distance x along the road in simplest radical form.

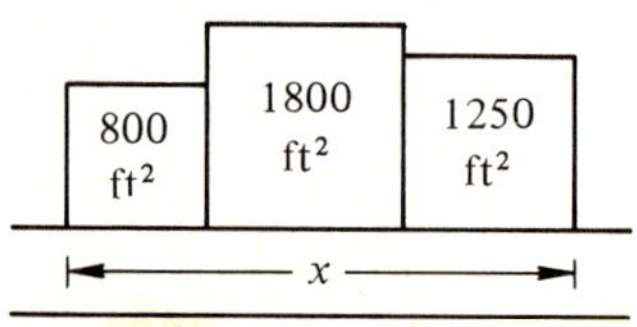

Figure 8-1

52. Find the area of a rectangle whose length is $\sqrt{7}$ and whose width is $2\sqrt{7} - 3$.

53. In the theory of waves in wires, one encounters the expression

$$\frac{\sqrt{d_1} - \sqrt{d_2}}{\sqrt{d_1} + \sqrt{d_2}}$$

Evaluate this expression for $d_1 = 10$ and $d_2 = 3$.

54. When dealing with vibratory motion, the equation

$$a^2 - 2al + k^2 = 0$$

is found. Show that $a = l + \sqrt{l^2 - k^2}$ and $a = l - \sqrt{l^2 - k^2}$ both satisfy the equation.

8-4 THE PYTHAGOREAN THEOREM

One of the important uses of square roots is in the determination of one of the sides of a right triangle when the other two sides are known. *A* **right triangle** *is a triangle that contains a* **right angle,** *or an angle of* 90°. *The side opposite this right angle is the* **hypotenuse.**

This application of square roots is based upon the very important property of a right triangle known as the **Pythagorean theorem.** In this section, we shall develop the Pythagorean theorem and show some of its applications.

The Pythagorean theorem states that

in a right triangle, the square of the length of the hypotenuse equals the sum of the squares of the lengths of the other two sides.

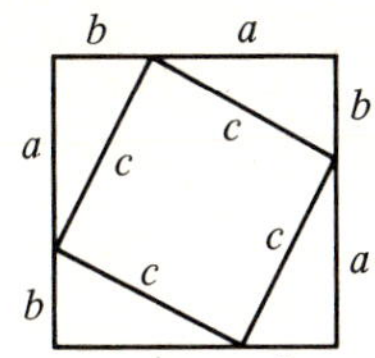

Figure 8-2

First, we shall show how this theorem may be found by the use of areas of geometric figures. In Fig. 8-2, a square of side c is inscribed in a square of side $a + b$ as shown. The area of the outer square $(a + b)^2$ minus the area of the four triangles of sides a, b, and c (the area of each of these triangles is $\frac{1}{2}ab$) equals the area of the inner square. This leads to

$$(a + b)^2 - 4\left(\frac{1}{2}ab\right) = c^2$$

$$a^2 + 2ab + b^2 - 2ab = c^2$$

$$\boxed{a^2 + b^2 = c^2} \qquad (8\text{-}9)$$

We see that this result is the Pythagorean relation for each of the four triangles of sides a, b, and c. Therefore, in any right triangle with sides a, b, and c, where c is the hypotenuse, Eq. (8-9) is valid.

We present now four examples illustrating the use of the Pythagorean theorem.

Example A For a right triangle ABC, with the right angle at C, $AC = 5$ (AC indicates the distance from A to C), and $BC = 12$, find AB. See Fig. 8-3.

Letting AC, BC, and AB denote the sides, we find that the Pythagorean theorem applied to this triangle is

$$(AC)^2 + (BC)^2 = (AB)^2$$

Substituting the values for AC and BC, we have

$$5^2 + 12^2 = (AB)^2$$
$$25 + 144 = (AB)^2$$
$$169 = (AB)^2$$
$$13 = AB$$

We find the final value, $AB = 13$, by taking the square roots of both sides.

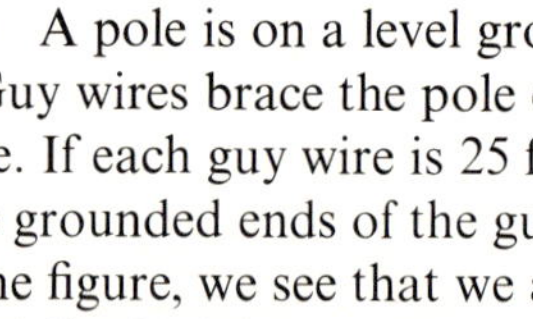

Figure 8-3

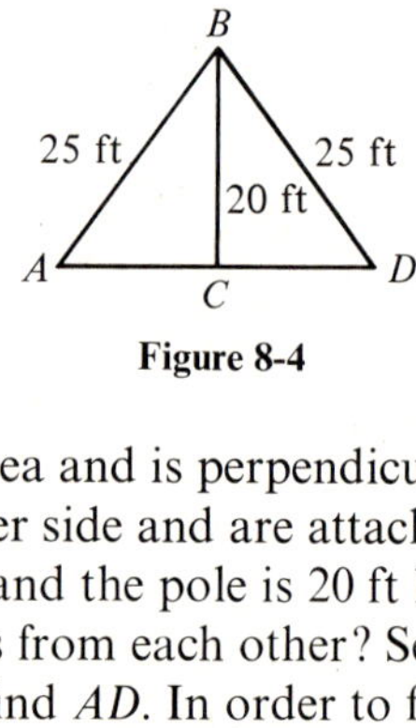

Figure 8-4

Example B A pole is on a level ground area and is perpendicular to the ground. Guy wires brace the pole on either side and are attached at the top of the pole. If each guy wire is 25 ft long and the pole is 20 ft high, how far are the grounded ends of the guy wires from each other? See Fig. 8-4.

From the figure, we see that we are to find AD. In order to find AD, we shall find AC, which in turn equals CD, and therefore $AD = 2AC$. From the Pythagorean theorem, we have $(AC)^2 + (BC)^2 = (AB)^2$. Substituting the known values of AB and BC, we have

$$(AC)^2 + (20)^2 = (25)^2$$
$$(AC)^2 + 400 = 625$$
$$(AC)^2 = 225$$
$$AC = 15 \text{ ft}$$

Therefore, $AD = 30$ ft.

Example C Find the general formula for the length of a diagonal of a square of side s. See Fig. 8-5.

From the statement of the problem, we wish to find d in terms of s. From the Pythagorean theorem, we have

$$d^2 = s^2 + s^2 = 2s^2$$
$$d = \sqrt{2s^2} = s\sqrt{2}$$

The diagonal, therefore, is equal to the value of a side of the square times $\sqrt{2}$.

Note that each of the right triangles created by the diagonal has two of its sides equal. The diagonal is the hypotenuse of these right triangles. Therefore, we may say that the hypotenuse of a right triangle with two sides equal is equal to the value of a side times $\sqrt{2}$.

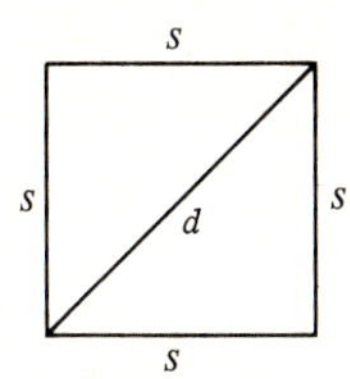

Figure 8-5

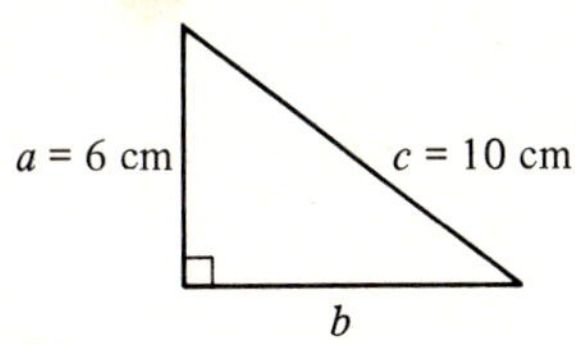

Figure 8-6

Example D If the hypotenuse of a triangle is 10 cm and one of the sides is 6 cm, how long is the other side?

We use the Pythagorean theorem to find the other side. See Fig. 8-6.

$$c^2 = a^2 + b^2$$
$$10^2 = 6^2 + b^2$$
$$100 = 36 + b^2$$
$$64 = b^2$$
$$b = 8 \text{ cm}$$

Any set of three numbers which satisfies the Pythagorean relation $a^2 + b^2 = c^2$ is known as a **Pythagorean triplet.** Often multiples of the basic set of integers which form a Pythagorean triplet are referred to as **Pythagorean numbers.** Consider the illustrations in the following example.

Example E We note in Example D that 6, 8, and 10 are each divisible by 2. Dividing gives us 3, 4, and 5, respectively. Since $3^2 + 4^2 = 5^2$, the set $\{3, 4, 5\}$ is a Pythagorean triplet. Also, since $6^2 + 8^2 = 10^2$, the set $\{6, 8, 10\}$ can be called a Pythagorean triplet, although it would usually be called a set of Pythagorean numbers since each number is the same multiple of the numbers of the basic set $\{3, 4, 5\}$.

In the same way, the set $\{5, 12, 13\}$ is a Pythagorean triplet, since $5^2 + 12^2 = 13^2$, and the set $\{15, 36, 39\}$ is a set of Pythagorean numbers since $15^2 + 36^2 = 39^2$.

Some applications of the use of the Pythagorean theorem require that the equation be applied more than once. The following example illustrates such an application.

Example F A rope is attached to the top of a 22-ft pole. A man holding the rope moves a certain distance from the pole and notes that there are 30 ft of rope from the ground to the top of the pole. He then moves another 10 ft from the pole. How long must the rope be in order to reach the ground at his feet? See Fig. 8-7.

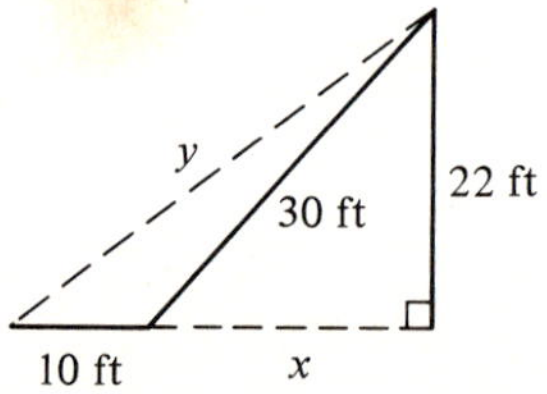

Figure 8-7

From the figure, we see that if we first find his original distance x from the pole, we can then determine the required distance y. First applying the Pythagorean theorem to the triangle with sides x, 22, and 30, we have

$$
\begin{aligned}
x^2 + (22)^2 &= 30^2 \\
x^2 + 484 &= 900 \\
x^2 &= 416 \\
x &= \sqrt{416} = 4\sqrt{26} \\
&= 20.4 \text{ ft}
\end{aligned}
$$

Here we approximated the square root of 26 to obtain the result. Now we add 10.0 to 20.4 and apply the Pythagorean theorem to the triangle with sides 30.4, 22, and y. This gives

$$(30.4)^2 + (22)^2 = y^2$$
$$924 + 484 = y^2$$
$$y^2 = 1408$$
$$y = 37.5 \text{ ft}$$

which means that the rope must be at least 37.5 ft long.

Exercises 8-4

In Exercises 1 through 8, use the Pythagorean theorem to solve for the unknown side of the triangle. In each case, c is the hypotenuse. Where appropriate, express answers in simplest radical form.

	a	b	c		a	b	c
1.	3	4	?	**2.**	9	12	?
3.	6	?	10	**4.**	2	?	4
5.	?	12	16	**6.**	?	10	18
7.	?	15	30	**8.**	6	?	38

In Exercises 9 through 16, use the Pythagorean theorem to solve for the unknown side in each triangle.

9.

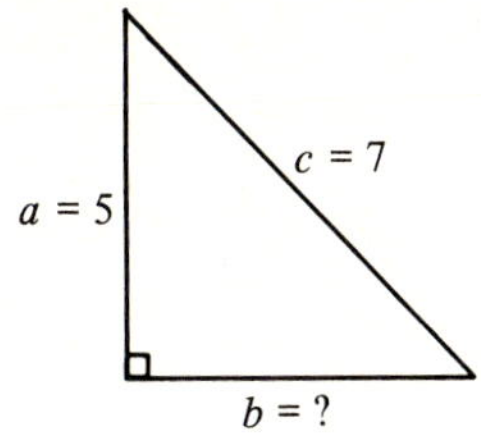

10.

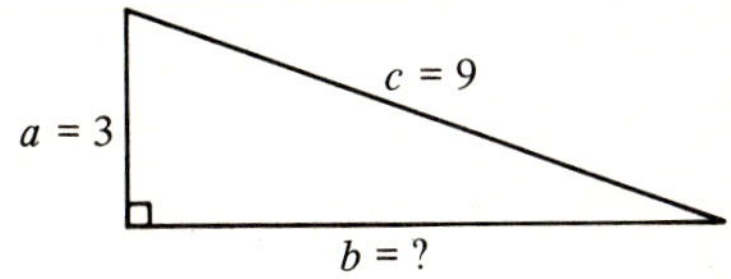

11.

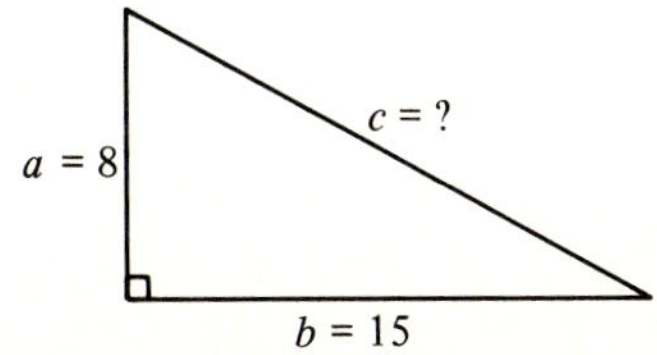

12.

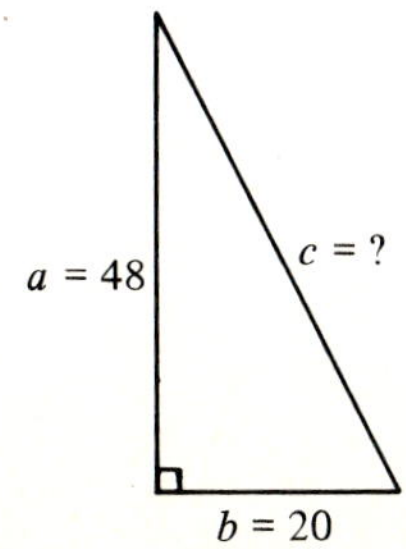

13.

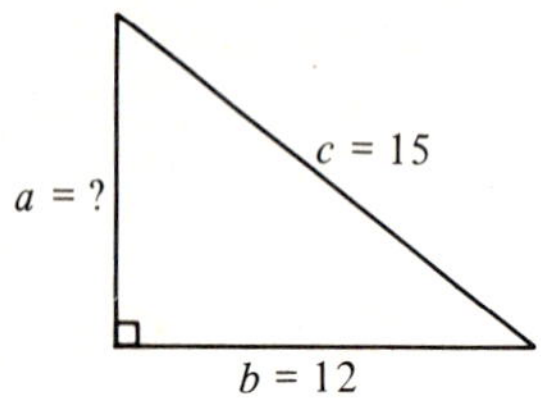

14.

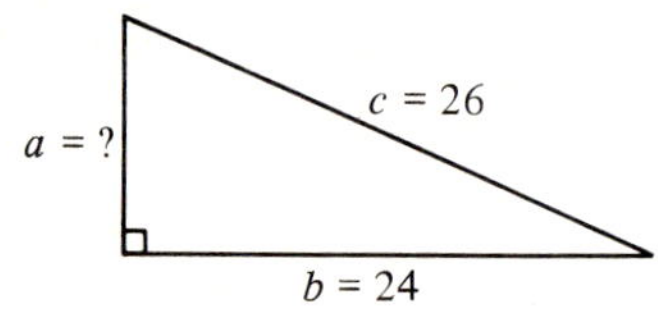

15.

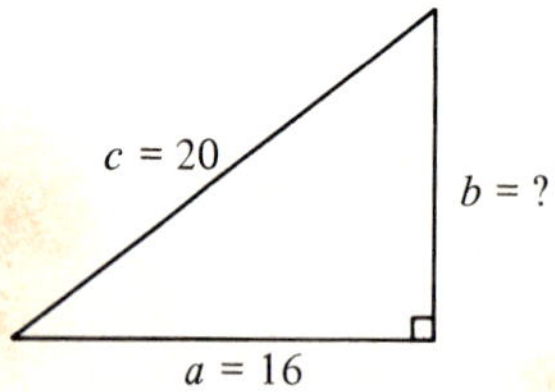

16.

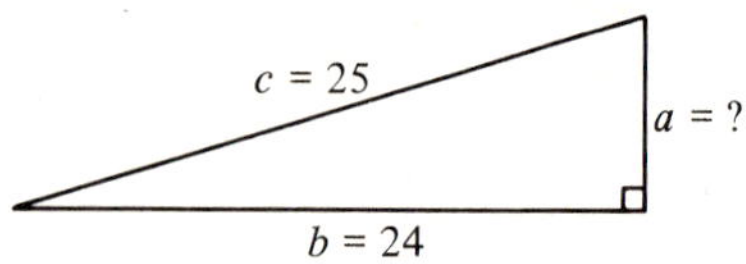

In Exercises 17 through 20, use the Pythagorean theorem to solve for the answers. If radicals are involved in the answers, leave the answers in simplest radical form.

17. If two sides (neither of which is the hypotenuse) of a right triangle are 10 cm and 24 cm, 2 times what number is the length of the hypotenuse of that triangle?

18. If two sides (neither of which is the hypotenuse) of a right triangle are 9 cm and 12 cm, 3 times what number is the length of the hypotenuse of that triangle?

19. If $a = 3$ and $c = 6$, what is the value of side b if c is the hypotenuse?

20. If $a = 3$ and $b = 3$, what is the value of side c if c is the hypotenuse?

In Exercises 21 through 24, determine whether or not the given sets of numbers are Pythagorean triplets or Pythagorean numbers.

21. $\{7, 40, 41\}$ **22.** $\{16, 20, 24\}$ **23.** $\{10, 24, 30\}$ **24.** $\{14, 48, 50\}$

In Exercises 25 through 40, set up the given problems and solve by use of the Pythagorean theorem. Except where noted, determine values to the nearest tenth.

25. Find the length of a diagonal of a square whose side is 15.5 in.

26. The shortest side of a triangle whose angles are 30°, 60°, and 90° (the shortest side is one-half the hypotenuse) is 8 m. Find the perimeter of the triangle.

27. A man rows across a river which is 600 ft wide. The current carries him downstream 40 ft from the point directly across from his starting point. How far to the nearest foot did he actually travel?

28. The guy wires bracing a telephone pole on a level ground area and the line along the ground between the grounded ends of the wires form an equilateral triangle whose side is 20 ft. Find the height of the point where the wires are attached to the pole.

29. The hypotenuse of a right triangle is 24 m, and one side is twice the other. Find the perimeter of the triangle.

30. The hypotenuse of a right triangle is three times as long as one of the sides. If the other side is 56 cm, what is the length of the hypotenuse?

31. One side of a right triangle is 15 ft long and the area of the triangle is 180 ft^2. What is the length of the hypotenuse? (The area is one-half the product of the two sides.)

32. The hypotenuse of a right triangle is 8 ft longer than one of the sides. The other side is 20 ft long. What is the length of the hypotenuse?

33. A rectangular room is 12 ft long, 10 ft wide, and 8 ft high. What is the distance from a corner on the floor to the opposite corner at the ceiling?

34. A square is inscribed in a circle (all four vertices of the square touch the circle). Find the length of the side of the square if the radius of the circle is 8 cm.

35. A ship 28 km due west of a lighthouse travels 18 km toward the lighthouse and then turns. After traveling another 14 km in a straight line, it is due north of the lighthouse. How far, on a direct line, is the ship from its starting point?

36. Two jets left the same airport at the same time, one traveling due south at 600 mi/h and the other traveling due east at 1000 mi/h. How far apart (to the nearest mile) are they after a half-hour?

37. Solve for the distance x in Fig. 8-8.

38. Solve for distances x and y in Fig. 8-9.

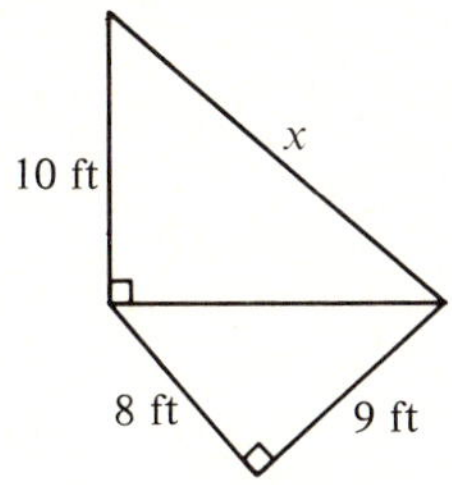

Figure 8-8

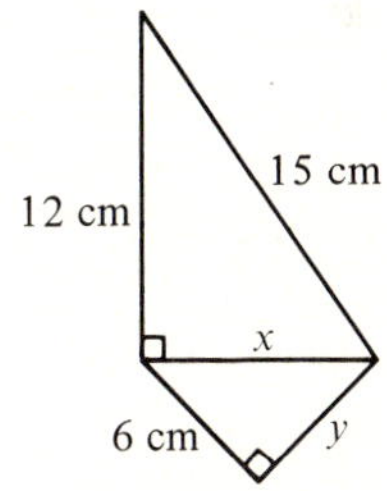

Figure 8-9

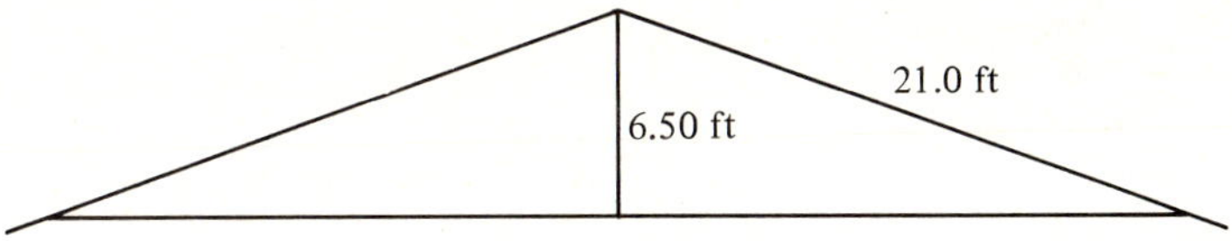

Figure 8-10

39. Figure 8-10 shows a roof truss. The rafters are 21.0 ft long, including a 1.5-ft overhang, and the height of the truss is 6.50 ft. Determine the length of the base of the truss.

40. A repairman places his ladder against the side of a house at a distance of 10 ft. If the ladder makes an angle of 60° with the ground and an angle of 30° with the wall of the house, how far up the house does the ladder reach? How long is the ladder? (see Exercise 26.)

8-5 FRACTIONAL EXPONENTS

Thus far, the only numbers we have used for exponents have been the positive integers, the negative integers, and zero. A natural question arises: can fractions be used as exponents? The answer to this question is yes; and as it turns out, these fractional exponents are very convenient in certain topics of mathematics. We discuss this subject at this time because, as we shall see, an expression raised to a fractional exponent can be interpreted as a radical.

For fractional exponents to be meaningful, they must satisfy the basic laws of exponents already established. The significance of fractional exponents is established in Example A.

Example A If $a^{1/3}$ is to have meaning, then

$$a^{1/3}a^{1/3}a^{1/3} = a^{1/3+1/3+1/3} = a$$

and

$$(a^{1/3})^3 = a$$

However, we have already established that

$$(\sqrt[3]{a})^3 = a$$

This means that $a^{1/3}$ must be interpreted as $\sqrt[3]{a}$ in order that the basic laws of exponents hold.

Generalizing on the conclusion of Example A, *we define $a^{1/n}$ to mean the principal nth root of a, or*

$$\boxed{a^{1/n} = \sqrt[n]{a}} \qquad (8\text{-}10)$$

Example B gives several illustrations of the meaning of Eq. (8-10)

Example B

$$4^{1/2} = \sqrt{4} = 2 \qquad (27)^{1/3} = \sqrt[3]{27} = 3$$

$$(-32)^{1/5} = \sqrt[5]{-32} = -2 \qquad (x^4)^{1/2} = \sqrt{x^4} = \sqrt{(x^2)^2} = x^2$$

Normally it is not necessary, and in fact it is often cumbersome, to include the radical interpretation when evaluating expressions with fractional exponents. Thus, the above results would usually be written as

$$4^{1/2} = 2 \qquad (27)^{1/3} = 3 \qquad (-32)^{1/5} = -2 \quad \text{and} \quad (x^4)^{1/2} = x^2$$

When an expression of the form $a^{1/n}$ is raised to some power, say m, the basic laws of exponents should still be valid. Thus,

$$(a^{1/n})^m = a^{(1/n)(m)} = a^{m/n}$$

The meaning of $a^{m/n}$ is shown in the following examples.

Example C $(a^{1/3})^2 = a^{2/3}$, which can be written as $(a^2)^{1/3}$. Thus, we have $(a^{1/3})^2 = (a^2)^{1/3}$, or $(\sqrt[3]{a})^2 = \sqrt[3]{a^2}$.

Generalizing on the result of Example C, we see that

$$\boxed{a^{m/n} = (\sqrt[n]{a})^m = \sqrt[n]{a^m}} \qquad (8\text{-}11)$$

Example D

$$8^{2/3} = (8^{1/3})^2 = 2^2 = 4 \qquad 4^{7/2} = (4^{1/2})^7 = 2^7 = 128$$

$$(27)^{4/3} = [(27)^{1/3}]^4 = 3^4 = 81 \qquad (x^4)^{3/2} = [(x^4)^{1/2}]^3 = (x^2)^3 = x^6$$

Example E $8^{2/3}$ written in radical form is $\sqrt[3]{8^2}$ or $(\sqrt[3]{8})^2$. Since $\sqrt[3]{8^2} = \sqrt[3]{64} = 4$ and $(\sqrt[3]{8})^2 = 2^2 = 4$, we see that $8^{2/3} = \sqrt[3]{8^2} = (\sqrt[3]{8})^2$. Also, $4^{7/2} = (\sqrt{4^7}) = (\sqrt{4})^7$.

The following examples illustrate the use of the basic laws of exponents with expressions that have fractional exponents.

Example F

$$8^{-2/3} = \frac{1}{8^{2/3}} = \frac{1}{(8^{1/3})^2} = \frac{1}{2^2} = \frac{1}{4}$$

$$(8^{4/5})^0 = 1$$

$$2^{1/2}2^{1/3} = 2^{1/2+1/3} = 2^{5/6}$$

$$\frac{2^{1/2}}{2^{1/3}} = 2^{1/2-1/3} = 2^{1/6}$$

$$\frac{4^{-1/2}}{2^{-5}} = \frac{2^5}{4^{1/2}} = \frac{32}{2} = 16$$

Example G

$$x^{1/2}x^{1/4} = x^{1/2+1/4} = x^{3/4}$$

$$\frac{a}{a^{1/2}b^{-2}} = a^{1-1/2}b^2 = a^{1/2}b^2$$

$$\frac{x^{-1/3}y^{2/3}}{x^{2/3}y^{-2/3}} = \frac{y^{2/3+2/3}}{x^{2/3+1/3}} = \frac{y^{4/3}}{x}$$

Exercises 8-5

In Exercises 1 through 8, change the given expressions to radical form.

1. $3^{1/2}$ 2. $6^{1/3}$ 3. $x^{1/4}$ 4. $y^{1/5}$
5. $a^{2/5}$ 6. $b^{3/5}$ 7. $x^{5/3}$ 8. $y^{7/3}$

In Exercises 9 through 40, evaluate the given expressions.

9. $9^{1/2}$ 10. $(36)^{1/2}$ 11. $(64)^{1/3}$ 12. $(125)^{1/3}$
13. $(16)^{1/4}$ 14. $(243)^{1/5}$ 15. $(-8)^{1/3}$ 16. $(-64)^{1/3}$
17. $4^{3/2}$ 18. $9^{3/2}$ 19. $8^{4/3}$ 20. $27^{2/3}$
21. $81^{3/4}$ 22. $16^{3/4}$ 23. $(-8)^{2/3}$ 24. $(-27)^{5/3}$
25. $36^{-1/2}$ 26. $100^{-1/2}$ 27. $16^{-1/4}$ 28. $32^{-2/5}$
29. $-(-32)^{-1/5}$ 30. $-64^{-2/3}$ 31. $(4^{1/2})(27^{2/3})$ 32. $(25^{1/2})(100^{3/2})$
33. $\dfrac{(64)^{1/2}}{(64)^{1/3}}$ 34. $\dfrac{81^{1/4}}{81^{1/2}}$ 35. $2^{-1/2} \cdot 2^{3/2}$ 36. $\dfrac{5^{1/2}}{5^{-1/2}}$
37. $(4^{-2})(27^{-1/3})$ 38. $\dfrac{(25)^{-3/2}}{(81)^{3/4}}$
39. $49^{-1/2} - 3(14^{-1})$ 40. $(125)^{-2/3} + 4(25)^{-1/2}$

In Exercises 41 through 48, use the laws of exponents to simplify the given expressions. Express all answers with positive exponents.

41. $a \cdot a^{1/2}$ 42. $\dfrac{b}{b^{1/3}}$ 43. $\dfrac{ab}{a^{1/4}}$ 44. $\dfrac{xyz^2}{z^{-3/2}}$
45. $\dfrac{x(x^{1/3})^4}{x^{2/5}}$ 46. $\dfrac{xy^{1/2}}{(y^{2/3})^{-2}}$ 47. $\dfrac{x^{-1/2}x^2}{x^{2/3}}$ 48. $\dfrac{xy^{-1/2}z}{z^{-2/3}x^{-1/2}}$

In Exercises 49 through 52, solve the given problems.

49. Use fractional exponents to express the side x of a square in terms of the area A of the square.
50. An expression which arises in Einstein's theory of relativity is $\sqrt{(c^2 - v^2)/c^2}$. Express this with the use of fractional exponents, and then simplify.
51. When the volume of a gas changes very rapidly, an approximate relation is that the pressure p varies inversely as the $\frac{3}{2}$ power of the volume. Given that p is 3 atm when $V = 100$ in.^{3}, find p when $V = 25$ in.^{3}
52. Kepler's third law of planetary motion may be stated as follows: *The mean radius (about the sun) of any planet is proportional to the $\frac{2}{3}$ power of the period of that planet.* Using the facts that the earth has a mean radius of about 93,000,000 mi and that its period is 1 year, calculate the mean radius of Saturn, assuming that its period is 27 years (actually, it is about 29.5 years, but the answer obtained is only about 6% in error).

8-6 SUMMARY

In this chapter, we have developed the concept of the root of a number. We designated the principal *n*th root of a number by the radical

$$\sqrt[n]{N}$$

We defined this radical to represent an imaginary number if $n = 2$ and N is negative. We also noted the relationship of imaginary numbers, real numbers, and complex numbers.

Simplification of radicals was discussed. We define a radical to be simplified if all factors which are perfect powers of the same order as that of the radical have been removed from under the radical and if the denominator has been rationalized. This simplification was based upon the following formulas:

$$\sqrt{ab} = \sqrt{a}\sqrt{b} \tag{8-5}$$

$$\sqrt{a^2} = a \qquad (a > 0) \tag{8-6}$$

$$\sqrt{\frac{a}{b}} = \frac{\sqrt{ab}}{b} \qquad (b \text{ not zero}) \tag{8-7}$$

$$(\sqrt{a} - \sqrt{b})(\sqrt{a} + \sqrt{b}) = a - b \tag{8-8}$$

Equation (8-8) is used for rationalizing denominators that consist of the sum of terms including radicals.

The basic operations with radicals were seen to be the same as the equivalent algebraic operations. Any radicals in the final expressions must be expressed in simplest form, with like radicals being combined.

We discussed the Pythagorean theorem, which is a very important property of a right triangle and which makes use of square roots. If c is the hypotenuse and a and b are the other two sides of a right triangle, the Pythagorean theorem states that

$$c^2 = a^2 + b^2 \tag{8-9}$$

We also discussed the concepts of the Pythagorean triplet and Pythagorean numbers.

Finally, we saw that fractions can be used as exponents, where

$$a^{m/n} = (\sqrt[n]{a})^m = \sqrt[n]{a^m} \tag{8-10}$$

is the interpretation of fractional exponents in terms of radicals. These fractional exponents follow all the laws of exponents.

The reader should be familiar with the following terms that were either introduced or reviewed in this chapter.

Complex number	Pythagorean numbers
imaginary part	Pythagorean theorem
real part	Pythagorean triplet
Hypotenuse	Rationalization of denominator
Imaginary number	Rational number
Irrational number	Real number
Like radicals	Right angle
Principal nth root	Right triangle

The following exercises can be used as supplementary or review problems for the topics covered in this chapter.

Exercises 8-6

In Exercises 1 through 32, evaluate the given expressions.

1. 5^{-1} **2.** 3^{-4} **3.** $\dfrac{1}{2^{-3}}$ **4.** $\dfrac{1}{5^{-1}}$

5. $3^0 6^{-1}$ **6.** $\dfrac{9^0}{4^{-3}}$ **7.** $\sqrt{169}$ **8.** $-\sqrt{900}$

9. $\sqrt[3]{216}$ **10.** $-\sqrt[3]{-125}$ **11.** $-\sqrt[4]{256}$ **12.** $\sqrt[4]{0.0081}$

13. $\sqrt{\dfrac{1}{9}}$ **14.** $\sqrt{\dfrac{4}{25}}$ **15.** $-\sqrt{\dfrac{9}{121}}$ **16.** $-\sqrt{\dfrac{144}{169}}$

17. $(100)^{1/2}$ **18.** $(625)^{1/2}$ **19.** $(1000)^{1/3}$ **20.** $(8000)^{1/3}$

21. $(49)^{3/2}$ **22.** $(243)^{3/5}$ **23.** $(121)^{3/2}$ **24.** $(8)^{7/3}$

25. $\dfrac{(16)^{3/4}}{(27)^{2/3}}$ **26.** $\dfrac{(25)^{3/2}}{5^{-1}}$ **27.** $\dfrac{(-216)^{2/3}}{3^{-2}}$ **28.** $\dfrac{(81)^{-3/4}}{(32)^{2/5}}$

29. $\sqrt{-81}$ **30.** $\sqrt{-169}$ **31.** $-\sqrt{-0.64}$ **32.** $-\sqrt{-0.01}$

In Exercises 33 through 48, write the given expressions in simplest form, expressing all results with positive exponents.

33. $3a^{-2}b$ **34.** $2xy^{-1}$ **35.** $\dfrac{mn^{-2}}{m^{-3}}$ **36.** $\dfrac{2rs^{-1}}{t^{-5}}$

37. $\dfrac{2x^3y^{-1}}{3x^{-2}y^2}$ **38.** $\dfrac{(2a)^0(b^{-1}c)}{4a^2bc^{-3}}$ **39.** $\dfrac{(-a)^{-2}xy^0}{(-x)^{-1}y^{-3}}$ **40.** $\dfrac{3^{-1}(ab^{-1})^{-2}}{(ab^0)^{-1}}$

41. $a^{1/4}a^{1/3}$ **42.** $x^{1/5}x^{2/3}$ **43.** $\dfrac{a^{2/3}}{a^{-1/2}}$ **44.** $\dfrac{b^{-1}}{b^{-1/2}}$

45. $(xy^{-2})^{1/2}$ **46.** $(8x^{-3}y^{3/2})^{1/3}$ **47.** $\frac{(st^{1/2})^{2/3}}{t^{-2}}$ **48.** $\frac{(16c^2)^{3/4}}{ac^{-1/5}}$

In Exercises 49 through 60, use Table 1 to evaluate the indicated square roots.

49. $\sqrt{52}$ **50.** $\sqrt{37}$ **51.** $\sqrt{6900}$ **52.** $\sqrt{1300}$
53. $\sqrt{240}$ **54.** $\sqrt{950}$ **55.** $\sqrt{180}$ **56.** $\sqrt{130}$
57. $\sqrt{0.47}$ **58.** $\sqrt{0.60}$ **59.** $\sqrt{0.060}$ **60.** $\sqrt{0.0031}$

In Exercises 61 through 84, express the given radicals in simplest form.

61. $\sqrt{40}$ **62.** $\sqrt{90}$ **63.** $\sqrt{128}$ **64.** $\sqrt{75}$
65. $\sqrt{72}$ **66.** $\sqrt{54}$ **67.** $\sqrt{108}$ **68.** $\sqrt{192}$
69. $\sqrt{\frac{1}{5}}$ **70.** $\sqrt{\frac{6}{11}}$ **71.** $\sqrt{\frac{4}{7}}$ **72.** $\sqrt{\frac{36}{37}}$
73. $\sqrt{4a^2}$ **74.** $\sqrt{28a}$ **75.** $\sqrt{125b^2c}$ **76.** $\sqrt{90b^3}$
77. $\sqrt{\frac{6}{a}}$ **78.** $\sqrt{\frac{3}{a^2}}$ **79.** $\sqrt{\frac{28}{3a}}$ **80.** $\sqrt{\frac{400000}{ab}}$
81. $\sqrt[3]{16a^3}$ **82.** $\sqrt[3]{81x^2y^4}$ **83.** $\sqrt[5]{64a^8}$ **84.** $\sqrt[7]{64a^8}$

In Exercises 85 through 108, perform the indicated operations and simplify.

85. $\sqrt{63} - 2\sqrt{28}$ **86.** $2\sqrt{45} - 3\sqrt{80}$ **87.** $3\sqrt{7} - 2\sqrt{6} + \sqrt{28}$
88. $5\sqrt{3} - \sqrt{27} - \sqrt{15}$ **89.** $2\sqrt{40} - 3\sqrt{90} + \sqrt{70}$ **90.** $\sqrt{12} - 3\sqrt{27} + 2\sqrt{80}$
91. $\sqrt{2a^2} + 3\sqrt{8} - \sqrt{32a^2}$ **92.** $\sqrt{20a} + 4\sqrt{5a^3} - 3\sqrt{45a}$
93. $\sqrt{2}(\sqrt{6} - 2\sqrt{24})$ **94.** $\sqrt{3}(3\sqrt{6} + \sqrt{54})$
95. $2\sqrt{5}(\sqrt{3} - 3\sqrt{5})$ **96.** $3\sqrt{2}(5\sqrt{2} + 2\sqrt{6})$
97. $\sqrt{a}(\sqrt{ab} - 3\sqrt{5b})$ **98.** $\sqrt{ab}(\sqrt{b} - 3\sqrt{a})$
99. $(\sqrt{6} - \sqrt{5})(2\sqrt{6} - 3\sqrt{5})$ **100.** $(3\sqrt{7} + 4\sqrt{2})(\sqrt{7} - 3\sqrt{2})$
101. $(2\sqrt{6} - 3\sqrt{7})(5\sqrt{6} - \sqrt{7})$ **102.** $(3\sqrt{11} - \sqrt{3})(\sqrt{11} + 2\sqrt{3})$
103. $(\sqrt{a} - 3\sqrt{b})(2\sqrt{a} + \sqrt{b})$ **104.** $(\sqrt{ab} - \sqrt{c})(2\sqrt{ab} + \sqrt{c})$
105. $\frac{\sqrt{2}}{\sqrt{5} - \sqrt{2}}$ **106.** $\frac{\sqrt{3}}{\sqrt{7} - 2\sqrt{3}}$
107. $\frac{\sqrt{2} - 1}{2\sqrt{2} + 3}$ **108.** $\frac{\sqrt{11} - \sqrt{5}}{2\sqrt{11} + \sqrt{5}}$

In Exercises 109 through 112, a, b, and c are the sides of right triangles, where c is the hypotenuse. Determine the values of the required sides. Where appropriate, express answers in simplest radical form.

	a	b	c
109.	8	?	17
110.	4	9	?
111.	36	?	60
112.	?	5	19

In Exercises 113 through 116, find the unknown side of each of the given right triangles (c is the hypotenuse) to determine whether or not each set of three numbers constitutes a Pythagorean triplet or a set of Pythagorean numbers.

113. $a = 16, b = 12, c = ?$

114. $a = 13, b = 84, c = ?$

115. $b = 56, c = 65, a = ?$

116. $a = 20, c = 52, b = ?$

In Exercises 117 through 126, solve the given problems.

117. In economic theory, when considering price and market demand, the expression $p = \sqrt{ad/b}$ arises. Rationalize the denominator of this expression.

118. In statistics, the expression $\sqrt{1 - n/N}$ is used. Combine the terms under the radical sign, and then rationalize the denominator.

119. A manufacturer found the standard deviation (see Exercise 56 of Section 8-2) of the the diameters of a certain machine part to be $\sqrt{2100/486}$ thousandths of an inch. Evaluate this standard deviation.

120. The velocity of sound in air, in meters/second, is proportional to the square root of $(273 + T)/273$, where T is the Celsius temperature. If $v = 331$ m/s when $T = 0°$C, find the equation relating the velocity of sound and temperature in rationalized form.

121. The density of an object equals its weight divided by its volume. A certain metal has a density of 1331 lb/ft^3. What is the edge of a cube of this metal which weighs 8 lb?

122. The absolute temperature of the filament of a light bulb equals approximately 1000 times the fourth root of the wattage. Find the Celsius temperature (273° less than the absolute temperature) of the filament of a 25-W bulb.

123. A tree casts a shadow of 12 m, and the distance from the end of the shadow to the top of the tree is 13 m. How high is the tree?

124. A rocket is taking off vertically at a distance of 1500 ft from an observer. If the rocket ascends at the rate of 500 ft/s, how far (to the nearest 10 ft) is the observer from the rocket after 10 s?

125. In psychology, when discussing the comparison of two stimuli, the expression

$$\frac{(b/a)^{3/5}}{(b/a)}$$

may arise. Simplify this expression using positive fractional exponents in the result.

126. In statistics, the expression

$$\frac{x_1 - x_2}{\sqrt{\dfrac{s_1^2}{m_1} + \dfrac{s_2^2}{m_2}}}$$

is used. Combine the terms under the radical sign, and then rationalize the denominator completely.

9

QUADRATIC EQUATIONS

9-1 QUADRATIC EQUATIONS

In Chapter 3, we introduced the basic methods of solving simple equations. Since then, additional algebraic operations have been discussed. We are now in a position to solve another important type of equation, the **quadratic equation.**

If a, b, and c are constants, the equation

$$\boxed{ax^2 + bx + c = 0} \tag{9-1}$$

is the standard form of **the general quadratic equation in *x*.** The purpose of this section is to identify quadratic equations and to discuss what is meant by the solution of the quadratic equation. The two sections which follow will discuss basic methods of solving such equations.

Example A The equation $2x^2 + 3x + 7 = 0$ is in the form of Eq. (9-1). Therefore, it is a quadratic equation, where $a = 2, b = 3$, and $c = 7$.

The equation $4x^2 - 7x + 9 = 0$ is also in the form of Eq. (9-1), despite the presence of the minus sign, since we may identify $a = 4$, $b = -7$, and $c = 9$.

The equation $x^3 - 9x + 8 = 0$ is *not* of the form of Eq. (9-1) due to the presence of x^3.

Since *it is the term in* x^2 *which distinguishes the quadratic equation from other types of equations*, if $a = 0$, the equation is not considered to be quadratic. However, either b, or c, or both may be zero in a quadratic equation.

Example B The equation $4x - 9 = 0$ is not a quadratic equation. It would fit the form of Eq. (9-1) only if $a = 0$, and such an equation is not quadratic.

The equation $3x^2 - 19 = 0$ is a quadratic equation, where $a = 3$, $b = 0$, and $c = -19$.

The equation $x^2 - 8x = 0$ is a quadratic equation, where $a = 1$, $b = -8$, and $c = 0$.

It is often necessary to perform basic operations on a quadratic equation to put it in the standard form of Eq. (9-1). Once this form is found, the values of a, b, and c may be determined. Consider the illustrations of Example C.

Example C The equation $3x^2 - 6 = 7x$ is not of the form of Eq. (9-1), but may easily be put in this form by subtracting $7x$ from each side of the equation. Performing this operation, we have

$$3x^2 - 7x - 6 = 0$$

which means $a = 3$, $b = -7$, and $c = -6$.

The equation $2x^2 = (x - 8)^2$ is not in the form of Eq. (9-1). In order to determine whether or not it may be put in this form, we must square the right-hand side as indicated. By then collecting terms on the left, we can establish the form of the equation. This leads to

$$2x^2 = x^2 - 16x + 64$$

or

$$x^2 + 16x - 64 = 0$$

This last equation is of the form of Eq. (9-1), where $a = 1$, $b = 16$, and $c = -64$.

The equation $3x^2 = (3x - 1)(x + 2)$ becomes

$$3x^2 = 3x^2 + 5x - 2$$

or

$$-5x + 2 = 0$$

We see that this last form is not the same as Eq. (9-1), and therefore the equation is not quadratic.

From Chapter 3, we recall that *a solution of an equation is a value of the unknown which, when substituted into the equation, makes the two sides of the equation equal.* Therefore, if a value of x is substituted into Eq. (9-1), the left-hand side must give zero if that value of x is a solution. As we shall see in the next section, *the solution of a quadratic equation is generally a pair of numbers, although occasionally only one number satisfies the equation.* We will again use set notation to denote those numbers that are members of the solution set. Example D illustrates checking possible values of x as solutions of a quadratic equation.

Example D Determine which, if any, of the values $x = 1$, $x = -3$, $x = \frac{1}{2}$, $x = 2$ are solutions of the equation $2x^2 + 5x - 3 = 0$.

Testing $x = 1$, we have

$$2(1)^2 + 5(1) - 3 = 2 + 5 - 3 = 4$$

Since the resulting value is not 0, the value $x = 1$ is not a solution.

Testing $x = -3$, we have

$$2(-3)^2 + 5(-3) - 3 = 2(9) - 15 - 3 = 18 - 15 - 3 = 0$$

Since the value is zero, the value $x = -3$ is a member of the set of solutions.

Testing $x = \frac{1}{2}$, we have

$$2\left(\frac{1}{2}\right)^2 + 5\left(\frac{1}{2}\right) - 3 = 2\left(\frac{1}{4}\right) + \frac{5}{2} - 3 = \frac{1}{2} + \frac{5}{2} - 3 = 3 - 3 = 0$$

Since two such values have been found, the solution set is $\{-3, \frac{1}{2}\}$. Any other value of x will prove not to be a solution.

Testing $x = 2$, we have

$$2(2)^2 + 5(2) - 3 = 2(4) + 10 - 3 = 15$$

which verifies that $x = 2$ is not a solution.

Our primary concern will be with quadratic equations that have real roots. However, it is possible that the solution of a quadratic equation can contain numbers with $i = \sqrt{-1}$. It is also possible that the real solutions of a quadratic equations are equal.

Example E The equation $x^2 + 4 = 0$ has the solution set $\{2i, -2i\}$. This can be verified by substitution:

$$(2i)^2 + 4 = 4i^2 + 4 = 4(-1) + 4 = -4 + 4 = 0$$
$$(-2i)^2 + 4 = (-2)^2 i^2 + 4 = 4(-1) + 4 = -4 + 4 = 0$$

The equation $x^2 + 4x + 4 = 0$ has the solution $x = -2$ only. The solution set is then $\{-2\}$. The reason why there is only one solution will be seen in the following section.

Exercises 9-1

In Exercises 1 through 12, determine whether or not the given equations are quadratic by performing algebraic operations that could put each in the form of Eq. (9-1). If the standard quadratic form is obtained, identify a, b, and c.

1. $x^2 - 7x = 4$

2. $3x^2 = 5 - 9x$

3. $x^2 = (x - 1)^2$

4. $2x^2 - x = 2x(x + 8)$

5. $(x + 2)^2 = 0$

6. $x(x^2 - 1) = x^3 + x^2$

7. $x^2 = x(1 - 6x)$

8. $x(1 + 2x) = 3x - 2x^2$

9. $x^2(1 - x) = 4$

10. $x(x^2 + 2x - 1) = 0$

11. $3x(x^2 - 2x + 1) = 1 + 3x^3$

12. $x(x^3 + 6) - 1 = x^2(x^2 + 1)$

In Exercises 13 through 24, if the equation is not of the form of Eq. (9-1), put it in that form. Then test the given values to determine which, if any, are members of the solution set of the equation.

13. $x^2 - x - 2 = 0, x = 1, x = -1, x = 2$

14. $x^2 + 4x + 3 = 0, x = -1, x = 1, x = -3$

15. $2x^2 - 3x + 1 = 0, x = -1, x = 0, x = \frac{1}{2}, x = 1$

16. $3x^2 - x - 2 = 0, x = 1, x = 2, x = -\frac{2}{3}, x = -1$

17. $V^2 + V = 12, V = -1, V = -3, V = 3, V = -4$

18. $y^2 = 3y + 10, y = 2, y = -1, y = 5$

19. $(3x - 4)(x - 2) = 0, x = -2, x = 1, x = 3$

20. $3t^2 + t + 2 = 2(t + 2), t = -\frac{2}{3}, t = -1, t = 1$

21. $s^2 = 4(s - 1), s = -2, s = 1, s = 2$

22. $x(x + 4) = 2x, x = -3, x = -2, x = 0, x = 1$

23. $n^2 + 6n - 3 = 6(n + 1), n = -3, n = -1, n = 3$

24. $x^2 + 16 = 0, x = -4, x = 2, x = 4$

In Exercises 25 through 28, solve the given problems.

25. A certain number is 42 less than its square. Show that the equation for this statement is quadratic.

26. A rectangle has a width w and a length which is 4 in. more than the width. If the area is 45 in.^{2}, show that the resulting equation involving the width is quadratic.

27. Under certain circumstances, the distance s that an object is above the ground is given by $s = 96t - 16t^2$, where t is the time. If $s = 144$ ft, find the resulting quadratic equation. Show that $t = 3$ s is a solution of the equation.

28. Show that the following problem leads to an equation which is quadratic and determine the equation: If the edge of a cube increases by 1 cm, the volume of the cube increases by 37 cm^3. Find the edge of the cube.

9-2 SOLUTION BY FACTORING

If a quadratic equation is in the form of Eq. (9-1) and the left-hand side is factorable, the solution is easily obtained. Once the factors are determined, each is set equal to zero; and the solution reduces to solving two simple equations of the type discussed in Chapter 3. The reason for this is that *the product of the factors is zero if at least one of the factors is zero.*

Example A Solve the equation $x^2 - x - 2 = 0$.

We first write the left-hand side in factored form. This gives

$$(x - 2)(x + 1) = 0$$

Next, each factor is set equal to zero, and the resulting equations are solved. Therefore, we have

$$x - 2 = 0 \qquad x + 1 = 0$$
$$x = 2 \qquad\qquad x = -1$$

Hence, the solution set is $\{2, -1\}$. Checking each of these values, we have

$$2^2 - 2 - 2 = 4 - 2 - 2 = 0 \quad \text{and} \quad (-1)^2 - (-1) - 2 = 1 + 1 - 2 = 0$$

Checking the solutions is always a good idea.

In Example A, if $x - 2 = 0$, the left-hand side may be written $0 \cdot (x + 1)$, and this product is zero, regardless of the value of $x + 1$. Therefore, we see that setting this factor equal to zero should give us a solution to the original equation. The same is true if $x + 1 = 0$. We must keep in mind, however, that the equation must be written in the form of Eq. (9-1) before we factor the left-hand side. This is necessary because *we must have zero on the right.* If any number other than zero appears on the right, this method is not generally applicable.

Example B Solve the equation $x^2 - 10x = -21$.

Many students would be tempted to factor the left-hand side into $x(x - 10)$ and then set $x = -21$ and $x - 10 = -21$. The "solution" set $\{-21, -11\}$ is not correct, as can be verified by substitution.

The correct procedure is first to write the equation as

$$x^2 - 10x + 21 = 0$$

which factors into $(x - 7)(x - 3) = 0$. In this manner, we obtain the correct solution set $\{3, 7\}$.

Example C Solve the equation $6x^2 = 5 - 7x$.

Following the procedure described above, we have

$$6x^2 + 7x - 5 = 0$$

$$(2x - 1)(3x + 5) = 0$$

$$2x - 1 = 0 \qquad 3x + 5 = 0$$

$$2x = 1 \qquad 3x = -5$$

$$x = \frac{1}{2} \qquad x = -\frac{5}{3}$$

Therefore, the solution set is $\{\frac{1}{2}, -\frac{5}{3}\}$.

In order to be certain no improper algebraic steps have been taken, *it is best to check each solution in the original equation*, even though it is not in the form of Eq. (9-1). Thus, we check and obtain

$$6\left(\frac{1}{2}\right)^2 \stackrel{?}{=} 5 - 7\left(\frac{1}{2}\right) \qquad 6\left(-\frac{5}{3}\right)^2 \stackrel{?}{=} 5 - 7\left(-\frac{5}{3}\right)$$

$$6\left(\frac{1}{4}\right) \stackrel{?}{=} 5 - \frac{7}{2} \qquad 6\left(\frac{25}{9}\right) \stackrel{?}{=} 5 + \frac{35}{3}$$

$$\frac{3}{2} = \frac{3}{2} \qquad \frac{50}{3} = \frac{50}{3}$$

Since the values on each side are equal, the solutions check.

In the preceding section, we mentioned that *the two solutions,* **roots** *of the equation, can be equal. This is true when the two factors are the same.* Example D illustrates this type of quadratic equation.

Example D Solve the equation $4x^2 - 12x + 9 = 0$.

Factoring this equation, we obtain

$$(2x - 3)(2x - 3) = 0$$

Setting each factor equal to zero gives

$$2x - 3 = 0$$

$$x = \frac{3}{2}$$

Since both factors are the same, the equation has a **double root** of $\frac{3}{2}$. Since each element of a set is written only once, the solution set is $\{\frac{3}{2}\}$. Checking verifies the solution.

Many students tend to solve a quadratic equation in which $c = 0$ improperly. Example E illustrates this type of quadratic equation and the error often made in its solution.

Example E Solve the equation $x^2 - 6x = 0$.

Noting that the two terms of this equation contain x, the student makes the error of dividing through by x. This results in the equation $x - 6 = 0$ and the solution $x = 6$. However, the solution $x = 0$ was lost by the division by x.

Instead of dividing through by x, we should factor the left side of the equation into

$$x(x - 6) = 0$$

By setting each factor equal to zero, we obtain

$$x = 0 \qquad x - 6 = 0$$

$$x = 6$$

Therefore, the solution set is $\{0, 6\}$. Checking verifies these roots.

A number of different kinds of verbally stated problems may lead to quadratic equations. Example F illustrates setting up and solving such a problem.

Example F The width of a certain box is 3 ft, and the height is 3 ft less than the length. If the volume of the box is 84 ft^3, find the length and height.

One good choice of unknowns is to

let h = the height of the box

Then, noting the second part of the first sentence of the statement of the problem, we have

$h + 3$ = the length of the box

(It would also be possible to let l = the length of the box and $l - 3$ equal its height.) Now using the fact that the volume is 84 ft^3, we have

$$3(h)(h + 3) = 84$$

$$3h^2 + 9h = 84$$

$$h^2 + 3h = 28$$

$$h^2 + 3h - 28 = 0$$

$$(h + 7)(h - 4) = 0$$

$$h + 7 = 0 \qquad h - 4 = 0$$

$$h = -7 \qquad h = 4$$

Since a negative value for the height has no practical significance, the height is 4 ft. This means that the length is 7 ft. Checking, we find that a box with dimensions 3 ft, 4 ft, and 7 ft has a volume of (3)(4)(7) = 84 ft^3.

Now that we are familiar with the factoring procedure to solve for the roots of a quadratic equation, we are able to find the quadratic equation if we know its roots, or its solution set. Consider the following example.

Example G Find the quadratic equation whose roots are -5 and $+7$; use x as the variable.

One of the given roots is -5. We know that if $x + 5 = 0$, then $x = -5$. Therefore, the factor for the root -5 must be $(x + 5)$. Since the other root is $+7$—and if $x - 7 = 0$, then $x = +7$—the second factor is $(x - 7)$. The product of these two factors gives us the required quadratic equation.

Thus, we form the quadratic equation as

$$(x + 5)(x - 7) = 0$$

or

$$x^2 - 2x - 35 = 0$$

Exercises 9-2

In Exercises 1 through 32, solve the given quadratic equations by factoring.

1. $x^2 - 4 = 0$
2. $4x^2 - 9 = 0$
3. $x^2 - 5x + 6 = 0$
4. $x^2 + x - 2 = 0$
5. $s^2 = 6s + 7$
6. $t^2 = 7t - 10$
7. $2x^2 + 3x = 2$
8. $3y^2 = 8 - 10y$
9. $2x^2 = 9 + 7x$
10. $2n^2 = 5n + 3$
11. $5x^2 + 4 = 21x$
12. $6x^2 - 7x - 20 = 0$
13. $t^2 + t = 12$
14. $3x^2 + 5x - 2 = 0$
15. $6n^2 - n = 2$
16. $4x^2 - 7x + 3 = 0$
17. $2v^2 + 30 = 17v$
18. $6x^2 + 13x - 5 = 0$
19. $8m^2 + 24m = 14$
20. $4x^2 + 49 = 28x$
21. $3 - x = 4x^2$
22. $x^2 + 4x + 4 = 0$
23. $x^2 + 9 = 6x$
24. $9p^2 + 20 = -27p$
25. $6x^2 + 7x = 10$
26. $x^2 - 5x = 0$
27. $s^2 + 6s = 0$
28. $4x^2 + 9 = 12x$
29. $9x^2 + 30x + 25 = 0$
30. $3r^2 = 21r$
31. $x^2 - 4a^2 = 0$ (a is constant)
32. $2x^2 - 8ax + 8a^2 = 0$ (a is constant)

In Exercises 33 through 40, find a quadratic equation for which the pair of roots is given. Use the variable given after each pair of roots.

33. $+1$ and $+1$; x
34. $+2$ and $+3$; x
35. -2 and -4; y
36. -3 and -5; y
37. -7 and $+6$; v
38. $+8$ and -5; v
39. $-\frac{1}{5}$ and $+\frac{1}{6}$; w
40. $+\frac{1}{3}$ and $-\frac{1}{2}$; w

In Exercises 41 through 50, set up quadratic equations and solve by factoring.

41. Three times a positive number subtracted from the square of the number results in 4. Find the number.
42. What is the approximate boiling point of water on the top of Mt. Baker in Washington (altitude is about 10,800 ft)? See Exercise 85 of Section 5-6.
43. The perimeter of a rectangular field is 100 m and the area is 600 m^2. Find the dimensions of the field.
44. The distance an object falls due to gravity is given by $s = 16t^2$, where s is the distance in feet and t is the time in seconds. How long does it take an object to fall 36 ft?
45. The floor of a certain room is covered with 324 square tiles. If the side of each tile were 1 in. longer, it would take only 256 tiles to cover the floor. Find the length of a side of each tile.
46. Under certain conditions, the motion of an object suspended by a helical spring requires the solution of the equation

$$D^2 + 8D + 12 = 0$$

Solve for D.

47. Under specified conditions, the deflection of a beam requires the solution of the equation

$$4Lx - x^2 - 4L^2 = 0$$

where x is the distance from one end and $2L$ is the length of the beam. Solve for x in terms of L.

48. Three times a number subtracted from twice the square of the number results in 2. Find the number.

49. The perimeter of a rectangular table is 200 cm and the area is 2500 cm^2. Find the dimensions of the table.

50. Four times a negative number added to the square of the number results in -4. Find the number.

9-3 SOLUTION BY FORMULA

Many quadratic equations cannot be solved by factoring. It is, however, possible to derive a formula by which any quadratic equation may be solved. The method of deriving this formula requires that we write the equation in a different form in the intermediate steps. The following example follows the basic line of reasoning we shall use to derive the formula.

Example A Solve the equation $x^2 + 4x - 21 = 0$.

This equation can easily be solved by factoring. However, there is an alternative *method* in which we are interested here.

The terms $x^2 + 4x$ are two of the terms we may obtain by expanding $(x + 2)^2$. In fact, if 4 is added to $x^2 + 4x$, we have $(x + 2)^2$ exactly. Therefore, let us write the given equation as

$$x^2 + 4x \qquad = 21$$

$$x^2 + 4x + 4 = 21 + 4$$

This means that another form of the equation is

$$(x + 2)^2 = 25$$

If we now take square roots of each side, we have

$$x + 2 = +5 \quad \text{or} \quad x + 2 = -5$$

Solving each of these equations gives us the solution set $\{3, -7\}$, which can be verified by checking.

The key to this solution is that the left-hand side is a perfect square, which allows us to solve simple equations after the square root is found.

The method used to find the solution in Example A is called **completing the square.** The basic idea of this method is to create the perfect square of a simple algebraic expression on the left, allowing the right to have any numerical value. Since $(x + a)^2 = x^2 + 2ax + a^2$, we see that the square of one-half the coefficient of x added to $x^2 + 2ax$ gives such a perfect square. We will now use this method to derive the general formula for solving quadratic equations.

First, we start with the general quadratic equation

$$ax^2 + bx + c = 0 \tag{9-1}$$

Next we divide through by a, obtaining

$$x^2 + \frac{b}{a}x + \frac{c}{a} = 0$$

This is done since we want the coefficient of the x^2-term to be 1. Now let us subtract c/a from each side, which gives

$$x^2 + \frac{b}{a}x \qquad = -\frac{c}{a}$$

Note that we have left space for the third term when we have completed the square on the left.

Now we take one-half of b/a, which is $b/2a$, square it, obtaining $b^2/4a^2$, and add this square to each side of the equation. Therefore, we have

$$x^2 + \frac{b}{a}x + \frac{b^2}{4a^2} = \frac{b^2}{4a^2} - \frac{c}{a}$$

The left-hand side is now the perfect square of $(x + b/2a)$. Indicating this and combining fractions on the right, we have

$$\left(x + \frac{b}{2a}\right)^2 = \frac{b^2 - 4ac}{4a^2}$$

Taking the square root of each side, we have

$$x + \frac{b}{2a} = \frac{\sqrt{b^2 - 4ac}}{2a} \quad \text{and} \quad x + \frac{b}{2a} = \frac{-\sqrt{b^2 - 4ac}}{2a}$$

Subtracting $b/2a$ from each side, combining fractions, and writing the result as one basic equation, we have

$$\boxed{x = \frac{-b \pm \sqrt{b^2 - 4ac}}{2a}} \tag{9-2}$$

which is known as the **quadratic formula.** If an equation is in the form of Eq. (9-1), then direct substitution of the appropriate quantities into Eq. (9-2) gives the solutions. Remember, *there are two solutions indicated in Eq. (9-2), one each for the* $+$ *sign and* $-$ *sign of the* $\pm$ *sign.* Therefore, the solution set is

$$\left\{\frac{-b + \sqrt{b^2 - 4ac}}{2a}, \frac{-b - \sqrt{b^2 - 4ac}}{2a}\right\}$$

The following examples illustrate the use of Eq. (9-2).

Example B Solve the equation $x^2 - 5x + 6 = 0$ by the quadratic formula.

Since the equation is in the proper form, we recognize that $a = 1$, $b = -5$, and $c = 6$. Therefore,

$$x = \frac{-(-5) \pm \sqrt{(-5)^2 - 4(1)(6)}}{2(1)}$$

$$= \frac{5 \pm \sqrt{25 - 24}}{2} = \frac{5 \pm 1}{2}$$

Therefore,

$$x = \frac{5 + 1}{2} = 3 \quad \text{and} \quad x = \frac{5 - 1}{2} = 2$$

and the solution set is $\{2, 3\}$.

These roots could have been found by factoring, but we see that the same results are obtained by use of the formula. The roots are easily checked in the original equation.

Example C Solve $2x^2 + 7x = 3$ by the quadratic formula.

Before the formula can be used, the equation must be put in the form of Eq. (9-1). This gives

$$2x^2 + 7x - 3 = 0$$

and therefore $a = 2$, $b = 7$, and $c = -3$. The solutions are

$$x = \frac{-7 \pm \sqrt{49 - 4(2)(-3)}}{2(2)}$$

$$x = \frac{-7 \pm \sqrt{49 + 24}}{4} = \frac{-7 \pm \sqrt{73}}{4}$$

therefore, the solution set is

$$\left\{\frac{-7 + \sqrt{73}}{4}, \frac{-7 - \sqrt{73}}{4}\right\}$$

This form of the result is generally acceptable. However, if decimal approximations are required, then using $\sqrt{73} = 8.544$, we have

$$x = \frac{-7 + 8.544}{4} = 0.386 \quad \text{and} \quad x = \frac{-7 - 8.544}{4} = -3.886$$

Example D Solve the equation $3x^2 = 2x - 5$ by the quadratic formula.

Putting the equation in the proper form as $3x^2 - 2x + 5 = 0$, we have $a = 3$, $b = -2$, and $c = 5$. Therefore,

$$x = \frac{-(-2) \pm \sqrt{(-2)^2 - 4(3)(5)}}{2(3)} = \frac{2 \pm \sqrt{4 - 60}}{6}$$

$$= \frac{2 \pm \sqrt{-56}}{6} = \frac{2 \pm 2\sqrt{-14}}{6} = \frac{1 \pm \sqrt{-14}}{3}$$

Since $\sqrt{-14} = i\sqrt{14}$, we see that the result contains imaginary numbers, and the solution set is

$$\left\{\frac{1 + i\sqrt{14}}{3}, \frac{1 - i\sqrt{14}}{3}\right\}$$

Example E A uniform strip is mowed around the four sides of a field 80 m by 100 m. A rectangular section of 6000 m^2 remains unmowed. Find the width of the strip.

A figure will help us visualize the situation. In Fig. 9-1, we have let x = the width of the strip. We see that the area which remains unmowed is given by $(80 - 2x)(100 - 2x) = 6000$.

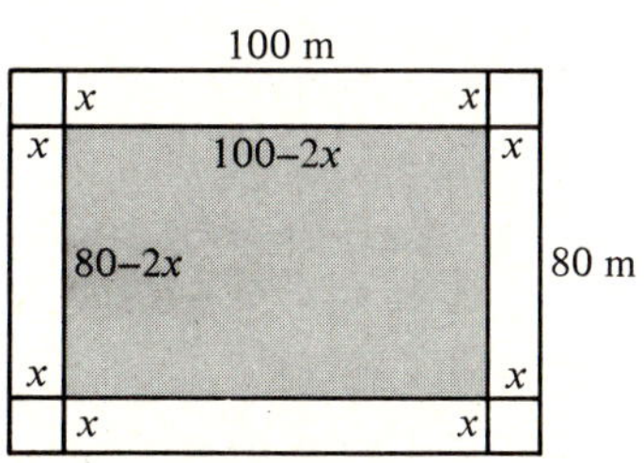

Figure 9-1

Simplifying, we have

$$8000 - 200x - 160x + 4x^2 = 6000$$
$$4x^2 - 360x + 2000 = 0$$
$$x^2 - 90x + 500 = 0$$

Solving this last equation by the quadratic formula, we have

$$x = \frac{90 \pm \sqrt{8100 - 2000}}{2} = \frac{90 \pm \sqrt{6100}}{2} = \frac{90 \pm 10\sqrt{61}}{2} = 45 \pm 5\sqrt{61}$$

Since $\sqrt{61} = 7.810$, it follows that $5\sqrt{61} = 39.05$. Thus, $x = 5.95$ and $x = 84.05$. However, the strip mowed would be wider than the field if $x = 84.05$, which means this answer cannot be true. If the strip is 5.95 m, the unmowed portion of the field is 88.1 m by 68.1 m, and this area is 6000 m^2. This means that the strip is about 5.95 m wide.

Exercises 9-3

In Exercises 1 through 32, solve the given quadratic equations by use of the quadratic formula.

1. $x^2 - x - 6 = 0$ **2.** $x^2 + 2x - 15 = 0$ **3.** $s^2 - 7s + 12 = 0$

4. $n^2 + 10n + 9 = 0$ **5.** $2x^2 + x - 1 = 0$ **6.** $3y^2 - 5y + 2 = 0$

7. $x^2 + 5x + 3 = 0$ **8.** $x^2 - 3x - 1 = 0$ **9.** $4t^2 = 8t - 3$

10. $3x^2 = x + 10$ **11.** $4x^2 - 9 = 0$ **12.** $3p^2 = 16$

13. $4x^2 - 12x = 7$ **14.** $2x^2 + 7x + 2 = 0$ **15.** $x^2 + x + 2 = 0$

16. $r^2 - 2r + 5 = 0$ **17.** $4x^2 = 8 - 2x$ **18.** $3u^2 = 18 - 6u$

19. $3t^2 - 5t = 0$ **20.** $2v^2 + 3v = 0$ **21.** $2t^2 - 3t = -8$

22. $4x^2 = 9x - 6$ **23.** $3q^2 = 10q$ **24.** $x^2 - 8x = 0$

25. $x^2 - 8 = 5x$ **26.** $t^2 - t = 8$ **27.** $7 - 15x = -2x^2$

28. $10t + 8 = 3t^2$ **29.** $n^2 - 2n = 6$ **30.** $s^2 = 4s + 2$

31. $6a^2x^2 + 11ax + 3 = 0$ (a is constant) **32.** $2x^2 + (a + 2)x + a = 0$ (a is constant)

Solve the equations of Exercises 33 through 38 by completing the square. (In Exercise 38, use p, q, and r as constants.)

33. $x^2 + 2x - 3 = 0$ **34.** $x^2 + 6x - 7 = 0$ **35.** $x^2 - 4x + 3 = 0$

36. $x^2 = 8x - 5$ **37.** $2x^2 + 3x + 1 = 0$ **38.** $px^2 + qx + r = 0$

Solve the quadratic equations obtained in Exercises 39 through 44 by means of the quadratic formula.

39. The side of one square exceeds the side of another square by 3 ft. The total area of the two squares is 65 ft^2. Find the side of each square.

40. Find two consecutive odd integers whose product is 195.

41. Under certain conditions, the distances s that an object is above the ground is given by $s = 100t - 16t^2$, where t is the time in seconds. After what time is the object 80 ft above the ground?

42. An object is sold for \$10, which includes a profit of as many percent as the number of dollars of original cost. Find the original cost.

43. Under specified conditions, the power developed in an element of an electric circuit is $P = EI - RI^2$, where P is the power, E is a specified voltage, and R is a specified resistance. Assuming that P, E, and R are constants, solve for I.

44. How wide a strip around the outside of a field 400 m by 600 m must a farmer mow in order to mow half the area?

9-4 SUMMARY

In this chapter, we have discussed the solution of the general quadratic equation, which has the standard form

$$ax^2 + bx + c = 0 \tag{9-1}$$

The two methods that were used primarily to solve this equation were (1) factoring and (2) the quadratic formula. In Section 9-3, we derived the quadratic formula as

$$x = \frac{-b \pm \sqrt{b^2 - 4ac}}{2a} \tag{9-2}$$

A third method for solving quadratic equations, completing the square, was used principally for deriving the quadratic formula.

We saw that the solution of this type of equation is in general a set of two numbers. It is possible that these numbers contain imaginary values. Also, it is possible that the numbers are the same, which means the equation has a double root and the solution set has one member. We also saw how to write a quadratic equation when the roots are known.

The reader should be familiar with the following terms introduced in this chapter.

Completing the square
Double root
Quadratic equation
Quadratic formula
Roots of an equation

The following exercises provide a means of reviewing the methods of solution developed in this chapter.

Exercises 9-4

In Exercises 1 through 12, solve the given quadratic equations by factoring.

1. $x^2 + 7x + 6 = 0$
2. $x^2 + 11x - 12 = 0$
3. $2x^2 - 7x + 5 = 0$
4. $4x^2 = 8x - 3$
5. $6n^2 = 35n + 6$
6. $9s^2 - 5s = 0$
7. $16x^2 - 24x + 9 = 0$
8. $4p^2 + 49 = 28p$
9. $18x^2 + 7x = 0$
10. $9r^2 - 55r + 6 = 0$
11. $t^2 + t = 110$
12. $2x^2 + 39x = 20$

In Exercises 13 through 20, solve the given quadratic equations by (a) factoring and (b) the quadratic formula.

13. $x^2 + 4 = 5x$
14. $2x^2 + 3 = 5x$
15. $6m^2 + 5 = 11m$
16. $7x^2 + 24x = 16$
17. $3x^2 - 6x + 3 = 0$
18. $5x^2 + 40x = 0$
19. $9x^2 - 25 = 0$
20. $9x = 9x^2 + 2$

In Exercises 21 through 32, solve the given quadratic equations by use of the quadratic formula.

21. $x^2 + x + 1 = 0$
22. $x^2 + x - 1 = 0$
23. $2x^2 - 5x + 1 = 0$
24. $3r^2 + 4r - 2 = 0$
25. $y^2 - 6y = 6$
26. $5t^2 = 5t + 3$
27. $2x^2 + 4x + 3 = 0$
28. $3u^2 - u = -9$
29. $x^2 + 4 = 0$
30. $4y^2 = -25$
31. $4x^2 = 36$
32. $5p^2 = 35p$

In Exercises 33 through 36, solve the given quadratic equations by completing the square.

33. $x^2 + 2x + 2 = 0$ **34.** $x^2 - 4x = 8$ **35.** $2x^2 - 5x = 3$ **36.** $3x^2 + 6x + 1 = 0$

Many equations containing algebraic fractions may be solved as quadratic equations. By first clearing fractions (see Section 6-5), the quadratic form may be derived. Exercises 37 through 40 illustrate this type of equation. In each, solve for x. (≠ means "does not equal.")

37. $\frac{1}{x} + x = 4 \quad (x \neq 0)$

38. $\frac{1}{3} + \frac{1}{x} = x \quad (x \neq 0)$

39. $\frac{1}{x} - \frac{2}{x - 1} = 2 \quad (x \neq 0, x \neq +1)$

40. $\frac{1}{2}(x + 2) - \frac{3}{x} = \frac{3}{2} \quad (x \neq 0)$

In Exercises 41 through 48, find a quadratic equation for which the pair of roots is given. Use the variable given after each pair of roots.

41. $+1$ and $+3$; x **42.** $+3$ and $+5$; y **43.** -3 and $+6$; x **44.** $+4$ and -7; y

45. $+\frac{1}{2}$ and $-\frac{1}{3}$; p **46.** $-\frac{1}{5}$ and $+\frac{1}{7}$; q **47.** $+4$ and 0; t **48.** -2 and 0; s

In Exercises 49 through 58, solve the resulting quadratic equations by any appropriate method.

49. Two numbers differ by 6. Their product is 91. Find the numbers.

50. Two consecutive positive even integers have a product of 224. Find these integers.

51. A company found that its profit for producing x units weekly was $p = 25x - x^2 - 100$. For which values of x is the profit zero?

52. If the speed of a car is increased by 20 mi/h, it can travel 300 mi in 4 h less time. Find its speed.

53. The length of a rectangle is 1 more than the width. If 3 is added to the length and 1 to the width, the new area is twice the original area. Find the dimensions of the original rectangle.

54. An object is sold for \$24, which includes a profit of twice as many percent as the number of dollars of original cost. Find the original cost.

55. If the radius of a circle is doubled, its area increases by 192π m^2. Find the radius of the circle.

56. Under certain frictional conditions, the motion of an object on a spring requires the solution of the equation $D^2 + 2rD + k^2 = 0$, where r and k are constants of the system. Solve for D.

57. A general formula for the distance s traveled by an object, given an initial velocity v and acceleration a in time t, is

$$s = vt + \frac{1}{2}at^2$$

Solve for t.

58. In the theory concerning the motion of biological cells and viruses, the equation $n = 2.5p - 12.6p^2$ is used. Solve for p in terms of n.

10
FUNCTIONS AND GRAPHS

10-1 FUNCTIONS

In the earlier chapters, we established the basic operations of algebra and discussed the solution of certain types of equations. Furthermore, we introduced a few basic formulas from geometry and other fields. In these formulas and equations, one quantity can be expressed in terms of one or more other quantities by the use of the basic operations. Therefore we can see that the various quantities of a formula have a specific interrelation as denoted by the formula.

If a scientific experiment were to be performed to determine whether or not a relationship exists between the distance an object falls and the time of its fall, observation of the results should indicate (approximately at least) that $s = 16t^2$, where s is the distance (in feet) and t is the time (in seconds) of fall. The experiment thus would show that the distance and the time were related.

In general, the reaction time of an individual depends on his age. Reaction times decrease until a person is about 20 years old, and then increase as he ages. Thus, it is possible to set up an approximate equation or a chart to show the relation of reaction time and age.

Considerations such as these lead us to the important mathematical concept of a **function.** In general,

> ***if two variables are so related that for each value of the first variable a single value of the second variable can be determined, the second variable is a function of the first variable.***

The first variable is called the **independent variable,** *and the second is called the* **dependent variable.** We shall restrict our attention to functions that are defined by formulas, equations, and graphical representations.

Example A The formula for the area of a circle in terms of its radius is $A = \pi r^2$. Here A is a function of r. The variable r is the independent variable, and A is the dependent variable. We can see that the words "dependent" and "independent" are appropriately chosen. As the formula is written, we choose a value of r; once this is done, the value of A depends upon that choice. Therefore, in this example, we say that the area of a circle is a function of the radius of the circle.

Example B If a boy is 22 years younger than his father, the boy's age y is a function of his father's age x, given by $y = x - 22$. Here x is the independent variable, and y is the dependent variable.

It may be well to ask: What is the advantage of introducing the word function, since the formulas we have been dealing with seem sufficient? *In mathematics, the word "function" specifically refers to the operation that is performed on the independent variable to find the dependent variable.* Consider Example C.

Example C The distance s (in feet) which an object falls in t seconds is given by $s = 16t^2$. The electric power developed in a 16-ohm resistor by a current of i amperes is given by $P = 16i^2$. Here, s is a function of t and P is a function of i. However, the function is the same. That is, in order to evaluate the dependent variable, the independent variable is squared and this result is multiplied by 16. Even though the letters are different, the *operation* is the same.

For convenience of notation and to emphasize the importance of the operational meaning of a function, the phrase "function of x" is written $f(x)$. (*This is a special notation, and does not mean f times x.*) Therefore, to state that "y is a function of x," we write $y = f(x)$.

Example D For the equation $y = x^2 - 5$, we say that $y = f(x)$, where $f(x) = x^2 - 5$. Thus, $y = x^2 - 5$ and $f(x) = x^2 - 5$ are different ways of writing the same function.

A particularly important use of the functional notation $f(x)$ is to designate the value of a function for a specified value of the independent variable. *Thus, "the value of the function $f(x)$ when $x = a$" is expressed as $f(a)$.*

Example E If $f(x) = 5 - 2x$, $f(0)$ is the value of the function for $x = 0$. Thus,

$$f(0) = 5 - 2(0) = 5$$

To find $f(-1)$, we have

$$f(-1) = 5 - 2(-1) = 5 + 2 = 7$$

Thus, $f(0) = 5$ and $f(-1) = 7$. We note that to find the function of the value specified on the left, we substitute this value into the function on the right. Also, if we state that $y = f(x)$, then $y = 5$ for $x = 0$ and $y = 7$ for $x = -1$.

Example F
If $f(x) = 4x - 2x^2$, find $f(-3)$.
To find $f(-3)$, we substitute -3 for x in the function. This gives us

$$\begin{aligned} f(-3) &= 4(-3) - 2(-3)^2 \\ &= -12 - 2(9) \\ &= -12 - 18 \\ &= -30 \end{aligned}$$

It is possible to establish many functions from verbal statements. Such functions may be based upon known formulas or a proper interpretation of a given statement.

Example G The length of a rectangle is three times its width. Express the perimeter as a function of the width.

The perimeter of a geometric figure is the distance around it. Therefore, the perimeter of a rectangle is twice the width plus twice the length. This may be expressed as $p = 2l + 2w$, where l is the length and w is the width. For the rectangle under consideration, we know that $l = 3w$. This means that

$$\begin{aligned} p = 2l + 2w &= 2(3w) + 2w \\ &= 6w + 2w \\ &= 8w \end{aligned}$$

Thus, the perimeter as a function of its width, under the given condition, is $p = 8w$.

Another comment regarding functions is in order at this time. Not all functions are valid for all values of the independent variable, nor are all values of the dependent variable always possible. Example H illustrates this point.

Example H In Example B, there is no real meaning for negative values of the boy's age or that of his father. Thus, we would restrict values of x to those 22 and above, and the values of y to 0 and above. (That is, $x \geq 22$ and $y \geq 0$.)

If we restrict our attention to real numbers, $f(x) = \sqrt{x - 1}$ is valid only for values of x greater than or equal to 1 ($x \geq 1$).

If $f(x) = 1/x$, all values of x are possible except $x = 0$, for $x = 0$ would necessitate division by zero.

Finally, we again note in the definition of a function that *only a single value of y can be determined for a given value of x. If for any value of the first variable x more than one value of the second variable y were possible, we would have what is known as a* **relation,** *not a function.* Consider the following example.

Example I For $y^2 = x$, we note that if $x = 4$, then $y = 2$ or $y = -2$. Since more than one value of y is possible for a given value of x, this equation defines a relation, not a function.

Exercises 10-1

In Exercises 1 through 4, identify the dependent variable and the independent variable of each of the given functions. In Exercises 3 and 4, c and a are constants.

1. $y = 3x^4$ **2.** $s = -16t^2$ **3.** $p = c/V$ **4.** $v = a(1.05)^t$

In Exercises 5 through 8, state the basic operation of the function that is to be performed on the independent variable. (See Example C.)

5. $f(x) = 3x$ **6.** $f(y) = y + 3$ **7.** $f(r) = 2 - r^2$ **8.** $f(s) = s + s^2$

In Exercises 9 through 12, indicate the function expressed by each of the given equations by replacing the dependent variable by the $f(x)$ type of notation. (See Example D.)

9. $y = 5 - x$ **10.** $F = 6q^2$ **11.** $v = t^2 - 3t$ **12.** $s = \dfrac{5}{1 - r}$

In Exercises 13 through 28, find the indicated values of the given functions.

13. $f(x) = x; f(0), f(1), f(-1)$

14. $f(x) = x + 2; f(0), f(1), f(-2)$

15. $f(t) = 2t - 1; f(4), f(-2)$

16. $f(r) = 2 - r; f(3), f(-3)$

17. $f(x) = 3x^2 - 2; f(0), f\left(\frac{1}{2}\right)$

18. $f(x) = 3 - x^2; f(2), f(0.3)$

19. $f(s) = s^3; f(-1), f(2)$

20. $f(p) = p^3 + 2p - 1; f(-1), f(a)$

21. $f(x) = x - 2x^2; f(a^2), f\left(\frac{1}{a}\right)$

22. $f(v) = v + \frac{1}{v}; f\left(\frac{1}{5}\right), f(5)$

23. $f(q) = \frac{q}{q - 3}; f(-3), f(3)$

24. $f(x) = x^4; f(-2), f\left(\frac{1}{2}\right)$

25. $f(z) = z^2 - z; f(2), f(-2)$

26. $f(x) = 2x - 3x^2; f(-1), f(-3)$

27. $f(t) = 3t - t^3; f(1), f(-2)$

28. $f(x) = x^3 - x^4; f(3), f(-3)$

In Exercises 29 through 32, determine whether the equation illustrates a function or a relation.

29. $y = x^2 + 1$ **30.** $y = x^3$ **31.** $y = 1 - x^3$ **32.** $y = x^4$

In Exercises 33 through 44, determine the required functions.

33. Express the perimeter of a square as a function of a side s.

34. Express the volume of a cube as a function of an edge e.

35. The number of tons t of air pollutants emitted by burning a certain fuel is 0.02 times the number of tons n of the fuel which are burned. Write the function given by this statement.

36. Express the simple interest I on \$100 at an interest rate of 5% per year as a function of the number of years, t.

37. Express the distance in miles s traveled for 3 h as a function of the velocity v in miles per hour.

38. Express the net profit P made on selling 10 items at a price p, if they cost a dollars each, as a function of p.

39. The voltage V across an electric resistor varies directly as the current i in the resistor. If the voltage is 10 V when the current is 2 A, express the current as a function of the voltage.

40. The stopping distance d of a car varies directly as the square of its speed s. Express d as a function of s if $d = 120$ ft when $s = 40$ mi/h.

41. Express a taxi fare F as a function of the distance d traveled. The charge is 35¢ for the first $\frac{1}{5}$ km and 10¢ for each $\frac{1}{5}$ km thereafter.

42. One electronic data-processing card-sorter can sort 1500 more cards per minute than a second sorter. Express the rate R_1 of the first sorter as a function of the rate R_2 of the second sorter.

43. The digits of a number written in base b ($b > 3$) are 132. Express the number N as a function of b.

44. Express the sum S of three consecutive integers as a function of n if n is the first of these integers.

10-2 TYPES OF GRAPHS

One of the most valuable ways of representing data and functions is by means of graphical methods. Graphs present a way of "seeing" the relationship that may exist for sets of data or between sets of numbers. A graph usually allows one to obtain a quick and meaningful understanding of a relationship; for this reason, graphs are used extensively in many fields, including business, science, and social studies.

We shall discuss two important types of graphs. In this section, we shall discuss those that represent statistical sets of data. In the remaining sections of the chapter, we shall develop the basic graphical representation of functions.

Of the numerous graphical methods used for representing data, we shall discuss three: the **circle graph,** the **bar graph,** and the **broken-line graph.** *The first of these, the circle graph, is particularly useful for showing the relationship of a whole category to the various parts of the category* by determining the percentage that each part is of the whole. Since there are 360° in a circle, each percent multiplied by 360° will give the number of degrees of the circle to be used for each part. Consider Example A.

Example A A family has a yearly income of $24,000. Of this, $3600 is paid out in taxes, $7200 is spent for food, $4800 is spent on shelter, $3600 is spent on clothing, $2400 is spent on entertainment, and $2400 is saved. (Some family!) Now, 3600 is 15% of 24,000, and 15% of 360° is 54°. Therefore, the part of the circle used to represent taxes will have an angle of 54°. In this way, we may set up Table 10-1, which in turn is used for the graph (see Fig. 10-1).

TABLE 10-1

Item	% of Income	Angle
Taxes	15	54°
Food	30	108°
Shelter	20	72°
Clothing	15	54°
Entertainment	10	36°
Savings	10	36°

Figure 10-1

A type of graph valuable for showing the relative magnitude of discrete data is the bar graph. In constructing a bar graph, one chooses a scale suitable for the data and then determines the length of each bar. The bars are then constructed either horizontally or vertically and are clearly labeled. The following is an example of a bar graph.

Example B Among the more important canals in the world are the Panama Canal (51 mi), the Suez Canal (103 mi), the Kiel Canal (61 mi), and the King Albert Canal (80 mi). Draw a bar graph of these data.

For the graph to be easily interpreted, the scale should be as large as is reasonable. In this case, we shall choose a scale of $\frac{1}{4}$ in. for 10 mi. With this scale, we obtain Table 10-2 for the approximate lengths of the bars to be used to represent the lengths of the canals (see Fig. 10-2).

TABLE 10-2

Canal	Length, mi	Length of bar, in.
Panama	51	1.3
Suez	103	2.6
Kiel	61	1.5
Albert	80	2.0

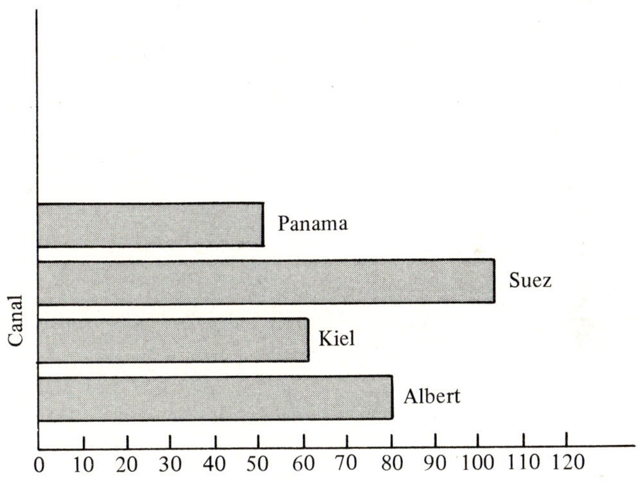

Figure 10-2

The last type of graph we shall consider in this section is the broken-line graph. This type of graph is constructed by connecting a series of points by straight-line segments. *It is used primarily to show how one variable changes as another variable (the equivalent to the independent variable of a function) changes continuously.* Appropriate scales should be chosen for each of the variables and properly labeled. Usually, the scale of the independent variable is placed horizontally, and the scale of the other variable (the dependent variable) is placed vertically. The points are then located and joined.

Example C The temperature recorded for a certain city from noon to midnight is shown in the table below. The broken-line graph of these data is shown in Fig. 10-3.

Temperature (°C)	15	18	20	26	25	23	20	17	12	13	10	8	6
Time (p.m.)	12	1	2	3	4	5	6	7	8	9	10	11	12

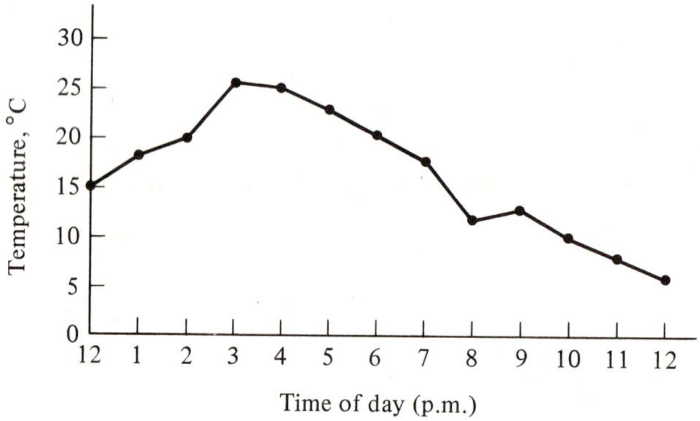

Figure 10-3

Exercises 10-2

In Exercises 1 through 6, make a circle graph to show the given data.

1. A school has a student body of which 40% are freshmen, 25% are sophomores, 20% are juniors, and 15% are seniors.
2. Of the surface area of the earth, 71% is covered by water, 9% by the continent of Asia, 6% by the continent of Africa, 5% by the continent of North America, 3% by the continent of South America, and 6% by the remaining land areas, including the continents of Europe, Antarctica, and Australia.
3. A certain type of brass is made with 85% copper, 5% zinc, 5% tin, and 5% lead.
4. In the Presidential election of 1912, Woodrow Wilson received 6,300,000 votes, Theodore Roosevelt received 4,100,000 votes, William Howard Taft received 3,500,000 votes, and the combined total of other candidates was 1,100,000 votes.
5. Of accidental deaths in the United States in a typical year, about 50,000 are due to motor vehicle accidents, 30,000 are due to home accidents, 15,000 are due to work accidents, and 15,000 are due to other types of accidents.
6. Of each 100 persons, the approximate numbers having each of the four blood groups are as follows: group O (45), group A (40), group B (10), group AB (5).

In problems 7 through 12, make a bar graph to represent the given data.

7. The number of calories per one-cup serving of various beverages is: apple juice (125), orange juice (110), ginger ale (80), milk (165), cocoa (230).

8. In a series of tests given to entering college students, it was found that the following numbers had their greatest apparent aptitude in the indicated areas: art and music (110), business (240), languages (100), social sciences (280), science and engineering (80).

9. The sales (in millions of dollars) of a corporation during a given year were found to be from: heavy machinery (58), light machinery (30), chemicals (14), special tools (10), and other sources (15).

10. In testing a set of electric resistors, it was found that (to the nearest ohm) 5 had a resistance of 5 ohms, 8 a resistance of 6 ohms, 17 a resistance of 7 ohms, 4 a resistance of 8 ohms, and 2 a resistance of 9 ohms.

11. The approximate frequencies of use of the vowels of the alphabet are as follows (based on e as 100): a (73), e (100), i (70), o (67), u (30).

12. List the prime integers between 1 and 60. Represent the number of primes having a final digit of 0, 1, 2, 3, 4, 5, 6, 7, 8, and 9.

In Exercises 13 through 18, construct a broken-line graph to represent the given data.

13. A thermocouple is a special type of voltage source for an electric circuit. The voltage is a function of the temperature. An experiment on a particular thermocouple gave the following results.

Voltage, volts	0	2.9	5.9	9.0	12.3	15.8
Temperature, °C	0	10	20	30	40	50

14. The value of a \$100 deposit earning 5% interest compounded annually is given in the following table.

Value, \$	100	105	110	116	122	128	134	141
t, years	0	1	2	3	4	5	6	7

15. Of Shakespeare's major plays, 6 were completed by 1595, 7 from 1596 to 1600, 5 from 1601 to 1605, 3 from 1606 to 1610, and 3 after 1610.

16. In an experiment in which pressure and volume of air were measured, the following results were obtained.

Volume, in.^{3}	200	180	160	140	120	100
Pressure, atm	2.0	2.2	2.6	3.0	3.6	4.5

17. A travel agency advertised that a two-week vacation to England would cost the amount shown in the table, if the vacation started on the given date:

Date	April 15	May 15	June 15	July 15	August 15	Sept. 15	Oct. 15
Cost	\$650	\$690	\$760	\$780	\$750	\$700	\$740

18. The circulation of a given magazine is given in the following table.

Year	1940	1950	1960	1970	1980
Circulation	200,000	350,000	700,000	900,000	950,000

In Exercises 19 through 22, make or obtain the required graphs.

19. Toss 6 pennies 50 times, each time recording the number of heads which appear. Represent on a bar graph the total number of times which 0, 1, 2, 3, 4, 5, and 6 heads appear.

20. Toss two dice 100 times, recording at each toss the sum which appears. Draw a broken-line graph of each sum and the total number of times it appears.

21. From the financial section of a recent newspaper find and clip out a graph which is of a type (at least approximately) discussed in this section.

22. From a newspaper or magazine (from other than the financial section), find and clip out a graph which is of a type discussed in this section.

10-3 THE RECTANGULAR COORDINATE SYSTEM

In this section, we shall develop the basic system on which we will be able to represent the graph of a function. Just as the graphical methods shown in the previous section help us to visualize the relation in statistical data, the graph of a function is often very useful as a means of depicting the basic characteristics of the function.

In Chapter 1, we showed how numbers can be represented on a line. Since a function represents two related sets of numbers, it is necessary to use such a number line for each of these sets. This is most conveniently done if the lines intersect at right angles in the same manner as on a broken-line graph.

Therefore, one line is placed horizontally and labeled the **x-axis.** *It is labeled just as the number line in Chapter 1 was labeled, with a point chosen as* 0 (*zero*), *called the* **origin,** *and with the positive values to the right and the negative values to the left of* 0. It is customary to use the numbers on the x-axis for the values of the *independent* variable.

Through the origin and perpendicular to the x-axis, another line is placed and labeled the **y-axis.** *The origin is also* 0 *for values on the y-axis with positive values above and negative values below.* The y-axis is customarily used for values of the *dependent* variable.

The x-axis and the y-axis as described constitute what is known as the **rectangular coordinate system.** *The four parts into which the plane is divided by the axes are known as* **quadrants,** and are numbered as shown in Fig. 10-4.

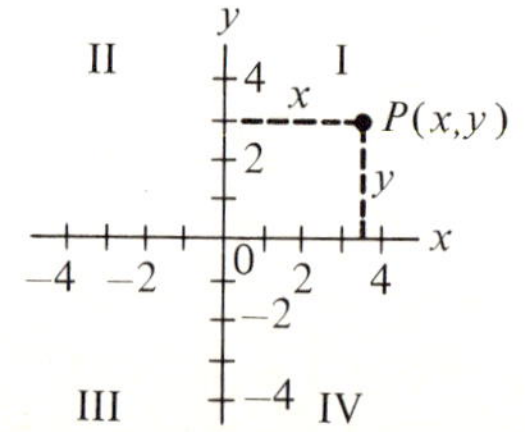

Figure 10-4

Any point P in the plane of the rectangular coordinate system is designated by the pair of numbers (x, y), always written in this order. Here, x is the value of the independent variable and y is the corresponding value of the dependent variable. *It can be seen that the x-value (called the* **abscissa**) *is the perpendicular distance of P from the y-axis, and the y-value (called the* **ordinate**) *is the perpendicular distance of P from the x-axis. The values of x and y are called the* **coordinates** *of the point P.*

Example A The positions of points $P(4, 1)$, $Q(-2, 5)$, $R(-3, -2)$, $S(4, -3)$, $T(-5, 0)$, and $V(0, 2)$ are shown in Fig. 10-5. Note that this representation of points allows *only one point for any pair of values* (x, y). Note also that this illustrates the rectangular formation of coordinates.

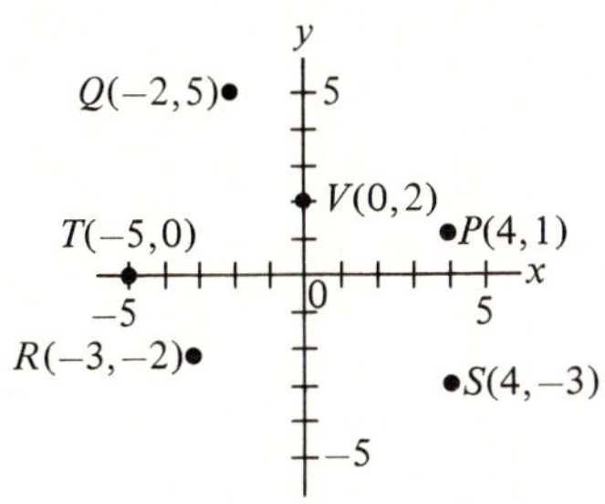

Figure 10-5

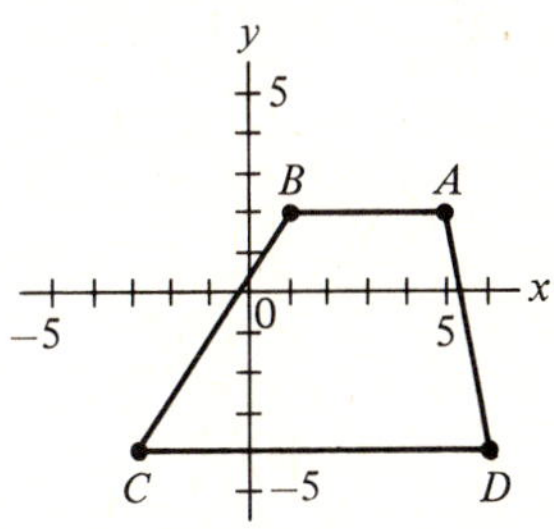

Figure 10-6

Example B If we join the points $A(5, 2)$, $B(1, 2)$, $C(-3, -4)$, and $D(6, -4)$ in order, we see that we have a four-sided geometric figure. The upper side is seen to be parallel to the x-axis since the y-coordinates are the same. This is also true of the lower side. Since they are both parallel to the x-axis, they are parallel to each other. Such a figure is known as a *trapezoid.* See Fig. 10-6.

Example C What is the sign of the ratio of the abscissa to the ordinate of a point in the second quadrant?

Since any point in the second quadrant is above the x-axis, the y-value (ordinate) is positive. Also, since any point in the second quadrant is to the left of the y-axis, the x-value (abscissa) is negative. The ratio of a negative number to a positive number is negative, which is the required answer.

Example D In what quadrant does a point (a, b) lie if $a < 0$ and $b < 0$?

A point for which the abscissa is negative (which is the meaning of $a < 0$) must be in either the second or third quadrant. A point for which the ordinate is negative is in either the third or fourth quadrant. Therefore, the required answer is the third quadrant. (We can see that this is the intersection of the two sets of quadrants.)

Exercises 10-3

In Exercises 1 through 4, determine (at least approximately) the coordinates of the points specified in Fig. 10-7.

1. A, B
2. C, D
3. E, F
4. G, H

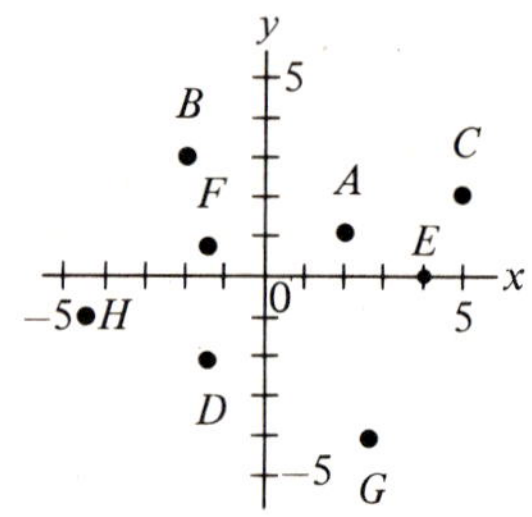

Figure 10-7

In Exercises 5 through 8, plot (at least approximately) the given points.

5. $A(0, 3)$, $B(3, 0)$
6. $A(-3, -6)$, $B(2, -4)$
7. $A(-1, 4)$, $B\left(\frac{1}{2}, 5\right)$
8. $A\left(-\frac{2}{3}, \frac{2}{3}\right)$, $B(0, -4)$

In Exercises 9 through 12, draw the geometric figures, the coordinates of whose vertices are given (do not cross lines).

9. $A(0, 3)$, $B(5, -1)$, $C(2, 3)$
10. $A(-1, -2)$, $B(0, -3)$, $C(1, 3)$, $D(-2, 4)$
11. $A(-2, -2)$, $B(5, -2)$, $C(5, 4)$, $D(-2, 4)$
12. $A(-1, -2)$, $B(2, 2)$, $C(7, 2)$, $D(4, -2)$

In Exercises 13 through 24, answer the given questions relevant to the rectangular coordinate system.

13. What is the abscissa of each point on the y-axis?
14. What is the ordinate of each point on the x-axis?
15. In which quadrants is the ratio of the abscissa to the ordinate positive?
16. In which quadrants is the ratio of the abscissa to the ordinate negative?
17. In what quadrant does a point (a, b) lie if $a > 0$ and $b < 0$?
18. In what quadrant does a point (a, b) lie if $a < 0$ and $b > 0$? Show how this point can be described as the intersection of two sets of quadrants.
19. A square has one vertex at the origin, one on the x-axis, one on the y-axis, and the other in the first quadrant. What are the coordinates of the vertices if the side of the square is 4?
20. What are the coordinates of the vertices of a square with sides of 2.5 in., whose center is at the origin and whose sides are parallel to the axes?

21. Three vertices of a rectangle are $(1, -3)$, $(7, -3)$, and $(7, 2)$. What are the coordinates of the fourth vertex?

22. Three vertices of a rectangle are $(0, 4)$, $(-2, 4)$, and $(-2, -1)$. What are the coordinates of the fourth vertex?

23. Describe the location of all points three units from the origin.

24. Describe the location of all points whose abscissas equal the ordinates.

10-4 THE GRAPH OF A FUNCTION

Having introduced the rectangular coordinate system in the preceding section, we are now in a position to find the graph of a function. In this way, we shall be able to get a "picture" of the function that will allow us to see its behavior. In this section, we shall restrict our attention to functions whose graphs are straight lines and functions whose graphs are what are known as **parabolas.**

The graph of a function is the set of all points whose coordinates (x, y) *satisfy the functional relationship* $y = f(x)$. By choosing a specific value for x, we can then find the corresponding value for y by evaluating $f(x)$. In this manner, we obtain the coordinates of one point. Repeating this process, we obtain as many points as necessary to plot the graph of the function. Usually we need to find only enough points to get a good approximation to the graph by joining these points by a smooth curve (as opposed to a broken-line graph).

We shall first consider the graph of a **linear function,** which is a function of the form

$$\boxed{f(x) = ax + b} \tag{10-1}$$

where a and b are constants. *It is called linear since the graph of such a function is always a straight line.* Example A illustrates the basic technique used.

Example A Graph the function $3x + 2$.

Since $f(x) = 3x + 2$, by letting $y = f(x)$ we may write $y = 3x + 2$. Now, by choosing numerical values for x and then substituting these values into the function, we can find the corresponding values for y. If $x = 0$, then

$$y = f(0) = 3(0) + 2 = 2$$

Therefore, the point $(0, 2)$ is on the graph of $3x + 2$. If $x = 1$, then

$$y = f(1) = 3(1) + 2 = 5$$

Accordingly, the point $(1, 5)$ is on the graph of $3x + 2$. This information will be most helpful when it appears in tabular form. We strongly recommend that, in preparing the table, you list the values of x in ascending order. In this way, the points indicated can be joined in order. After the table has been set up, the indicated points are plotted on a rectangular coordinate system and then joined. The table for this function is given below. The graph is shown in Fig. 10-8. Note that the line is extended beyond the points found in the table. This extension indicates that we know it continues in each direction.

x	-3	-2	-1	0	1	2
y	-7	-4	-1	2	5	8

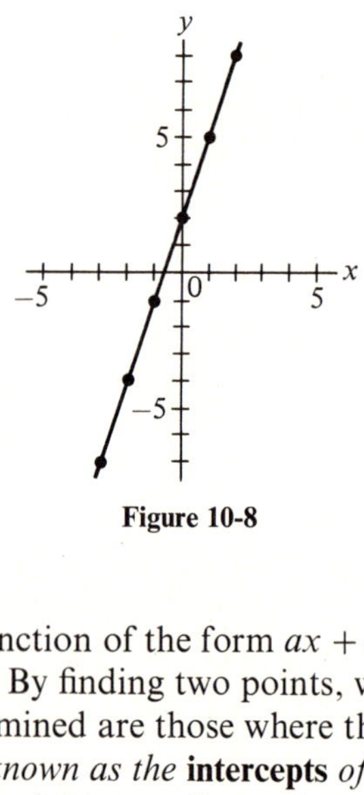

Figure 10-8

The knowledge that a function of the form $ax + b$ is a straight line can be used to definite advantage. By finding two points, we can draw the line. Two points that are easily determined are those where the curve crosses each of the axes. *These points are known as the* **intercepts** *of the line.* The reason these are easily found is that one of the coordinates is zero. *By setting $x = 0$ and $y = 0$, in turn, and determining the corresponding value of the other respectively, we obtain the coordinates of the intercepts.* A third point should be found as a check.

Example B Graph the function $-2x - 5$.

By setting $y = -2x - 5$ and then letting $x = 0$, we obtain $y = -5$. This means that the graph passes through the point $(0, -5)$. Now setting $y = 0$, we have $x = -\frac{5}{2}$. This means the graph passes through the point $(-\frac{5}{2}, 0)$.

As a check, we let $x = 1$ and find $y = -7$, meaning that the line passes through $(1, -7)$. The table is given below, and the graph is shown in Fig. 10-9.

x	$-\frac{5}{2}$	0	1
y	0	-5	-7

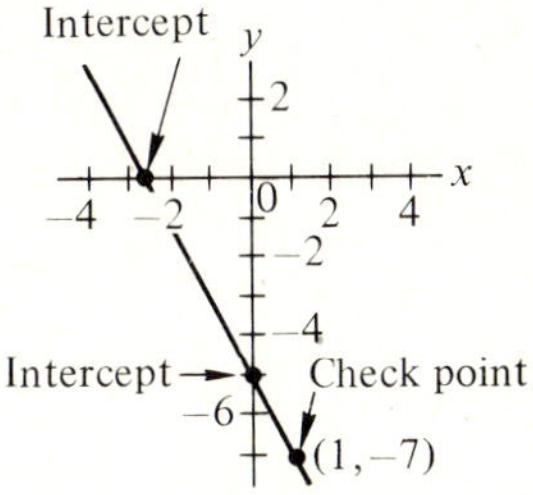

Figure 10-9

If a line passes through the origin, it has only the one intercept (0, 0). In this case, it is necessary to find at least one point other than the intercept in order to plot the graph. However, the origin and two other points (one as a check) are always easily found. Exercises 1 through 4 at the end of this section illustrate this type of straight line.

We now turn our attention to the graph of the quadratic function

$$f(x) = ax^2 + bx + c \qquad (10\text{-}2)$$

where a, b, and c are constants. *The graph of this function is a* **parabola,** *which is illustrated in the examples that follow.* Note that the right-hand side of Eq. (10-2) is the same expression as that appearing in the quadratic equations of Chapter 9.

Example C Graph the function $y = \dfrac{x^2}{2}$.

The method of plotting this type of function is precisely the same as that explained in Example A. We choose values of x and then obtain the values of y, thus determining and tabulating the coordinates of a set of representative points on the graph. In finding the y-values, we must be careful with negative values of x. Remember that the square of a negative number is a positive number. With these ideas in mind, we obtain the following table for $y = \dfrac{x^2}{2}$. The curve is shown in Fig. 10-10.

x	-3	-2	-1	0	1	2	3
y	$\frac{9}{2}$	2	$\frac{1}{2}$	0	$\frac{1}{2}$	2	$\frac{9}{2}$

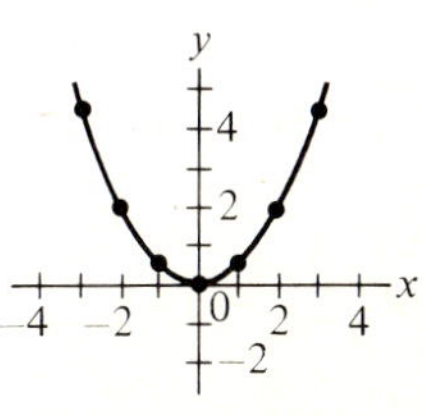

Figure 10-10

Note that the points in Fig 10-10 are connected with a smooth curve, not straight-line segments.

Figure 10-10 shows the basic shape of a parabola, which is the graph of a quadratic function. However, the parabola may be shifted right or left and up or down, or it may open right or left or down. The following examples illustrate the graphs of other quadratic functions.

Example D Graph the function $y = 2x^2 - 4x - 7$.

Again, we set up a table and find the y-values by assuming certain values for x. The question arises: Exactly how many values of x should be chosen? There is no explicit answer. However, we should determine enough points so that we can easily connect the points and indicate the true basic shape of the curve. Also for this function, we must again be careful in handling the negative values of x. For example, if $x = -2$, then

$$y = 2(-2)^2 - 4(-2) - 7 = 2(4) + 8 - 7 = 9.$$

The table for this function is given below, and the graph is shown in Fig. 10-11.

x	-2	-1	0	1	2	3	4
y	9	-1	-7	-9	-7	-1	9

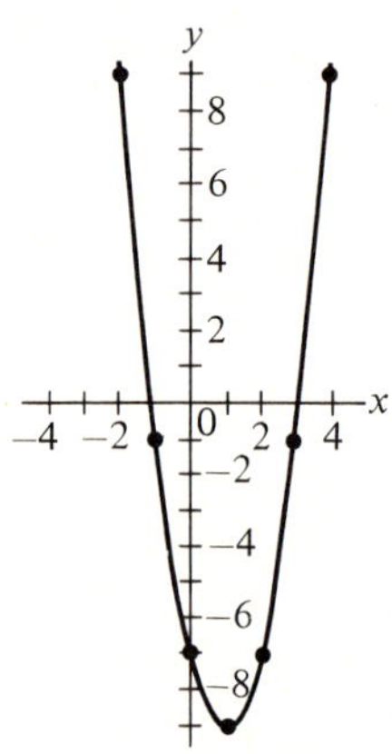

Figure 10-11

Example E Graph the function $y = 4 + 2x - x^2$.
We set up the table as shown

x	-3	-2	-1	0	1	2	3	4	5
y	-11	-4	1	4	5	4	1	-4	-11

and graph the function in Fig. 10-12. Once again, special care must be used in finding y-values for negative values of x. For example, if $x = -2$, we have

$$y = 4 + 2(-2) - (-2)^2 = 4 - 4 - (4) = -4$$

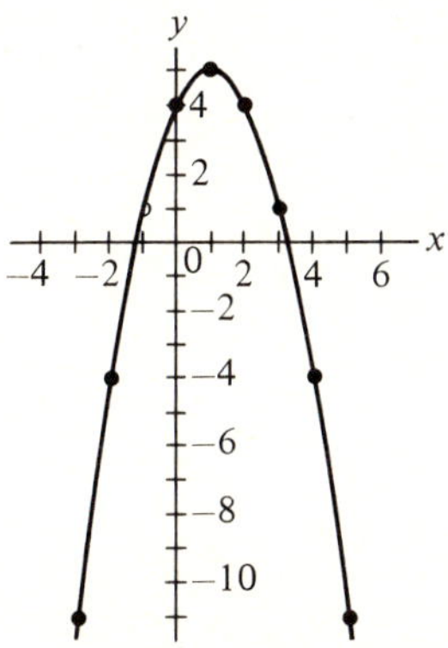

Figure 10-12

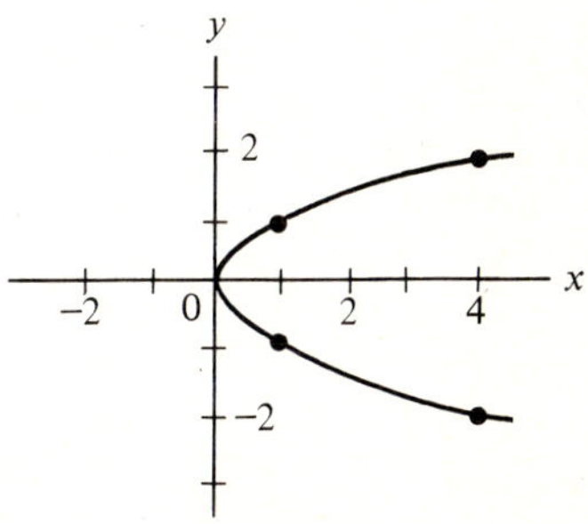

Figure 10-13

Example F Graph the equation $y^2 = x$.
We set up the table as shown below.

x	0	1	1	4	4
y	0	1	-1	2	-2

We recall from Section 10-1 that for a function there may be only one value of y for any given value of x. From the table we see that for each value of x, except $x = 0$, there are two values of y. Therefore, we know that this equation represents a relation, not a function.

Figure 10-13 shows the graph of this relation. From the graph we readily see that for each value of x there is more than one value of y.

Exercises 10-4

In Exercises 1 through 28, graph the given functions.

1. $y = x$

2. $y = -x$

3. $y = -2x$

4. $y = 4x$

5. $y = 3x - 5$

6. $y = 2x + 14$

7. $y = 9 - 4x$

8. $y = 8 - 2x$

9. $y = 6x - 1$

10. $y = -5x + 3$

11. $y = \frac{1}{2}x + 1$

12. $y = \frac{1}{3}x - 2$

13. $y = -x + \frac{1}{3}$

14. $y = 2x - \frac{1}{2}$

15. $y = \frac{2 - x}{3}$

16. $y = \frac{4 + 3x}{2}$

17. $y = x^2$

18. $y = -x^2$

19. $y = x^2 - 4$

20. $y = 4 - x^2$

21. $y = 2x - x^2$

22. $y = x^2 - 2x$

23. $y = \frac{x^2}{3}$

24. $y = \frac{x^2}{4}$

25. $y = 1 - \frac{x^2}{2}$

26. $y = 2 + \frac{x^2}{3}$

27. $y = 2x^2 - 3x + 2$

28. $y = 3 - x - x^2$

In Exercises 29 through 36, graph the functions by plotting the dependent variable along the y-axis and the independent variable along the x-axis.

29. A board 10 m long is cut into two pieces. Express the length y of one piece as a function of the length x of the other piece. Plot the graph.

30. Under certain conditions, the price p for each of x units of a commodity is given by $p = -\frac{3}{4}x + 6$. Plot the graph.

31. A firm can produce up to 500 units per day of an item which sells for \$2. Fixed costs are \$200 daily and each item costs \$0.50 to produce. Find the equation relating profit p and the number x of units produced daily. Plot the graph.

32. If a spring is stretched x in. by a force F, the equation relating x and F is $F = kx$, where k is a constant. A force of 10 lb stretches a given spring 2 in. Plot the graph for the spring.

33. Express the area A of a circle as a function of its diameter d and plot the graph.

34. Express the total surface area A of a cube as a function of its edge e and plot the graph.

35. Let d equal a number x minus the square of x. Plot the graph.

36. The distance h (in feet) above the surface of the earth of an object as a function of the time (in seconds) is given by $h = 60t - 16t^2$ if the object is given an initial upward velocity of 60 ft/s. Plot the graph.

In Exercises 37 and 38, plot the graphs of the given equations. Determine whether the equation represents a function or a relation.

37. $y^2 = x + 4$

38. $y^3 = x + 1$

In Exercises 39 and 40, state whether the graph represents a function or a relation.

39.

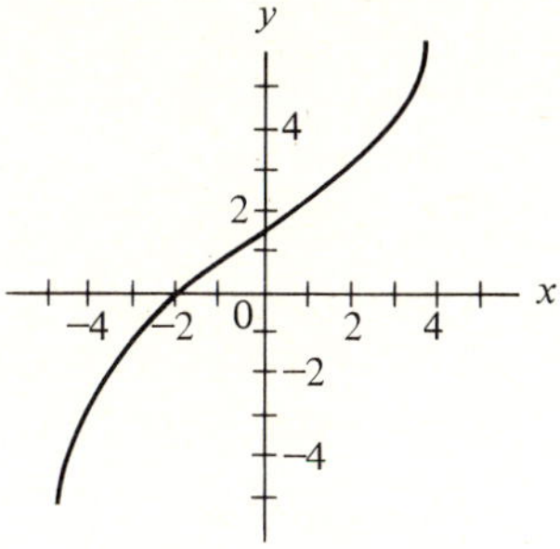

Figure 10-14

40.

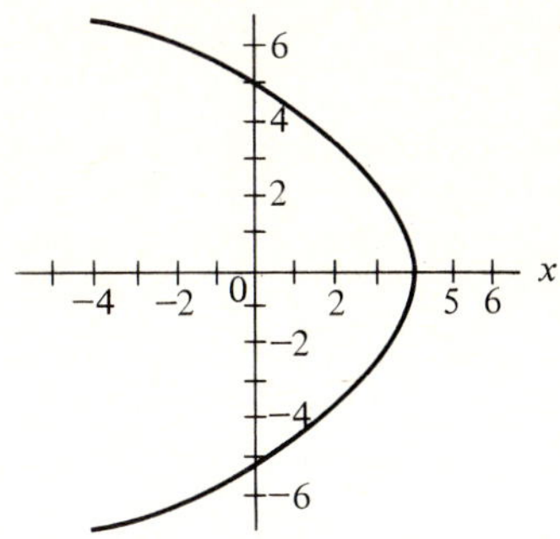

Figure 10-15

It is possible to find approximate solutions to quadratic equations by means of the graph of the quadratic function. Plotting the graph of the quadratic function, we find that the solutions to the equation $ax^2 + bx + c = 0$ are those values of x where the curve crosses the x-axis. Using this method, graphically solve the quadratic equations given in Exercises 41 through 44.

41. $2x^2 - x - 4 = 0$

42. $x^2 + 3x - 5 = 0$

43. $5 - 7x - x^2 = 0$

44. $2 - 3x - x^2 = 0$

10-5 GRAPHS OF OTHER FUNCTIONS

In the preceding section, we discussed the method of plotting the graph of a function. However, we limited our discussion to two important types of functions. In this section, we shall introduce the graphs of several other types of functions.

The basic procedure for plotting the graphs of functions is that presented in Section 10-4. That is, we select specific values of x and then find the corresponding values of y in order to find particular points which lie on the graph of the function. Following are four examples of plotting the graphs of functions that are not linear or quadratic.

Example A Graph the function $y = 2^x$.

We note for this function that the independent variable x is an exponent. Therefore, in order to plot the graph, it is necessary to handle powers of 2—including both positive and negative exponents—properly. For example, if $x = 3$, we have $y = 2^3 = 8$. If $x = -2$, we have

$$y = 2^{-2} = \left(\frac{1}{2}\right)^2 = \frac{1}{4}$$

The table below gives the values from which the points are plotted in Fig. 10-16.

x	-2	-1	0	1	2	3
y	$\frac{1}{4}$	$\frac{1}{2}$	1	2	4	8

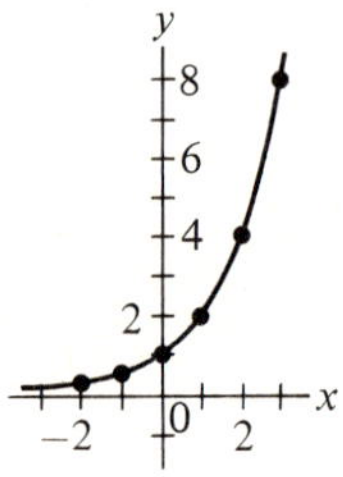

Figure 10-16

This is the graph of an *exponential function*. A discussion of this function is taken up in the next chapter.

Example B Graph the function $y = 2 + \dfrac{1}{x}$.

In finding the points for this graph, we must be careful not to set $x = 0$, which would necessitate divisions by zero. However, in order to get an accurate graph, we must choose values of x near zero. Therefore, you will find fractional values of x in the table below. For example, if $x = \frac{1}{2}$, then

$$y = 2 + \frac{1}{\frac{1}{2}} = 2 + 2 = 4$$

This graph is known as a **hyperbola.** See Fig. 10-17.

x	-3	-2	-1	$-\frac{1}{2}$	$-\frac{1}{4}$	$\frac{1}{4}$	$\frac{1}{2}$	1	2	3
y	$\frac{5}{3}$	$\frac{3}{2}$	1	0	-2	6	4	3	$\frac{5}{2}$	$\frac{7}{3}$

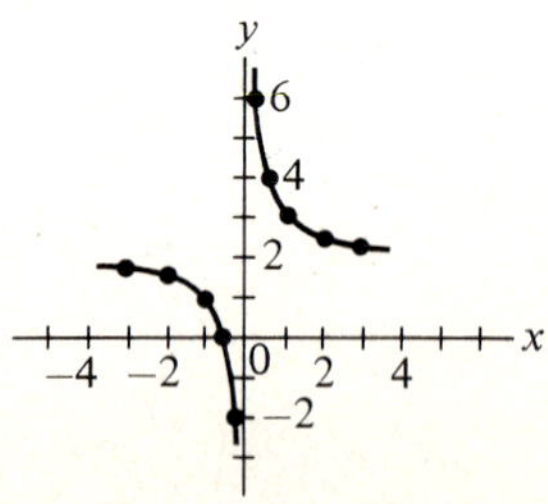

Figure 10-17

Example C Graph the function $y = x^3 - 3x$.

Proper use of signed numbers and their powers is essential in finding the values of y for values of x, particularly negative values. For example, if $x = -2$, then

$$y = (-2)^3 - 3(-2) = -8 + 6 = -2$$

The following table gives the points used to plot the graph in Fig. 10-18.

x	-3	-2	-1	0	1	2	3
y	-18	-2	2	0	-2	2	18

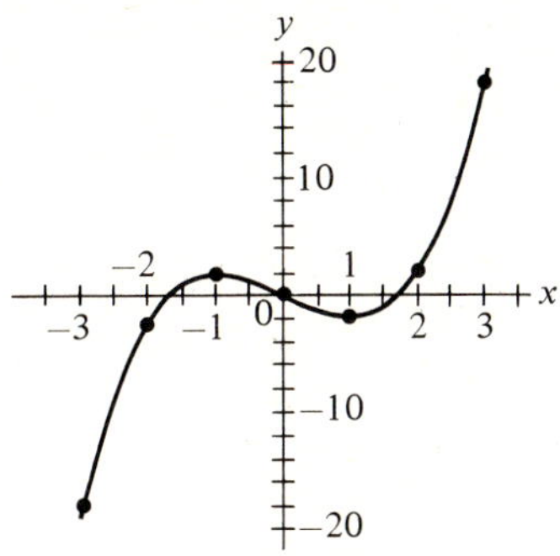

Figure 10-18

This curve is known as a **cubic curve.** Note that the scale is different for both x and y. Scales may differ when the range of values used differs considerably.

Example D Graph the function $y = \sqrt{x + 1}$.

In finding the y-values of this graph, we must be careful to use only values of x which will give us real values for y. This means that the value under the radical sign must be zero or greater. For example, if $x = -5$, we would have $y = \sqrt{-4}$, which is an imaginary number. Therefore, x cannot be -5. In fact, we cannot have any value of x which is less than -1. Thus, we find the following points for the graph:

x	-1	0	1	2	3	4	6	8
y	0	1	1.4	1.7	2	2.2	2.6	3

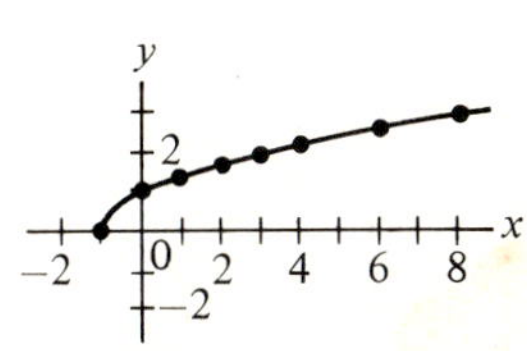

Figure 10-19

The graph is plotted in Fig. 10-19. This graph is *half of a parabola.*

Exercises 10-5

In Exercises 1 through 28, graph the given functions.

1. $y = 3^x$
2. $y = 2^{2x}$
3. $y = \left(\frac{1}{3}\right)^x$
4. $y = 2^{-x}$
5. $y = \frac{1}{x}$
6. $y = 1 + \frac{2}{x}$
7. $y = 2 - \frac{1}{x}$
8. $y = \frac{3}{x} - 4$
9. $y = \sqrt{x}$
10. $y = 3\sqrt{x}$
11. $y = \sqrt{x + 4}$
12. $y = \sqrt{1 + x}$
13. $y = \sqrt{4 - x}$
14. $y = \sqrt{2 - x}$
15. $y = \sqrt{x^2 + 1}$
16. $y = \sqrt{25 + x^2}$
17. $y = \frac{2}{x^2}$
18. $y = \frac{1}{x^2}$
19. $y = x^3$
20. $y = 6x - x^3$
21. $y = -\frac{1}{2}x^3$
22. $y = -\frac{1}{3}x^3$
23. $y = \frac{1}{x^2 + 1}$
24. $y = \frac{1}{x^2 - 1}$
25. $y = \frac{1}{4}x^4$
26. $y = x^4 - 2x^2$
27. $y = \frac{1}{\sqrt{1 + x}}$
28. $y = \frac{1}{\sqrt{4 + x}}$

In Exercises 29 through 34, graph the given functions by plotting the dependent variable along the y-axis and the independent variable along the x-axis.

29. The total profit P a manufacturer makes in producing x units of a commodity is given by $P = x^3 - 3x^2 - 5x - 150$. Plot the graph, using values of 0 through 10 for x.
30. If the number of bacteria in a certain culture increases by 50% each hour, the number N of bacteria present after t hours is $N = 1000(\frac{3}{2})^t$, if 1000 were originally present. Graph the function.
31. Under the condition of constant temperature, it was found that the pressure p (in atmospheres) and the volume V (in cubic inches) in an experiment on air were related by $p = 10/V$. Plot the graph.
32. The gravitational force of attraction between two objects is inversely proportional to the square of the distance between their centers. Assuming the constant of proportionality to be 8 units, plot the graph.
33. An object is p in. from a lens of a focal length of 5 in. The distance from the lens to the image is given by $q = \frac{5p}{p - 5}$. Plot the graph.
34. The average unit cost C a manufacturer has in producing y units of a commodity is given by $C = \frac{500}{y + 50}$. Plot the graph, using values of 0 through 100 for y. Calculate C for $y = 0, 10, 20, 30$, etc.

10-6 GRAPHS OF INEQUALITIES

In the previous sections, we have discussed the graphs of certain types of functions. In this section, we consider graphs of inequalities. The algebraic solution of inequalities was introduced in Chapter 3.

We shall now demonstrate the graphs of inequalities through the following examples.

Example A Graph the inequality $x > 2$ on the rectangular coordinate system.

We first draw the line $x = 2$. This is a line parallel to the y-axis and two units to the right. We note that for every point on this line $x = 2$, regardless of the y-coordinate.

Now, all values of x which are greater than 2 lie to the right of the line $x = 2$. Thus, *the graph of $x > 2$ consists of all points to the right of $x = 2$.* In Fig. 10-20, we indicate this by shading the area to the right of $x = 2$. We note $x = 2$ as a *dashed line* to indicate that it is *not* part of the graph of the inequality.

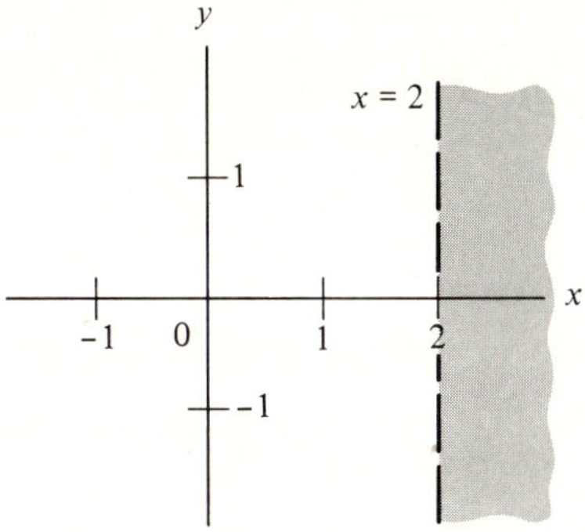

Figure 10-20

Example B Graph the following on the rectangular coordinate system: $x \geq -3$.

We draw the line $x = -3$ (Fig. 10-21). Then *all points on the line $x = -3$ and to the right of $x = -3$* describe the graph of $x \geq -3$. This is shown by shading the area to the right of $x = -3$ and by drawing in $x = -3$ as a *solid line.*

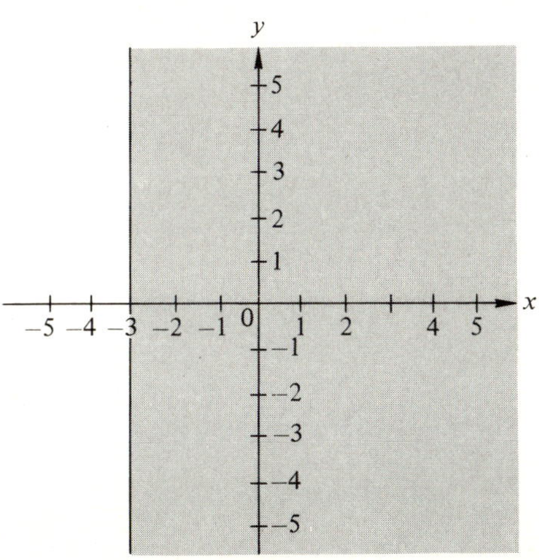

Figure 10-21

Example C Graph $y \le 0$ on the rectangular coordinate system.

We draw the line $y = 0$ (Fig. 10-22); this is the x-axis. Then all points *on the x-axis and below the x-axis* constitute the graph of $y \le 0$.

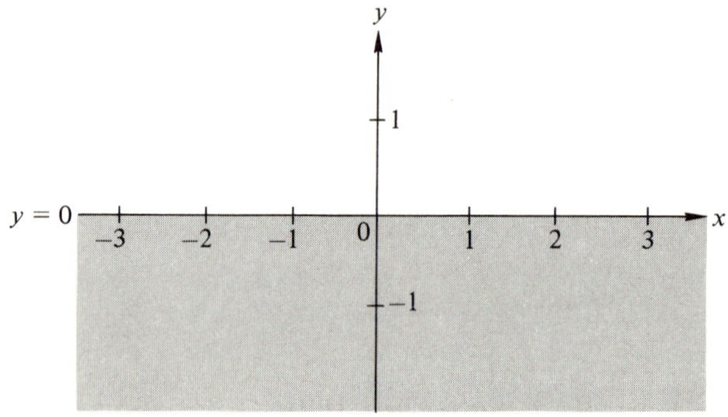

Figure 10-22

Example D Graph $y > 2x + 4$.

We first graph the line $y = 2x + 4$, using the intercepts (0, 4) and (−2, 0) as shown in Fig. 10-23. We draw the line segment to represent this line.

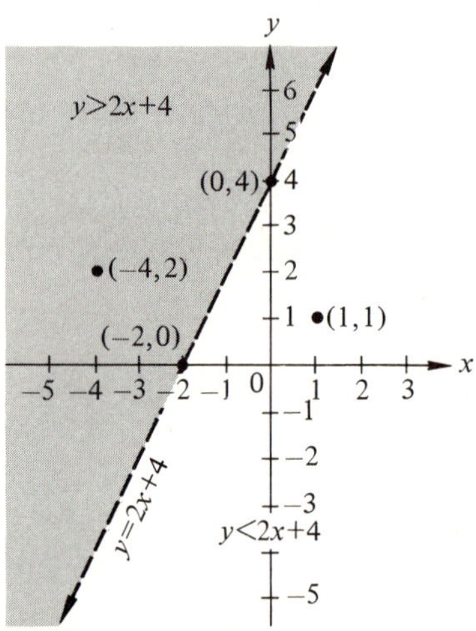

Figure 10-23

The question now arises concerning where the points are that satisfy the inequality $y > 2x + 4$. To determine this, we test a point that is not on the line $y = 2x + 4$. Using the point (1, 1), which is under the line, we have

$$y > 2x + 4$$

$$1 > 2(1) + 4$$

$$1 > 6$$

This is a false statement. This tells us that points in the area under the line must represent "less than" values. Now, testing the point $(-4, 2)$, which is above the line, we have

$$y > 2x + 4$$
$$2 > 2(-4) + 4$$
$$2 > -8 + 4$$
$$2 > -4$$

This is a correct statement, which tells us that the points in the area above the line represent the "greater than" values, and these are the ones we want. We must be careful not to include points on the line, and we therefore show it as a dashed line.

Therefore, we shade the area above the line $y = 2x + 4$ to indicate the graph of the inequality $y > 2x + 4$.

Example E Graph the inequality $y \leq 2 - x$.

First we graph the line $y = 2 - x$ by using the intercepts $(0, 2)$ and $(2, 0)$, as shown in Fig. 10-24. Next we test one representative point above the line, and one representative point below the line. Testing $(1, 3)$, which is above the line, we have

$$y \leq 2 - x$$
$$3 \leq 2 - 1$$
$$3 \leq 1$$

which is false. Now testing $(1, -1)$, which is below the line, we have

$$-1 \leq 2 - 1$$
$$-1 \leq 1$$

which is true. Therefore, points below the line and on the line satisfy the inequality. Thus, to graph the inequality $y \leq 2 - x$, we shade the area below the line, and draw the line as a solid line, indicating that it is part of the graph.

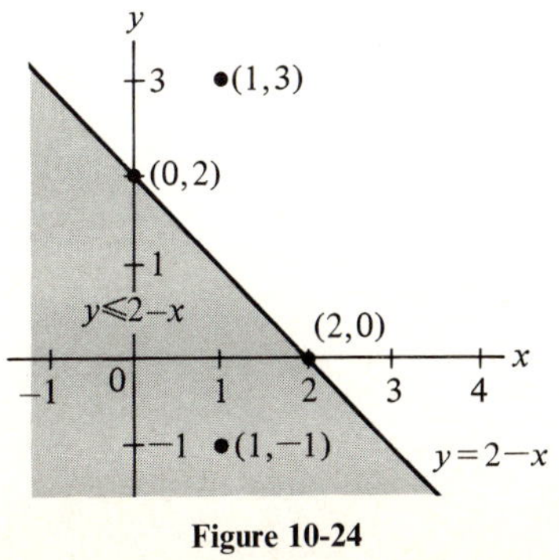

Figure 10-24

Example F Graph the inequality $2x - 3y \le 6$.

We first graph the line $2x - 3y = 6$, as shown in Fig. 10-25, by using the intercepts $(0, -2)$ and $(3, 0)$. To find the points that satisfy the inequality $2x - 3y \le 6$, we test two points, one above the line and the other below the line. To represent points under the line, we use the point $(2, -3)$. This gives us

$$2x - 3y \le 6$$
$$2(2) - 3(-3) \le 6$$
$$4 + 9 \le 6$$
$$13 \le 6$$

which is a false statement. This means that points under the line do not satisfy the inequality. Now, testing $(3, 2)$, which is above the line, we have

$$2(3) - 3(2) \le 6$$
$$6 - 6 \le 6$$
$$0 \le 6$$

which is true. Therefore, we shade the area above the line, and draw the line as a solid line to represent the graph of the inequality $2x - 3y \le 6$.

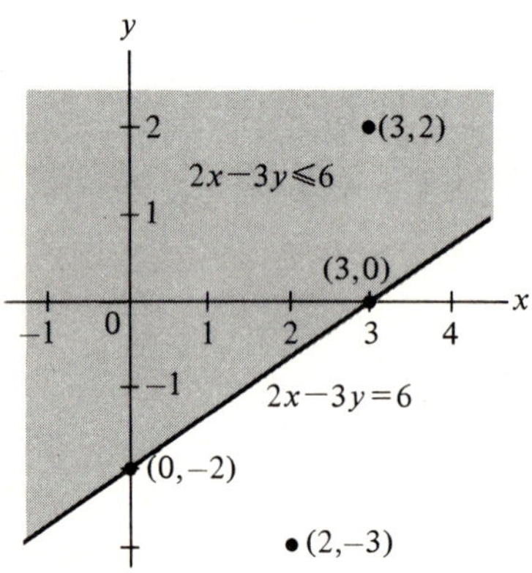

Figure 10-25

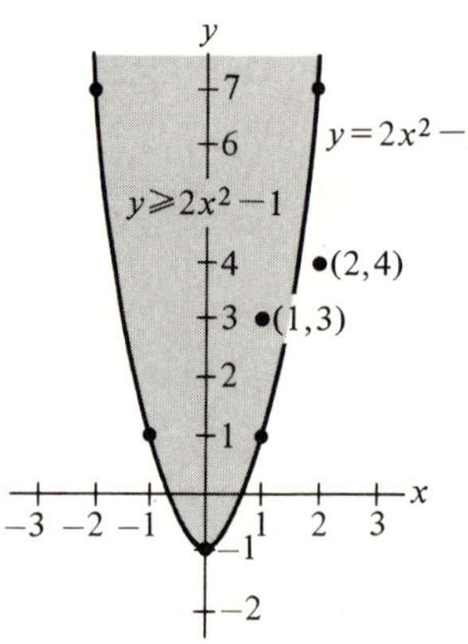

Figure 10-26

Example G Graph the inequality $y \ge 2x^2 - 1$.

We graph the parabola $y = 2x^2 - 1$ in Fig. 10-26 by using the values shown in the table.

x	y
−2	7
−1	1
0	−1
1	1
2	7

We draw the curve to separate the two areas involved. To find the points that satisfy the inequality $y \geq 2x^2 - 1$, we test two points, one above the curve and the other below the curve. One of these points will represent those points which satisfy the inequality. Below the curve, we choose (2, 4).

$$y \geq 2x^2 - 1$$
$$4 \geq 2(2^2) - 1$$
$$4 \geq 7$$

This is a false statement. Now, choosing (1, 3) above the curve, we have

$$3 \geq 2(1^2) - 1$$
$$3 \geq 1$$

which is a true statement. Therefore, the graph of the inequality $y \geq 2x^2 - 1$ consists of those points above the parabola and those points on the parabola. Thus, the area above the parabola is shaded, and the parabola is shown as a solid curve.

We also note that the graph of the inequality $y < 2x^2 - 1$ consists of the points below the parabola, and that the boundary for the two inequalities is the curve $y = 2x^2 - 1$.

Exercises 10-6

In Exercises 1 through 28, graph the given inequalities on the rectangular coordinate system.

1. $x > 0$	**2.** $x < 3$	**3.** $y \leq 3$	**4.** $y \geq 0$
5. $x \leq -2$	**6.** $y \leq -4$	**7.** $y > -5$	**8.** $x > -8$
9. $y > 1 - x$	**10.** $y < x + 1$	**11.** $y < 2x - 1$	**12.** $y > 3x + 4$
13. $y \leq 2x + 3$	**14.** $y \geq 3x - 5$	**15.** $\frac{y}{2} < 4$	**16.** $\frac{y}{3} < 6$
17. $3y < 8 - 6x$	**18.** $2y \geq 3 + 2x$	**19.** $y - 3 > x$	**20.** $y - 1 \leq 2x$
21. $3x + 4y < 3$	**22.** $4x + 3y \geq 5$	**23.** $2x - 5y \geq 10$	**24.** $6x - y < 6$
25. $y \leq x^2$	**26.** $y \geq x^2 + 1$	**27.** $y < 2 - x^2$	**28.** $y > x^3 + 1$

In Exercises 29 through 32, set up and graph the appropriate inequalities.

29. What average speed will allow a traveler to drive a car at least 124 km in 4 h?

30. If a person is to earn more than $18,000 in a year, what must be his or her average monthly income?

31. If one number is less than six more than twice another number, what are the numbers?

32. It takes 2 h for a company to produce each of x medium-quality parts and 5 h to produce each of y high-quality parts. How many of each type can be produced in 200 h or less?

10-7 THE SLOPE OF A LINE

In Section 10-4, we introduced the linear function $y = f(x) = ax + b$. In this section, we shall show that useful specific meanings can be given to the coefficient of the x-term and also to the b-term.

Consider the line which passes through points A and B in Fig. 10-27.

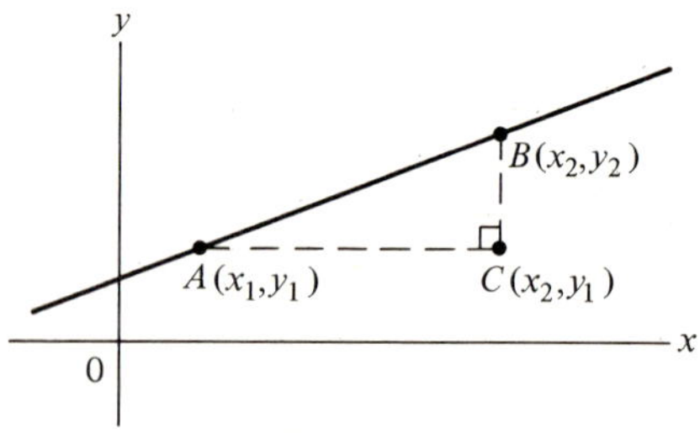

Figure 10-27

Point A has coordinates (x_1, y_1), and point B has coordinates (x_2, y_2). Point C is directly horizontal from point A and directly vertical from point B. Therefore, point C has coordinates (x_2, y_1).

As a point moves along the line AB, it is moving horizontally and it is also moving vertically. Along the line, from one point to another, there is a constant and continuous change in the y-coordinate with a corresponding change in the x-coordinate. The relationship between these changes in y and x determines the steepness of the line AB.

Mathematically then, *the ratio of the change in the y-coordinate to the corresponding change in the x-coordinate from one point to another point is known as the* **slope** *of the line through the points.* Thus, using the letter m to represent slope, we have

$$m = \frac{y_2 - y_1}{x_2 - x_1} \tag{10-3}$$

The following examples illustrate the use of Eq. (10-3) and certain important properties of the slope of a line.

Example A Find the slope of the line that passes through the points $(-1, 0)$ and $(0, 2)$.

In Fig. 10-28, we draw the line through the two given points. We consider $(0, 2)$ as (x_2, y_2). This means that $(-1, 0)$ is (x_1, y_1). Although we may choose either point as (x_1, y_1), once the choice is made *the correct order must be maintained.* Using Eq. (10-3), the slope is

$$m = \frac{2 - 0}{0 - (-1)} = \frac{2}{1} = 2$$

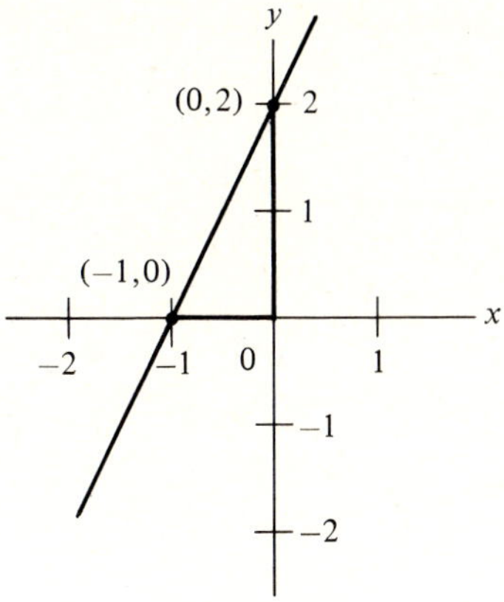

Figure 10-28

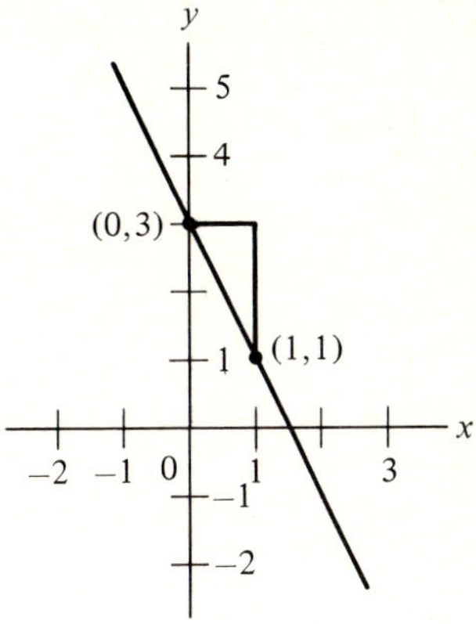

Figure 10-29

Example B Find the slope of the line that passes through the points (1, 1) and (0, 3).

In Fig. 10-29, we draw the line through the two given points. We consider 3 as the y_2; then 0 is the x_2. Also, we must now consider 1 as y_1 and 1 as x_1. Using Eq. (10-3), the slope is

$$m = \frac{3-1}{0-1} = \frac{2}{-1} = -2$$

Note in Example A that *as x increases, y also increases.* Therefore, *the slope is* **positive.** In Example B, the reverse is true. *As x increases, y decreases.* Here, *the slope is* **negative.**

We shall now see how the slope m relates to the linear function $y = ax + b$. If we choose $(0, b)$ as one point and (x, y) as another point on a line, then the slope is

$$m = \frac{y-b}{x-0} = \frac{y-b}{x}$$

Now, if we solve the equation $y = ax + b$ for a, we have

$$a = \frac{y-b}{x}$$

Thus, we see that *the coefficient a of the x-term is actually the slope of the line.*

Again, recall the equation of a straight line $y = ax + b$. If we let $x = 0$, then $y = b$, and we have the point $(0, b)$. This is the point where the straight line crosses (intercepts) the y-axis. Thus, *b is called the* **y-intercept** *of the line.* Since it is customary to use m as the slope of the line, we have

$$\boxed{y = mx + b} \tag{10-4}$$

where m is the slope and b is the y-intercept of the line. *Eq.* (10-4) *is known as the* **slope-intercept** *form of the equation of a straight line.*

Example C

1. Find the slope of the line $y = -\frac{2}{3}x + 5$.

Since the equation is solved for y, the coefficient of x is the slope. Therefore, we simply read the slope as $-\frac{2}{3}$.

2. Find the slope of the line $2x + 3y = 4$.

Here, *we must write the equation of the line in slope-intercept form.* Thus, we solve for y; this gives us

$$y = -\frac{2}{3}x + \frac{4}{3}$$

Thus, the slope is $-\frac{2}{3}$.

Example D Graph $y = 2x - 3$ by determining the slope and y-intercept of the line.

Since the equation is solved for y, the coefficient of x is the slope. Therefore, the slope of this line is 2. Also, since the b-term is the y-intercept, we see that the y-intercept of this line is -3.

We can use this information to sketch the line as shown in Fig. 10-30.

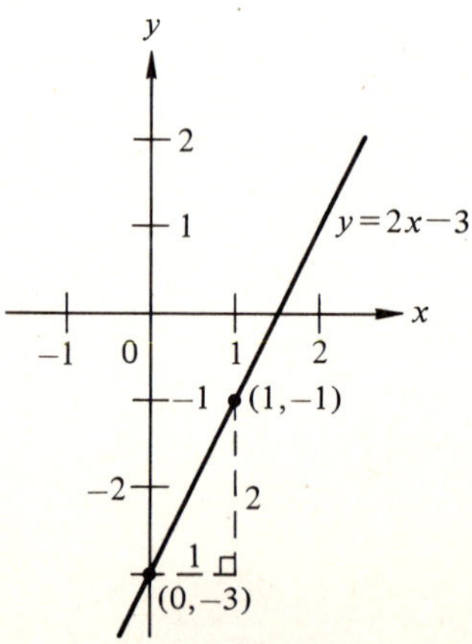

Figure 10-30

Since the y-intercept of the line is -3, we know it passes through $(0, -3)$. Also, since the slope is 2, we know that y increases 2 units for each unit increase of x. Thus, starting at $(0, -3)$, if x increases 1 unit and y increases 2 units, we have the point $(1, -1)$. Thus, the line must pass through $(1, -1)$ as well as through $(0, -3)$. Therefore, we draw the line through these two points.

Example E Given the two points (2, 3) and $(-1, 4)$, find the slope and the y-intercept associated with the line through these points, and write the equation of this line.

We first find the slope of the line represented by the points. Thus,

$$m = \frac{4-3}{-1-2} = \frac{1}{-3} = -\frac{1}{3}$$

Now, to find the y-intercept, we use *either one* of the two points and the slope-intercept form of the equation of a line. Thus, using the point (2, 3) and the known slope of $-\frac{1}{3}$ in the slope-intercept form, we solve for b.

$$3 = -\frac{1}{3}(2) + b$$

$$3 = -\frac{2}{3} + b$$

$$\frac{11}{3} = b$$

This means that the required equation is $y = -\frac{1}{3}x + \frac{11}{3}$. This line is shown in Fig. 10-31.

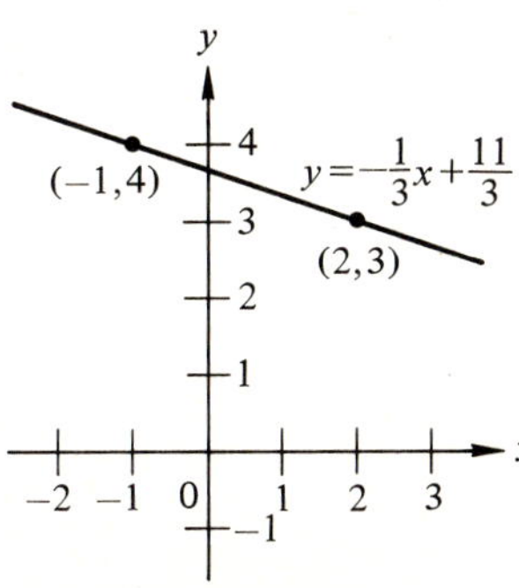

Figure 10-31

Example F Find the equation of the line that has the slope -3 and passes through the point $(-1, 3)$.

We use the slope-intercept form of the equation of a line to find the y-intercept. Thus,

$$3 = (-3)(-1) + b$$

$$3 = 3 + b$$

$$b = 0$$

Then the equation required is $y = -3x + 0$, or $y = -3x$. Note that the y-intercept here is 0. The line passes through the origin, as shown in Fig. 10-32.

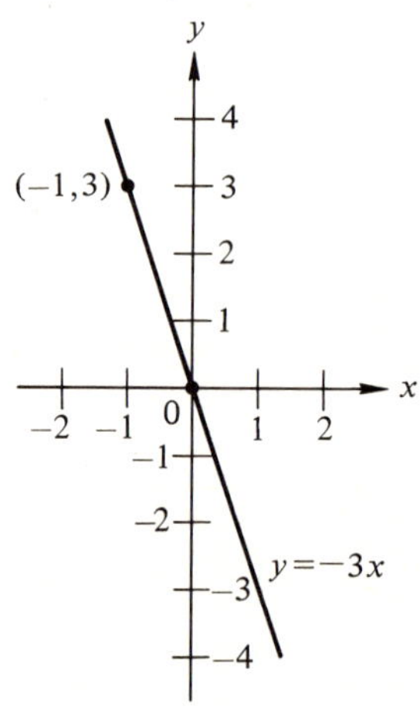

Figure 10-32

Exercises 10-7

In Exercises 1 through 8, find the slope of the line that passes through the given points.

1. $(-2, 0)$, and $(0, 3)$

2. $(-5, 0)$ and $(1, 3)$

3. $(2, 2)$ and $(0, 4)$

4. $(-1, 4)$ and $(5, -1)$

5. $(-1, -3)$ and $(2, -1)$

6. $(4, -2)$ and $(-5, 1)$

7. $(-7, 3)$ and $(-1, -6)$

8. $(5, 4)$ and $(-3, 8)$

In Exercises 9 through 20, find the slope and y-intercept of each line.

9. $y = 3x + 1$

10. $y = 4x + 2$

11. $y = -x + 6$

12. $y = -5x - 3$

13. $y = \frac{5}{3}x$

14. $y = -\frac{2}{5}x$

15. $x = -\frac{1}{4}y$

16. $x = \frac{6}{5}y$

17. $y - 2 = x + 1$

18. $y - 3 = 2x + 3$

19. $6y - 2x + 3 = 0$

20. $3x - 5y + 2 = 0$

In Exercises 21 through 24, find the slope and y-intercept of each line and then sketch the line.

21. $y + 6 = x - 3$

22. $y - 2x + 1 = 0$

23. $2y + 3x - 6 = 0$

24. $5x - 2y + 4 = 0$

In Exercises 25 through 28, find the equation of the line that passes through the given point and has the given slope.

25. $(2, 1), m = 4$

26. $(1, 4), m = 3$

27. $(-1, -1), m = -5$

28. $(-3, 0), m = -2$

In Exercises 29 through 32, find the equation of the line which passes through the two given points.

29. $(1, 1)$ and $(0, 2)$

30. $(2, 4)$ and $(0, 4)$

31. $(1, -2)$ and $(2, -1)$

32. $(-1, -3)$ and $(3, -2)$

In Exercises 33 through 36, solve the given problems.

33. A ladder is placed against a building, and the slope of the ladder is -2. If the foot of the ladder is 10 ft from the building, how many feet above the ground does the ladder touch the building?

34. An old tree is braced on either side with a guy wire. To keep it in a vertical position, each wire is fastened to the tree 9 m above the ground. If each wire is pinned to the ground 6 m from the tree, what are the slopes of the guy wires?

35. The total fixed cost for a company to operate a certain plant is \$400 per day. It also costs \$2 for each unit produced in the plant. Find the equation relating the total cost C of operating the plant and the number of units n produced each day. Interpret the fixed cost and the cost per unit in terms of the slope and the y-intercept of the resulting linear equation.

36. For temperature readings the metric system uses degrees Celsius (°C) and the traditional English system uses degrees Fahrenheit (°F). If 0°C corresponds to 32°F and 100°C corresponds to 212°F, find the equation relating the Celsius temperature C and the Fahrenheit temperature F.

10-8 INVERSE FUNCTIONS

We introduced the concept of a function in Section 10-1. There have been frequent occasions when we were given a function $y = f(x)$ and were asked to determine the value of y for a corresponding value of x. However, there are occasions when we are interested in finding the value of the independent variable corresponding to a given value of the function. For example, the function relating the perimeter p of a square and its side s is $p = 4s$. There will be occasions when we know the value of the perimeter and wish to find the value of the side. Establishing functions for the purpose of interchanging the dependent and independent variables is the topic of this section.

Under certain conditions it is possible to obtain a new function from a given function by interchanging the values of x and $f(x)$ in all pairings $(x, f(x))$. *Such a function is called the* **inverse** *of the given function and is denoted by* $f^{-1}(x)$. (The -1 is not a negative exponent, but is a notation for the inverse function.)

Example A If $f(x) = 3x + 1$, some of the $(x, f(x))$ pairings that we may determine are (0, 1), (1, 4), and (2, 7). If we interchange these pairings to (1, 0), (4, 1), and (7, 2), as well as all other possible pairings, we have the pairings for the inverse function.

We recall from Section 10-1 that *for a function* $y = f(x)$, *there may be only one value of y for each value of x*. If the pairings $(x, f(x))$ are reversed, and *for these new pairings to properly define a function, there also may only be one value of x for each value of y. Such a function, one for which we have only one value of x for any given value of y, is termed* **one-to-one.**

Example B The function $f(x) = 3x + 1$ is one-to-one, since for each x we choose, we obtain a different value for y. In fact, *all linear functions are one-to-one*.

The function $f(x) = x^2$ is not a one-to-one function. If $x = -1$, $y = 1$; and if $x = 1$, $y = 1$. Thus, we have two values of x for the value of $y = 1$. Actually, we have two values of x for all valid values of y, except $y = 0$.

From the preceding discussion, we can see that *a function can have an inverse only if it is a one-to-one function*. Only such a function will lead to a properly defined function when the pairings $(x, f(x))$ are reversed.

We can now determine how to find the inverse function of a given function. From our definition of an inverse function, we see that x and $f(x)$, or x and y, are interchanged when we find the pairings for the inverse function. This tells us that x and y must be interchanged to find the inverse function. Therefore, the steps necessary to determine the inverse function of $y = f(x)$ are

(1) ***interchange the x and y variables*, and**
(2) ***solve the resulting equation for the new y variable.***

Example C Find the inverse function of the function $y = 3x + 2$.

Since this is a linear function, it is one-to-one and therefore has an inverse (see Example B). Now, interchanging x and y, we have

$$x = 3y + 2$$

We now solve this last equation for y.

$$x - 2 = 3y$$

$$y = \frac{x - 2}{3}$$

Therefore, the inverse function is $y = \frac{x-2}{3}$. Using functional notation, the original function is $f(x) = 3x + 2$, and its inverse is

$$f^{-1}(x) = \frac{x-2}{3}$$

Note that the operations in the function and its inverse are reversed. In the original function we were to multiply x by 3 and then add 2. In the inverse we subtract 2 from x and then divide by 3.

The graphs of inverse functions are related. In Fig. 10-33, we see that the points (a, b) and (b, a) are reflections of each other across the line $y = x$. Since in finding the inverse of a function, x and y are interchanged (as in these coordinates), *the graphs of inverse functions are reflections of each other across the line* $y = x$.

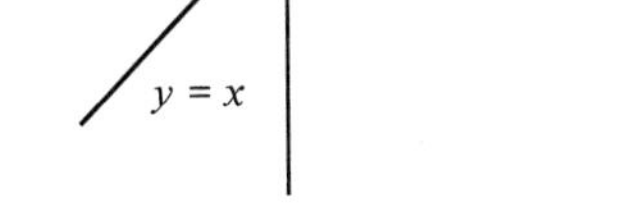

Figure 10-33

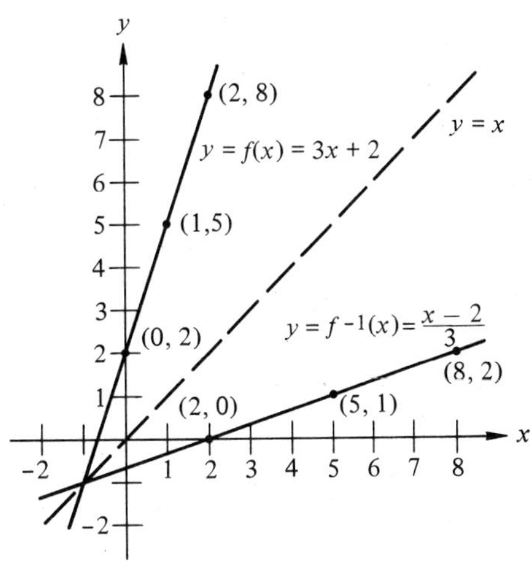

Figure 10-34

Example D In Example C, for the function $y = f(x) = 3x + 2$, we found that the inverse function is $y = f^{-1}(x) = \frac{x-2}{3}$. Graphing each function, we find that the points (0, 2), (1, 5), and (2, 8) are on the graph of $y = 3x + 2$. We also find that the points (2, 0), (5, 1), and (8, 2) are on the graph of $y = f^{-1}(x) = \frac{x-2}{3}$. Graphing these two lines, and the line $y = x$, in Fig. 10-34, we see that each is the reflection of the other across the line $y = x$.

Another important point regarding inverse functions is that *each function is the inverse of the other*. In Example C, if we started by saying that $f(x) = \frac{x-2}{3}$, we would find that $f^{-1}(x) = 3x + 2$.

We can find a method of checking inverse functions by carefully considering the meaning of an inverse function. We can see that *the inverse function of the function of x is x*. This is written as $f^{-1}[f(x)] = x$. This is seen by noting that if $f(a) = b$, then $f^{-1}(b) = a$, which is consistent with our definition of an inverse function.

Example E In Example C, we had the function

$$f(x) = 3x + 2$$

which had as its inverse function

$$f^{-1}(x) = \frac{x-2}{3}$$

To show that these two functions are inverses, we show that $f^{-1}[f(x)] = x$. Thus,

$$\begin{aligned} f^{-1}[f(x)] &= \frac{(3x+2)-2}{3} \\ &= \frac{3x+2-2}{3} \\ &= \frac{3x}{3} = x \end{aligned}$$

This verifies that we have the proper inverse function. Since each function is the inverse of the other, we could also show that $f[f^{-1}(x)] = x$.

Example F Find the inverse of the function $y = f(x) = \frac{-3x+2}{4}$, sketch the graph of the function and its inverse, and then verify that the functions are inverses.

Since the function is linear, it is one-to-one, and therefore has an inverse function. Next, we interchange x and y and solve for the new y.

$$\begin{aligned} x &= \frac{-3y+2}{4} \\ 4x &= -3y + 2 \\ 3y &= -4x + 2 \\ y &= \frac{-4x+2}{3} \end{aligned}$$

Therefore, the inverse function is

$$f^{-1}(x) = \frac{-4x + 2}{3}$$

The graphs of $f(x) = \dfrac{-3x + 2}{4}$ and $f^{-1}(x) = \dfrac{-4x + 2}{3}$ are shown in Fig. 10-35.

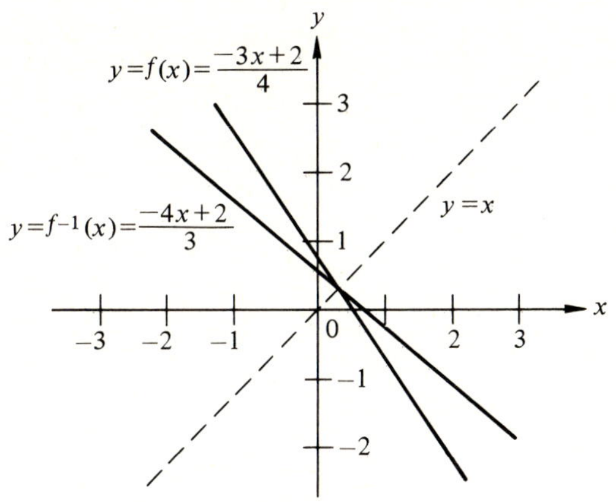

Figure 10-35

Now we find $f^{-1}[f(x)]$.

$$f^{-1}[f(x)] = \frac{-4\left(\dfrac{-3x + 2}{4}\right) + 2}{3} = \frac{-(-3x + 2) + 2}{3}$$

$$= \frac{3x - 2 + 2}{3} = \frac{3x}{3} = x$$

Therefore, the functions are inverses.

Many functions which are not linear are one-to-one functions and therefore have inverse functions. The following example illustrates one such case.

Example G Find the inverse of the function $y = \dfrac{3}{x-2}$.

Following the procedure previously established, we interchange x and y, and solve for the new y.

$$x = \frac{3}{y-2}$$

$$xy - 2x = 3$$

$$xy = 2x + 3$$

$$y = \frac{2x+3}{x}$$

This means that the inverse function is

$$f^{-1}(x) = \frac{2x+3}{x}$$

The above function is one-to-one, and we can verify that these functions are inverses by finding $f^{-1}[f(x)]$.

Exercises 10-8

In Exercises 1 through 12, find the inverse functions of the given functions. Each given function is one-to-one.

1. $y = f(x) = x + 4$ **2.** $y = f(x) = x - 3$ **3.** $y = f(x) = 3x - 1$

4. $y = f(x) = 2x + 3$ **5.** $y = f(x) = 7 - 5x$ **6.** $y = f(x) = 5 - 6x$

7. $y = f(x) = \dfrac{2x-3}{2}$ **8.** $y = f(x) = \dfrac{x+3}{3}$ **9.** $y = f(x) = \dfrac{1}{x}$

10. $y = f(x) = -\dfrac{4}{x}$ **11.** $y = f(x) = x^3$ **12.** $y = f(x) = \dfrac{2}{x+1}$

In Exercises 13 through 20, find the inverse functions of the given functions and sketch the graphs of $f(x)$ and $f^{-1}(x)$.

13. $y = f(x) = x - 1$ **14.** $y = f(x) = x + 2$

15. $y = f(x) = 4x + 3$ **16.** $y = f(x) = 2x - 7$

17. $y = f(x) = -\dfrac{2x+1}{8}$ **18.** $y = f(x) = \dfrac{5x+4}{4}$

19. $y = f(x) = -\dfrac{2}{x}$ **20.** $y = f(x) = 8x^3$

In Exercises 21 through 28, verify that the given function and its inverse are inverse functions by finding $f^{-1}[f(x)]$. These functions are the same as those in Exercises 1–4 and 7–10.

21. $y = f(x) = x + 4$

22. $y = f(x) = x - 3$

23. $y = f(x) = 3x - 1$

24. $y = f(x) = 2x + 3$

25. $y = f(x) = \dfrac{2x - 3}{2}$

26. $y = f(x) = \dfrac{x + 3}{3}$

27. $y = f(x) = \dfrac{1}{x}$

28. $y = f(x) = -\dfrac{4}{x}$

In Exercises 29 through 32, show that the given functions are not one-to-one functions, and therefore do not have inverses.

29. $y = f(x) = x^2 + 1$

30. $y = f(x) = 3 - x^2$

31. $y = f(x) = x^3 - 4x$

32. $y = f(x) = \dfrac{1}{(x - 1)^2}$

In Exercises 33 through 36, solve the given problems involving inverse functions.

33. Express the perimeter p of a square as a function of its side x, and then find the inverse function.

34. Express the circumference C of a circle as a function of its radius x, and then find the inverse function.

35. A firm finds that its profit p from selling x items is given by $p = 10x - 50$. Find the inverse function. What use can be made of the inverse function?

36. The velocity v (in meters per second) of an object which has fallen for x seconds is given by the function $v = 9.8x$. Find the inverse function, and explain its use.

10-9 SUMMARY

In this chapter, we have introduced the mathematical concepts of a function and of a relation as well as the basic method of plotting the graph of a function. Graphs applicable to statistical types of data were also shown.

We pointed out that a function is essentially a general way of expressing an operation that is to be performed on the independent variable. The functional notation that was introduced is a general way of expressing the value of the function $f(a)$ for a given value a of the independent variable.

The circle graph is a valuable means of demonstrating the relationship of a whole category to the parts of the category. The bar graph is used to show the relative magnitude of discrete data. The broken-line graph shows how one variable changes as another variable changes continuously (usually).

The basic technique for plotting the graph of a function on the rectangular coordinate system is to assume values of the independent variable and find the corresponding values of the dependent variable. Care must be taken when handling signed numbers, and one must be sure that enough points are found to represent the graph properly.

The procedure for graphing inequalities was presented. The slope of a line was introduced, and the slope-intercept form of a straight line was discussed.

Finally, the inverse function was introduced. Along with the meaning of the inverse function we defined a one-to-one function and showed procedures for deriving and verifying the inverse function. The relationship between the graphs of the function and its inverse was demonstrated.

Below is a list of the terms introduced in this chapter. The reader should be certain that all the terms in the list are now familiar to him or her.

Abscissa
Bar graph
Broken-line graph
Circle graph
Coordinates
Dependent variable
Function
  inverse
  linear
  one-to-one
Graph of a function
Independent variable
Intercept
Ordinate
Origin
Parabola
Quadrant
Quadratic function
Rectangular coordinate system
Relation
Slope
Slope-intercept form
x-axis
y-axis
y-intercept

The exercises given below can be used to review the material of the chapter.

Exercises 10-9

In Exercises 1 through 12, find the indicated values of the given functions.

1. $f(x) = -x$; $f(-1)$, $f\left(-\frac{1}{2}\right)$
2. $f(x) = 1 - 3x$; $f(-2)$, $f\left(\frac{1}{3}\right)$
3. $f(y) = 3y - 5$; $f(-5)$, $f(-2)$
4. $f(x) = 2x^2 - 1$; $f(\sqrt{2})$, $f\left(-\frac{1}{2}\right)$
5. $f(t) = -t^2$; $f(3)$, $f(-4)$
6. $f(z) = z^2 - 2z - 3$; $f(-3)$, $f(0.2)$
7. $f(r) = -r^3$; $f(0)$, $f\left(\frac{r}{2}\right)$
8. $f(v) = 7 - v - 4v^2$; $f(-1)$, $f(-v)$
9. $f(w) = 2w^2 + 2$; $f(-2)$, $f(w^2)$
10. $f(p) = p^2 + p - 1$; $f(0)$, $f\left(\frac{p^2}{4}\right)$
11. $f(q) = -q^3 - 3$; $f(-1)$, $f(-2q)$
12. $f(x) = -2x^2 + 3x - 1$; $f(-1)$, $f(-x)$

In Exercises 13 through 16, plot the given points and then connect them to form a geometric figure.

13. $(0, 4), (-1, 2), (3, -2)$
14. $(-1, 0), (0, 1), (1, 2)$
15. $(2, 4), (3, 6), (-1, 2), (0, 0)$
16. $(4, 1), (0, -1), (2, 2), (1, -1)$

In Exercises 17 through 36, graph the given functions.

17. $y = 4x - 4$
18. $y = 4 - 4x$
19. $y = 3 - 2x$
20. $y = 3x + 2$
21. $y = 3 - \frac{1}{2}x$
22. $y = 4x - \frac{2}{3}$
23. $y = \frac{5x - 1}{4}$
24. $y = \frac{1}{5}x + \frac{1}{2}$
25. $y = 3x^2 - 4$
26. $y = 3 - 2x^2$
27. $y = \frac{x^2 - 2}{4}$
28. $y = 1 - 3x - x^2$
29. $y = 8 - x^3$
30. $y = 2^{x-1}$
31. $y = \left(\frac{1}{4}\right)^{2x}$
32. $y = 2x^2 - x^3$
33. $y = 4 - \frac{1}{x}$
34. $y = \frac{1}{x - 1}$
35. $y = 2\sqrt{x} - 3$
36. $y = \frac{1}{\sqrt{x}}$

In Exercises 37 through 40, find an equation of the line that passes through the two given points, and graph the line.

37. $(0, 0)$ and $(-1, -4)$
38. $(-1, 2)$ and $(1, -2)$
39. $(3, -4)$ and $\left(\frac{1}{3}, \frac{1}{2}\right)$
40. $\left(-\frac{2}{3}, 0\right)$ and $\left(0, \frac{2}{3}\right)$

In Exercises 41 through 44, find the slope and the y-intercept, and graph the line.

41. $y = -\frac{2}{3}x - 5$
42. $2x = y + 15$
43. $-2y + 3x + 2 = 0$
44. $4y - 5x - 7 = 0$

In Exercises 45 through 52, graph the given inequalities.

45. $x \geq 0$
46. $y \leq -1$
47. $y > 2 - 3x$
48. $y < 2x - 5$
49. $8y \leq 4 - x$
50. $4y \geq -2x + 1$
51. $y \geq -x^2 + 2$
52. $y \leq 4 + 3x^2$

In Exercises 53 through 56, graph each function and determine whether the function has an inverse.

53. $f(x) = -3x + 4$
54. $f(x) = -2x - 5$
55. $f(x) = -4x^2$
56. $f(x) = 5x^2$

In Exercises 57 through 64, find the inverse functions of the given functions, verify them, and graph the original function and the inverse function.

57. $y = f(x) = \frac{1}{2}x - 3$
58. $y = f(x) = \frac{1}{4}x - 7$
59. $y = f(x) = \frac{-3x + 11}{3}$

60. $y = f(x) = \dfrac{-4x + 12}{4}$ **61.** $y = f(x) = -\dfrac{3}{x}$ **62.** $y = f(x) = \dfrac{6}{x + 1}$

63. $y = f(x) = x^3 + 1$ **64.** $y = f(x) = x^5$

In Exercises 65 through 70, draw the appropriate graphs.

65. In a certain mathematics class, 5% of the class received A's, 15% received B's, 40% received C's, 25% received D's, and 15% received F's. Use a circle graph to represent these data.

66. In a high school senior class, it was found that 15 students had I.Q.'s below 100; 63 had I.Q.'s ranging from 100 to 109; 46 had I.Q.'s ranging from 110 to 119; 13 had I.Q.'s from 120 to 129; and 6 had I.Q.'s of 130 and above. Use a circle graph to represent these data.

67. Of the mountains in the world which are higher than 15,000 ft, there are 42 in Asia, 7 in Europe, 14 in North America, 28 in South America, and 7 elsewhere. Use a bar graph to represent these data.

68. A study of a certain musical sound yielded the following data for the frequency (a measure of the pitch) and the amplitude (a measure of loudness).

Frequency (cycles/s)	220	440	660	880	1100
Amplitude (units)	10	2	3	1	0.5

Use a bar graph to represent these data.

69. The approximate percentage of the population of the United States of persons 17 years of age or younger is shown in the following table. Use a broken-line graph to represent these data.

Year	1900	1910	1920	1930	1940	1950	1960	1970	1980
Percentage	42	40	38	35	30	31	36	35	30

70. The increase in length of a certain metal rod was measured as a function of the temperature with the following results.

Increase in length, mm	0.5	2.2	4.7	8.2	12
Increase in temperature, °C	50	100	150	200	250

Use a broken-line graph to represent these data.

In Exercises 71 through 78, determine the required functions.

71. A rectangle has a width of 4 in. and a length of x in. Express its perimeter as a function of x.

72. The sales tax in a certain state is 6%. Express the tax T on a certain item as a function of the cost C of the item.

73. The I.Q. of a person is defined as 100 times the ratio of his mental age M to his chronological age A. Express the I.Q. of a 15-year-old person as a function of M.

74. The rate H at which heat is developed in a filament of an electric light bulb is proportional to the square of the electric current I. A current of 0.5 A produces heat at the rate of 60 W. Express H as a function of I.

75. A kitchen exhaust fan should remove each minute a volume of air that is proportional to the floor area. A properly operating fan removes 200 ft^3 per min for a kitchen of 100 ft^2 of floor area. Express the volume of air that is removed per minute for a properly operating fan as a function of the floor area A.

76. The vertical side of the rectangle of a Norman window (a semicircle surmounted on a rectangle) is 2 m. Express the area of the window as a function of the radius of the circular part.

77. A number of white mice were subjected to various doses of radioactivity. It was found that the time T each lived afterward was inversely proportional to the square of the intensity I of the dose. Assuming that, on the average, a mouse lived 10 h after having received a dose of 4 units, express T as a function of I.

78. Express the perimeter of a square as a function of its area.

In Exercises 79 through 84, plot the dependent variable along the y-axis and the independent variable along the x-axis.

79. An art painting is 2 m high by 3 m wide. A museum wishes to make reproductions of differing sizes, keeping the ratio of height to width the same. Express h as a function of w and plot the graph.

80. A temperature in degrees Fahrenheit F equals 32 more than $\frac{9}{5}$ the number of degrees Celsius C. Express F as a function of C and plot the graph. From the graph determine the point at which $F = C$.

81. Express a number N which equals twice the square of a number x as a function of x. Plot the graph.

82. The path of a projectile is approximately parabolic. If y is the vertical distance from the ground to the projectile and x is the horizontal distance traveled, the equation $y = x - 0.0004x^2$ is the equation of the path of a given projectile, where the distances are measured in feet. Plot the graph, using units of 50 ft for x.

83. A rectangular tract of land is to have a perimeter of 800 ft. Express the area as a function of its width and plot the graph.

84. A rectangular table is to have a perimeter of 600 cm. Express the area as a function of its length and plot the graph.

11

EXPONENTIAL AND LOGARITHMIC FUNCTIONS

11-1 DEFINITION OF A LOGARITHM

In Chapters 4 and 8, we dealt in some detail with exponents. There is a special use of exponents which is important in computational work and for theoretical purposes. Exponents used in this manner are given the name **logarithms.** Today, with the extensive use of computers and electronic calculators, logarithms are used much less for computation than in the past. However, their usefulness in advanced mathematics and in applications in technical fields remains of great importance.

To illustrate the importance of logarithms, many applications may be cited. The basic units used to measure the intensity of sound, and those used to measure the intensity of earthquakes, are defined in terms of logarithms. In chemistry, the distinction between a base and an acid is defined in terms of logarithms. In electrical transmission lines, power gains and losses are measured in terms of logarithmic units.

Chapters 4 and 8 dealt with exponents in expressions of the form x^n, where we showed that n could be any rational number. Here we shall deal with expressions of the form b^x, where x is any real number. When we look at these expressions, we note the primary difference is that in the second expression *the exponent is variable. We therefore define*

$$\boxed{y = b^x} \qquad (11\text{-}1)$$

to be the **exponential function.** Here *x is called the* **logarithm** *of the number y to the base b.*

In our work with logarithms we shall restrict all numbers to the real number system. This leads us to choose the base as a positive number other than 1. We know that 1 raised to any power will result in 1, which would make y a constant regardless of the value of x. Negative numbers for b would result in imaginary values for y if x were any fractional exponent with an even integer for its denominator.

Example A $y = 2^x$ is an exponential function, where x is the logarithm of y to the base 2. This means that 2 raised to a given power gives us the corresponding value of y.

If $x = 2$, $y = 2^2 = 4$; this means that 2 is the logarithm of 4 to the base 2. If $x = 4$, $y = 2^4 = 16$; this means that 4 is the logarithm of 16 to the base 2. If $x = \frac{1}{2}$, $y = 2^{1/2} = 1.41$; this means that $\frac{1}{2}$ is the logarithm of 1.41 to the base 2.

Using the definition of a logarithm, Eq. (11-1) may be solved for x, and is written in the form

$$\boxed{x = \log_b y} \qquad (11\text{-}2)$$

This equation is read in accordance with the definition of x in Eq. (11-1): *x equals the logarithm of y to the base b*. Generally, then,

the logarithm of a number is that power to which the base b must be raised in order to obtain the number.

Now, since a logarithm is an exponent, and x is a logarithm, Eqs. (11-1) and (11-2) state the same relationship, but in a different manner. Equation (11-1) is the *exponential form*, and Eq. (11-2) is the *logarithmic form*.

Example B The equation $y = 2^x$ would be written as $x = \log_2 y$ if we put it in logarithmic form. When we choose values of y to find the corresponding values of x from this equation, we ask ourselves, "2 raised to what power gives y?" Hence, if $y = 4$, we know that 2^2 is 4, and x would be 2. If $y = 8$, $2^3 = 8$, or $x = 3$.

Example C $3^2 = 9$ in logarithmic form is $2 = \log_3 9$; $4^{-1} = \frac{1}{4}$ in logarithmic form is $-1 = \log_4(\frac{1}{4})$. Remember, the exponent may be negative. The base must be positive.

Example D

$(64)^{1/3} = 4$ in logarithmic form is $\frac{1}{3} = \log_{64} 4$

$(32)^{3/5} = 8$ in logarithmic form is $\frac{3}{5} = \log_{32} 8$

Example E

$\log_2 32 = 5$ in exponential form is $32 = 2^5$

$\log_6(\frac{1}{36}) = -2$ in exponential form is $\frac{1}{36} = 6^{-2}$

Example F Find b, given that $-4 = \log_b(\frac{1}{81})$.

Writing this in exponential form, we have $\frac{1}{81} = b^{-4}$, and using our knowledge of exponents, we solve for b:

$$\frac{1}{81} = b^{-4}$$

$$\frac{1}{3^4} = \frac{1}{b^4}$$

Since, in the equality, both b and 3 have the same exponent 4, $b = 3$.

Example G Find y, given that $\log_4 y = \frac{1}{2}$.

In exponential form we have $y = 4^{1/2}$, or $y = (2^2)^{1/2} = 2$.

We see that the exponential form is very useful for determining values written in logarithmic form. For this reason, it is important that you learn to transform readily from one form to the other.

Exercises 11-1

In Exercises 1 through 12, express the given equations in logarithmic form.

1. $3^3 = 27$ **2.** $5^2 = 25$ **3.** $4^4 = 256$ **4.** $8^2 = 64$

5. $4^{-2} = \frac{1}{16}$ **6.** $3^{-2} = \frac{1}{9}$ **7.** $2^{-6} = \frac{1}{64}$ **8.** $(12)^0 = 1$

9. $8^{1/3} = 2$ **10.** $(81)^{3/4} = 27$ **11.** $(\frac{1}{4})^2 = \frac{1}{16}$ **12.** $(\frac{1}{2})^{-2} = 4$

In Exercises 13 through 24, express the given equations in exponential form.

13. $\log_3 81 = 4$ **14.** $\log_{11} 121 = 2$ **15.** $\log_9 9 = 1$ **16.** $\log_{15} 1 = 0$

17. $\log_{25} 5 = \frac{1}{2}$ **18.** $\log_8 16 = \frac{4}{3}$ **19.** $\log_{243} 3 = \frac{1}{5}$ **20.** $\log_{1/32}(\frac{1}{8}) = \frac{3}{5}$

21. $\log_{10} 0.1 = -1$ **22.** $\log_7(\frac{1}{49}) = -2$ **23.** $\log_{0.5} 16 = -4$ **24.** $\log_{1/3} 3 = -1$

In Exercises 25 through 40, determine the value of the unknown.

25. $\log_4 16 = x$ **26.** $\log_5 125 = x$ **27.** $\log_{10} 0.01 = x$ **28.** $\log_{16}(\frac{1}{4}) = x$

29. $\log_7 y = 3$ **30.** $\log_8 N = 3$ **31.** $\log_8 y = -\frac{2}{3}$ **32.** $\log_7 y = -2$

33. $\log_b 81 = 2$ **34.** $\log_b 625 = 4$ **35.** $\log_b 4 = -\frac{1}{3}$ **36.** $\log_b 4 = \frac{2}{3}$

37. $\log_{10} 10^{0.2} = x$ **38.** $\log_5 5^{1.3} = x$ **39.** $\log_3 27^{-1} = x$ **40.** $\log_b(\frac{1}{4}) = -\frac{1}{2}$

In Exercises 41 through 44, perform the indicated operations.

41. If there are initially 1000 bacteria in a culture, and the number of bacteria then doubles each hour, the number of bacteria as a function of time is $N = 1000(2^t)$. By writing this equation in logarithmic form, solve for t.

42. In 1975 it was estimated that the total annual world demand for copper was $C = 9e^{0.08t}$, where C is in billions of kilograms of copper and t is the number of years after 1975. By rewriting this equation in logarithmic form, solve for t.

43. An equation relating the number N of atoms of radium at any time t in terms of the number of atoms at $t = 0$, N_0, is $\log_e(N/N_0) = -kt$, where k is a constant. By expressing this equation in exponential form, solve for N.

44. When dealing with the optical brightness of objects, the equation $D = \log_{10}(I_0/I)$ is found. By writing this equation in exponential form, solve for I.

11-2 GRAPHS OF $y = b^x$ AND $y = \log_b x$

When we are working with functions, we must keep in mind that a function is defined by the operation being performed on the independent variable, and not by the letter chosen to represent it. However, for consistency, it is normal practice to let y represent the dependent variable and x represent the independent variable. Therefore, *the* **logarithmic function** *is*

$$y = \log_b x \tag{11-3}$$

We note that in forming the logarithmic function, in effect we interchanged the x and y variables in Eq. (11-1) and solved for y, using the definition of a logarithm. This is precisely the method we developed in Section 10-8 to find an inverse function. Therefore, *the exponential function and the logarithmic function are inverse functions.*

Graphical representation of functions is often valuable when we wish to demonstrate their properties. We shall now show the graphs of the exponential function [Eq. (11-1)] and the logarithmic function [Eq. (11-3)].

Example A Plot the graph of $y = 2^x$.

Assuming values for x and then finding the corresponding values for y, we obtain the following table. We might recall that this is the same graph of Example A in Section 10-5.

x	-3	-2	-1	0	1	2	3
y	$\frac{1}{8}$	$\frac{1}{4}$	$\frac{1}{2}$	1	2	4	8

From these values we plot the curve, as shown in Fig. 11-1.

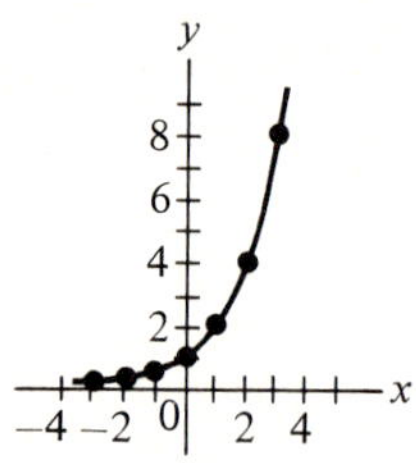

Figure 11-1

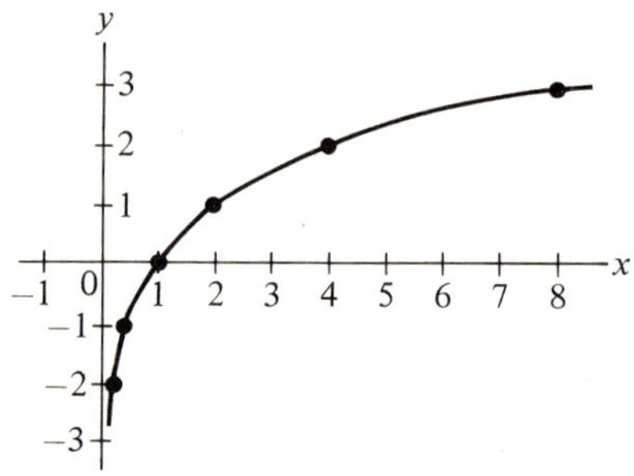

Figure 11-2

Example B Plot the graph of $y = \log_2 x$.

We can find the points for this graph more easily if we first write the equation in its exponential form $x = 2^y$. Then, by assuming values for y, we find the corresponding values for x.

x	$\frac{1}{4}$	$\frac{1}{2}$	1	2	4	8
y	-2	-1	0	1	2	3

Using these values, we construct the graph seen in Fig. 11-2.

If we compare the tables of values used in Examples A and B, we see that the x and y values are interchanged. This is further confirmation of the fact that the exponential and logarithmic functions are inverse functions. Also, as seen in Fig. 11-3, when we graph $y = 2^x$ and $y = \log_2 x$ on the same graph, their graphs are reflections of each other across the line $y = x$. Again, this is a basic characteristic of the graphs of inverse functions, as shown in Section 10-8.

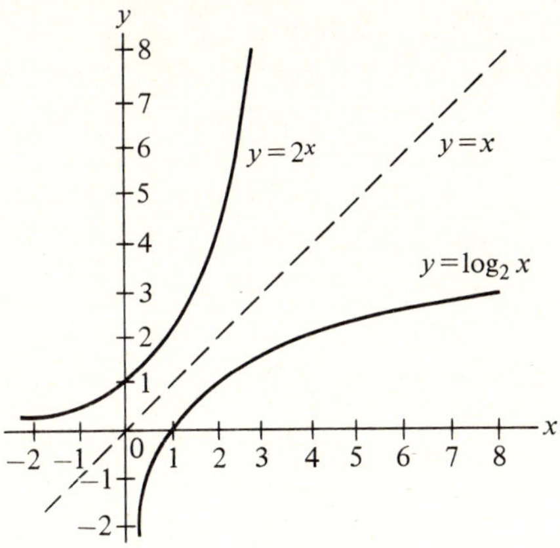

Figure 11-3

Any exponential or logarithmic curve, where $b > 1$, will be similar in shape to those of Examples A and B. From these curves we can draw certain conclusions:

(1) ***If $x < 1$, $\log_b x < 0$; if $x = 1$, $\log_b 1 = 0$; if $x > 1$, $\log_b x > 0$. Also, $\log_b x$ exists only for $x > 0$.***
(2) ***If $x > 1$, x increases more rapidly than $\log_b x$.***
(3) ***For all values of x, $b^x > 0$.***
(4) ***If $x > 1$, b^x increases more rapidly than x.***

Although the bases important to applications are greater than 1, to understand how the curve of the exponential function differs somewhat if $b < 1$, let us consider the following example.

Example C Plot the graph of $y = (\frac{1}{2})^x$.

The values are found for the table below; the graph is plotted in Fig. 11-4.

x	−3	−2	−1	0	1	2	3	4
y	8	4	2	1	$\frac{1}{2}$	$\frac{1}{4}$	$\frac{1}{8}$	$\frac{1}{16}$

We note that as the values of x increase, the values of $(\frac{1}{2})^x$ decrease. This is different from the behavior of $y = b^x$, where $b > 1$.

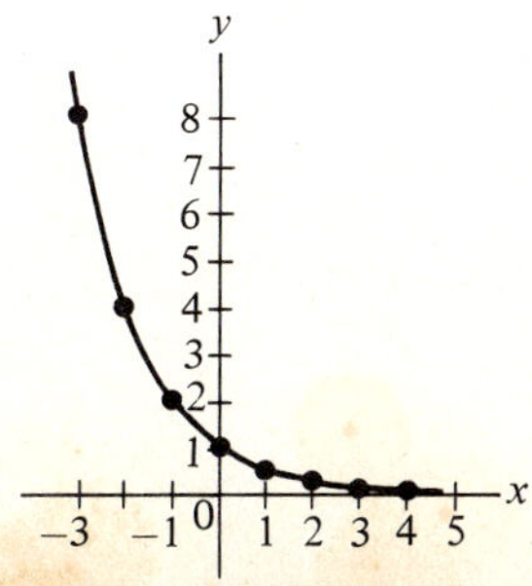

Figure 11-4

Example D Plot the graph of $y = 2 \log_3 x$.

As in Example B, it is generally more convenient to work with the exponential form. In order to change this equation to exponential form, we first divide each side by 2. Thus, we have

$$\frac{y}{2} = \log_3 x \quad \text{or} \quad x = 3^{y/2}$$

Now, choosing values of y, we then calculate corresponding values of x and obtain the following table.

x	0.3	0.6	1	1.7	3	5.2
y	−2	−1	0	1	2	3

See Fig. 11-5.

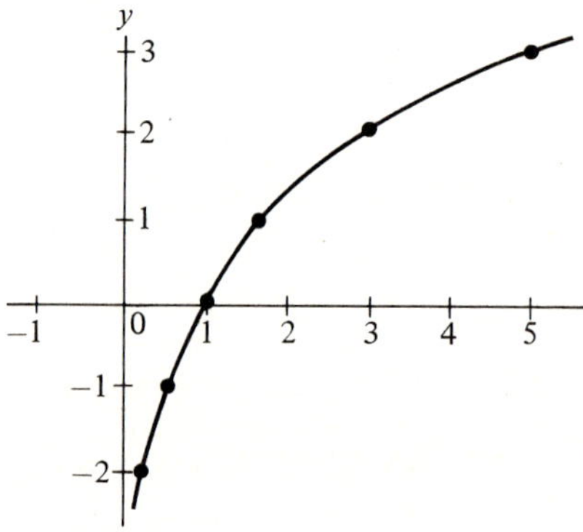

Figure 11-5

Exercises 11-2

In Exercises 1 through 16, plot the graphs of the given functions. In Exercises 1 through 8, values of x from −3 or −2 through 2 or 3 are appropriate, and in Exercises 9 through 16, values of y from −3 or −2 through 2 or 3 are appropriate.

1. $y = 3^x$ **2.** $y = 4^x$ **3.** $y = 10^x$ **4.** $y = 2.7^x$

5. $y = (\frac{1}{3})^x$ **6.** $y = (\frac{1}{4})^x$ **7.** $y = 2(2^x)$ **8.** $y = 1.5(4^x)$

9. $y = \log_3 x$ **10.** $y = \log_4 x$ **11.** $y = \log_{10} x$ **12.** $y = \log_{2.7} x$

13. $y = \log_{1.5} x$ **14.** $y = \log_{0.5} x$ **15.** $y = 3 \log_2 x$ **16.** $y = 0.5 \log_4 x$

In Exercises 17 and 18, plot the graphs of the given functions on the same graph. What conclusion can be seen from the graph?

17. $y = 3^x$ and $y = \log_3 x$ (See Exercises 1 and 9.)

18. $y = 4^x$ and $y = \log_4 x$ (See Exercises 2 and 10.)

In Exercises 19 and 20, plot the indicated graphs.

19. If \$1000 is placed in a bank account in which interest is compounded daily with an effective 8% annual interest, the amount in the account after t years is $A = 1000(1.08)^t$. Plot A as a function of t for $0 \le t \le 5$ years.

20. Under certain conditions the temperature T (in degrees Celsius) of a cooling object as a function of time (in minutes) is $T = 50.0(10^{-0.1t})$. Plot T as a function of t.

11-3 PROPERTIES OF LOGARITHMS

Since a logarithm is an exponent, it must follow the laws of exponents. Those laws which are of the greatest importance at this time are listed here for reference.

$$b^u \cdot b^v = b^{u+v} \tag{11-4}$$

$$b^u/b^v = b^{u-v} \tag{11-5}$$

$$(b^u)^n = b^{nu} \tag{11-6}$$

We shall now show how these laws for exponents give certain useful properties to logarithms.

If we let $u = \log_b x$ and $v = \log_b y$ and write these equations in exponential form, we have $x = b^u$ and $y = b^v$. Therefore, forming the product of x and y, we obtain

$$xy = b^u b^v = b^{u+v} \quad \text{or} \quad xy = b^{u+v}$$

Writing this last equation in logarithmic form yields

$$u + v = \log_b xy$$

or

$$\log_b x + \log_b y = \log_b xy \tag{11-7}$$

Equation (11-7) states the property that *the logarithm of the product of two numbers is equal to the sum of the logarithms of the numbers.*

Using the same definitions of u and v to form the quotient of x and y, we then have

$$\frac{x}{y} = \frac{b^u}{b^v} = b^{u-v} \quad \text{or} \quad \frac{x}{y} = b^{u-v}$$

Writing this last equation in logarithmic form, we have

$$u - v = \log_b\left(\frac{x}{y}\right)$$

or

$$\boxed{\log_b x - \log_b y = \log_b\left(\frac{x}{y}\right)} \qquad (11\text{-}8)$$

Equation (11-8) states the property that *the logarithm of the quotient of two numbers is equal to the logarithm of the numerator minus the logarithm of the denominator*.

If we again let $u = \log_b x$ and write this in exponential form, we have $x = b^u$. To find the nth power of x, we write

$$x^n = (b^u)^n = b^{nu}$$

Expressing this equation in logarithmic form yields

$$nu = \log_b(x^n)$$

or

$$\boxed{n \log_b x = \log_b(x^n)} \qquad (11\text{-}9)$$

This last equation states that *the logarithm of the nth power of a number is equal to n times the logarithm of the number*. The exponent n may be integral or fractional and, therefore, we may use Eq. (11-9) for finding powers and roots of numbers.

Example A Using Eq. (11-7), we may express $\log_4 15$ as a sum of logarithms as follows:

$$\log_4 15 = \log_4(3 \cdot 5) = \log_4 3 + \log_4 5$$

Using Eq. (11-8), we may express $\log_4(\frac{5}{3})$ as the difference of logarithms as follows:

$$\log_4(\tfrac{5}{3}) = \log_4 5 - \log_4 3$$

Using Eq. (11-9), we may express $\log_4 9$ as twice $\log_4 3$ as follows:

$$\log_4 9 = \log_4(3^2) = 2 \log_4 3$$

Example B Using Eqs. (11-7) through (11-9), we may express a sum or difference of logarithms as the logarithm of a single quantity.

$$\log_4 3 + \log_4 x = \log_4(3 \cdot x) = \log_4 3x$$

$$\log_4 3 - \log_4 x = \log_4\left(\frac{3}{x}\right)$$

$$\log_4 3 + 2\log_4 x = \log_4 3 + \log_4(x^2) = \log_4 3x^2$$

$$\log_4 3 + 2\log_4 x - \log_4 y = \log_4\left(\frac{3x^2}{y}\right)$$

We may use Eq. (11-9) to find another important property of logarithms. Since $b = b^1$ in logarithmic form is $\log_b b = 1$, we have

$$\log_b(b^n) = n\log_b b = n(1) = n$$

or

$$\boxed{\log_b(b^n) = n} \qquad (11\text{-}10)$$

Equation (11-10) may be used to evaluate certain logarithms when the exact values may be determined.

Example C We may evaluate $\log_3 9$ in the following manner. Using Eq. (11-10), we have

$$\log_3 9 = \log_3(3^2) = 2$$

We can establish the exact value since the base of logarithms and the number being raised to the power are the same. Of course, this could have been evaluated directly from the definition of a logarithm.

However, we cannot establish an exact value of $\log_4 9$ in this way. We have

$$\log_4 9 = \log_4(3^2) = 2\log_4 3$$

We must leave it in this form until we are able to evaluate $\log_4 3$. This can be done, and we shall consider this type of expression later in the chapter.

Example D Calculate $\log_3(3^{0.4})$.

Using Eq. (11-10), we may calculate this logarithm. This gives us

$$\log_3(3^{0.4}) = 0.4$$

Although we did not calculate $3^{0.4}$ at this point, we were able to calculate $\log_3 3^{0.4}$.

Example E We may also use Eq. (11-10) in combination with other properties of logarithms. The following illustration shows how it can be used alone or in combination with Eq. (11-8) to evaluate the indicated logarithm.

$$\log_5(\tfrac{1}{25}) = \log_5 1 - \log_5 25 = 0 - \log_5(5^2) = -2$$

$$\log_5(\tfrac{1}{25}) = \log_5(5^{-2}) = -2$$

Either method is appropriate.

In the following examples, the properties of logarithms of Eqs. (11-7) through (11-10) are further illustrated.

Example F

$$\log_2 6 = \log_2(2 \cdot 3) = \log_2 2 + \log_2 3 = 1 + \log_2 3$$

$$\log_3(\tfrac{2}{9}) = \log_3 2 - \log_3 9 = \log_3 2 - \log_3(3^2) = \log_3 2 - 2$$
$$= -2 + \log_3 2$$

Example G

$$\log_{10}\sqrt{7} = \log_{10}(7^{1/2}) = \tfrac{1}{2}\log_{10} 7$$

This demonstrates the property which is especially useful for finding roots of numbers. We see that we need merely to multiply the logarithm of the number by the fractional exponent representing the root to obtain the logarithm of the root.

Similarly,

$$\log_{10}\sqrt[3]{10x} = \log_{10}(10x)^{1/3} = \tfrac{1}{3}\log_{10}(10x)$$
$$= \tfrac{1}{3}(\log_{10} 10 + \log_{10} x)$$
$$= \tfrac{1}{3}(1 + \log_{10} x)$$

Exercises 11-3

In Exercises 1 through 12, express each as a sum, difference, or multiple of logarithms. (See Example A.)

1. $\log_5 xy$ **2.** $\log_3 7y$ **3.** $\log_7\left(\frac{5}{a}\right)$ **4.** $\log_3\left(\frac{r}{s}\right)$

5. $\log_2(a^3)$ **6.** $\log_8(n^5)$ **7.** $\log_6 abc$ **8.** $\log_2\left(\frac{xy}{z}\right)$

9. $\log_5\sqrt[4]{y}$ **10.** $\log_4\sqrt[7]{x}$ **11.** $\log_2\left(\frac{\sqrt{x}}{a^2}\right)$ **12.** $\log_3\left(\frac{\sqrt{y}}{8}\right)$

In Exercises 13 through 20, express each as the logarithm of a single quantity. (See Example B.)

13. $\log_b a + \log_b c$ **14.** $\log_2 3 + \log_2 x$ **15.** $\log_5 9 - \log_5 3$
16. $\log_8 6 - \log_8 a$ **17.** $\log_b x^2 - \log_b \sqrt{x}$ **18.** $\log_4 3^3 + \log_4 9$
19. $2 \log_e 2 + 3 \log_e n$ **20.** $\frac{1}{2} \log_b a - 2 \log_b 5$

In Exercises 21 through 28, determine the exact value of each of the given logarithms.

21. $\log_2(\frac{1}{32})$ **22.** $\log_3(\frac{1}{81})$ **23.** $\log_2(2^{2.5})$ **24.** $\log_5(5^{0.1})$
25. $\log_7 \sqrt{7}$ **26.** $\log_6 \sqrt[3]{6}$ **27.** $\log_3 \sqrt[4]{27}$ **28.** $\log_5 \sqrt[3]{25}$

In Exercises 29 through 40, express each as a sum, difference, or multiple of logarithms. In each case, part of the logarithm may be determined exactly.

29. $\log_3 18$ **30.** $\log_5 75$ **31.** $\log_2(\frac{1}{6})$ **32.** $\log_{10}(0.05)$
33. $\log_3 \sqrt{6}$ **34.** $\log_2 \sqrt[3]{24}$ **35.** $\log_2(4^2 \cdot 3^3)$ **36.** $\log_7(7^4 \cdot 3^5)$
37. $\log_{10} 3000$ **38.** $\log_{10}(40^2)$ **39.** $\log_{10}(\frac{27}{100})$ **40.** $\log_5(\frac{4}{125})$

In Exercises 41 through 44, using $\log_{10} 2 = 0.301$, *evaluate each of the given expressions.*

41. $\log_{10} 4$ **42.** $\log_{10} 20$ **43.** $\log_{10}(0.5)$ **44.** $\log_{10} 8000$

In Exercises 45 and 46, plot the indicated graphs.

45. Plot the graphs of $y = 2 \log_2 x$ and $y = \log_2 x^2$, and show that they are the same.
46. Plot the graphs of $y = \log_2 4x$ and $y = 2 + \log_2 x$, and show that they are the same.

11-4 POWERS OF TEN; COMMON LOGARITHMS

In Section 11-1, we stated that a base of logarithms must be a positive number, not equal to one. In the examples and exercises of the previous sections we used a number of different bases. There are, however, only two bases which are used extensively. They are 10 and e, where e equals approximately 2.718.

Base 10 logarithms are used primarily for calculations and in certain types of applications. Base e logarithms are used extensively in technical and scientific work. In this section, we discuss how to determine the base 10 logarithm of any given number. Base e logarithms are considered in Section 11-7.

Logarithms to the base 10 are known as **common logarithms.** Since we know that a logarithm is the power to which the base must be raised in order to equal a given number, we first briefly consider powers of ten.

From our previous work with exponents, we know that

$$10^0 = 1 \quad \text{and} \quad 10^1 = 10$$

Also, recalling the relationship between roots of numbers and fractional exponents, we can establish other powers of 10 as shown in the following example.

Example A From Table 1, we find that $\sqrt{10} = 3.162$. Since $\sqrt{10} = 10^{1/2}$, we therefore know that

$$10^{1/2} = 3.162$$

From other available tables, we find that $\sqrt[3]{10} = 2.154$. Therefore, since $\sqrt[3]{10} = 10^{1/3}$,

$$10^{1/3} = 2.154$$

Let us now tabulate the results we have found. Also, instead of writing the exponents as fractions, we shall write the exponents in their equivalent decimal form. For example, instead of writing $10^{1/4}$, we shall write $10^{0.2500}$. This will prove to be convenient as we go on. Therefore, we have

$$10^{0.0000} = 1.000$$
$$10^{0.3333} = 2.154$$
$$10^{0.5000} = 3.162$$
$$10^{1.0000} = 10.000$$

Thus, we see that it is possible to raise 10 to a power to obtain a given number. However, this is the definition of a logarithm to the base 10. Therefore, *these powers of 10 are the common logarithms of the numbers* on the right.

In general, given a number N,

$$\boxed{10^x = N} \quad (11\text{-}11)$$

means that

$$\boxed{\log N = x} \quad (11\text{-}12)$$

where the symbol $\log N$, without any base being noted, denotes the common logarithm of N.

Example B $10^{0.2500} = 1.778$ means that the logarithm to the base 10, or the common logarithm, of 1.778 is 0.2500. We therefore write this as $\log 1.778 = 0.2500$.

10^2 means that $\log 100 = 2$.

We have now established the meaning of the common logarithm of a number, and that these logarithms are used in making calculations. We will now discuss certain relevant aspects about the numbers that we shall use in these calculations.

In the discussions which follow in this chapter, we shall be using numbers that have no more than three **significant digits** (except for Section 11-6, where numbers with four significant digits will be used). By this statement, we mean that *the first nonzero digit and the next two digits which appear in any given number are the only ones we will be able to consider* (*other than any zeros used to locate the decimal point*) *in any calculation. If* a given number contains additional digits, it must be **rounded off.** Consider the illustrations in the following examples.

Example C The number 58,300 has three significant digits: 5, 8, and 3. The two zeros are used only to locate the decimal place properly.

The number 0.00106 has three significant digits: 1, 0, and 6. The first three zeros are used to locate the decimal point properly.

The number 8.9136 has five significant digits. For our purposes in this chapter, it must be rounded off to three significant digits. (See Example D.)

The following numbers have the indicated number of significant digits:

0.114 (three), 16,000 (two), 1.034 (four), 10.7 (three), 0.1008 (four)

103,000,000 (three) 1,000,000,000,000 (one)

Example D The number 8.9136 has five significant digits. In rounding it off to three significant digits, we replace it with that number with three significant digits to which it is closest. The numbers 8.91 and 8.92 are the three-significant-digit numbers that are the nearest to 8.9136. However, 8.91 is closer than 8.92. This means that 8.9136 rounded off to three significant digits is 8.91.

For simplicity, we shall adopt the following policy of rounding off:

1. Drop all appropriate digits and replace them with zeros.
2. Keep only enough zeros to locate the decimal point properly.
3. If the first digit dropped was 0, 1, 2, 3, or 4, the rounding off is completed.
4. If the first digit dropped was 5, 6, 7, 8, or 9, add 1 to the last digit not dropped.

Example E The following numbers are shown along with their three significant-digit equivalents.

67,428—67,400 36,870,800—36,900,000

0.1055—0.106 0.0010073—0.00101

827.45—827 918,400—918,000

24,045—24,000 (The first zero is significant since we rounded off the number to *three* significant digits.)

We return now to finding logarithms of numbers. Table 2 in Appendix C gives four-place logarithms (the logarithms are written with four significant digits) for numbers with three significant digits from 1.00 to 9.99. From our discussion, we know that these are decimals ranging from 0.0000 to 0.9996. We also know that these logarithms are powers to which 10 must be raised in order to equal the given number N. The method of finding the logarithm of a number from 1.00 to 9.99 from this table is illustrated in the following example.

Example F To find the logarithm of 8.36, we first look under N to find 83 (the first two significant digits). To the right of 83 and under 6 (the third significant digit) we find the digits 9222. Therefore,

$$\log 8.36 = 0.9222$$

This means that $10^{0.9222} = 8.36$.

To find the logarithm of 4.08, we look to the right of 40 and under 8 and find the digits 6107. Thus,

$$\log 4.08 = 0.6107$$

which means that $10^{0.6107} = 4.08$.

Now the question arises as to how one may find the logarithm of a positive number not between 1.00 and 9.99. Let us first consider numbers greater than 10.

By writing a number greater than 10 as the product of a number between 1.00 and 9.99 and an integral power of 10, we can use the basic laws of exponents along with the meaning of the logarithm to find the logarithms of such numbers. Consider the following example.

Example G To find the logarithm of 5340, we first write

$$5340 = 5.34 \times 10^3$$

We find the logarithm of 5.34 to be 0.7275, which means that $5.34 = 10^{0.7275}$. Thus,

$$5340 = 5.34 \times 10^3 = (10^{0.7275})(10^{3.0000}) = 10^{3.7275}$$

Since 3.7275 is the required power to which 10 must be raised to equal 5340, we have

$$\log 5340 = 3.7275$$

Generalizing on this example, if we express a positive number N as

$$\boxed{N = M \times 10^C} \tag{11-13}$$

we note that the logarithm of N is given by the equation

$$\boxed{\log N = C + \log M} \tag{11-14}$$

Now, in Eq. (11-13) we have expressed the number N in **scientific notation.** *That is, it has been written as the product of a number M between 1 and 10 and a power of 10.*

The following examples show how to change a number from ordinary notation to scientific notation.

Example H

$$56{,}000 = 5.6(10{,}000) = 5.6 \times 10^4$$

$$0.143 = \frac{1.43}{10} = 1.43 \times 10^{-1}$$

$$0.000804 = \frac{8.04}{10{,}000} = 8.04 \times 10^{-4}$$

$$2.97 = 2.97(1) = 2.97 \times 10^0$$

From Example H we can establish a basic method of changing numbers from ordinary notation to scientific notation. *The decimal point is moved so that only one nonzero digit is to its left. The number of places moved is the value of C. It is positive if the decimal point is moved to the left, and it is negative if it is moved to the right.*

Example I

$$56000 = 5.6 \times 10^4$$

4 places

$$0.143 = 1.43 \times 10^{-1}$$

1 place

$$0.000804 = 8.04 \times 10^{-4}$$

4 places

$$2.97 = 2.97 \times 10^0$$

0 places

To change a number from scientific notation to ordinary notation, the procedure described above is reversed. The following example illustrates the method.

Example J To change 4.18×10^7 to ordinary notation, we must move the decimal point seven places to the right. Therefore, additional zeros must be included for the proper location of the decimal point. Thus, $4.18 \times 10^7 = 41{,}800{,}000$.

To change 7.28×10^{-2} to ordinary notation, we must move the decimal point two places to the left. Again, additional zeros must be included. Thus, $7.28 \times 10^{-2} = 0.0728$.

In Eq. (*11-14*), which is $\log N = C + \log M$, *C is known as the* **characteristic** *of the logarithm of N, and* $\log M$ *is the* **mantissa** *of the logarithm of N. Thus: The characteristic is the integral power of 10 of the number in scientific notation, and the mantissa is the logarithm of the number between* 1 *and* 10 *of this form.*

Example K $69{,}200 = 6.92 \times 10^4$. Therefore,

$$\log 69{,}200 = 4 + \log 6.92 = 4 + 0.8401 = 4.8401$$

$20.6 = 2.06 \times 10^1$. Therefore,

$$\log 20.6 = 1 + \log 2.06 = 1 + 0.3139 = 1.3139$$

To find the logarithm of a positive number less than 1, we follow the same method. The only apparent difference is that the characteristic is negative. Consider the following example.

Example L $0.0272 = 2.72 \times 10^{-2}$. Therefore,

$$\log 0.0272 = -2 + \log 2.72 = -2 + 0.4346$$

We shall find that for some purposes this form of the logarithm is somewhat inconvenient. Thus, we shall express the characteristic as $8 - 10$. In fact, we shall follow the procedure of expressing all negative characteristics in a similar manner: subtracting 10 from a positive number. Thus,

$$\begin{aligned} \log 0.0272 &= -2 + 0.4346 = 8 - 10 + 0.4346 = 8 + 0.4346 - 10 \\ &= 8.4346 - 10 \end{aligned}$$

(Note that the form $-2 + 0.4346$ *cannot* be written as -2.4346, for this would indicate that the entire number is negative. Only the 2 is negative, but the 0.4346 is positive.)

Example M

Number	Form in Scientific Notation	Indicated Movement of Decimal Point	Characteristic	Mantissa	Logarithm
12	(1.2)(10)	1.2	+1	0.0792	1.0792
30,400	$(3.04)(10^4)$	3.0400	+4	0.4829	4.4829
0.123	$(1.23)(10^{-1})$	0.123	−1	0.0899	9.0899 − 10
0.00065	$(6.5)(10^{-4})$	0.00065	−4	0.8129	6.8129 − 10
7.68	$(7.68)(10^0)$	7.68	0	0.8854	0.8854

As mentioned in Section 11-1, some of the important measurements in scientific and technical work are defined in terms of logarithms. The following example illustrates one such application, and another is found in the exercises.

Example N It has been found that the human ear responds to sound on a scale which is approximately proportional to the logarithm of the intensity of the sound. Thus, the loudness of a sound, measured in decibels (dB), is defined by the equation $b = 10 \log(I/I_0)$, where I is the intensity of the sound and I_0 is the minimum intensity detectable.

A busy city street has traffic and other noise with an intensity of 10^7 times the minimum value of I_0. What is the loudness of this noise in decibels? From this statement, we know that $I = 10^7 I_0$. Thus,

$$\frac{I}{I_0} = \frac{10^7 I_0}{I_0} = 10^7$$

Therefore, the loudness is given by

$$b = 10 \log 10^7$$

Since $\log 10^7 = 7$, we have

$$b = 10 \log 10^7 = (10)(7)$$
$$= 70 \text{ dB}$$

This section has demonstrated how to find the logarithm of a positive number. It might be pointed out that more errors occur in the proper determination of the characteristic than in any other step.

Exercises 11-4

In Exercises 1 through 4, rewrite the given expressions so that the powers of 10 *are written in decimal, not fractional, form. Express each power to four decimal places.*

1. $10^{2/3} = 4.642$ **2.** $10^{2/7} = 1.931$ **3.** $10^{9/10} = 7.943$ **4.** $10^{11/20} = 3.548$

In Exercises 5 through 8, rewrite the given expressions so that decimal powers of 10, *expressed to four decimal places, are substituted for the given roots of* 10.

5. $\sqrt[6]{10} = 1.468$ **6.** $\sqrt[7]{10} = 1.390$ **7.** $\sqrt[5]{10^2} = 2.512$ **8.** $\sqrt[8]{10^7} = 7.499$

In Exercises 9 through 12, identify the logarithm of the number on the left.

9. $1000 = 10^3$ **10.** $10{,}000 = 10^4$ **11.** $2 = 10^{0.3010}$ **12.** $3 = 10^{0.4771}$

In Exercises 13 through 16, determine the number of significant digits in each number.

13. 845.2; 0.0071 **14.** 60.248; 0.1002 **15.** 4000; 0.004 **16.** 0.000207; 3.40

In Exercises 17 through 20, round off each of the given numbers to three significant digits.

17. 7.856; 0.95672 **18.** 88327; 4.3949 **19.** 34.55; 9070 **20.** 30050; 30005

In Exercises 21 through 36, determine the characteristic of the logarithm of each given number. Write any negative characteristics in the -10 *notation.*

21. 27.5 **22.** 356 **23.** 4090 **24.** 750000
25. 1.04 **26.** 6.92 **27.** 0.106 **28.** 0.725
29. 0.00509 **30.** 0.000752 **31.** 89.2 **32.** 63300000
33. 0.0395 **34.** 0.00000632 **35.** 0.0649 **36.** 0.998

In Exercises 37 through 56, determine the logarithm of each given number by using Table 2.

37. 6.93 **38.** 3.17 **39.** 5.02 **40.** 6.10
41. 5800 **42.** 79,000 **43.** 105 **44.** 900,000
45. 29 **46.** 85.7 **47.** 879 **48.** 707,000,000
49. 0.0642 **50.** 0.00163 **51.** 0.303 **52.** 0.990
53. 0.000863 **54.** 0.0101 **55.** 0.00000884 **56.** 0.0909

In Exercises 57 through 60, determine the logarithm of the numbers given in each problem.

57. A certain radio station broadcasts at a frequency of 1340000 Hz.

58. The Grand Coulee Dam contains about 10,600,000 yd^3 of concrete.

59. The mass of the Milky Way has been estimated to be about 6×10^{41} lb.

60. The coefficient of thermal expansion for aluminum is 2.2×10^{-5} per degree Celsius.

In Exercises 61 and 62, solve the given applied problems. (*See Example N.*)

61. Riveting has a sound intensity 10^{10} times that of the minimum level. What is the loudness of riveting in decibels? How many times greater is the intensity of the sound of riveting than the noise of the city street in Example N?

62. Measured on the Richter scale, the magnitude of an earthquake of intensity I is defined as $R = \log(I/I_0)$, where I_0 is a minimum level for comparison. What is the Richter scale reading of an earthquake whose intensity is 10^8 times I_0? (This is approximately the value of the 1906 San Francisco earthquake.)

11-5 COMPUTATIONS USING LOGARITHMS

Logarithms were developed in the seventeenth century for the purpose of making tedious and complicated calculations which arose in astronomy and navigation. Until recently (the mid 1970s), when calculators came into extensive use, logarithms were used commonly for calculational purposes. The slide rule was also used extensively, but its accuracy was generally limited to three significant digits. (The construction of a slide rule is based on logarithms.) Also, logarithms are used for some computations performed by calculators.

Logarithms can be used to make more complicated calculations by means of additions, subtractions, and basic multiplications and divisions. By means of logarithms and the basic arithmetic operations, we can determine the answer to such problems as $\sqrt[5]{7.60}$ or $(89.1)^{0.3}$. In this section, we show how logarithms are used to make such calculations to an accuracy up to four significant digits. (Greater accuracy is obtainable with more extensive tables.) Even if logarithms are not required for such calculations, performing them with logarithms provides an opportunity to better understand the meaning and properties of logarithms. The following examples illustrate the use of logarithms in computations.

Example A By the use of logarithms, calculate the value of (2.15)(3.84).

Equation (11-7) tells us that $\log xy = \log x + \log y$. If we find log 2.15 and log 3.84 and add them, we shall have the logarithm of the required product. Using this result, we look up the **antilogarithm,** which is the desired product.

$$\begin{aligned} \log 2.15 &= 0.3324 \\ \log 3.84 &= 0.5843 \\ \hline \log(2.15)(3.84) &= 0.9167 \end{aligned}$$

If we now find the number whose logarithm is 0.9167, which is the antilogarithm of 0.9167, we will have the desired product. Looking in the body of Table 2, we find that the number closest to 9167 is 9165, which we shall use to determine our answer. Since 0.9165 is the logarithm of 8.25, we have

$$(2.15)(3.84) = 8.25$$

Generalizing on the method used in Example A, we have the following procedure for multiplication by use of logarithms.

(1) ***Determine the logarithms of the numbers to be multiplied.***
(2) ***Find the sum of these logarithms.***
(3) ***The required product is the antilogarithm of this sum.***

The following examples illustrate the method.

Example B Find the product of 1.30 and 120 by use of logarithms.

$$\begin{aligned}\log 1.30 &= 0.1139\\ \log 120 &= \underline{2.0792}\\ &\ 2.1931\end{aligned}$$

From the table, a mantissa of 0.1931 is the logarithm of 1.56. Since the characteristic of the logarithm is 2, we have

$$(1.30)(120) = (1.56)(10^2) = 156$$

Example C Find the product of 0.00472 and 60.8 by logarithms.

$$\begin{aligned}\log 0.00472 &= 7.6739 - 10\\ \log 60.8 &= \underline{1.7839}\\ &\ 9.4578 - 10\end{aligned}$$

From the table, the mantissa nearest to 0.4578 is 0.4579. Using this value, we have the logarithm of 2.87. Since the characteristic is $9 - 10 = -1$, the antilogarithm of 9.4578 − 10 is 0.287. Thus,

$$(0.00472)(60.8) = (2.87)(10^{-1}) = 0.287$$

The next type of calculation we consider is that of the division of one number by another number. Consider the following example.

Example D Find the quotient $\dfrac{629}{27.0}$ by using logarithms.

From Eq. (11-8) we know that $\log(x/y) = \log x - \log y$. Therefore, by subtracting log 27.0 from log 629, we shall have the logarithm of the quotient. The antilogarithm gives the desired result.

$$\begin{aligned}\log 629 &= 2.7987\\ \log 27.0 &= \underline{1.4314}\\ &\ 1.3673\end{aligned}$$

From the table, the mantissa nearest to 0.3673 is 0.3674, which is the logarithm of 2.33. The characteristic is 1. Therefore,

$$\frac{629}{27.0} = (2.33)(10) = 23.3$$

From Example D, we see that *to divide one number by another using logarithms, we subtract the logarithm of the denominator from the logarithm of the numerator and then find the antilogarithm of the result.* The following example illustrates the method.

Example E Find the quotient $\frac{39.2}{0.0512}$ by using logarithms.

To avoid any possible confusion in the subtraction and the results, we shall set up the logarithms such that the result of subtracting the parts of the logarithms containing the mantissa will be positive. To do this for this problem, we express the characteristic of the logarithm of 39.2 as $11 - 10$. Therefore, we have the following steps.

$$\begin{aligned} \log 39.2 &= 11.5933 - 10 \\ \log 0.0512 &= \underline{\ \ 8.7093 - 10} \\ & \quad\ \ 2.8840 \end{aligned}$$

From the table, the mantissa nearest to 0.8840 is 0.8842, which is the logarithm of 7.66. The characteristic is 2. (Note that in the subtraction of the logarithms the -10's are subtracted out.) Therefore,

$$\frac{39.2}{0.0512} = (7.66)(10^2) = 766$$

We shall now see how to raise numbers to powers and take roots of numbers by using logarithms. The basic technique is developed in the following examples.

Example F Evaluate $(2.05)^3$ by logarithms.

From Eq. (11-9), we know that $\log x^n = n \log x$. Therefore, by multiplying log 2.05 by 3, we shall have the logarithm of the power. The antilogarithm gives the desired result.

$$\log 2.05 = 0.3118$$

$$3(\log 2.05) = 3(0.3118) = 0.9354$$

In the table, the mantissa nearest to 0.9354 is 0.9355, which is the logarithm of 8.62. Therefore,

$$(2.05)^3 = 8.62$$

Example G Evaluate $\sqrt{86.7}$ by logarithms.

By writing $\sqrt{86.7} = 86.7^{1/2}$, we can again make use of Eq. (11-9) to find this square root. By multiplying log 86.7 by $\frac{1}{2}$, we have the logarithm of the square root. Again, the antilogarithm gives the desired square root.

$$\log 86.7 = 1.9380$$

$$\frac{1}{2}(\log 86.7) = \frac{1}{2}(1.9380) = 0.9690$$

In the table, the mantissa nearest to 0.9690 is 0.9689, which is the logarithm of 9.31. Therefore,

$$\sqrt{86.7} = 9.31$$

From Examples F and G, we see that *to raise a number to a power by using logarithms, we multiply the logarithm of the number by the exponent and find the antilogarithm of the result. In finding the root of a number, we interpret the root as a fractional exponent and proceed as with the power of a number.* The following examples further illustrate the method.

Example H Evaluate $(6840)^5$ by logarithms.

$$\log 6840 = 3.8351$$

$$5(\log 6840) = 5(3.8351) = 19.1755$$

In the table, the mantissa nearest to 0.1755 is 0.1761, which is the logarithm of 1.50. The characteristic is 19. Therefore,

$$(6840)^5 = 1.50(10^{19})$$

This is the most convenient form of the result. If we write it in ordinary notation, we have 15,000,000,000,000,000,000.

Example I Evaluate $\sqrt[3]{0.916}$ by logarithms.

$$\sqrt[3]{0.916} = (0.916)^{1/3}$$

$$\log 0.916 = 9.9619 - 10$$

In order to have 10 subtracted from the term containing the mantissa after the division, we write the characteristic as 29 − 30. Therefore,

$$\log 0.916 = 29.9619 - 30$$

$$\frac{1}{3}(\log 0.916) = \frac{1}{3}(29.9619 - 30) = 9.9873 - 10$$

In the table, the mantissa nearest to 0.9873 is 0.9872, which is the logarithm of 9.71. The characteristic is $9 - 10 = -1$. Therefore,

$$\sqrt[3]{0.916} = 9.71(10^{-1}) = 0.971$$

When evaluating an expression in which several multiplications and divisions are indicated, the basic properties of logarithms indicate that the following procedure should be followed.

(1) ***Find the logarithm of each number.***
(2) ***Find the sum of the logarithms of the numbers in the numerator.***
(3) ***Find the sum of the logarithms of the numbers in the denominator.***
(4) ***Subtract the result of Step 3 from that of Step 2.***
(5) ***The desired result is the antilogarithm of the difference found in Step 4.***

Example J Evaluate $\dfrac{(65.1)\sqrt{804}}{(3.07)(567)}$ by logarithms.

To find the logarithm of the numerator, we add log 65.1 to $\frac{1}{2}$ log 804. (Remember, $\log 804^{1/2} = \frac{1}{2}\log 804$.) To find the logarithm of the denominator, we add log 3.07 to log 567. We then subtract these results and find the antilogarithm.

$$\begin{array}{lrr} & \log 65.1 = 1.8136 & \log 3.07 = 0.4871 \\ \log 804 = 2.9053 & \frac{1}{2}(\log 804) = \underline{1.4527} & \log 567 = \underline{2.7536} \\ & 3.2663 & 3.2407 \\ & \underline{3.2407} \text{ (subtract)} & \\ & 0.0256 & \end{array}$$

In the table, the mantissa nearest to 0.0256 is 0.0253, which is the logarithm of 1.06. Therefore,

$$\frac{(65.1)\sqrt{804}}{(3.07)(567)} = 1.06$$

Calculations such as those shown in this section can be quickly and easily performed with an electronic calculator. However, if a calculator is not available, logarithms provide a method by which such calculations can be made. It can be seen that powers and roots of numbers can easily be determined by use of logarithms. It should also be pointed out that calculators make use of logarithms in performing many of these types of calculations.

Exercises 11-5

In each of the following exercises, perform the indicated calculations by means of logarithms.

1. $(5.20)(1.83)$ **2.** $(31.9)(4.92)$ **3.** $(0.831)(92.1)$

4. $(0.0726)(452)$ **5.** $(6000)(3.18)$ **6.** $(0.00709)(0.847)$

7. $(78{,}000)(1.47)$ **8.** $(8080)(0.0715)$ **9.** $\dfrac{8.12}{3.07}$ **10.** $\dfrac{84{,}600}{45.2}$

11. $\dfrac{0.841}{0.617}$ **12.** $\dfrac{0.00721}{8.15}$ **13.** $\dfrac{1.83}{92.5}$ **14.** $\dfrac{58.5}{9060}$

15. $\dfrac{0.0984}{0.0076}$ **16.** $\dfrac{156{,}000}{79{,}400}$ **17.** $(6.45)^2$ **18.** $(34.1)^3$

19. $(1.45)^{10}$ **20.** $(21{,}900)^2$ **21.** $(0.531)^2$ **22.** $(0.0874)^4$

23. $(9.08)^7$ **24.** $(0.00806)^5$ **25.** $\sqrt{26.8}$ **26.** $\sqrt{268}$

27. $\sqrt[3]{56.9}$ **28.** $\sqrt[3]{1950}$ **29.** $\sqrt[4]{31.9}$ **30.** $\sqrt[5]{250}$

31. $\sqrt{0.592}$ **32.** $\sqrt[3]{0.0828}$ **33.** $(9.84)^{0.3}$ **34.** $(482)^{0.6}$

35. $(2.56)^{2.1}$ **36.** $(89.5)^{1.3}$ **37.** $\dfrac{(45.1)(6.12)}{73.8}$ **38.** $\dfrac{3.14}{(0.184)(61.3)}$

39. $\dfrac{(5.03)(89.1)}{(46.1)(2.36)}$ **40.** $\dfrac{(0.0187)(816)}{(51.5)(0.0845)}$ **41.** $(\sqrt{56.9})(\sqrt[3]{18.8})$ **42.** $\dfrac{\sqrt[4]{8540}}{12.4}$

43. $\dfrac{(73.8)(\sqrt{0.854})}{0.308}$ **44.** $\dfrac{(0.544)^{0.3}(\sqrt{12900})}{(17.8)^3(0.0441)}$

45. Find the area of a rectangle whose length is 8.50 m and whose width is 2.75 m.

46. A bank pays 5.25% annual interest on deposits. How much is the annual interest on a deposit of \$3580 in this bank?

47. A person's I.Q. is defined to be 100 times the ratio of his or her mental age to his or her chronological age. If a boy 13 years old is found to have a mental age of 14 years and 6 months, what is his I.Q.?

48. The area of the United States is about 3,620,000 mi^2. The population in 1970 was approximately 205,000,000. On the average, how many persons were there per square mile in the United States in 1970?

49. What is the side of a square whose area is 308 cm^2?

50. Under certain conditions, the population of a bacterial culture equaled the fourth power of the number P_0 originally present. Determine the population if $P_0 = 1500$.

51. If \$100 is invested at 4% interest compounded quarterly, it will grow in 10 years to an amount whose value is given by the expression $100(1.01)^{40}$. Calculate this amount.

52. The velocity of sound in air is given by $v = \sqrt{1.410p/d}$, where p is the pressure and d is the density. Given that $p = 1.01 \times 10^6$ units and $d = 1.29 \times 10^{-3}$ units, find v (in cm/s).

11-6 INTERPOLATION

In the calculations of the preceding section, we have been using the mantissa in the table which is closest to the obtained value. It is, however, possible to use the obtained value and at the same time achieve greater accuracy. We shall now show how we may use the four-place table of logarithms with numbers having four significant digits.

The method of finding mantissas which lie between those listed is known as **linear interpolation.** *The basic assumption of this method is this: Differences between mantissas are in the same proportion as the difference between the associated numbers.* This assumption is not strictly correct, although it is an excellent approximation. The method of finding values by interpolation may be used with a great many tables of values. The following example develops the method and its reasoning with the table of logarithms, and an example of its use with another table is illustrated at the end of the section. In each case, interpolation allows us to use the table to obtain values between those which are directly listed.

Example A Find the logarithm of 3.267.

Checking Table 2, we find that the logarithm of 3.267 lies between log 3.260 and log 3.270. Since 3.267 lies at $\frac{7}{10}$ of the way from 3.260 to 3.270, we assume that log 3.267 lies $\frac{7}{10}$ of the way from log 3.260 to log 3.270. The value of log 3.267 can be found by the use of the following diagram.

$$10\left[\begin{array}{l} 7\left[\begin{array}{l}\log 3.260 = 0.5132 \\ \log 3.267 = \quad \vdots \end{array}\right] x \\ \quad \log 3.270 = 0.5145 \end{array}\right] 13$$

First, we note that 13 is the **tabular difference** (the actual position of the decimal point is not of importance here) between the values of the logarithms. Next, we set up the proportion

$$\frac{7}{10} = \frac{x}{13}$$

which gives the value of $x = 9.1$. Since the mantissa of the required logarithm may have only four significant digits, we round off the value of x to $x = 9$. This is then added to the logarithm of 3.260 to yield

$$\log 3.267 = 0.5141$$

Example B Find the logarithm of 67.28.

The solution is indicated in the following diagram.

$$10\left[\begin{array}{l} 8\left[\begin{array}{l}\log 67.20 = 1.8274\\ \log 67.28 = \quad\vdots\end{array}\right]x \\ \quad\log 67.30 = 1.8280\end{array}\right]6$$

Setting up the proportion, we have

$$\frac{8}{10}=\frac{x}{6}$$

which gives $x = 4.8$. Rounding off, we have $x = 5$, and this means

$$\log 67.28 = 1.8279$$

When calculations are performed by means of logarithms, it is necessary to interpolate in order to find the values of antilogarithms. The procedure is the same as that previously shown, as the following examples illustrate.

Example C Find the number whose logarithm is 0.4554.

The nearest mantissas in the table are 0.4548 and 0.4564. This leads to the following diagram.

$$10\left[\begin{array}{l} x\left[\begin{array}{l}\log 2.850 = 0.4548\\ \log\ldots\quad = 0.4554\end{array}\right]6 \\ \quad\log 2.860 = 0.4564\end{array}\right]16$$

We now have the proportion

$$\frac{x}{10}=\frac{6}{16}$$

which leads to $x = 3.75$. Rounding this off to the nearest integer, we have $x = 4$. Thus, $\log 2.854 = 0.4554$, and the required result is 2.854.

Example D Find the number whose logarithm is 3.9301.

From the table, we set up the following diagram.

$$10\left[\begin{array}{l} x\left[\begin{array}{l}\log 8510 = 3.9299\\ \log\ldots\quad = 3.9301\end{array}\right]2 \\ \quad\log 8520 = 3.9304\end{array}\right]5$$

We now set up the proportion

$$\frac{x}{10} = \frac{2}{5}$$

which leads to $x = 4$. Therefore,

$$\log 8514 = 3.9301$$

and the required result is 8514.

The following examples illustrate calculations in which interpolation is used for determining logarithms and antilogarithms.

Example E Evaluate (6.481)(376.7) by means of logarithms.

$$\begin{aligned} \log 6.481 &= 0.8117 \\ \log 376.7 &= \underline{2.5760} \\ &\ 3.3877 \end{aligned}$$

The antilogarithm of 3.3877 is 2442, which means that

$$(6.481)(376.7) = 2442$$

Example F Evaluate $\frac{\sqrt{21840}}{(3.146)^3}$ by means of logarithms.

$$\log 21840 = 4.3393, \qquad \frac{1}{2}(\log 21840) = 2.1697$$

$$\log 3.146 = 0.4977, \qquad 3(\log 3.146) = \underline{1.4931}$$

$$0.6766$$

The antilogarithm of 0.6766 is 4.749, which is the desired result.

As mentioned at the beginning of the section, we may use interpolation with a great many tables of values. In the following example, interpolation is used with a table which lists the velocity of sound in water at certain temperatures.

Example G In an experiment designed to measure the velocity of sound in water, the values in the following table were found. By use of interpolation, determine the velocity of sound in water, in meters per second, at a temperature of 23°C.

v (m/s)	T(°C)
1430	0
1455	10
1487	20
1528	30
1580	40
1645	50
1722	60

Since 23°C is $\frac{3}{10}$ of the way from 20°C to 30°C, we assume that the velocity of sound at 23°C is $\frac{3}{10}$ of the way between 1487 m/s and 1528 m/s. The difference between these velocities is 41 m/s, which means we have the proportion

$$\frac{3}{10} = \frac{x}{41} \quad \text{or} \quad 10x = 123$$

or

$$x = 12.3$$

Thus, we add 12 m/s (rounded off) to 1487 m/s to find that the velocity is 1499 m/s at 23°C, according to the results of this experiment.

Exercises 11-6

In Exercises 1 through 8, find the logarithm of each of the given numbers.

1. 6.942 **2.** 8514 **3.** 0.3167 **4.** 0.001053

5. 29.11 **6.** 100.5 **7.** $(3.142)(10^3)$ **8.** $(7.081)(10^{-4})$

In Exercises 9 through 16, find the antilogarithm of each of the given logarithms.

9. 0.6407 **10.** 0.9209 **11.** 1.9412 **12.** 3.2475

13. 2.4585 **14.** 9.8706 − 10 **15.** 5.4567 − 10 **16.** 19.1049 − 20

In Exercises 17 through 30, perform the indicated calculations by means of logarithms.

17. $(67.08)(0.4921)$ **18.** $\dfrac{16.72}{6.893}$ **19.** $(51.82)^3$ **20.** $\sqrt{93.12}$

21. $(75.18)\sqrt{3805}$ **22.** $\dfrac{\sqrt[3]{0.9105}}{45.19}$

23. $\dfrac{(0.6112)(0.1052)^4}{\sqrt{0.08737}}$ **24.** $\dfrac{\sqrt{23.75}\sqrt[5]{56.18}}{\sqrt[4]{456.6}\sqrt[3]{1.764}}$

25. A certain stock sells for $62.75 per share. If there are 756,000 shares of stock in the company, what is the total value of the stock?

26. A person's annual retirement income is $9065. How much per week is this income?

27. Find the side of a cube whose volume is 893.4 cm^3.

28. If an amount of $1845 is deposited in a bank which pays 5.5% interest, compounded annually, the value of this deposit in 8 years is $1845(1.055)^8$. Calculate this value.

29. The surface area of a sphere is given by $A = 4\pi r^2$, where r is the radius of the sphere. Calculate the surface area of the earth, assuming that $r = 3960$ mi and $\pi = 3.142$.

30. If a pendulum is to beat seconds, the formula for its length is $L = g/\pi^2$. Find the length, in feet, of a pendulum which is to beat seconds if $g = 32.16$ units and $\pi = 3.142$.

In Exercises 31 and 32, use the table in Example G. Find by interpolation the indicated values of v and T.

31. Find v for $T = 16°C$. **32.** Find T for $v = 1610$ m/s.

11-7 NATURAL LOGARITHMS

The logarithms we have used in the previous three sections have 10 as their base. That is, each is actually an exponent of 10. Since the base of our number system is 10 (see Section 1-7) and we can write numbers conveniently with powers of 10 in scientific notation, these logarithms are very useful for calculation. This was shown in previous sections.

Historically, logarithms were developed for the purpose of making calculations easier. As we noted in Section 11-1, with the advent of calculators and computers this use of logarithms is not as important today as it was. As mathematics developed, however, other uses of logarithms arose, and in more advanced mathematics they are of considerable importance.

We cannot develop here the logarithms of significance in mathematics and in numerous applications. We can, however, describe them and demonstrate a few of their applications in science and technology.

In our development of logarithms to the base ten—common logarithms—we noted that each is an exponent of 10. The number 10 was chosen because it is a convenient base for logarithms to be used for calculation. Actually, as we stated in Section 11-1, it is possible to use almost any positive number as the base of logarithms. Regardless of the base chosen, *the logarithm of a number is the exponent to which the base must be raised to obtain the number.* The following example briefly reviews this concept using base two.

Example A We could choose 2 as a base of logarithms. If we did so, since

$$2^3 = 8$$

we would say that 3 is the logarithm of 8 to the base two. That is, 3 is the exponent to which 2 must be raised to obtain 8.

In the same way, since

$$2^4 = 16$$

4 is the logarithm of 16 to the base two.

As it turns out, there is no specific reason for choosing 2 as a base of logarithms, such as we chose 10 for its use in calculations. There is another number, however, which arises in advanced mathematics, that is very useful as a base of logarithms. This number is e, an irrational number equal approximately to 2.718. *Logarithms to the base e are called* **natural logarithms.** We denote the natural logarithm of x as $\ln x$. That is, by definition,

$$\boxed{\log_e x = \ln x} \tag{11-15}$$

Example B Since $e^0 = 1$, we have $\log_e 1 = 0$, or $\ln 1 = 0$. Since $e^1 = e$, we have $\log_e e = 1$, or $\ln e = 1$. Since $\log_e e^{1/2} = \frac{1}{2}$, we have $\ln e^{1/2} = \frac{1}{2}$, or $\ln e^{0.5} = 0.5$. In the same way, we have $\ln e^2 = 2$.

Using the approximate value of $e = 2.718$, we have

$$\ln e = \ln 2.718 = 1$$

$$\ln e^{0.5} = \ln 2.718^{0.5} = \ln 1.649 = 0.5$$

$$\ln e^2 = \ln 2.718^2 = \ln 7.389 = 2$$

Tables of values of natural logarithms are available. Table 4 in Appendix C is such a table. Since we cannot express numbers in terms of powers of e in the same way we can express them in terms of powers of 10, this is not a table of mantissas. The natural logarithms listed may be used only with the given numbers.

Example C In Table 4, to find ln 7.5, we look opposite 7.5 in the N column and find 2.0149 in the $\ln N$ column. Thus,

$$\ln 7.5 = 2.0149$$

Also,

$$\ln 60 = 4.0943$$

and

$$\ln 0.7 = 9.6433 - 10 = -0.3567$$

The meanings of these logarithms are

$$e^{2.0149} = 7.5$$
$$e^{4.0943} = 60$$
$$e^{-0.3567} = 0.7$$

It is possible to use interpolation in Table 4 to obtain reasonable approximations of natural logarithms which are not listed. Such values are correct to about three or four significant digits. It is also possible to obtain values of natural logarithms by use of the LN key on a scientific calculator. Values obtained in this way are generally correct to seven or more significant digits.

In our development of logarithms, we obtained certain important properties of logarithms. The properties are not dependent on the base which is used and are therefore valid for any base, including e. The first of these properties is based on Eq. (11-7), which is $\log_b x + \log_b y = \log_b xy$. Using base e natural logarithms, we have

$$\boxed{\ln xy = \ln x + \ln y} \qquad (11\text{-}16)$$

Therefore, as with logarithms to other bases: *The natural logarithm of the product of two numbers is equal to the sum of the natural logarithms of the numbers.*

Example D Using the fact that $600 = 60 \times 10$, find $\ln 600$.

From Eq. (11-16), we have

$$\ln 600 = \ln(60 \times 10) = \ln 60 + \ln 10$$

From Table 4 we have $\ln 60 = 4.0943$ and $\ln 10 = 2.3026$. Thus,

$$\begin{aligned}\ln 600 &= 4.0943 + 2.3026\\ &= 6.3969\end{aligned}$$

The properties of logarithms associated with division and powers of numbers are also valid for natural logarithms. Therefore, referring to Eq. (11-8), which is $\log_b(x/y) = \log_b x - \log_b y$, we have

$$\boxed{\ln \frac{x}{y} = \ln x - \ln y} \qquad (11\text{-}17)$$

Therefore: *The natural logarithm of the quotient of two numbers is equal to the natural logarithm of the numerator minus the natural logarithm of the denominator*. Also, referring to Eq. (11-9), which is $\log_b x^n = n \log_b x$, we have

$$\boxed{\ln x^n = n \ln x} \tag{11-18}$$

This means: *The natural logarithm of the nth power of a number is equal to n times the natural logarithm of the number.*

Example E Using Eq. (11-17), we find that

$$\begin{aligned}\ln \frac{7}{4} &= \ln 7 - \ln 4\\ &= 1.9459 - 1.3863\\ &= 0.5596\end{aligned}$$

Thus,

$$\ln \frac{7}{4} = \ln 1.75 = 0.5596$$

Using Eq. (11-18), we find that

$$\begin{aligned}\ln 1.5^2 &= 2 \ln 1.5\\ &= 2(0.4055)\\ &= 0.8110\end{aligned}$$

Thus,

$$\ln 1.5^2 = \ln 2.25 = 0.8110$$

We can see from Example E that the properties of natural logarithms given in Eqs. (11-16) through (11-18), along with values listed in Table 4, allow us to determine the natural logarithms of numbers which are not listed. It is also possible to obtain the value of a natural logarithm from a common logarithm. If a number $N = e^u$, where $u = \ln N$, by finding common logarithms of N, we get

$$\begin{aligned}\log e^u &= \log N\\ u \log e &= \log N\\ u &= \frac{\log N}{\log e}\end{aligned}$$

$$\boxed{\ln N = \frac{\log N}{\log e}} \tag{11-19}$$

Since $\log e = 0.43429$,

$$\ln N = \frac{\log N}{0.43429}$$

or

$$\boxed{\ln N = 2.3026 \log N} \qquad (11\text{-}20)$$

Example F From Table 2, we find that log 436 = 2.6395. Therefore, by using Eq. (11-20), we get

$$\begin{aligned} \ln 436 &= 2.3026(\log 436) \\ &= 2.3026(2.6395) \\ &= 6.0777 \end{aligned}$$

The following example illustrates the use of natural logarithms in an applied problem.

Example G If Q_0 mg of radium decays through radioactivity, an equation relating the amount Q which remains after t years is

$$t = \frac{\ln Q_0 - \ln Q}{4.10 \times 10^{-4}}$$

Determine the number of years for 500 mg of radium to decay such that 400 mg remains.

From the given information $Q_0 = 500$ mg and $Q = 400$ mg. Thus,

$$t = \frac{\ln 500 - \ln 400}{4.10 \times 10^{-4}}$$

We can evaluate ln 500 and ln 400 or we can use Eq. (11-19). Using Eq. (11-19), we get

$$\begin{aligned} t &= \frac{\ln \frac{500}{400}}{4.10 \times 10^{-4}} = \frac{\ln 1.25}{4.10 \times 10^{-4}} \\ &= \frac{0.2231}{4.10 \times 10^{-4}} \\ &= 5.44 \times 10^2 \text{ years} \end{aligned}$$

Thus, it takes 544 years for 500 mg of radium to decay such that 400 mg remains.

Exercises 11-7

In Exercises 1 through 4, determine the required natural logarithms. (See Example B.)

1. $\ln e^{1/3}$ **2.** $\ln e^5$ **3.** $\ln 2.718^{-1}$ **4.** $\ln 2.718^{-1/2}$

In Exercises 5 through 8, find the indicated natural logarithms from Table 4.

5. $\ln 8.4$ **6.** $\ln 18$ **7.** $\ln 0.9$ **8.** $\ln 0.2$

In Exercises 9 through 20, use the properties of natural logarithms (Eqs. 11-16, 11-17, 11-18) and the given values to find the required natural logarithms. The necessary values may be obtained from Table 4. (See Examples D and E.)

9. Given that $200 = 20 \times 10$, find $\ln 200$.

10. Given that $700 = 7 \times 100$, find $\ln 700$.

11. Given that $54 = 3 \times 18$, find $\ln 54$.

12. Given that $560 = 8 \times 70$, find $\ln 560$.

13. Given that $3.25 = \frac{13}{4}$, find $\ln 3.25$.

14. Given that $4.375 = \frac{35}{8}$, find $\ln 4.375$.

15. Given that $0.06 = \frac{6}{100}$, find $\ln 0.06$.

16. Given that $0.45 = \frac{45}{100}$, find $\ln 0.45$.

17. Given that $1.44 = 1.2^2$, find $\ln 1.44$.

18. Given that $7.84 = 2.8^2$, find $\ln 7.84$.

19. Given that $343 = 7^3$, find $\ln 343$.

20. Given that $1024 = 4^5$, find $\ln 1024$.

In Exercises 21 through 28, use Eq. (11-20) and logarithms from Table 2 to find the indicated natural logarithms.

21. $\ln 2.65$ **22.** $\ln 8.05$ **23.** $\ln 673$ **24.** $\ln 5230$

25. $\ln 47.1$ **26.** $\ln 92.5$ **27.** $\ln 0.850$ **28.** $\ln 0.0235$

In Exercises 29 and 30, solve the applied problem.

29. How many years does it take for 1000 mg of radium to decay such that 750 mg remains? (See Example G.)

30. If interest is compounded continuously (daily compounded interest closely approximates this), a bank account can double in t years according to the equation

$$i = \frac{\ln 2}{t}$$

where i is the interest rate. What interest rate is required for an account to double in 8.5 years?

11-8 SUMMARY

In this chapter, we have defined the exponential function and the logarithmic function. In the exponential function

$$\boxed{y = b^x} \tag{11-1}$$

we defined x as the logarithm of the number y to the base b. The logarithmic function

$$y = \log_b x \tag{11-3}$$

and the exponential function were shown to be inverse functions through an analysis of their definitions and their graphs.

The important properties of logarithms were developed. They are shown in Eqs. (11-7), (11-8), and (11-9).

$$\log_b xy = \log_b x + \log_b y \tag{11-7}$$

$$\log_b \frac{x}{y} = \log_b x - \log_b y \tag{11-8}$$

$$\log_b x^n = n \log_b x \tag{11-9}$$

These properties are the basis for the use of base 10 common logarithms in calculations. Thus, multiplication is performed by the addition of logarithms, division is performed by the subtraction of logarithms, and powers of numbers are found by the multiplication of logarithms.

In finding common logarithms, we wrote numbers in scientific notation, which is the product of a number between 1 and 10 and an integral power of 10. To find the common logarithm of a number from a table requires that we find the sum of the logarithm of the number between 1 and 10 (the mantissa) and the integral power of 10 (the characteristic).

We also showed how interpolation can be used to obtain greater accuracy from tables and in calculations with logarithms. Finally, we introduced natural logarithms, which are logarithms to the irrational base number e, which equals approximately 2.718.

The reader should be familiar with the following terms introduced in this chapter.

Antilogarithm
Characteristic
Exponential function
Linear interpolation
Logarithmic function
Logarithms
  base e
  base 10
  common
  natural
Mantissa
Round off
Scientific notation
Significant digits
Tabular difference

The following exercises can be used to review the material presented in the chapter.

Exercises 11-8

In Exercises 1 through 8, determine the value of x.

1. $\log_{10}x = 4$ **2.** $\log_9 x = 3$ **3.** $\log_2 64 = x$ **4.** $\log_{12}144 = x$

5. $\log_x 36 = 2$ **6.** $\log_x 243 = 5$ **7.** $\log_5 x = -1$ **8.** $\log_x 8 = \frac{3}{4}$

In Exercises 9 through 16, express each as a sum, difference, or multiple of logarithms. Wherever possible, evaluate logarithms of the result.

9. $\log_3 2x$ **10.** $\log_5\left(\frac{7}{a}\right)$ **11.** $\log_3(t^2)$ **12.** $\log_6\sqrt{5}$

13. $\log_2 28$ **14.** $\log_7 98$ **15.** $\log_4\sqrt{48}$ **16.** $\log_6\left(\frac{5}{36}\right)$

In Exercises 17 through 20, graph the given equations.

17. $y = 0.5(5^x)$ **18.** $y = 2(6^x)$ **19.** $y = 0.5 \log_4 x$ **20.** $y = 10 \log_{16}x$

In Exercises 21 through 48, perform all calculations by means of logarithms. If the letter i *appears after the number of the exercise, interpolation is required.*

21. (4.15)(8.39) **22.** (0.941)(7.11) **23.** (0.0181)(0.00721)

24. (16,200)(531) **25i.** (84.12)(6.947) **26i.** (0.01361)(9.103)

27. $\frac{67.1}{4.11}$ **28.** $\frac{0.985}{2.31}$ **29.** $\frac{7.14}{9310}$ **30.** $\frac{0.0862}{0.174}$

31i. $\frac{89.15}{9.176}$ **32i.** $\frac{15010}{0.9007}$ **33.** $(10.8)^3$ **34.** $(1.08)^{10}$

35. $(76.1)^4$ **36.** $(0.405)^5$ **37i.** $(6.184)^4$ **38i.** $(1.034)^{0.3}$

39. $\sqrt{23.1}$ **40.** $\sqrt{84.9}$ **41.** $\sqrt[3]{1.17}$ **42.** $\sqrt[3]{789}$

43i. $\sqrt{1308}$ **44i.** $\sqrt[3]{0.9006}$ **45.** $(61.2)(\sqrt{128})$ **46.** $\frac{\sqrt{86000}}{45.8}$

47. $\frac{(0.721)(98.4)}{\sqrt[3]{8.17}}$ **48.** $\frac{(6.10)(\sqrt[5]{38.1})}{(8.01)^3}$

In Exercises 49 through 52, by using logarithms to the base 10, *find the natural logarithms of the given numbers.*

49. 8.86 **50.** 33.0 **51.** 2.07 **52.** 0.542

In Exercises 53 through 66, perform any necessary calculations by logarithms. Exercise 61 requires interpolation.

53. Find the area of a square 35.7 m on a side.

54. Find the area of a rectangle whose sides are 45.7 ft and 85.3 ft.

55. In 1976 there were 149,000,000 Americans of voting age. Of these, 56.5% voted in the presidential election. How many people voted?

56. A person with an income of \$18,500 in a given year found that about 31.7% of the income was paid out in taxes. How much was paid in taxes by this person?

57. The radiator of a certain car contains 6.25 L of alcohol and 9.45 L of water. What percent of this mixture is alcohol?

58. A woman drove 556 mi in 11 h and 12 min. What was her average speed?

59. Light travels at the rate of 186,000 mi/s. The earth is about 92,900,000 mi from the sun. How long does it take the light from the sun to reach the earth?

60. The population of a city was 67,800 in 1960. By 1970, it had increased in population by 15.5%. By 1980, it had increased in population by 23.1% over the 1970 population. What was the 1980 population?

61i. A formula for the area of a triangle with sides a, b, and c is

$$A = \sqrt{s(s - a)(s - b)(s - c)}$$

where s is half the perimeter. Calculate the area of a triangle with sides 45.78, 56.81, and 32.17, respectively.

62. If an object is falling due to gravity, its velocity v in terms of the distance fallen h is given by $v = \sqrt{64.4h}$. What is the velocity, in feet per second, of an object which has fallen 655 ft?

63. The number of bacteria in a certain culture after t hours is given by $N = 1000(10^{0.0451t})$. How many are present after 3 h?

64. In the study of population growth of a certain area with a present population of 10,000, it is estimated that the population will increase 15% each year for t years. This means that the population in t years is given by $10{,}000(1.15)^t$. What will be the population in 2.5 years?

65. Calculate (456)(1.88)(5.19)(0.994) by means of logarithms and by direct multiplication. Compare the methods.

66. Show that $\log(1/P) = -\log P$.

12
SIMULTANEOUS LINEAR EQUATIONS

12-1 PROBLEMS LEADING TO TWO EQUATIONS

In Chapter 7, we discussed in detail some of the methods of analyzing stated problems. In each case, we solved a given problem by setting up an equation in which one unknown was present. Many of those problems had only one unknown, but a number of them had two or more unknowns.

Example A Consider these two problems:

1. The larger of two numbers is 10 more than the smaller. Their sum is 4 less than 3 times the smaller. What are the numbers?

2. Two books together cost \$22, with one costing \$2 less than twice the other. What is the cost of each?

In the first problem, there are *two* unknown numbers, and in the second problem there are *two* unknown costs. We can solve each of these problems by methods discussed in Chapter 7 by expressing one of the unknowns in terms of the other. However, many such problems can be solved as easily, if not more easily, by the use of two equations with two unknowns. Two such equations can be set up for each of these problems, as is shown in Examples B and C.

When a given problem gives rise to two separate equations involving two unknowns and we must solve for these unknowns, the equations are referred to as **simultaneous equations.** Not only do certain stated problems give rise to simultaneous equations, but many problems from science result in simultaneous equations. When forces on an object are analyzed in physics, simultaneous equations often result. In electricity, when determining the electric currents in a circuit, one encounters and then solves several equations with as many unknown currents. Other similar situations could be cited.

Many problems that result in simultaneous equations result in more than two equations. We shall restrict our attention in this chapter to two simultaneous equations in two unknowns. Some of the methods introduced, however, are applicable to systems with more equations.

The purpose of this section is to analyze problems and set up simultaneous equations with two unknowns. The sections which follow will show how such systems of equations are solved for the unknowns they contain. We now present four examples of determining systems of two equations. Since the numerical solutions are not of direct concern here, they are not included with the examples, but they are included as Exercises 67 through 70 in Section 12-5.

Example B Let us set up the first of the problems introduced in Example A. Since the two numbers are the unknowns,

let x = the larger number

and

let y = the smaller number

The first sentence of the problem tells us that $x = 10 + y$. The second sentence tells us that $x + y = 3y - 4$. We now have the required two equations in the two unknowns x and y.

Example C We shall now set up the second of the problems introduced in Example A.

Since the costs of the books are the unknowns,

let x = cost of the less expensive book

and

let y = cost of the more expensive book

The first part of the first sentence of the statement of the problem tells us that $x + y = 22$. The second part of the first sentence of the statement of the problem tells us that $2x - 2 = y$. We now have our two equations.

Example D If part of \$7000 is invested at 6% and the remainder at 7%, how much is invested at each rate if the total annual income from the investments is \$460?

We are asked to find the amount invested at each percentage; thus,

let $x =$ the amount invested at 6%

and

let $y =$ the amount invested at 7%

We know that the total amount invested is \$7000, which means that $x + y = 7000$. The income from the 6% investment is $0.06x$, and that from the 7% investment is $0.07y$. Since the total income is \$460, it follows that $0.06x + 0.07y = 460$. We now have the necessary two equations in the two unknowns.

Example E Two jet planes start at cities A and B 6400 mi apart and travel the same route toward each other. If the jets pass each other 2 h later and the jet that started from city A travels 200 mi/h faster than the other, how far from city A do they pass each other?

Since we wish to determine the distance from city A to the place where they pass,

let $x =$ the distance from city A to where they pass

and

let $y =$ the distance from city B to where they pass

Since the cities are 6400 mi apart, we have $x + y = 6400$. We are also informed that one jet travels 200 mi/h faster than the other. Therefore, it is necessary to set up the other equation in terms of speeds. The speed of each is the distance traveled divided by the time. Since the jet from city A travels x mi in 2 h, its speed is $x/2$. In the same way, the speed of the other jet is $y/2$. Since the jet from city A travels 200 mi/h faster, we have

$$\frac{x}{2} = \frac{y}{2} + 200$$

We now have the necessary two equations.

An alternate method would be to

let $u =$ the speed of the jet from city A

and

let $v =$ the speed of the jet from city B

Even though the required distance from city A is not one of the unknowns, if we find u, we may find the distance by multiplying u by 2, the time elapsed until the planes meet. Using these unknowns, we have the first equation, $u = v + 200$, from the fact that the jet from city A travels 200 mi/h faster. Also, since distance equals speed times time, the distance from city A to the meeting place is $2u$, and the distance from city B is $2v$. The total of these distances is 6400 mi. Therefore, $2u + 2v = 6400$ is the other equation.

Exercises 12-1

In each of the following exercises, designate the unknown quantities by appropriate letters and form two equations involving these unknowns. (The completed numerical solutions for these problems have been included as part of the review exercises at the end of the chapter.)

1. The sum of two numbers is 19 and their difference is 11. Find the numbers.
2. One number is five times as large as another number. If 4 is subtracted from the smaller number and 16 is subtracted from the larger number, the results are equal. What are the numbers?
3. In a certain school there are 74 more boys than girls. There are 500 students in the school. How many boys and girls are there in the school?
4. On a certain mathematics test, two-thirds as many students received a grade below C as those who received C or better. If there are 25 students in the class, how many received a grade below C and how many received C or better?
5. A boy has \$3.50 in dimes and quarters, and there are as many dimes as quarters. How many of each has he?
6. One hundred twenty children and 140 adults attended a show for which the total receipts were \$265. If a father and his two young sons paid \$2.75, how much were adult tickets and children's tickets?
7. A woman pays \$490 each year to insure her car and her home. The car insurance is \$10 more than twice the home insurance. How much does she pay to insure each?
8. Based upon estimates of sources of air pollution, fuel consumption for transportation contributes a percent of pollution which is 16% less than all other sources combined. What percent of air pollution is a result of this one source?
9. A line segment is 12 cm long. Where must a point be located on the line such that the longer part is $\frac{3}{2}$ cm longer than the shorter part?
10. A board is 23 ft long. Where must it be cut such that the longer piece is 2 ft longer than the shorter piece?
11. A man saves \$850 each year. Part of this amount is earned during the winter and the other part is earned during the summer. The amount of his savings earned during the winter is \$70 more than three times the amount of his savings earned during the summer. How much does he save during each period?

12. The difference between two numbers is 13. If two times one-third the sum of the numbers is 78, find the two numbers.

13. An amount of $5000 is invested partly at 6% and partly at 7%. How much is invested at each rate if the total annual income from both of the investments is $330?

14. A man's will reads: My wife and son are each to receive at least $4000. Any remaining portion of my estate is to be divided such that my wife receives twice as much as my son. If the total estate was $14,000, how much did the wife and son receive?

15. The speed of one train is 5 mi/h more than that of another train. If they start at the same time at stations 425 mi apart and arrive at the same station at the same time 5 h later, what is the speed of each?

16. Two trains travel at 40 km/h and 36 km/h, respectively. If the faster train travels for a certain length of time and the slower train takes 2 h more, the total distance traveled by the two trains is 680 km. Find the time each travels.

17. The voltage across an electric resistor equals the current times the resistance. The sum of two resistances is 14 ohms. If 3 A (the current) passes through the smaller resistor and 5 A through the larger resistor, the sum of the voltages is 60 V. What is the value of each resistor?

18. The sum of two electric voltages is 100 V. If the higher voltage is doubled and the other is halved, the sum becomes 155 V. What are the voltages?

19. The relation between Celsius and Fahrenheit temperature scales is that the number of Fahrenheit degrees equals 32 more than $\frac{9}{5}$ the number of Celsius degrees. When will the readings be equal?

20. When will a Fahrenheit temperature reading be 100° more than a Celsius reading? (See Exercise 19.)

21. From 1921 through 1964, the New York Yankees won one American League pennant less than twice the number won by the other teams combined. How many pennants did the Yankees win, and how many did the other teams win?

22. The Roman emperor Nero was 31 yr old when he died. Twice the year of his birth less 6 was the year of his death. In what year was he born and in what year did he die?

23. Find a two-digit number such that the sum of the digits is 9 and the number itself is increased by 9 if the digits are reversed.

24. When a pulley belt makes one complete revolution, one of the pulley wheels makes 5 revolutions and the other makes 4 revolutions. If the smaller wheel is replaced by another wheel which has a circumference of 4 ft less, the replacement wheel makes 10 revolutions for each revolution of the belt. What are the circumferences of the two original wheels?

12-2 GRAPHICAL SOLUTION OF SIMULTANEOUS EQUATIONS

We shall now take up the problem of solving for the unknowns when we have two simultaneous equations with two unknowns. In all, three methods will be discussed in this chapter. In this section, we shall show how the solution may be found graphically. The sections that follow will discuss two basic algebraic methods of solution.

In Chapter 10, we graphed several kinds of functions, among them the linear function. *We saw that the graph of $y = ax + b$ is always a straight line.* However, there are a limitless number of points on the line. If a second line were to cross a given line, there would be only one point common to the two lines. The coordinates of this point, and only this point, would satisfy the equations of both lines.

Example A For the lines $y = 3x - 3$ and $2x + 3y = 6$, we find that the intercepts of the first line are $(0, -3)$ and $(1, 0)$ and that those of the second line are $(3, 0)$ and $(0, 2)$. These points allow us to graph each of the lines, as shown in Fig. 12-1. The actual coordinates of the point of intersection are $(\frac{15}{11}, \frac{12}{11})$. This result can be shown to be correct since these coordinates satisfy the equation of each line. For the first line, we have

$$\frac{12}{11} = 3\left(\frac{15}{11}\right) - 3 = \frac{45}{11} - \frac{33}{11} = \frac{12}{11}$$

and for the second line, we have

$$2\left(\frac{15}{11}\right) + 3\left(\frac{12}{11}\right) = \frac{30}{11} + \frac{36}{11} = \frac{66}{11} = 6$$

Even though we cannot obtain this accuracy for the coordinates of the point of intersection from the graph, we can see that the point of intersection is about (1.4, 1.1) and that this is the type of solution we shall attempt to obtain in this section.

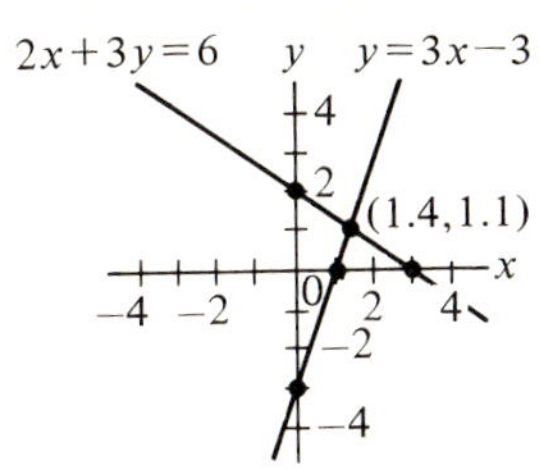

Figure 12-1

In the terminology associated with sets, all the points which satisfy the first equation constitute a set as do all the points which satisfy the second equation. The point of intersection is the only point which satisfies both equations, and this point is the only member of both sets; that is, it is the intersection of the two sets.

Therefore, in order to solve a pair of simultaneous linear equations, we shall graph the equations and determine the point of intersection. The x-coordinate of this point is the desired value of x, and the y-coordinate of this point is the desired value of y. Together they are the solution to the system of two equations in two unknowns. Since the coordinates of no other point satisfy both equations, the unique solution has been determined.

Example B Solve the system of equations

$$x - y = 6$$
$$2x + y = 3$$

graphically.

We determine the intercepts and one check point for each of the equations. Then we graph each equation, as shown in Fig. 12-2.

$x - y = 6$

x	y
0	-6
6	0
1	-5

$2x + y = 3$

x	y
0	3
$\frac{3}{2}$	0
1	1

From the graph, we see that the lines cross at about $(3, -3)$. Therefore, the solution to this set of equations is $x = 3$, $y = -3$. (Actually, this solution is exact, although this cannot be shown without substitution of the values into the given equations.)

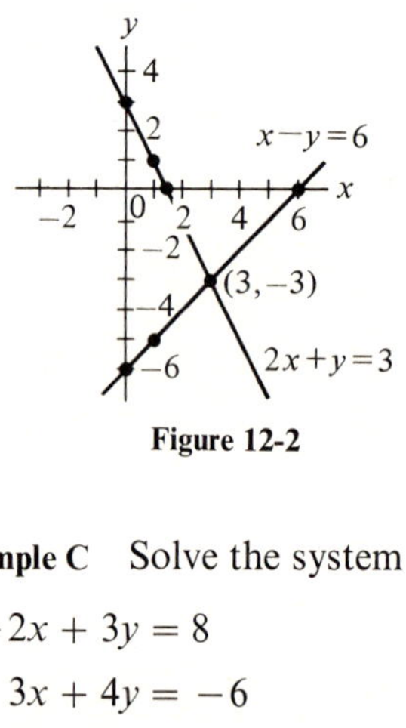

Figure 12-2

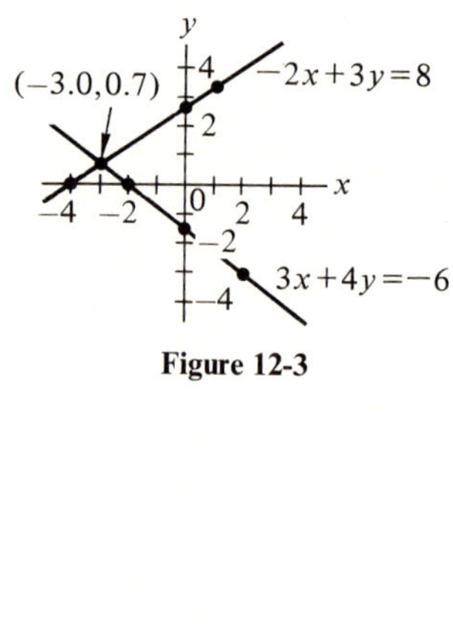

Figure 12-3

Example C Solve the system of equations

$$-2x + 3y = 8$$
$$3x + 4y = -6$$

graphically.

First we determine the intercepts and one check point for each equation. The lines are then graphed, as shown in Fig. 12-3.

$-2x + 3y = 8$

x	y
0	$\frac{8}{3}$
-4	0
1	$\frac{10}{3}$

$3x + 4y = -6$

x	y
0	$-\frac{3}{2}$
-2	0
2	-3

From the graph, we can estimate the coordinates of the point of intersection to be $(-3.0, 0.7)$. [Actually, they are $(-\frac{50}{17}, \frac{12}{17})$.] Thus, the required solution is $x = -3.0$ and $y = 0.7$.

There are special circumstances in which two lines will not intersect at just one point. This is the case (a) if the lines are parallel and do not intersect at all or (b) if the lines are coincident, all the points of one being the same as all of the points of the other. *When the lines are parallel, the system is called* **inconsistent,** *and when they are coincident, the system is called* **dependent.** The following examples illustrate these cases.

Example D Solve the system of equations

$$x + y = 3$$
$$2x + 2y = 9$$

graphically.

We set up the following tables to indicate the intercepts and check point for each line. The lines are then graphed in Fig. 12-4.

$x + y = 3$

x	y
0	3
3	0
1	2

$2x + 2y = 9$

x	y
0	$\frac{9}{2}$
$\frac{9}{2}$	0
2	$\frac{5}{2}$

We observe that, within the limits of accuracy of the graphing, the lines appear to be parallel. (They are, in fact, parallel; and they are always parallel when one equation of the system can be multiplied through by a constant such that the coefficients of the variables are the same, respectively, as those in the other equation and the constant is different. In this case, if the first equation is multiplied by 2, we have $2x + 2y = 6$, which is the same as the second equation except for the constant.) *Since the lines are parallel, there is no solution. The system is inconsistent.*

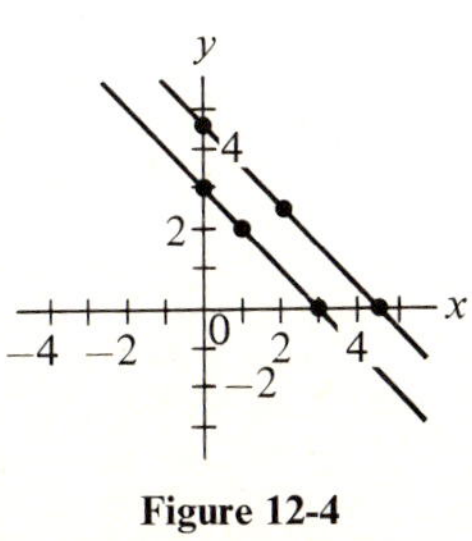

Figure 12-4

Example E Solve the system of equations

$$x + 2y = 6$$
$$3x + 6y = 18$$

graphically.

We set up the following tables to indicate the intercepts and check points for each line. The lines are then graphed in Fig. 12-5.

$x + 2y = 6$

x	y
0	3
6	0
2	2

$3x + 6y = 18$

x	y
0	3
6	0
2	2

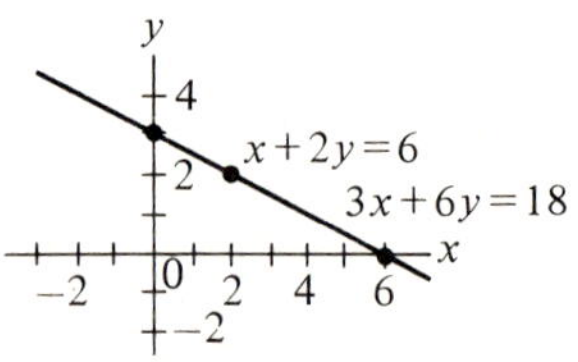

Figure 12-5

We note that both intercepts are the same. Therefore, the two lines are coincident. Hence, *every point on the lines is a solution in that the coordinates satisfy both equations.* This, in turn, means that there is an infinite number of solutions. However, *since no unique solution may be determined, we call the system dependent.*

Exercises 12-2

In Exercises 1 through 24, solve the given systems of equations graphically. Where possible, estimate the coordinates of the point of intersection to the nearest $\frac{1}{10}$ of a unit. If variables other than x and y are used, plot the first one along the x-axis and the second along the y-axis.

1. $x + y = 4,$ $x - y = 4$

2. $x - y = 2,$ $2x + y = 6$

3. $y - x = 2,$ $2y + x = 8$

4. $2x + y = 4,$ $3x + 2y = 6$

5. $r - s = 1,$ $2r + s = 5$

6. $m + n = 6,$ $3m + n = 2$

7. $x + 3y = 8,$ $x - y = 0$

8. $x + y = 1,$ $2x - 8y = 1$

9. $2y + x = 8,$ $4y + 2x = 16$

10. $2k + m = 4,$ $4k + 2m = 6$

11. $x + 4y = 4$ $-2x + 2y = 7$

12. $3x - 2y = 4$ $-9x - 4y = 13$

13. $a + 2b = -3,$ $3a - 5b = -3$

14. $2u - 3v = 8,$ $u + 4v = 6$

15. $3p - q = 6,$ $2p + 2q = 7$

16. $2s - 5t = 10,$ $s + 2t = 8$

17. $3x + 4y = 4,$ $2x - y = 6$

18. $2x + 3y = 9,$ $4x - y = 8$

19. $x - 3y = 2,$ $3x - 9y = 6$

20. $s + t = 7$ $2s + 2t = 7$

21. $3x + 7y = 12,$ $3x + 5y = 0$

22. $2x - 4y = 7,$
$3x + 2y = 3$

23. $\frac{1}{2}x + y = 4$
$x + \frac{1}{3}y = 2$

24. $\frac{1}{3}x - 2y = 8$
$3x - \frac{1}{5}y = 2$

In Exercises 25 through 30, set up the appropriate equations and solve them graphically.

25. The sum of two numbers is 18 and their difference is 6. Find the numbers.

26. The value of 9 coins—nickels and pennies—is 25¢. How many of each are there?

27. The perimeter of a rectangle is 28 in. The length is 4 in. longer than the width. Find the dimensions of the rectangle.

28. The perimeter of a rectangular garden is 42 m. The width is 6 m less than the length. Find the dimensions of the rectangle.

29. A man can row downstream 10 mi in 2 h, and he can row 8 mi upstream in 4 h. Find the rate at which he rows in still water and the rate of the current of the stream.

30. If a current of 2 A passes through a resistor R_1 and a current of 3 A passes through a resistor R_2, the total voltage across the resistors is 8 V. If the current in the first resistor is changed to 4 A and that in the second resistor is changed to 1 A, the total voltage is 11 V. Find the resistances. (See Exercise 17 of Section 12-1.)

12-3 SOLUTION BY ALGEBRAIC SUBSTITUTION

We have just seen how a system of two linear equations can be solved graphically. This technique is good for obtaining a "picture" of the solution of two equations. One of the difficulties of graphical solutions is that the solutions are, in general, approximate. If exact solutions are required, we must turn to algebraic methods. In this section, we shall present a basic algebraic method of solution, and we shall discuss another one in Section 12-4.

The method of this section is called algebraic **substitution.** *Using this method, we first solve one of the equations for one of the two unknowns and then substitute the result into the other equation, thus obtaining a simple equation in one unknown.* The following examples illustrate the method.

Example A Solve the system of equations

$$2x + y = 4$$
$$3x - y = 1$$

by substitution.

The first step is to solve one of the equations for one of the unknowns. About the only consideration on which the choice of equation and unknown is based is the ease of manipulation. For example, in this system, it is slightly easier to solve for y than for x, for fractions can be avoided. Inspection also shows that one algebraic step is all that is required to solve the first equation for y. Therefore, the first equation would be a good choice. It should be emphasized, however, that either equation can be solved for either unknown, and the final result will be the same. Thus, solving the first equation for y, we have $y = 4 - 2x$. This solution is now substituted into the second equation, giving

$$3x - (4 - 2x) = 1$$

This equation is now solved for x.

$$\begin{aligned} 3x - 4 + 2x &= 1 \\ 5x &= 5 \\ x &= 1 \end{aligned}$$

Therefore, the value of y that corresponds to $x = 1$ is found by substituting $x = 1$ into either equation. Since the first has already been solved for y in terms of x, we have

$$y = 4 - 2(1) = 2$$

Thus, the solution is $x = 1$ and $y = 2$. We check it by substituting both values into the other equation. This gives $3(1) - 2 = 1$. Therefore, the solution checks.

Example B Solve the system of equations

$$\begin{aligned} 3x + 2y &= 7 \\ x - 3y &= 6 \end{aligned}$$

by substitution.

Inspection of the equations shows that the solution for x in the second equation appears best. Thus, solving the second equation for x, we have $x = 3y + 6$. Substituting this expression for x into the first equation, we have

$$3(3y + 6) + 2y = 7$$

This equation is now solved for y:

$$\begin{aligned} 9y + 18 + 2y &= 7 \\ 11y &= -11 \\ y &= -1 \end{aligned}$$

Substituting $y = -1$ into $x = 3y + 6$, we have

$$x = 3(-1) + 6 = -3 + 6 = 3$$

The solution is $x = 3$ and $y = -1$. Checking this solution by substituting these values into the first equation, we have

$$3(3) + 2(-1) = 9 - 2 = 7$$

These values therefore satisfy the first equation, and the solution is confirmed.

Example C Solve the system of equations

$$2x - 4y = 23$$
$$3x + 5y = -4$$

by substitution.

There is very little choice as to which variable to solve for or which equation to use. Therefore, simply choosing to solve the first equation for x, we have $2x = 23 + 4y$, or

$$x = \frac{23 + 4y}{2}$$

Substituting this expression into the second equation and then solving for y, we have

$$3\left(\frac{23 + 4y}{2}\right) + 5y = -4$$
$$3(23 + 4y) + 10y = -8$$
$$69 + 12y + 10y = -8$$
$$22y = -77$$
$$y = -\frac{7}{2}$$

Substituting this value into the solution for x, we have

$$x = \frac{23 + 4\left(-\frac{7}{2}\right)}{2} = \frac{23 - 14}{2} = \frac{9}{2}$$

Thus, the solution is $x = \frac{9}{2}$ and $y = -\frac{7}{2}$. Checking this solution in the second equation, we have

$$3\left(\frac{9}{2}\right) + 5\left(-\frac{7}{2}\right) = \frac{27}{2} - \frac{35}{2} = -\frac{8}{2} = -4$$

Thus, the solution checks.

When attempting to solve a system that turns out to be dependent, one can reduce the equation to $0 = 0$ *after substitution.* This result may be deemed reasonable because $0 = 0$ regardless of the value of x and because the coordinates of all points that satisfy one equation of a dependent system also satisfy the other equation, regardless of the value of x. *If the system is inconsistent, the equation after substitution may be reduced to* $0 = a$, *where* a *is not zero.* This result also may be deemed reasonable because zero does not equal a nonzero number and because there are no points whose coordinates satisfy both equations.

Example D Solve the system of equations

$$2x - y = 3$$
$$4x - 2y = 6$$

by substitution.

Solving the first equation for y and substituting this expression into the second equation, we have

$$y = 2x - 3$$
$$4x - 2(2x - 3) = 6$$
$$4x - 4x + 6 = 6$$
$$6 = 6$$

By subtracting 6 from each side, we have $0 = 0$. Therefore, the system is dependent.

Example E Solve the system of equations

$$3x - 2y = 4$$
$$9x - 6y = 2$$

by substitution.

Solving the first equation for x, we have

$$x = \frac{2y + 4}{3}$$

Substituting this expression into the second equation, we have

$$9\left(\frac{2y + 4}{3}\right) - 6y = 2$$
$$3(2y + 4) - 6y = 2$$
$$6y + 12 - 6y = 2$$
$$12 = 2$$
$$10 = 0$$

Since this cannot be true, the system is inconsistent.

Exercises 12-3

In Exercises 1 through 24, solve the given systems of equations by substitution.

1. $y = 5 + x$
$2x - y = 1$

2. $s = 1 - t$
$2s - 8t = 1$

3. $x - y = 2$
$2x + y = 8$

4. $m - n = 5$
$m + 2n = -1$

5. $u + v = 10$
$u - v = 4$

6. $x = y - 1$
$x + y = 5$

7. $2x + y = 0$
$3x + 2y = 1$

8. $2x = 3y$
$x - 4y = 0$

9. $x - 2y = 0$
$2x - 3y = 1$

10. $a - 2b = 4$
$2a + b = 6$

11. $3x = 2y$
$-x + 4y = 0$

12. $2x + 3y = 6$
$x - y = 4$

13. $x - 3y = 6$
$3x = 6y + 18$

14. $x = y + 1$
$x - y = 5$

15. $k - 8u = 4$
$2k + u = 2$

16. $z + 2u = 3$
$3z - u = 1$

17. $\frac{1}{3}x + y = 9$
$x - \frac{1}{3}y = 6$

18. $\frac{1}{2}x + y = 8$
$x + 4y = 7$

19. $2x - 5y = 1$
$3x - 2y = -4$

20. $2u + 3w = 6$
$4u + 2w = 8$

21. $2r + 3s = 2$
$3r - 2s = 16$

22. $3m + 2n = 7$
$5m + 3n = 10$

23. $3x - 4k = 12$
$4x + 5k = -7$

24. $3y - 2z = 6$
$2y + 9z = 8$

In Exercises 25 through 30, set up the appropriate equations and solve by substitution.

25. One number minus twice another is 3. The sum of the two numbers is 12. Find the numbers.

26. On a spelling test of 100 words, a student spelled correctly 4 more than 3 times the number she misspelled. How many words did she spell correctly?

27. A basketball team scored twice as many field goals (2 points each) as foul shots (1 point each). If they scored 90 points, how many field goals and how many foul shots did they make?

28. On a particular day on the stock exchange, 1716 issues were traded, of which 346 were unchanged in price. If the gainers were 18 less than 3 times the losers, how many stocks gained and how many lost in the day's transactions?

29. The specific gravity of an object may be defined as its weight in air divided by the difference between its weight in air and its weight when submerged in water. If the sum of the weights in water and in air of an object of specific gravity equal to 10 is 30 lb, what is its weight in air?

30. Is it possible that the difference of two numbers is 8 and also that two times one of the numbers is 3 less than twice the other?

12-4 SOLUTION BY THE ADDITION-SUBTRACTION METHOD

The second algebraic method of solving two simultaneous equations that we shall discuss is known as the **addition-subtraction** method. *The basis of this method can be described as follows: If one of the unknowns appears on the same side of each equation and if the coefficients of this unknown are numerically the same, it is possible to add (or subtract) the left-hand sides as well as the right-hand sides in order to obtain an equation with only one of the unknowns remaining.* Example A illustrates the basic method.

Example A Solve the system of equations

$$2x - y = 1$$
$$3x + y = 9$$

by the addition-subtraction method.

We note that, if the left-hand sides of the two equations are added, there will be $+y$ plus $-y$, which means that the sum will not contain y. The two sides of each equation are equal. Hence, if the left-hand sides are added, the sum will equal the sum of the right-hand sides. Since y is not present, the resulting equation will contain only x, and thus can be solved for x. We may then find y by substituting the value of x into either equation and then solving for y. Thus, adding the left-hand sides and equating this sum to the sum of the right-hand sides, we have

$$\begin{array}{r} 2x - y = 1 \\ 3x + y = 9 \\ \hline 5x \quad = 10 \\ x = 2 \end{array}$$

Substituting $x = 2$ into the second equation, we have

$$3(2) + y = 9$$
$$6 + y = 9$$
$$y = 3$$

The solution is $x = 2$ and $y = 3$. Checking this solution in the first equation, we have

$$2(2) - 3 = 1$$

which means that the solution is correct.

Example B Solve the system of equations

$$2x + 3y = 7$$
$$2x - y = -5$$

by the addition-subtraction method.

If we subtract the left-hand side of the second equation from the left-hand side of the first equation, x will not appear in the result. Thus, we may obtain the solution by subtracting the respective sides of the equations. Performing this operation, we have

$$\begin{array}{r} 2x + 3y = 7 \\ 2x - y = -5 \\ \hline 4y = 12 \\ y = 3 \end{array} \qquad \text{(subtracting)}$$

Substituting $y = 3$ into the first equation, we have

$$\begin{aligned} 2x + 3(3) &= 7 \\ 2x + 9 &= 7 \\ 2x &= -2 \\ x &= -1 \end{aligned}$$

The solution is $x = -1$, $y = 3$. Checking this solution in the second equation, we have

$$2(-1) - (3) = -2 - 3 = -5$$

In many equations it is necessary to multiply one (or both) of the equations by constants such that the resulting coefficients of one unknown are numerically the same. The following two examples illustrate this procedure.

Example C Solve the system of equations

$$\begin{aligned} 2x + 3y &= -5 \\ 4x - y &= 4 \end{aligned}$$

by the addition-subtraction method.

If each term of the second equation is multiplied by 3, it will be possible to eliminate y by addition and continue to the solution. Thus, we have

$$\begin{aligned} 2x + 3y &= -5 \\ 12x - 3y &= 12 \end{aligned}$$

Adding the equations, we have

$$\begin{aligned} 14x \qquad &= 7 \\ x &= \frac{1}{2} \end{aligned}$$

Substituting $x = \frac{1}{2}$ into the first equation, we have

$$\begin{aligned} 2\left(\frac{1}{2}\right) + 3y &= -5 \\ 1 + 3y &= -5 \\ 3y &= -6 \\ y &= -2 \end{aligned}$$

The solution is $x = \frac{1}{2}$, $y = -2$. Checking this solution by substituting into the second equation, we have

$$4\left(\frac{1}{2}\right) - (-2) = 2 + 2 = 4$$

Example D Solve the system of equations

$$2x - 4y = 5$$
$$3x + 5y = 8$$

by the addition-subtraction method.

It is not possible to multiply just one of these equations by an integer to make the coefficients of one of the variables numerically the same. Therefore, we multiply both equations by appropriate integers. We can multiply the first equation by 3 and the second equation by 2 in order to make the coefficients of x equal, or we may multiply the first equation by 5 and the second one by 4 in order to eliminate y. Choosing this second method, we obtain the new set of equations

$$\begin{aligned} 10x - 20y &= 25 \\ 12x + 20y &= 32 \\ \hline 22x \qquad &= 57 \\ x &= \frac{57}{22} \end{aligned}$$

$$2\left(\frac{57}{22}\right) - 4y = 5$$

$$-4y = 5 - \frac{57}{11}$$

$$-4y = \frac{55 - 57}{11} = -\frac{2}{11}$$

$$y = \frac{-2}{-44} = \frac{1}{22}$$

Checking, we have

$$3\left(\frac{57}{22}\right) + 5\left(\frac{1}{22}\right) = \frac{171}{22} + \frac{5}{22} = \frac{171 + 5}{22} = \frac{176}{22} = 8$$

Therefore, the solution is $x = \frac{57}{22}$, $y = \frac{1}{22}$.

If the system is dependent or inconsistent, we have the same type of result as that mentioned in the preceding section. That is, if the system is dependent, the addition-subtraction method will result in $0 = 0$ after the equations are combined; the result will be $0 = a$ (a not zero) if the system is inconsistent.

Example E Solve the system of equations

$$3x - 6y = 8$$
$$-x + 2y = 3$$

by the addition-subtraction method.

Multiplying the second equation by 3 and adding equations, we have

$$\begin{aligned} 3x - 6y &= 8 \\ -3x + 6y &= 9 \\ \hline 0 &= 17 \end{aligned}$$

Since this result is not possible, the system is inconsistent.

Exercises 12-4

In Exercises 1 through 24, solve the given systems of equations by the addition-subtraction method.

1. $x + y = 4$, $x - y = 2$

2. $x + y = 5$, $x - 2y = 1$

3. $2x + y = 16$, $x - y = 2$

4. $x + y = 11$, $2x - y = 1$

5. $m + n = 9$, $2m - n = 0$

6. $2x + y = 7$, $2x - y = 1$

7. $d + t = 3$, $2d + 3t = 10$

8. $2r + s = 7$, $r + 2s = -1$

9. $x + 2n = -3$, $3x - 5n = -3$

10. $2x + y = 4$, $5x + y = 13$

11. $x - 2y = 5$, $2x + y = 20$

12. $4t - x = 5$, $3t - x = 1$

13. $7x - y = 5$, $7x - y = 4$

14. $3x - y = 4$, $-6x + 2y = -8$

15. $a + 7b = 15$, $3a + 2b = 7$

16. $-8x + 7y = 2$, $3x - 5y = 9$

17. $13p - 18q = 21$, $4p + 12q = 5$

18. $11k + 15t = 1$, $2k + 3t = 1$

19. $\frac{2}{7}x - \frac{1}{3}y = -3$, $\frac{5}{14}x + \frac{5}{3}y = -10$

20. $\frac{7}{8}x + \frac{5}{12}y = -4$, $x + y = 8$

21. $2m = 10 - 3n$, $3m = 12 - 4n$

22. $62x = 4y + 43$, $15x + 12y = 17$

23. $\frac{1}{x} + \frac{2}{y} = 3$, $\frac{1}{x} - \frac{2}{y} = -1$

24. $\frac{4}{x} - \frac{1}{y} = 1$, $\frac{6}{x} + \frac{2}{y} = 5$

In Exercises 25 through 30, set up the appropriate equations and solve by the addition-subtraction method.

25. The sum of two numbers is 28. Three times the smaller number is 16 more than the larger number. Find the numbers.

26. The sum of the numerator and denominator of a fraction is 18. Twice the numerator is 3 less than the denominator. Find the fraction.

27. Three times the number of counties in Wyoming is 19 less than twice the number of counties in Idaho. Also, twice the number of counties in Wyoming is two more than the number in Idaho. How many counties has each state?

28. A woman has \$50 to spend in buying pieces for her china set. For this amount, she can buy 5 dinner plates and 10 salad plates or 8 dinner plates and 6 salad plates. How much is each of these pieces?

29. If part of \$6000 is invested at 6% and the remainder at 8%, the total annual income is \$400. How much is invested at each rate?

30. By weight, one alloy consists of 60% copper and 40% zinc. Another is 30% copper and 70% zinc. How many grams of each alloy are needed to make 120 g of an alloy which consists of 50% of each metal?

12-5 SUMMARY

In this chapter, we have shown how systems of simultaneous equations in two unknowns can arise from certain stated problems. We also mentioned that such systems arise naturally in many applied situations. Hence, we established the need to be able to solve such systems.

Three basic methods for solving systems of two linear equations were discussed. The first of these was the graphical method. We solved simultaneous equations by drawing the graph of each of the equations and determining the point where the lines crossed. The coordinates of this point gave us the solution, which was the set of values of x and y satisfying both equations.

The other methods of solution were algebraic in nature. The first was the method of substitution, where we solved for one of the unknowns in terms of the other unknown in one of the equations and then substituted this expression into the other equation, thereby creating an equation with only one unknown. We saw that the solution was readily completed when this equation was solved. The other method was based upon either the addition or subtraction of each side of one equation from the corresponding side of the other. When the coefficients of one of the unknowns were numerically the same, the resulting equation after addition or subtraction had but one unknown. The solution then followed essentially in the same manner as in the method of substitution.

We made a definite point to show that all solutions should be checked in the original equations. Doing this ensures that any errors that might have been made in the solution will in general be discovered.

We also discussed dependent and independent systems. In a dependent system, the coordinates of any point which satisfies one equation also satisfies the other. In an inconsistent system, there is no set of values which satisfies both equations.

The terms listed below have been introduced in the chapter. If any of these terms are not familiar to you, you should review them before you attempt to do the exercises which follow.

Addition-subtraction method
Dependent system
Graphical method
Inconsistent system
Point of intersection
Simultaneous equations
Substitution method

Exercises 12-5

In Exercises 1 through 34, solve the given systems of equations by one of the algebraic methods and check by solving by the other.

1. $2x + y = 6$
$3x - y = 6$

2. $3x + y = 12$
$3x - y = 24$

3. $x + 2y = 12$
$x + 3y = 16$

4. $x + 2y = 12$
$2x - y = 4$

5. $2x - y = 4$
$x + 3y = 16$

6. $y + 2x = -3$
$3y - 5x = -3$

7. $2p + 7q = 10$
$3p - 2q = 10$

8. $m = n + 3$
$m = 5 - 3n$

9. $2u - 3v = 5$
$-u + 4v = -5$

10. $4x + 3y = -1$
$5x + 2y = 4$

11. $3a + 7b = 15$
$2a - 5b = 39$

12. $2p = 3x + 1$
$2p = 2x - 3$

13. $3x = -16 - y$
$7x = -8 - 5y$

14. $3y = 11 - 5h$
$15y - 15h = 7$

15. $-6y + 4z = -8$
$5y + 6z = 2$

16. $7x + 9y = 3$
$5x + 4y = 1$

17. $3n + d = 20$
$6n + 2d = 40$

18. $x - 2y = 6$
$-2x + 4y = 5$

19. $3x + y = -5$
$6x + 8y = 2$

20. $4x - y = 7$
$8x + 3y = -26$

21. $2x - 3y = 8$
$5x + 2y = 9$

22. $8x - 9y = 3$
$3x + 4y = 2$

23. $8s + 5t = 6$
$7s - 9t = 8$

24. $17x - 9y = 10$
$8x + 7y = -13$

25. $7x + 2y = 3$
$5x + 3y = 10$

26. $3x + 2y = 5$
$2x - 6y = 7$

27. $\frac{2}{3}x - \frac{3}{4}y = 26$
$\frac{1}{3}x + \frac{3}{4}y = -14$

28. $\frac{4x + 5y}{20} = 3$
$\frac{1}{2}x - \frac{1}{3}y = 2$

29. $\frac{9}{r} - \frac{5}{s} = 2$
$\frac{7}{r} - \frac{3}{s} = 6$

30. $\frac{2}{x} - \frac{2}{y} = 4$
$\frac{1}{x} - \frac{3}{y} = -2$

31. $0.03x + 2y = 1$
$0.02x - 3y = 5$

32. $0.10x - 0.05y = 3$
$0.03x - 0.20y = 2$

33. $\frac{x - 1}{9} + \frac{y - 3}{6} = 3$
$2x - \frac{2y - 5}{4} = 0$

34. $\frac{x + 3}{7} - y = \frac{1}{2}$
$x + \frac{y - 2}{8} = \frac{2}{3}$

In Exercises 35 through 42, solve the given systems of equations graphically.

35. $x + 3y = 2$
$6x - 3y = 5$

36. $2x + 3y = 0$
$x + y = 2$

37. $3x - y = 6$
$x - 2y = 4$

38. $4x - y = 9$
$4x + 3y = 0$

39. $4u + v = 2$
$3u - 4v = 30$

40. $r + 4s = 6$
$r + 6s = 7$

41. $3x - y = 5$
$-6x + 2y = 3$

42. $m + 3n = 7$
$3m - 2n = 5$

For Exercises 43 through 66, complete the solutions of Exercises 1–24 of Exercises 12-1. Use any appropriate method.

For Exercises 67 through 70, complete the solutions of Examples B–E of Section 12-1. Use any appropriate method.

Exercises 71 through 74 illustrate scientific problems which give rise to simultaneous equations. Solve these by any appropriate method.

71. In analyzing a certain electric circuit, you obtain the following equations:

$$4i_1 - 10i_2 = 3 \qquad -10i_2 - 5(i_1 + i_2) = 6$$

Solve for the indicated electric currents i_1 and i_2 (in amperes).

72. The equation $s = v_0 t + \frac{1}{2}at^2$ is used to relate the distance s traveled by an object in time t if the initial velocity is v_0 and the acceleration is a. For an object moving down an inclined plane, it is noted that $s = 32$ ft when $t = 2$ s, and $s = 63$ ft when $t = 3$ s. Find v_0 (in ft/s) and a (in ft/s^2).

73. Two ropes support a given weight. The equations used to find the tensions in the ropes are

$$0.866T_2 - 0.500T_3 = 0$$
$$0.500T_2 + 0.866T_3 = 50$$

Find T_2 and T_3 (in pounds).

74. Nitric acid is produced from air and nitrogen compounds. To determine the size of the equipment required, one often uses a relationship between the air flow rate m (in moles/h) and exhaust nitrogen rate n. For one particular operation, one encounters the following equations:

$$1.58m + 41.5 = 38.0 + 2.00n$$
$$0.424m + 36.4 = 189 + 0.0728n$$

Solve for m and n.

In Exercises 75 through 86, set up the appropriate equations and solve.

75. A man has \$1.10 in nickels and dimes. If there are 15 coins in all, how many of each has he?

76. The legislative body of a certain county has 37 members, all of whom belong to one or the other of two political parties. There is one more than twice the number of members from one party than of members from the other party. How many belong to each of the political parties?

77. A man invested part of \$5000 at 5% and the remainder at 6%. If he earns \$272 each year from the investments, find the amount invested at each rate.

78. A student received 16 more points on her second mathematics test than on her first. She then determined she must get an 88 on the third test to raise her average to 80. What did she get on each of her first two tests?

79. Two thermometer readings differ by 48°C. Four times the smaller reading exceeds twice the larger reading by 2°C. What are the readings?

80. The cost of booklets at a printing firm consists of a fixed charge and a charge for each booklet. The total cost of 1000 booklets is \$550; the total cost of 2000 booklets is \$800. Find the fixed charge and the cost of each booklet.

81. A plumber worked for 3 h on a particular job and her helper worked on it for 4 h. Together they earned \$25.50. If the hours worked were reversed, they would have earned \$27.00 together. What is the hourly rate each receives?

82. The circumference of a circular lawn area is 5.60 ft more than the perimeter of a nearby square area. The diameter of the circle is 10.0 ft more than the side of the square. Find the diameter of the circle and the side of the square. (Use $\pi = 3.14$ and three significant digits in working this problem.)

83. A car travels 355 mi by going at one speed for 3 h and at another speed for 4 h. If it had traveled 3 h at the second speed and 4 h at the first speed, it would have gone 345 mi. Find the two speeds.

84. How many cubic centimeters of each of a 10% solution and a 25% solution of nitric acid must be used to make 95 cm^3 of a 20% solution?

85. The sum of the digits of a two-digit number is 7. With the digits reversed, the number is 9 more than the original number. Find the number.

86. The sum of the digits of a two-digit number is 9. With the digits reversed the number obtained in 9 less than 3 times the original number. Find the number.

In Exercises 87 and 88, determine the indicated sets.

87. What set describes the set of solution points of two parallel (noncoincident) lines?

88. What is the relation of the set P of all points of the plane and the set S of points of the solution of two nonparallel, noncoincident lines?

13
TOPICS FROM GEOMETRY

13-1 ANGLES AND TRIANGLES

In several of the examples and exercises in the previous chapters, we used a few basic formulas from geometry, assuming that you have some knowledge of certain basic geometric figures and that you are able to use the formulas given in Sections 3-2 and 8-4. In this chapter, we shall formally present some basic topics from geometry.

This section will be devoted to introducing terminology related primarily to angles and triangles. The remaining sections will be devoted to introducing other important geometric figures, along with topics pertinent to the study of geometry. The study of geometry included in this chapter is not complete, but it should be sufficient to give you a good idea of the basic methods and topics of the subject.

Geometry deals with the properties and measurements of angles, lines, surfaces, and volumes and the basic figures they form. In a more complete course in geometry, many properties are shown through proofs. A few examples of this procedure are shown in Section 13-7. However, it is not possible here or anywhere to define every word and prove every statement, and therefore certain words and concepts must be accepted without definition. *Generally, in geometry, the concepts of a* **point,** *a* **line,** *and a* **plane** *are accepted as being known intuitively.* We can then define other terms with the help of these undefined terms.

If two lines meet at point P (see Fig. 13-1), *the amount of rotation necessary to bring one line together with the other is called the* **angle** *through which the first line was rotated.* Notation and terminology associated with angles are illustrated in the following example.

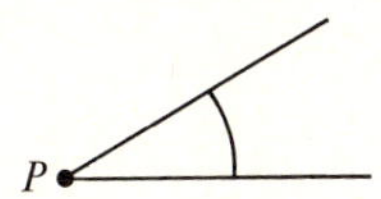

Figure 13-1

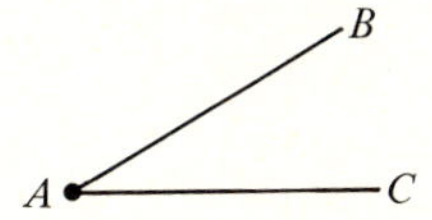

Figure 13-2

Example A In Fig. 13-2, the angle formed by lines AB and AC is denoted by $\angle BAC$. *The* **vertex** *of the angle is the point A, and the* **sides** *of the angle are AB and AC.*

One complete rotation (see Fig. 13-3a) *of a line about a point is defined to be an angle of 360 degrees, written as 360°. A* **straight angle** *contains 180°* (see Fig. 13-3b), *and a* **right angle** *(denoted, as in* Fig. 13-3c, *by* ∟) *contains 90°.* If two lines meet so that the angle between them is a right angle, the lines are said to be **perpendicular.**

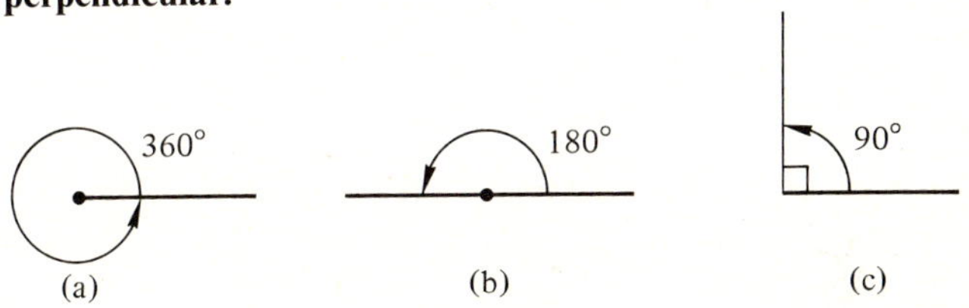

Figure 13-3

A common method of stating the magnitude of an angle is to give the **measure** *of the angle.* That is, the measure of a straight angle is 180°, and the measure of a right angle is 90°. However, in our discussions we shall refer to the angle rather than the measure of the angle. Also, we shall use the same symbol for the angle and the measure of the angle.

A device that can be used to measure angles approximately is a **protractor** (see Fig. 13-4).

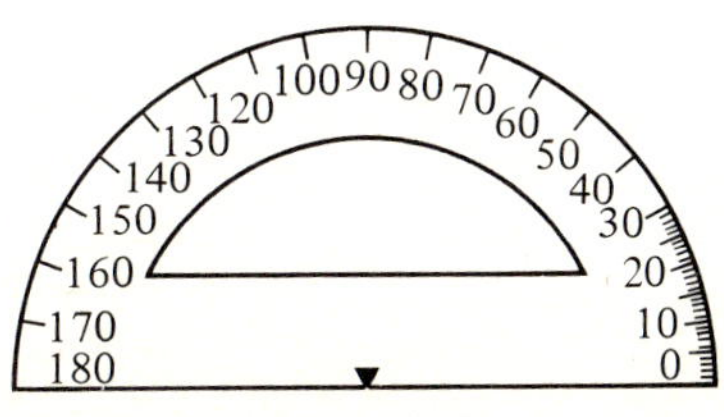

Figure 13-4

When it is necessary to measure angles to an accuracy of less than 1°, **decimal parts of a degree** *may be used.* The use of decimal parts of a degree has become common with the extensive use of electronic calculators. *Previously the common way of expressing angles to such accuracy was to divide each degree into 60 equal parts called* **minutes,** *and to divide each minute into 60 equal parts called* **seconds.** Surveyors and astronomers frequently measure angles to the nearest second.

The notation for minutes is ′ and that for seconds is ″. Thus, an angle of 32 degrees, 15 minutes, and 38 seconds is denoted as $32°15'38''$. Since $1° = 60'$, we find a decimal part of a degree in minutes by multiplying the decimal by $60'$. Also, since $1' = (\frac{1}{60})°$, we express minutes in terms of a decimal part of a degree by dividing by 60.

Example B

1. $0.2° = (0.2)(60') = 12'$. Thus, $17.2° = 17°12'$.
2. $36' = (\frac{36}{60})° = 0.6°$. Thus, $58°36' = 58.6°$.
3. $6.35° = 6°21'$, since $(0.35)(60') = 21'$.
4. $74°11' = 74.18°$, since $11' = (\frac{11}{60})° = 0.18°$ (to the nearest hundredth).

There are many special names given to angles and pairs of angles depending upon their size, relative position, and so on. **Adjacent angles** *have a common* **vertex** *and a side common between them* (see Fig. 13-5). *Two lines that cross form two pairs of equal angles "across" the point of intersection; these angles are called* **vertical angles** (see Fig. 13-6). **Perpendicular** *lines are defined to be a pair of lines that cross and form equal adjacent angles* (see Fig. 13-7).

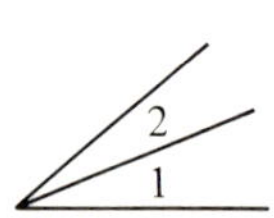

Angles 1 and 2 are adjacent

Figure 13-5

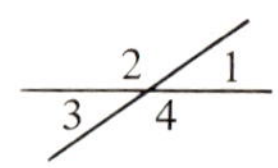

Angles 1 and 3 are vertical
Angles 2 and 4 are vertical

Figure 13-6

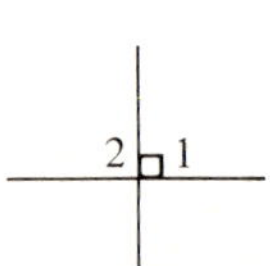

Equal adjacent angles 1 and 2

Figure 13-7

Supplementary angles *are two angles whose sum is 180°.* **Complementary angles** *are two angles whose sum equals 90°. An angle less than 90° is an* **acute angle** (see Fig. 13-8). *An angle greater than 90° but less than 180° is an* **obtuse angle** (see Fig. 13-9).

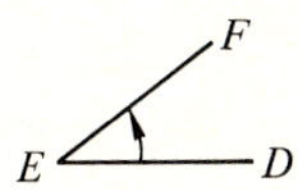

Acute angle *DEF*

Figure 13-8

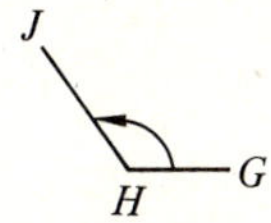

Obtuse angle *GHJ*

Figure 13-9

Example C In Fig. 13-10, $\angle AOB$ is an acute angle, $\angle AOC$ is a straight angle, $\angle BOC$ is an obtuse angle, and angles *AOB* and *BOC* are supplementary angles.

In Fig. 13-11, $\angle DOE$ is an acute angle, $\angle DOF$ is a right angle, and angles *DOE* and *EOF* are complementary angles.

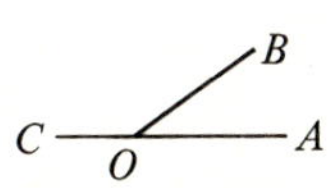

Figure 13-10

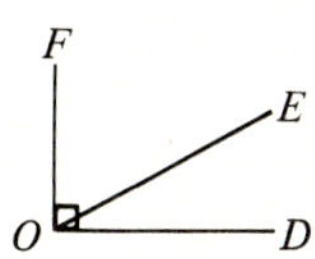

Figure 13-11

In a plane, if a line crosses two given **parallel** *lines (their extensions will not meet) or nonparallel lines, it is called a* **transversal.** *If a transversal crosses a pair of parallel lines, certain sets of equal angles result.* In Fig. 13-12, the **corresponding angles** *are equal.* (That is, $\angle 1 = \angle 5$, $\angle 2 = \angle 6$, $\angle 3 = \angle 7$, and $\angle 4 = \angle 8$.) Also the **alternate interior angles** *are equal* ($\angle 3 = \angle 6$ and $\angle 4 = \angle 5$), and the **alternate exterior angles** *are equal* ($\angle 1 = \angle 8$ and $\angle 2 = \angle 7$).

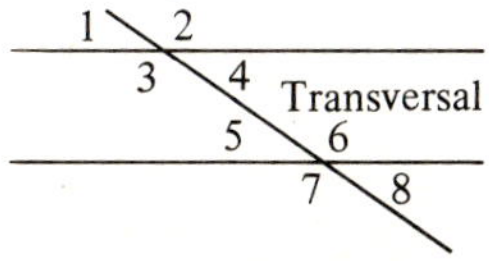

Figure 13-12

When a part of the plane is bounded and closed in by straight line segments, it is called a **polygon.** *In general, polygons are named according to the number of sides they contain. A* **triangle** *has three sides, a* **quadrilateral** *has four sides, and so on.* In this section, we shall consider only triangles. Other types of figures are covered in the next section.

In an **equilateral triangle,** *the three sides are equal in length and each angle is a 60° angle* (see Fig. 13-13a). *In an* **isosceles triangle,** *two of the sides are equal in length and the* **base angles** (*the angles opposite the equal sides*) *are also equal* (see Fig. 13-13b). *In a* **scalene triangle,** *no two sides are equal in length and no two angles have equal measure* (see Fig. 13-13c).

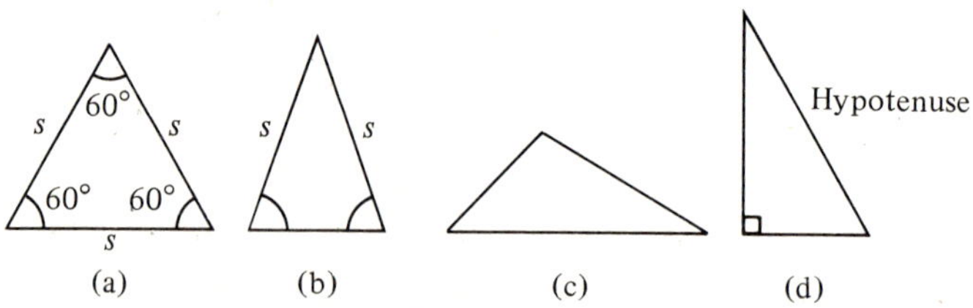

Figure 13-13

One of the most important triangles in scientific and technical applications is the **right triangle.** *In a right triangle, one of the angles is a right angle. The side opposite the right angle is called the* **hypotenuse,** *and the other two sides are called the* **legs.** See Fig. 13-13d.

Example D

1. The triangle in Fig. 13-14a, denoted as $\triangle ABC$, is equilateral. Since $AC = 5$, we also know that $AB = 5$ and $BC = 5$. All of the angles are 60° angles, and we note this by writing $\angle A = 60°$, $\angle B = 60°$, and $\angle C = 60°$.

2. The triangle in Fig. 13-14b is isosceles since $DF = 7$ and $EF = 7$. The base angles are $\angle D$ and $\angle E$. Therefore, since $\angle D = 70°$, we know that $\angle E = 70°$.

3. The triangle in Fig. 13-14c, $\triangle PQR$, is scalene since none of the sides is equal in length to another side. Therefore, none of the angles has the same measure as any other angle.

4. The triangle in Fig. 13-14d, $\triangle LMN$, is a right triangle since $\angle M = 90°$. The hypotenuse is the side LN, and the legs are MN and LM.

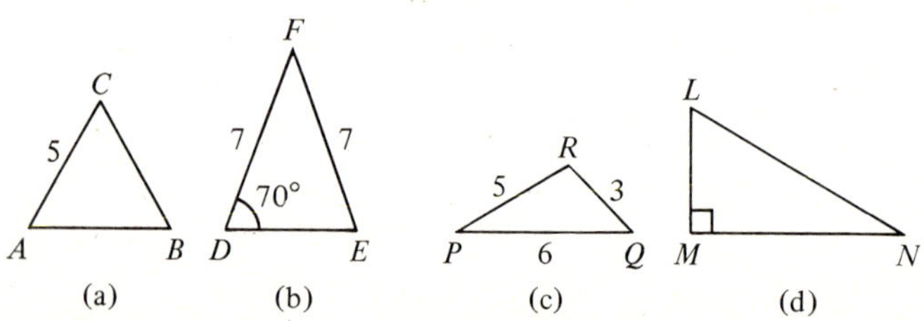

Figure 13-14

There are certain basic properties of triangles that we shall mention here. One very important property is that *the sum of the three angles of any triangle is 180°*. Also, *the three* **medians** (*line segments drawn from a vertex to the* **midpoint** *of the opposite side*) *meet at a single point* (see Fig. 13-15). *This point of intersection of the medians is called the* **centroid** *of the triangle. It is also true that the three* **angle bisectors** *meet at a common point* (see Fig. 13-16) *as do the three* **altitudes** (*heights*), *which are drawn from a vertex perpendicular to the opposite side* (*or its extension*). See Fig. 13-17.

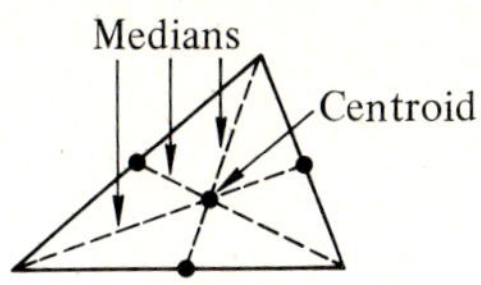

Figure 13-15

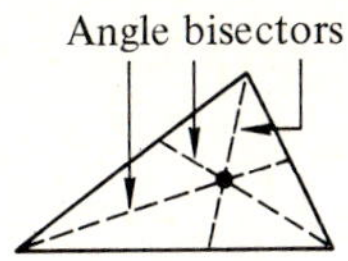

Figure 13-16

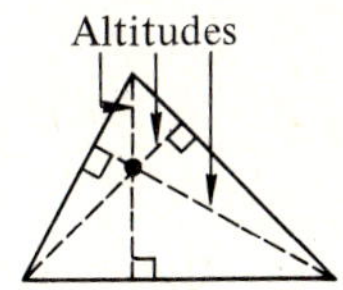

Figure 13-17

Example E In the triangle in Fig. 13-18, we know two of the angles: the right angle and the 42° angle. Since the sum of the three angles is 180°, we may determine the angle at A, $\angle A$. Therefore, since $42° + 90° = 132°$ and $180° - 132° = 48°$, we find that $\angle A = 48°$.

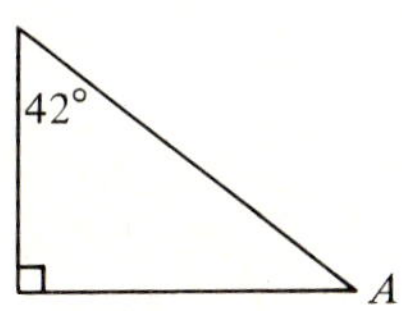

Figure 13-18

Exercises 13-1

In Exercises 1 through 4, use a protractor to measure the angles given in Fig. 13-19 to the nearest degree.

1. 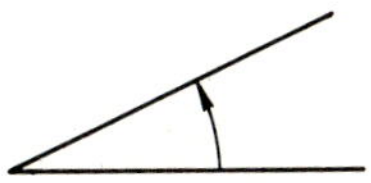

2. 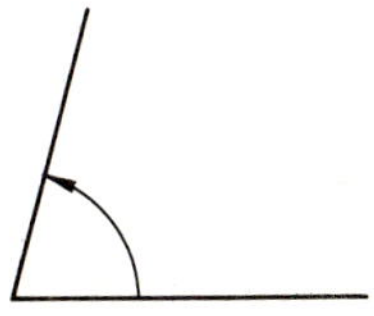

3. 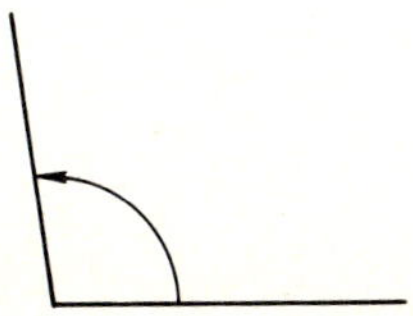

4. 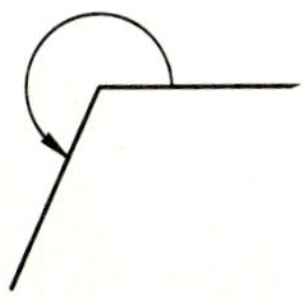

Figure 13-19

In Exercises 5 through 8, express each angle in degrees and decimal parts of a degree.

5. $156°15'$ **6.** $33°48'$ **7.** $67°6'$ **8.** $16°57'$

In Exercises 9 through 12, answer the given questions about the figures in Fig. 13-20.

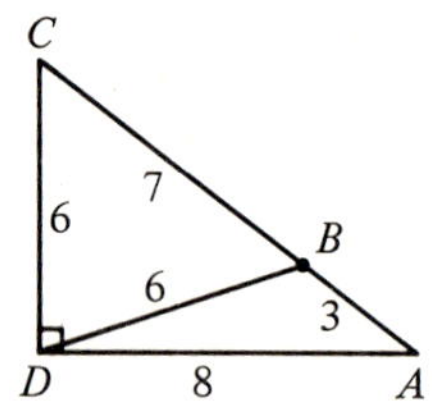

Figure 13-20

9. In Fig. 13-20, identify the right triangle.

10. In Fig. 13-20, identify the scalene triangle.

11. In Fig. 13-20, identify the isosceles triangle.

12. In Fig. 13-20, if $\angle DBC = 54°$, how many degrees are there in $\angle C$?

In Exercises 13 through 20, use Fig. 13-21 and identify all the listed types of angles and pairs of angles. CE is perpendicular to DA.

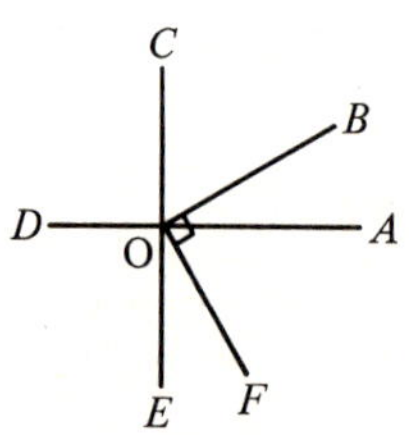

Figure 13-21

13. Straight angles

14. Right angles

15. Acute angles

16. Obtuse angles

17. Adjacent acute angles

18. Vertical angles

19. Complementary angles

20. Supplementary angles (start at A)

In Exercises 21 through 24, use Fig. 13-22 and identify the listed pairs of angles. Use only angles 1 through 5.

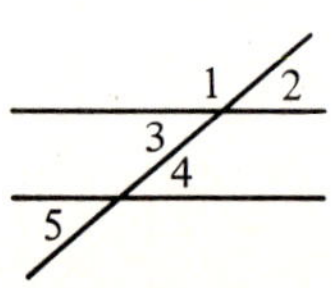

Figure 13-22

21. Alternate interior angles

22. Alternate exterior angles

23. Corresponding angles

24. Vertical angles

In Exercises 25 through 28, use Fig. 13-23, where $\angle ABD = 30^\circ$, $\angle SBK = 30^\circ$, *and* EK *is perpendicular to* AC. *Determine the indicated angles.*

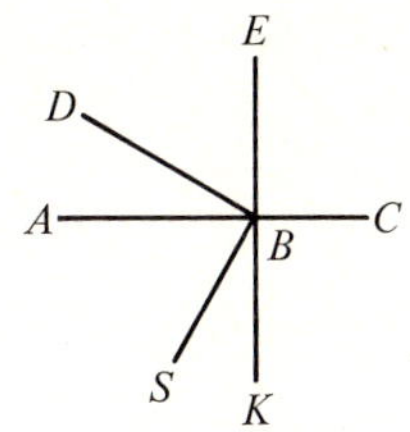

Figure 13-23

25. $\angle DBE$ **26.** $\angle DBC$ **27.** $\angle ABS$ **28.** $\angle EBS$

In Exercises 29 through 32, use Fig. 13-24, where AOP *and* KOT *are straight lines,* $\angle TOP = 40^\circ$, $\angle OBC = 90^\circ$, *and* $\angle AOL = 55^\circ$. *Determine the indicated angles.*

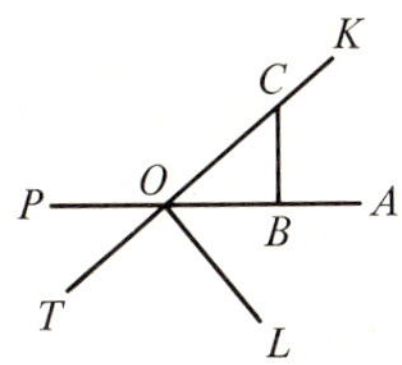

Figure 13-24

29. $\angle LOT$ **30.** $\angle POL$ **31.** $\angle KOP$ **32.** $\angle KCB$

In Exercises 33 through 36, determine the indicated angles. The triangle in Fig. 13-25 is isosceles and $\angle B = 70^\circ$.

Figure 13-25

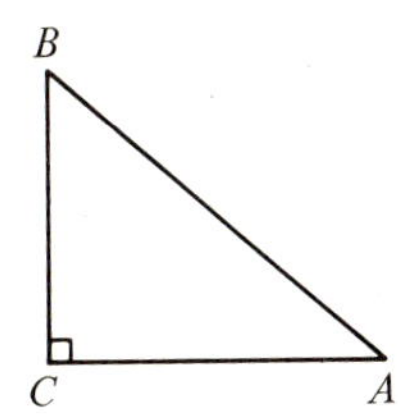

Figure 13-26

33. In Fig. 13-25, find $\angle C$.

34. In Fig. 13-25, find $\angle A$.

35. In Fig. 13-26, find $\angle A + \angle B$.

36. In Fig. 13-26, if $\angle A = 40^\circ$, $\angle B = ?$

In Exercises 37 through 40, answer the given questions and draw the appropriate figures.

37. What is the maximum number of right angles a triangle may contain?

38. How many acute angles are there in a triangle that contains an obtuse angle?

39. Draw a *scalene* triangle (no two sides are equal) along with its medians and altitudes. Is the point of intersection of the medians the same as that of the altitudes?

40. Draw a triangle with an obtuse angle along with its altitudes. Where do the altitudes meet?

13-2 OTHER GEOMETRIC FIGURES—PERIMETER

In the preceding section, we introduced the subject of geometry and discussed various kinds of angles and triangles. In this section, we will consider other geometric figures, primarily quadrilaterals and the circle, and the concept of the **perimeter** (*distance around*) of a geometric figure.

As we mentioned in Section 13-1, a polygon is a part of the plane closed in by straight line segments, and a quadrilateral is a polygon of four sides (see Fig. 13-27a). Considering now specific types of quadrilaterals, *the opposite sides of a* **parallelogram** *are parallel and equal in length, and its opposite angles are equal* (see Fig. 13-27b). *A* **rectangle** *is a parallelogram with four right angles* (see Fig. 13-27c). *Therefore, the opposite sides of a rectangle are equal and parallel. A* **square** *is a rectangle all sides of which are equal* (see Fig. 13-27d). *A* **trapezoid** *is a quadrilateral with two of the sides parallel* (see Fig. 13-27e). *These sides are called the* **bases** *of the trapezoid. A* **rhombus** *is a parallelogram all sides of which are equal* (see Fig. 13-27f).

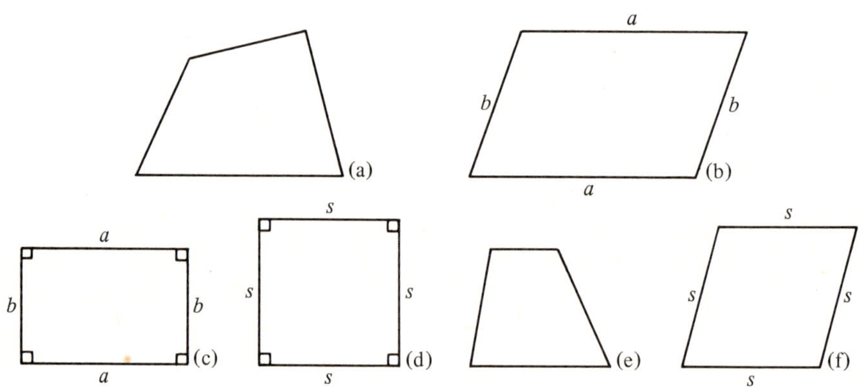

Figure 13-27

Sets are convenient for demonstrating relationships among the various types of geometric figures. See Example A.

Example A Let Q be the set of all quadrilaterals, P the set of all parallelograms, and R the set of all rectangles. Since a rectangle has all of the properties of a parallelogram, set R is a subset of P, or $R \subseteq P$. Since a parallelogram is one type of quadrilateral, set P is a subset of set Q, or $P \subseteq Q$. Since $R \subseteq P$ and $P \subseteq Q$, we also know that $R \subseteq Q$.

In a polygon, a line segment that joins any two nonadjacent vertices is called a **diagonal.** From this definition, we can see that a triangle cannot have any diagonals but that polygons with four or more sides do have diagonals.

Example B Figure 13-28 shows a **hexagon** (6 sides). We can see that three diagonals may be drawn from vertex A (AC, AD, and AE). Also, three may be drawn from vertex B (BD, BE, and BF). From vertex C, three diagonals may be drawn although only two (CE and CF) are different from those already drawn because diagonal AC has already been included. From vertex D, three diagonals may be drawn although only one is different from those already drawn (DF). No other diagonals not already included can be drawn from vertices E and F. Therefore, it is possible to draw three different diagonals from any vertex, and nine diagonals in all.

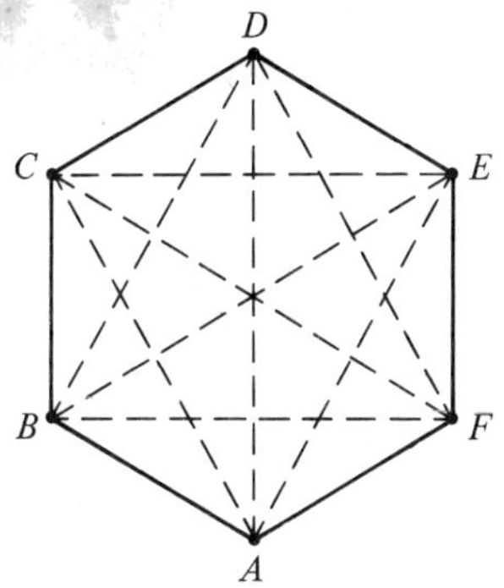

Figure 13-28

Turning our attention briefly to the **circle,** we see that *all of the points on a circle are the same distance from a fixed point in the plane* (see Fig. 13-29). *This point O is the* **center** *of the circle. The distance ON (or OM) from the center to a point on the circle is the* **radius** *of the circle. The distance MN between two points on the circle and on a line passing through the center of the circle is the* **diameter** *of the circle.* Thus, *the diameter is twice the radius.*

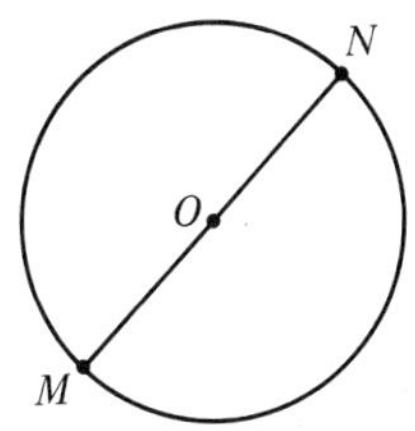

Figure 13-29

Also associated with the circle is the **chord,** *which is a line segment having its endpoints on points on the circle. A* **tangent** *is a line that touches a circle (does not pass through) at one point. A* **secant** *is a line that passes through two points of a circle. The* **circumference** *is the total distance around the circle, and an* **arc** *is part of the circumference. A* **central angle** *is the angle formed at the center by two radii.*

Example C In Fig. 13-30, line segment AB is a diameter, and line segments AO, CO, and BO are radii. Line segment DE is a chord. Line S, which passes through D and E, is a secant, and line T, which touches the circle at D, is a tangent. Radii CO and AO form a central angle at O, where arc AC is part of the circumference.

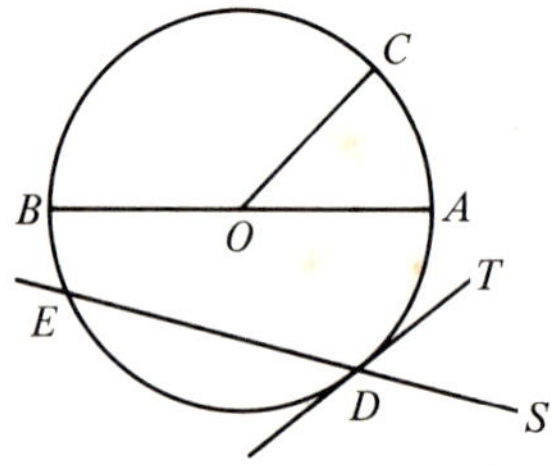

Figure 13-30

One of the basic concepts associated with a plane figure is that of its **perimeter.** *The perimeter of a figure is the distance around it.* From this definition, it is possible to derive the formulas for the perimeters of many plane figures. However, if one remembers the definitions of the figures and that of the perimeter, it is not necessary to memorize most of these perimeter formulas. Consider the formulas given in Example D.

Example D For the given figures, we have the following perimeter formulas.

Figure	Perimeter
1. Triangle with sides a, b, c	$p = a + b + c$
2. Equilateral triangle with side s	$p = 3s$
3. Quadrilateral with sides a, b, c, d	$p = a + b + c + d$
4. Parallelogram, or rectangle, with sides a and b	$p = 2a + 2b$
5. Square of side s	$p = 4s$

Example E A basketball court is a rectangle of length 94 ft and width 50 ft (these are the maximum allowable dimensions). Find its perimeter.

Since the court is a rectangle, its perimeter is twice its length plus twice its width. Thus,

$$\begin{aligned} p &= 2(94) + 2(50) \\ &= 188 + 100 \\ &= 288 \text{ ft} \end{aligned}$$

Thus, the perimeter of the basketball court is 288 ft.

The perimeter of a circle, which is the **circumference,** *cannot be found directly from the definition. Therefore, the number π, equal to about 3.14 and a member of the set of irrational numbers* (see Section 1-6), *is defined such that the circumference c is*

$$\boxed{c = 2\pi r} \tag{13-1}$$

where r is the radius of the circle. Another formula for the circumference is

$$\boxed{c = \pi d} \tag{13-2}$$

where d is the diameter of the circle.

Example F The circumference of a circle of radius 3 cm is

$$c = 2\pi(3) = 6\pi = 18.8 \text{ cm}$$

The circumference of a circle of diameter 8 m is

$$c = \pi(8) = 25.1 \text{ m}$$

Note: It is often more convenient to leave the results as a multiple of π than to replace π by its decimal equivalent.

It is possible to find the perimeters of geometric figures that are combinations of basic figures by the definition of perimeter. Consider Example G.

Example G A Norman window is one whose shape is a rectangle surmounted by a **semicircle** (half a circle). See Fig. 13-31. Find the perimeter of a Norman window if the vertical side h is 3 ft and the radius r of the circular part is 1 ft.

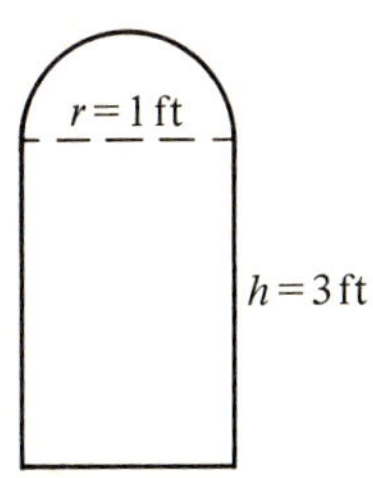

Figure 13-31

Let us first find a formula for the perimeter in terms of r and h. The perimeter will be the sum of the lengths of the two vertical sides, the base, and the semicircular portion at the top. Each of the two vertical sides is h;

the base is equal to the diameter of the circle at the top, or $2r$; the semicircular portion at the top is $2\pi r/2$, since it is half a circle. Thus,

$$p = 2h + 2r + \frac{2\pi r}{2}$$
$$= 2h + 2r + \pi r$$

Since, in this case, $h = 3$ ft and $r = 1$ ft, we have

$$p = 2(3) + 2(1) + \pi(1) = 6 + 2 + \pi$$
$$= 8 + \pi = 11.14 \text{ ft}$$

Exercises 13-2

In Exercises 1 through 26, find the perimeters (circumferences for circles) of the indicated figures for the given values.

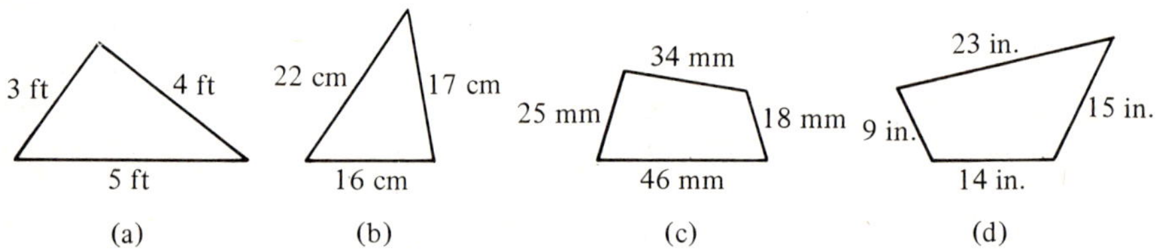

Figure 13-32

1. The triangle shown in Fig. 13-32a
2. The triangle shown in Fig. 13-32b
3. The quadrilateral shown in Fig. 13-32c
4. The quadrilateral shown in Fig. 13-32d
5. Triangle: $a = 320$ m, $b = 278$ m, $c = 298$ m
6. Triangle: $a = 52$ yd, $b = 49$ yd, $c = 64$ yd
7. Quadrilateral: $a = 16.5$ in., $b = 17.3$ in., $c = 21.8$ in., $d = 29.2$ in.
8. Quadrilateral: $a = 6.92$ cm, $b = 8.26$ cm, $c = 9.93$ cm, $d = 8.07$ cm
9. Equilateral triangle: $s = 64.6$ mm
10. Equilateral triangle: $s = 128$ ft
11. Isosceles triangle: equal sides of 15.3 in., third side of 26.5 in.
12. Isosceles triangle: equal sides of 36.2 cm, third side of 12.5 cm
13. Square: $s = 0.65$ m
14. Square: $s = 2.36$ yd
15. Rhombus: side of 0.15 mi
16. Rhombus: side of 178 mm
17. Parallelogram: $a = 47.2$ cm, $b = 36.8$ cm
18. Parallelogram: $a = 1.69$ in., $b = 1.46$ in.
19. Rectangle: $l = 68.7$ ft, $w = 46.6$ ft
20. Rectangle: $l = 4.57$ m, $w = 0.97$ m
21. Circle: $r = 15.1$ cm
22. Circle: $d = 42.0$ mm

23. A pentagon (5 sides) each of whose sides is 7 ft
24. A hexagon (6 sides) each of whose sides is 3 mm
25. A decagon (10 sides) each of whose sides is 4 cm
26. A dodecagon (12 sides) each of whose sides is 8 in.

In Exercises 27 through 30, find a formula for the perimeter of the given figures. (In the figures, dashed lines are shown only to indicate how the figures are constructed.)

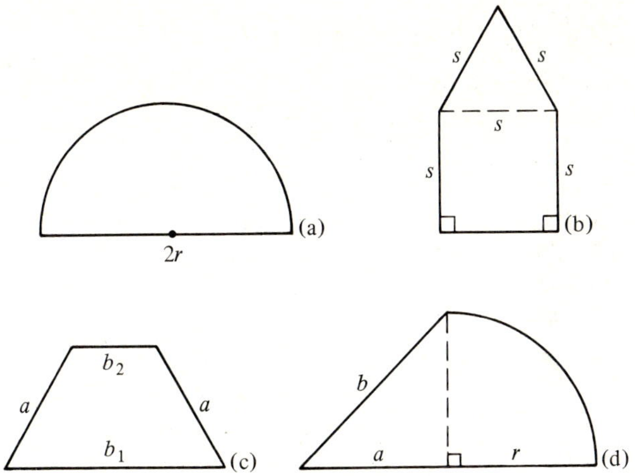

Figure 13-33

27. An isosceles triangle with equal sides s and a third side a
28. A rhombus with side s
29. The semicircle in Fig. 13-33a
30. The figure shown in Fig. 13-33b

In Exercises 31 through 34, find a formula for the perimeter of the indicated figure, and then solve for the given letter.

31. A triangle with sides a, b, and c, for a
32. The isosceles trapezoid in Fig. 13-33c, for a
33. A circle with radius r, for r
34. The figure in Fig. 13-33d, for b

In Exercises 35 through 42, solve the given applied problems involving perimeters.

35. A floor of a room is in the shape of a square 12 ft on a side. How many feet of floor molding are required for this room if the room has two doors 3 ft wide?
36. A soccer field is a rectangle 360 ft long and 240 ft wide. What is the perimeter of this soccer field?
37. If rug binding costs 15¢ per foot, what would be the cost of the binding for a rectangular rug 12 ft by 18 ft?

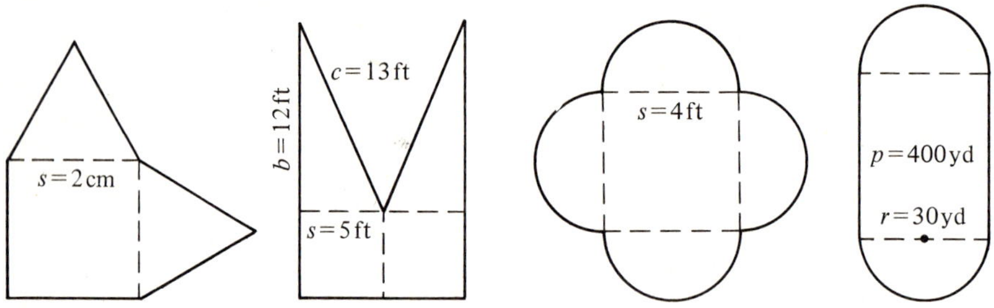

Figure 13-34

38. A certain machine part is in the shape of a square with equilateral triangles attached to two sides. See Fig. 13-34a. If the side of the square is 2 cm, what is the perimeter of the machine part?

39. Find the perimeter of the figure in Fig. 13-34b, which is made up of two squares surmounted by two equal right triangles.

40. A field is a square 500 m on a side with a rectangular area 200 m by 100 m removed from one corner. How much fencing is required to enclose the field? How much fencing would be required if the field were simply the square 500 m on a side?

41. A certain flower bed is made up of semicircular areas attached to a square. See Fig. 13-34c. If the side of the square is 4 ft, what is the perimeter of the flower bed?

42. The area within a race track is a rectangle with semicircles at each end (see Fig. 13-34d). If the radius of the circular parts is 30 yd and the perimeter of the area within the track is 400 yd, how long is each straight section of the track?

In Exercises 43 through 52, you are asked to draw certain conclusions regarding various properties of geometric figures. In each case, draw an appropriate figure to help in your determination.

43. What is the total number of diagonals in a quadrilateral?

44. What is the total number of diagonals from one vertex of a pentagon (five sides)?

45. What is the total number of diagonals in a pentagon?

46. How do the diagonals of a rectangle cut each other?

47. How do the diagonals of a rhombus cut each other?

48. How many quadrilaterals, bound by a decagon (10 sides), can be formed in the decagon, with the center of the decagon as a vertex common to the quadrilaterals? (Two sides of each quadrilateral extend from the center of the decagon to vertices of the decagon.)

49. Draw a circle and a radius of the circle. Draw a line tangent to the circle through the endpoint of the radius. What is the relation between the radius and the tangent line?

50. Draw a circle with two tangent lines that are perpendicular. What is your conclusion about the arc of the circle between the points of tangency?

51. If two chords meet perpendicularly at a point on the circumference of a circle, what is your conclusion about the arc of the circle between the other two points on the chords?

52. Draw a circle. Then draw another circle around the first one such that all points on the first circle are at a distance of 2 mm from the second circle. What is your conclusion about the centers of the two circles? Is it possible to make a general statement about such circles?

13-3 AREA OF GEOMETRIC FIGURES

In this section, we shall develop additional important formulas from geometry. These formulas will involve the concept of area for the geometric figures introduced in the previous sections.

We shall not attempt to define area, choosing instead to consider the concept intuitively. However, we can define the measure of an area. We begin by defining the area of a parallelogram; from this definition, the formulas for many other geometric figures follow easily. Thus, *the* **area** *of a parallelogram is defined as*

$$\boxed{A = bh} \tag{13-3}$$

where b is the length of one side, the **base,** *and h is the perpendicular distance, called the* **height,** *between the base and the opposite side.* See Fig. 13-35.

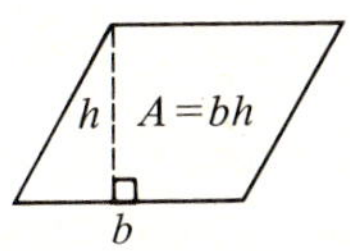

Figure 13-35

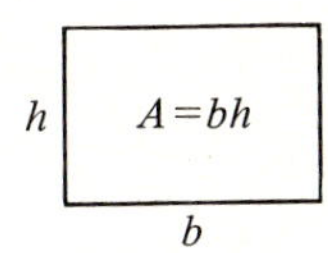

Figure 13-36

Since the sides of a rectangle intersect at right angles, its area is the product of the lengths of a pair of its intersecting sides. See Fig. 13-36.

Since a square is a rectangle in which the length and width are equal, the area of a square of side s is

$$\boxed{A = s^2} \tag{13-4}$$

See Fig. 13-37.

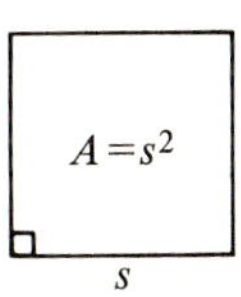

Figure 13-37

Example A The area of a parallelogram of base 6 in. and height 4 in. is

$$A = (6)(4) = 24 \text{ in}^2.$$

See Fig. 13-38a.

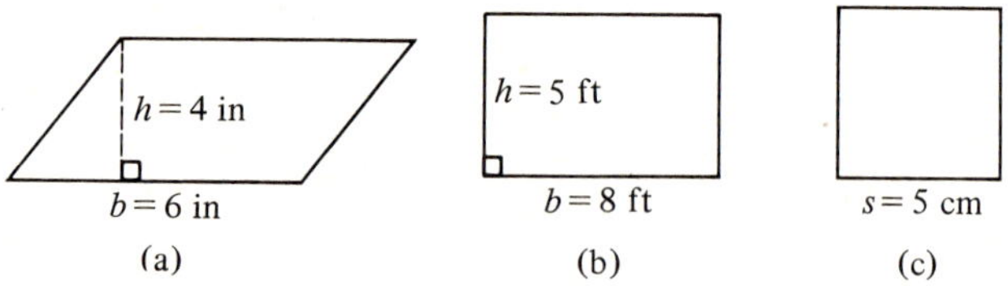

Figure 13-38

The area of a rectangle of base 8 ft and height 5 ft is

$$A = (8)(5) = 40 \text{ ft}^2$$

See Fig. 13-38b. This problem could also have been given in terms of the length and width of the rectangle. In this case, the length would have been 8 ft and the width 5 ft.

The area of a square of side 5 cm is

$$A = (5)^2 = 25 \text{ cm}^2$$

See Fig. 13-38c.

When we join opposite **vertices** (points of intersection of two sides) of a parallelogram, we see that two equal triangles are formed. Thus, *the area of a triangle is*

$$\boxed{A = \frac{bh}{2}} \qquad (13\text{-}5)$$

See Fig. 13-39.

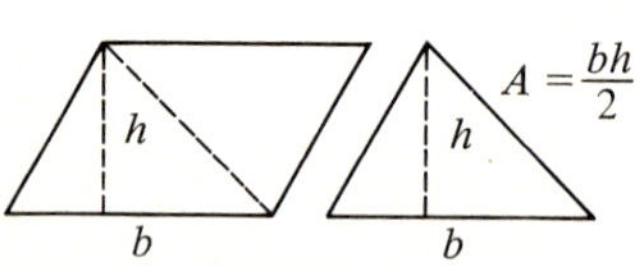

Figure 13-39

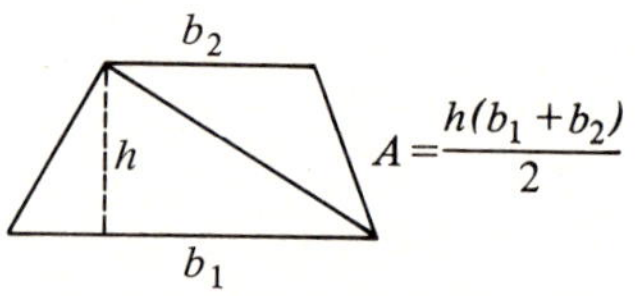

Figure 13-40

When we join opposite vertices of a trapezoid, we see that two triangles are formed. See Fig. 13-40. The area of the lower triangle is $b_1h/2$ and that of the upper triangle is $b_2h/2$. The sum of these areas is equal to

$$\boxed{A = \frac{h(b_1 + b_2)}{2}} \qquad (13\text{-}6)$$

which is the required formula for the area of the trapezoid.

Example B The area of a triangle of base 12 ft and height 7 ft is

$$A = \frac{(12)(7)}{2} = 42 \text{ ft}^2$$

See Fig. 13-41a.

Figure 13-41

The area of a trapezoid of bases 3 m and 6 m and height 4 m is

$$A = \frac{4(3 + 6)}{2} = 18 \text{ m}^2$$

See Fig. 13-41b.

Example C A painter charges 20¢/ft^2 for painting house exteriors. One end of a house is a rectangle surmounted by a triangle. If the base of the rectangle is 30 ft, the height of the rectangle is 9 ft, and the height of the triangle is 7 ft, what would be his charge for this part of the job? There are 3 windows, each 2 ft by 3 ft, in the side of the house. See Fig. 13-42.

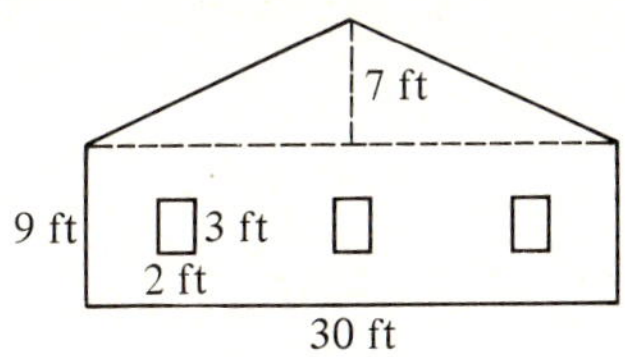

Figure 13-42

The area of the side of the house is the area of the rectangle plus that of the triangle minus the area of the windows. Thus,

$$\begin{aligned} A &= 30(9) + \frac{30(7)}{2} - 3(2)(3) \\ &= 270 + 105 - 18 = 357 \text{ ft}^2 \end{aligned}$$

The charge, in cents, is

$$\begin{aligned} C &= 20A \\ &= 20(357) = 7140¢ \quad \text{or} \quad C = \$71.40 \end{aligned}$$

The area of a circle cannot be found directly from the area of a parallelogram. It is expressed in terms of π, as is the circumference. *The area of a circle is*

$$\boxed{A = \pi r^2} \tag{13-7}$$

See Fig. 13-43.

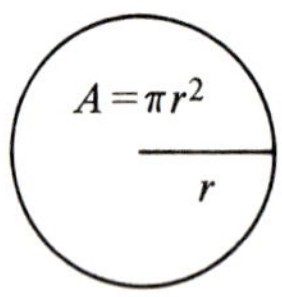

Figure 13-43

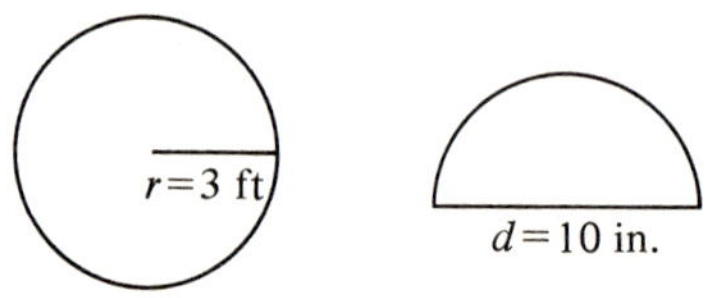

(a) (b)

Figure 13-44

Example D The area of a circle of radius 3 ft is

$$A = \pi(3)^2 = 9\pi = 28.3 \text{ ft}^2$$

See Fig. 13-44a.

The area of a semicircle of diameter 10 cm is

$$A = \frac{\pi(5)^2}{2} = \frac{25\pi}{2} = 39.3 \text{ cm}^2$$

See Fig. 13-44b. Here, since we were given the diameter, we first determined that the radius was 5 cm.

Exercises 13-3

In Exercises 1 through 16, find the areas of the indicated geometric figures.

1. Rectangle: $l = 60.0$ cm, $w = 45.0$ cm
2. Rectangle: $l = 152$ ft, $w = 85.0$ ft
3. Square: $s = 7.60$ in.
4. Square: $s = 0.160$ km
5. Parallelogram: $b = 72.0$ mm, $h = 34.0$ mm
6. Parallelogram: $b = 1.50$ yd, $h = 1.20$ yd
7. Rhombus: side $= 16.5$ in., altitude $= 6.40$ in.
8. Rhombus: side $= 240$ cm, altitude $= 150$ cm
9. Triangle: $b = 0.750$ m, $h = 0.640$ m
10. Triangle: $b = 64.0$ in., $h = 14.5$ in.
11. Right triangle: legs $= 16.5$ ft and 28.8 ft
12. Right triangle: legs $= 396$ mm and 250 mm

13. Trapezoid: $h = 0.0120$ km, $b_1 = 0.0250$ km, $b_2 = 0.0180$ km
14. Trapezoid: $h = 1.23$ yd, $b_1 = 3.74$ yd, $b_2 = 2.36$ yd
15. Circle: $r = 0.478$ ft
16. Circle: $d = 5.38$ cm

In Exercises 17 through 24, determine (a) the perimeter and (b) the area of the given figures.

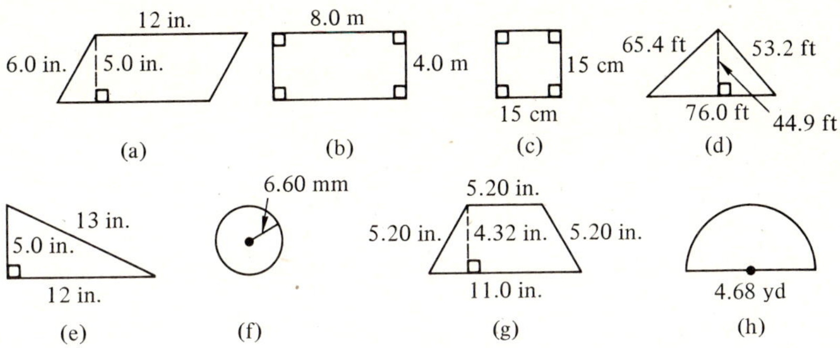

Figure 13-45

17. The figure in Fig. 13-45a
18. The figure in Fig. 13-45b
19. The figure in Fig. 13-45c
20. The figure in Fig. 13-45d
21. The figure in Fig. 13-45e
22. The figure in Fig. 13-45f
23. The figure in Fig. 13-45g
24. The figure in Fig. 13-45h

In Exercises 25 through 28, the solutions will require the use of the Pythagorean theorem (see Section 8-4). Find the required areas.

25. Find the area of a right triangle in which the hypotenuse is 17 in. and one leg is 8 in.
26. Find the area of a right triangle in which the hypotenuse is 12.5 m and one leg is 10 m.
27. Find the area of a rectangle in which the diagonal is 8 cm and the width is 3 cm.
28. Find the area of a rectangle in which the diagonal is 14 in. and the length is 10 in.

In Exercises 29 through 32, find a formula for the areas of the given figures.

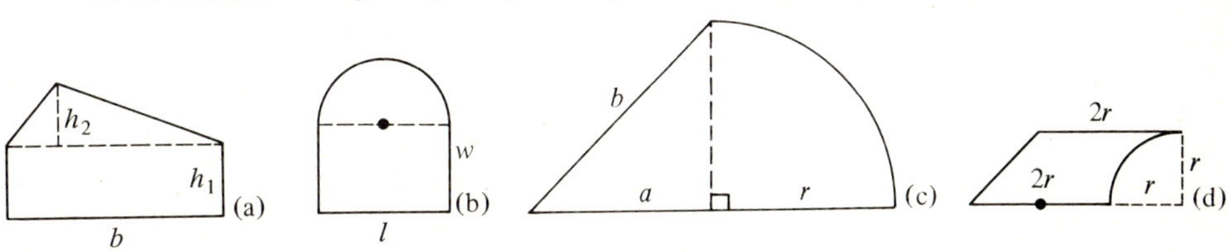

Figure 13-46

29. The figure in Fig. 13-46a
30. The figure in Fig. 13-46b
31. The figure in Fig. 13-46c
32. The figure in Fig. 13-46d

In Exercises 33 through 40, solve the given problems involving areas.

33. A man wishes to put screening in a window frame which has a square opening 60 cm on a side. If he must put a 2-cm-wide strip under the molding, how many square centimeters of screening does he need?

34. A tennis court is a rectangle 78 ft long and 36 ft wide. What is the area of the tennis court?

35. Atmospheric pressure is approximately 15 lb/in.^{2} in all directions. What is the total atmospheric pressure on the outside of a box 5 in. by 8 in. by 12 in.?

36. If a gallon of paint will cover 300 ft^2, how much paint is required to paint the walls of a room 12 ft by 16 ft by 8 ft, if the room has three windows 2 ft by 3 ft and two doors 3 ft by 6 ft?

37. What is the area of the flower bed in Fig. 13-34c?

38. If it costs 10¢/yd^2 to grass an area, how much would it cost to grass the area within the racetrack of Fig. 13-34d?

39. A metal plate is in the shape of a right triangle with sides of 3.56 cm, 4.08 cm, and 5.41 cm. Determine the area of the plate.

40. How many pipes 1.00 cm in radius are required to carry as much water as a pipe 3.00 cm in radius?

13-4 SIMILAR TRIANGLES

In our discussion of geometry in the preceding sections, we studied several important basic properties of geometric figures, such as the concepts of perimeter and area, which deal with the actual size of a figure. Also, we saw that the length of the sides and the size of the angles, respectively, are of importance in defining the properties of figures. In this section, we consider the properties of triangles that are of the same basic shape although not necessarily the same size.

At this point, we need some terminology for the development of this material. **Corresponding sides** *and* **corresponding angles,** *one from each of two different triangles, are those which have the same relative position with respect to the other parts of the triangle.*

Example A In Fig. 13-47, the triangles are lettered so that corresponding angles have the same letters. That is, angles A and A' are corresponding angles, B and B' are corresponding angles, and C and C' are corresponding angles. Also, the sides between these vertices are corresponding sides. That is, AB corresponds to side $A'B'$, and so forth. Even though the triangles are not drawn so that corresponding parts are in the same position relative to the page, both have the largest angle between the shortest and second shortest sides, with other relative positions the same.

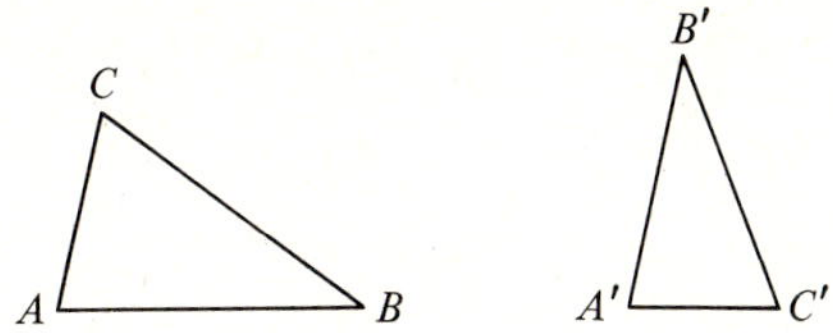

Figure 13-47

If the corresponding angles of two triangles are equal, or if the corresponding sides are proportional, the triangles are said to be **similar.**

Example B The two triangles in Fig. 13-47 have been drawn in such a manner that corresponding angles are equal. This means that the corresponding sides are also proportional:

$$\frac{AB}{A'B'} = \frac{BC}{B'C'} = \frac{CA}{C'A'}$$

If the corresponding angles and the corresponding sides of two triangles are equal, then the two triangles are said to be **congruent.** As a result of this definition, the areas and perimeters of congruent triangles are also equal.

Example C A right triangle with legs of 2 in. and 4 in. is congruent to any other right triangles with legs of 2 in. and 4 in. However, it is similar to any right triangle with legs 5 in. and 10 in., since the ratios of corresponding sides are equal.

One equilateral triangle of side 6 cm is congruent to any other equilateral triangle of side 6 cm; and it is similar to any other equilateral triangle, regardless of the length of the side. We know that corresponding angles are equal and that the ratios of corresponding sides must be equal.

From our discussion, we can see that similar triangles have two basic properties:

1. *Corresponding angles are equal.*
2. *Corresponding sides are proportional.*

If we wish to show that two triangles are similar, we must show that one of these conditions is valid. (If one is valid, so is the other in the case of triangles.) Also, if we know that two triangles are similar, we use these properties to determine unknown parts of one triangle from known parts of the other. The following examples illustrate the use of the properties of similar triangles.

Example D In Fig. 13-48, given triangle ABC where DE is parallel to BC, show that triangle ADE is similar to triangle ABC.

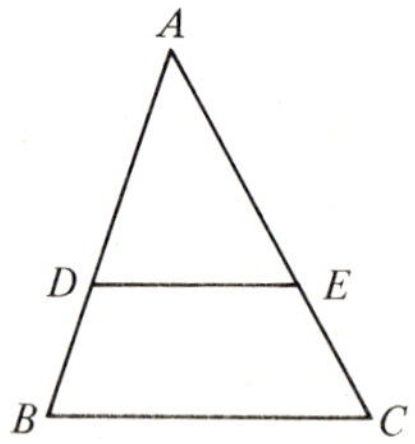

Figure 13-48

We shall review certain symbols that are commonly used in geometry. The symbol $\triangle$ denotes "triangle," $\sim$ denotes "similar," and as we noted earlier, $\|$ denotes "parallel." Therefore, the statement above can be stated as: Given $\triangle ABC$ where $DE \parallel BC$, show that $\triangle ADE \sim \triangle ABC$.

To show that the triangles are similar, we recall that corresponding angles of parallel lines cut by a transversal are equal. This means that $\angle ADE = \angle ABC$ and that $\angle AED = \angle ACB$. Since $\angle DAE$ is common to both triangles, we have one angle in each triangle equal to one angle in the other triangle. Since the corresponding angles are equal, $\triangle ADE \sim \triangle ABC$.

Example E In Fig. 13-49, in right $\triangle ABC$ with right angle at C, $CD \perp AB$. (As we noted earlier, $\perp$ means "perpendicular.") Show that $\triangle ADC \sim \triangle CDB$.

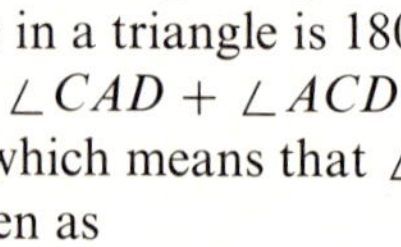

Figure 13-49

First, both triangles contain a right angle. That is, $\angle CDA = \angle CDB$. Since the sum of the angles in a triangle is 180°, the sum of the other two angles in $\triangle ADC$ is 90°, or $\angle CAD + \angle ACD = 90°$. Also, angles ACD and DCB are complementary, which means that $\angle ACD + \angle DCB = 90°$. These two equations can be written as

$$\angle CAD = 90° - \angle ACD \qquad \angle DCB = 90° - \angle ACD$$

Since the right-hand sides of these equations are equal, we conclude that $\angle CAD = \angle DCB$. We have now shown that two angles are respectively equal in the two triangles. Since, in any triangle, the sum of the angles is 180°, the remaining angles must also be equal. Therefore $\angle ACD = \angle CBD$.

The triangles are therefore similar. We do note, however, that the corresponding sides appear in different positions with respect to the page. This is indicated by writing the ratio of corresponding sides as

$$\frac{AD}{DC} = \frac{DC}{DB} = \frac{AC}{CB}$$

where the sides of $\triangle ADC$ are written in the numerators and the corresponding sides of $\triangle\ CDB$ are written as denominators.

Example F In Fig. 13-50, $\triangle ABC \sim \triangle DEF$, where $AC = 6$ cm, $DF = 4$ cm, $FE = 3$ cm, and $ED = 2$ cm. Find the lengths of sides CB and AB.

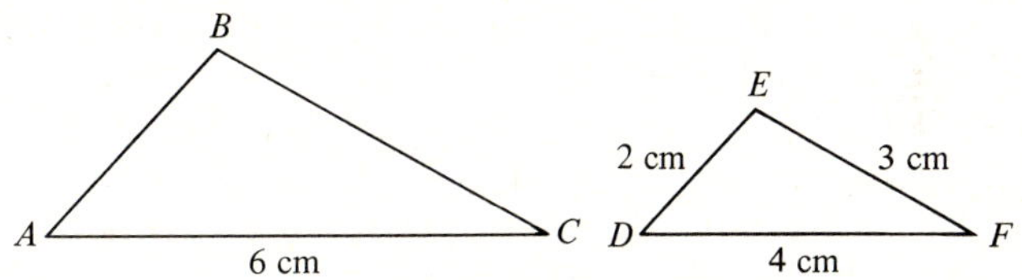

Figure 13-50

Since the triangles are similar, the corresponding sides are proportional. This means that

$$\frac{AC}{DF} = \frac{CB}{FE} = \frac{BA}{ED}$$

Substituting the known values, we get

$$\frac{6}{4} = \frac{CB}{3} = \frac{BA}{2}$$

Since the middle and right ratios are both equal to $\frac{6}{4}$, we can solve for the unknown side each contains. Therefore,

$$\frac{CB}{3} = \frac{6}{4} \qquad CB = \frac{6(3)}{4} = \frac{9}{2}\text{ cm}$$

$$\frac{BA}{2} = \frac{6}{4} \qquad BA = \frac{6(2)}{4} = 3\text{ cm}$$

Example G On level ground a tree casts a shadow 24 ft long. At the same time, a pole 4.0 ft high casts a shadow 3.0 ft long. How high is the tree? See Fig. 13-51.

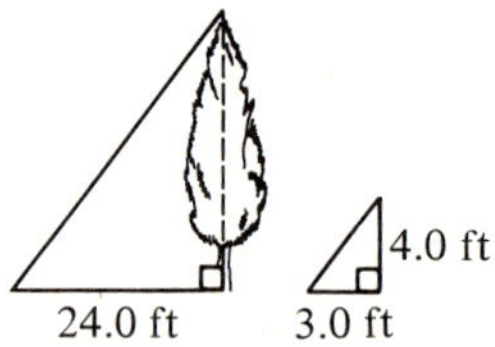

Figure 13-51

The rays of the sun are essentially parallel. Hence, the two triangles indicated in Fig. 13-51 are similar, since each has a right angle and the angles at the tops are equal. The lengths of the hypotenuses are of no importance in this problem. We shall therefore use only the other sides in stating the ratios of corresponding sides. Thus,

$$\frac{h}{4.0} = \frac{24}{3.0} \qquad h = 32 \text{ ft}$$

We conclude that the tree is 32 ft high.

One of the most practical uses of similar figures is that of **scale drawings.** Maps, charts, blueprints, and most drawings which appear in books are familiar examples of scale drawings. Actually, there have been numerous scale drawings used in this book in the previous sections.

In any scale drawing, all distances are drawn a certain ratio of the distances they represent and all angles equal the angles they represent. Consider the following example.

Example H In drawing a map of the area shown in Fig. 13-52, a scale of 1 cm = 200 km is used. In measuring the distance between Chicago and Toronto on the map, we find it to be 3.5 cm. Therefore, the actual distance x between Chicago and Toronto is found from the proportion

$$\frac{x}{3.5 \text{ cm}} = \frac{200 \text{ km}}{1 \text{ cm}} \qquad \text{or} \qquad x = 700 \text{ km}$$

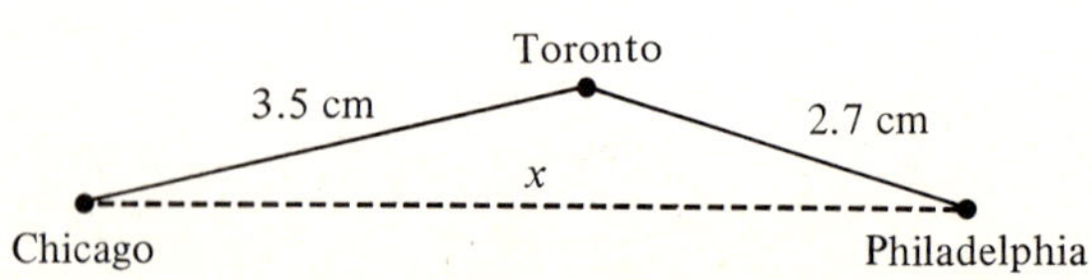

Figure 13-52

If we did not have the scale but knew that the distance between Chicago and Toronto is 700 km, then by measuring distances on the map between Chicago and Toronto (3.5 cm) and between Toronto and Philadelphia (2.7 cm), we could find the distance between Toronto and Philadelphia. It is found from the following proportion, determined by use of similar triangles:

$$\frac{700 \text{ km}}{3.5 \text{ cm}} = \frac{y}{2.7 \text{ cm}}$$

$$y = \frac{2.7(700)}{3.5} = 540 \text{ km}$$

Exercises 13-4

In Exercises 1 through 4, determine whether or not the given triangles are (a) similar and (b) congruent. Angles that are equal are marked in the same way, and so are equal sides.

1. The triangles in Fig. 13-53

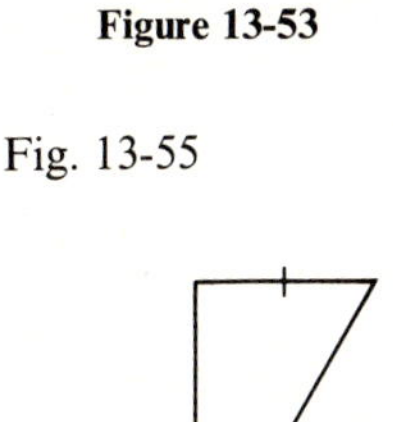

Figure 13-53

2. The triangles in Fig. 13-54

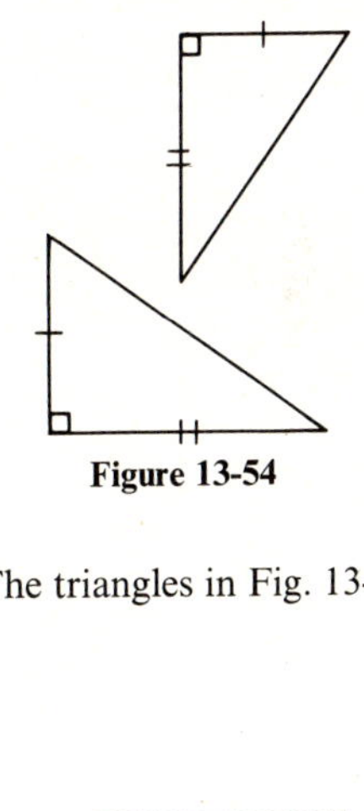

Figure 13-54

3. The triangles in Fig. 13-55

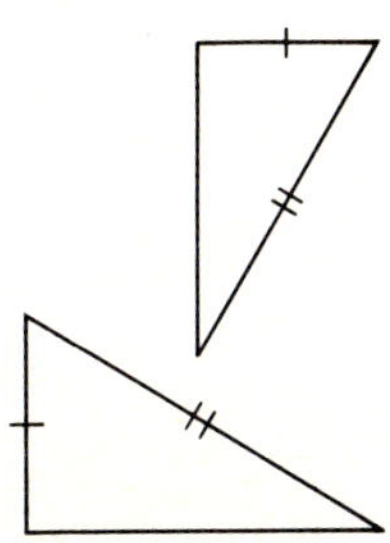

Figure 13-55

4. The triangles in Fig. 13-56

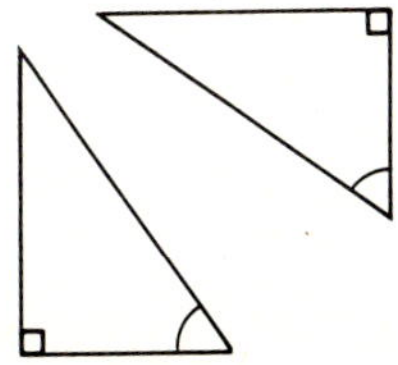

Figure 13-56

In Exercises 5 through 8, for the given pairs of triangles, identify the corresponding angles and corresponding sides to the given angles and sides. The triangles in each of the figures are similar.

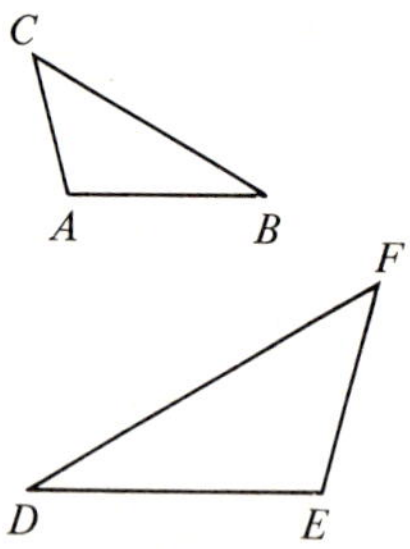

Figure 13-57

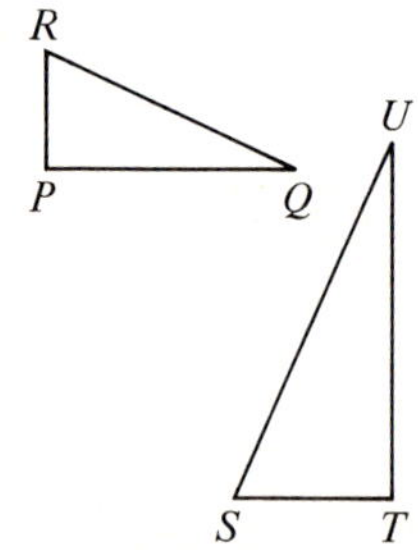

Figure 13-58

5. In Fig. 13-57, $\angle A$ corresponds to ______ and side AC corresponds to ______.

6. In Fig. 13-57, $\angle F$ corresponds to ______ and side DF corresponds to ______.

7. In Fig. 13-58, $\angle Q$ corresponds to ______ and side RP corresponds to ______.

8. In Fig. 13-58, $\angle S$ corresponds to ______ and side UT corresponds to ______.

In Exercises 9 through 12, for the given sides of the triangles of Figs. 13-57 and 13-58, find the indicated sides. In Fig. 13-57: $AB = 5$, $BC = 7$, $AC = 4$, *and* $FE = 8$. *In Fig. 13-58:* $RP = 6$, $PQ = 8$, $RQ = 10$, *and* $ST = 9$.

9. In Fig. 13-57, find DE.

10. In Fig. 13-57, find FD.

11. In Fig. 13-58, find TU.

12. In Fig. 13-58, find SU.

In Exercises 13 through 16, use Fig. 13-59. In this figure, $BD \parallel AE$, $AE = 18$, *and* $DB = 6$.

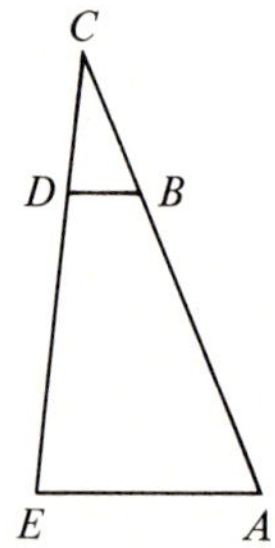

Figure 13-59

13. If $DC = 7$, find CE.

14. If $CA = 27$, find CB.

15. If $CB = 10$, find BA.

16. If $CE = 24$, find DE.

In Exercises 17 and 18, use $\triangle ABC$ in Fig. 13-60. In the figure, $AC \perp BC$ and $CD \perp AB$.

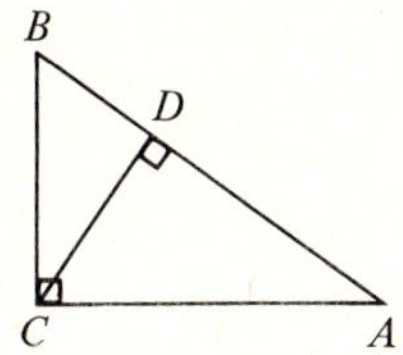

Figure 13-60

17. Given that $\triangle ABC \sim \triangle ACD$, find AB if $AD = 9$ and $AC = 12$.

18. Given that $\triangle ABC \sim \triangle CBD$, find BD if $BC = 6$ and $AB = 9$.

In Exercises 19 and 20, solve for the unknown side by use of an appropriate proportion. Parallel lines are marked with arrows.

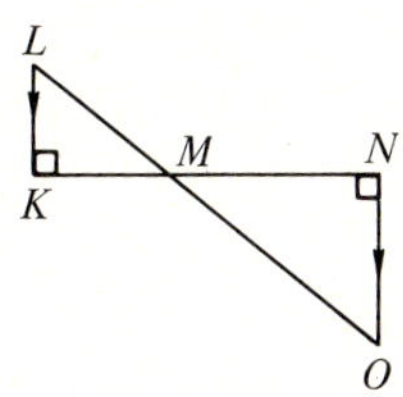

Figure 13-61

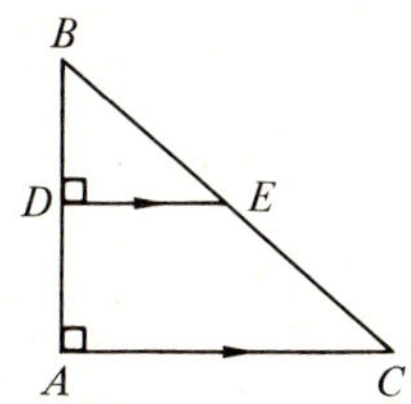

Figure 13-62

19. In Fig. 13-61, KM $= 6$, $MN = 9$, and $MO = 12$. Find LM.

20. In Fig. 13-62, $BD = 5$, $BE = 8$, and $BA = 10$. Find BC.

In Exercises 21 and 22, construct the appropriate figures.

21. Draw $\triangle ABC$ such that $AB \perp BC$ and $\angle ACB = \angle CAB$.

22. Draw $\triangle ABC$ with D on AB and E on AC such that $DE \parallel BC$.

In Exercises 23 and 24, find the required values.

23. Two triangles are similar. The sides of the larger triangle are 3.0 cm, 5.0 cm, and 6.0 cm, and the shortest side of the other triangle is 2.0 cm. Find the remaining sides of the smaller triangle.

24. Two triangles are similar. The angles of the smaller triangle are 50°, 100°, and 30°, and the sides of the smaller triangle are 7.00 in., 9.00 in., and 4.57 in. The longest side of the larger triangle is 15.0 in. Find the other two sides and the three angles of the larger triangle.

In Exercises 25 and 26, show that the required triangles are similar.

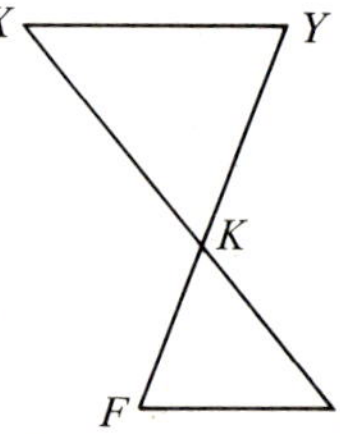

Figure 13-63

25. In Fig. 13-63, show that $\triangle XYK \sim \triangle NFK$. ($XY \parallel FN$)

26. In Fig. 13-60, show that $\triangle ACB \sim \triangle ADC$.

In Exercises 27 and 28, find the required values.

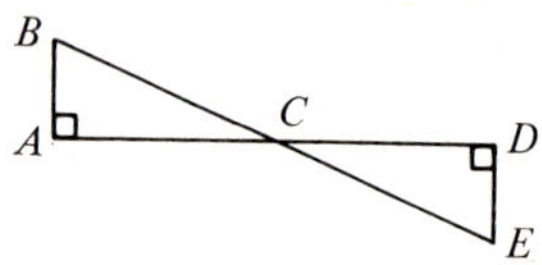

Figure 13-64

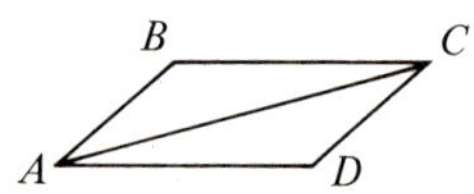

Figure 13-65

27. In Fig. 13-64, $\triangle ABC \cong \triangle EDC$ ($\cong$ means "congruent"). If $AD = 16$ in., how long is AC?

28. In Fig. 13-65, $\triangle ABC \cong \triangle ADC$. If $\angle CAD = 40°$, how many degrees are there in $\angle CAB + \angle ABC$?

In Exercises 29 through 40, solve the given problems. Draw figures as needed to aid in setting up the problem.

29. A man casts a shadow 5.0 ft long. His son casts a shadow 3.0 ft long. If the son is 4.0 ft tall, how tall is the father?

30. On level ground a tree casts a shadow 36 m long. At the same time, a pole 9 m high casts a shadow 12 m long. How high is the tree?

31. A 1-m stick is placed vertically in the shadow of a vertical pole such that the ends of their shadows are at the same point. If the shadow of the meter stick is 80 cm long and that of the pole is 280 cm long, how high is the pole?

32. In constructing a metal support in the form of $\triangle ABC$ as shown in Fig. 13-66, it is deemed necessary to strengthen the support with an added brace DE which is parallel to BC. How long must the brace be if $AB = 20$ in., $AD = 14$ in., and $BC = 25$ in.?

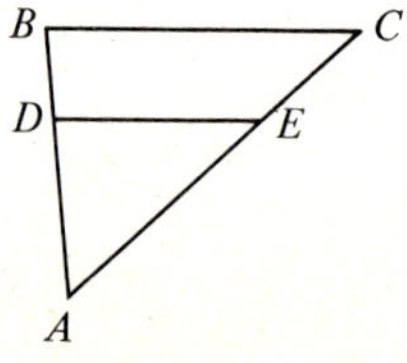

Figure 13-66

33. A certain house blueprint has a scale of $1\frac{1}{4}$ in. = 10 ft. The living room is 18 ft long. What distance on the blueprint represents this length?

34. On a map, 12 cm = 100 mi. What is the distance between two cities if the distance between them on the map is 7.5 cm?

35. From Fig. 13-52, find the distance from Chicago to Philadelphia (see Example H).

36. On a map the scale is 1 cm = 50 mi. If the actual distance between Albany and New York is 125 mi, find the distance between these cities on the map.

37. A 4.0-ft wall stands 2.0 ft from a building. The ends of a straight pole touch the building and the ground 6.0 ft from the wall. A point on the pole touches the wall's top. How high on the building does the pole touch?

38. To find the width *ED* of a river, a surveyor places markers at *A*, *B*, *C*, and *D* (see Fig. 13-67). She places them so that $AB \parallel ED$, $BC = 50.0$ m, $DC = 300$ m, and $AB = 80.0$ m. How wide is the river?

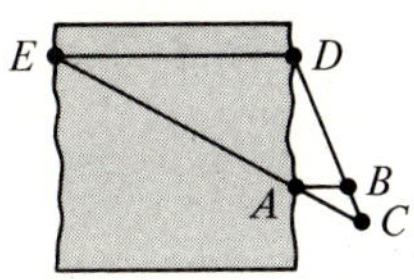

Figure 13-67

39. A 30.0-m pole on level ground is supported by two 60.0-m guy wires attached at its top. The guy wires are on opposite sides of the pole. How far is it between the grounded ends of the wires?

40. Town *B* is due east of town *A*, and town *E* is due east of town *D*. The direct routes from *A* to *E* and *B* to *D* cross at town *C*. The route *DC* is 15 mi, *DB* is 45 mi, and *DE* is 24 mi. How far is it from town *A* to town *B*?

13-5 SOLID GEOMETRIC FIGURES: PRISMS AND CYLINDERS

Until now we have restricted our attention to plane geometry. There are, however, a number of important solid geometric figures. The purpose of this and the following section is to discuss the determination of surface area and volume of the most important of these figures.

A **polyhedron** *is a solid geometric figure which is bounded by planes. The plane surfaces of the polyhedron are called* **faces,** *and the intersections of the faces are called* **edges.** In Fig. 13-68, a polyhedron of four faces and six edges is illustrated.

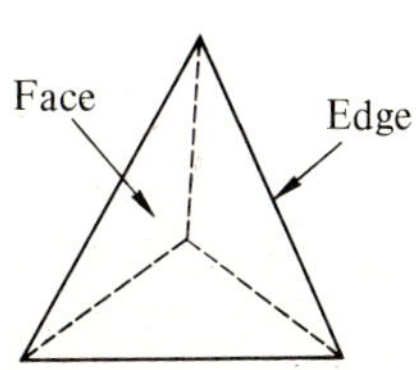

Figure 13-68

A **prism** *is a polyhedron whose bases are parallel and equal polygons and whose side faces are parallelograms.* In Fig. 13-69, a prism with triangular bases is shown. We shall consider only prisms in which the side edges are perpendicular to the bases **(right prisms).**

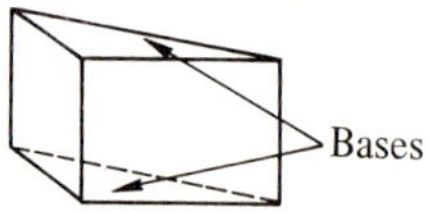

Figure 13-69

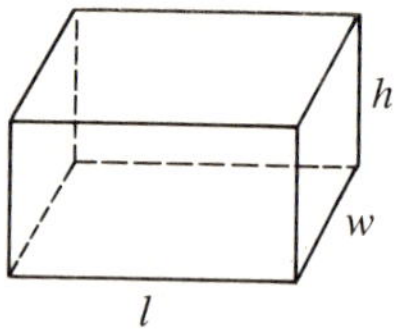

Figure 13-70

A prism of particular importance is a **rectangular solid,** which we now illustrate in Fig. 13-70. In Chapter 3, we defined the **volume** *of a rectangular solid* of length l, width w, and height h as

$$\boxed{V = lwh} \tag{13-8}$$

We now note that *the bottom face of the rectangular solid, which is of length* l *and width* w, *is the* **base** *of the solid.* If B is the area of the base, which means that $B = lw$ for the rectangular solid, we have

$$\boxed{V = Bh} \tag{13-9}$$

as the formula for the volume. Equation (13-9) can be used to find the volume of a rectangular solid, as well as the volume of any prism for which we know or can determine the base area B.

Example A The prism shown in Fig. 13-71 has a regular hexagon as a base. If we are able to determine the area of the hexagon, we can use Eq. (13-9) to find the volume. If, for example, $B = 12$ in.^{2} and $h = 5.0$ in., the volume is

$$V = (12)(5.0) = 60 \text{ in.}^3$$

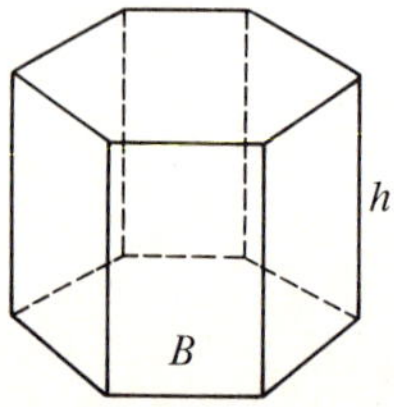

Figure 13-71

Example B The base of the prism shown in Fig. 13-72 is a right triangle with legs of 14.6 cm and 10.9 cm. The height of the prism is 12.3 cm. To find the volume, we first find the area of the base to be one-half the product of the lengths of the legs of the triangle. Thus,

$$B = \frac{1}{2}(14.6)(10.9)$$
$$= 79.6 \text{ cm}^2$$

This means that the volume is

$$V = Bh = (79.6)(12.3)$$
$$= 979 \text{ cm}^3$$

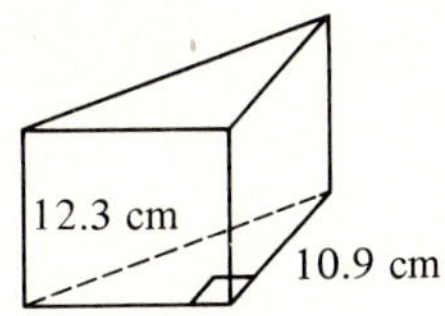

Figure 13-72

The **lateral area** *L of a prism is the area of the faces other than the bases. The* **total area** *A of a prism is the lateral area plus the area of the bases.* Consider the following example.

Example C For the prism in Fig. 13-72, the lateral area is the sum of the areas of the front face, right face, and back face. To find each of these areas, we multiply length by width. For the back face we use the Pythagorean theorem to find the length. Thus,

$$L = (14.6)(12.3) + (10.9)(12.3) + (\sqrt{14.6^2 + 10.9^2})(12.3)$$
$$= 179.6 + 134.1 + (18.22)(12.3) = 313.7 + 224.1$$
$$= 537.8 \text{ cm}^2$$

Thus, the lateral area is 538 cm^2 (rounded off to three significant digits). Noting that the length of each face is multiplied by the height of the prism, we see that: *The lateral area may be found by multiplying the perimeter of the base by the height.*

The total area is found by adding the area of the bases to the lateral area. Thus,

$$A = 537.8 + 2\left(\frac{1}{2}\right)(14.6)(10.9)$$
$$= 537.8 + 159.1$$
$$= 696.9 \text{ cm}^2.$$

Thus, the total area is 697 cm^2.

In the case of a rectangular solid (see Fig. 13-70), there are three pairs of equal rectangular faces. Thus, *a formula for the total area of a rectangular solid is*

$$\boxed{A = 2lw + 2wh + 2lh} \qquad (13\text{-}10)$$

For a cube, each of the six faces is a square. Thus, *the total area of a cube is*

$$\boxed{A = 6e^2} \tag{13-11}$$

Here e is the length of the edge of the cube.

Example D For a rectangular solid for which $l = 260$ mm, $w = 130$ mm, and $h = 150$ mm, the total area is

$$\begin{aligned} A &= 2(260)(130) + 2(130)(150) + 2(260)(150) \\ &= 67600 + 39000 + 78000 \\ &= 184{,}600 \text{ mm}^2 \end{aligned}$$

The total area of a cube 6.82 in. on an edge is

$$\begin{aligned} A &= 6(6.82)^2 = 6(46.5) \\ &= 279 \text{ in.}^2 \end{aligned}$$

A **right circular cylinder** *is generated by revolving a rectangle about one of its sides. In a right circular cylinder each base is a circle, and the* **cylindrical surface** *is perpendicular to the bases.* See Fig. 13-73.

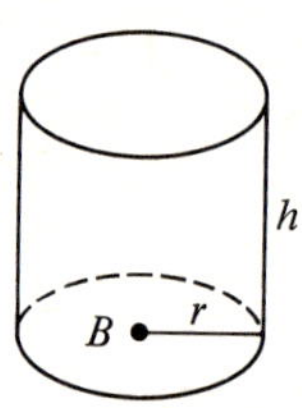

Figure 13-73

As with the prism, *the volume of a right circular cylinder is the product of the area of the base and the height, or altitude, of the cylinder.* Since the base is a circle, the area of the base is πr^2, and we have the formula

$$\boxed{V = \pi r^2 h} \tag{13-12}$$

for the volume, where r is the radius of the base and h is the height.

Example E Find the volume of a right circular cylinder for which $d = 16.5$ in. and $h = 6.75$ in.

In this case, we are given the diameter of the base, which means that the radius r is 8.25 in. Therefore,

$$\begin{aligned} V &= \pi(8.25)^2(6.75) \\ &= (3.14)(68.06)(6.75) \\ &= 1440 \text{ in.}^3 \end{aligned}$$

The lateral area of a prism is the product of the perimeter of a base and the height. In a similar manner, *the lateral area of a cylinder is the circumference of a base times the height.* We can think of this as the label on a can being removed and flattened out as demonstrated in Fig. 13-74. Since the circumference of a base is $2\pi r$, the lateral area of a cylinder is

$$\boxed{L = 2\pi rh} \qquad (13\text{-}13)$$

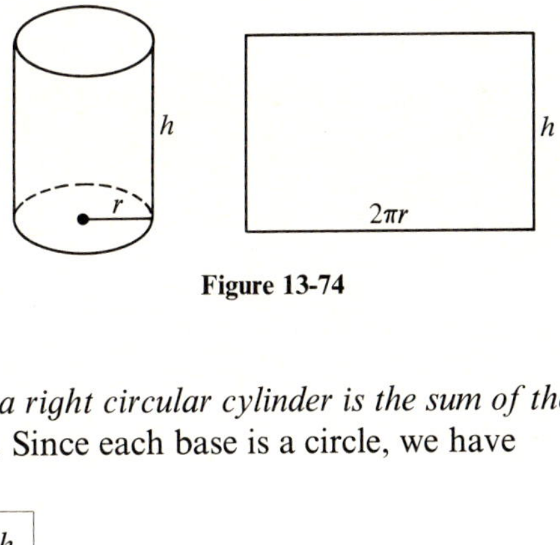

Figure 13-74

The total area of a right circular cylinder is the sum of the areas of the bases and the lateral area. Since each base is a circle, we have

$$\boxed{A = 2\pi r^2 + 2\pi rh} \qquad (13\text{-}14)$$

as the formula for the total area.

Example F An oil can in the shape of a right circular cylinder has a radius of 12.5 cm and is 30.0 cm high. What is (a) the lateral area and (b) the total area of the can?

Using Eq. (13-13), with $r = 12.5$ and $h = 30.0$, we get

$$L = 2\pi(12.5)(30.0) = 2(3.14)(12.5)(30.0)$$
$$= 2355 \text{ cm}^2$$

Using Eq. (13-14), we have

$$A = 2\pi(12.5)^2 + 2\pi(12.5)(30.0)$$
$$= 2(3.14)(156.3) + 2(3.14)(12.5)(30.0)$$
$$= 981 + 2355$$
$$= 3336 \text{ cm}^2$$

Therefore, to three significant digits, $L = 2360 \text{ cm}^2$ and $A = 3340 \text{ cm}^2$.

Exercises 13-5

In Exercises 1 through 8, determine the volumes of the prisms with the given dimensions.

1. Triangular base: $B = 460 \text{ ft}^2$, $h = 15.0$ ft
2. Triangular base: $B = 37.8 \text{ m}^2$, $h = 5.27$ m
3. Quadrilateral base: $B = 2750 \text{ cm}^2$, $h = 36.0$ cm
4. Polygon base: $B = 8630 \text{ in.}^2$, $h = 40.2$ in.
5. Base is a right triangle with legs 2.37 ft and 4.22 ft; $h = 1.77$ ft.
6. Base is a right triangle with legs 36.3 mm and 58.2 mm; $h = 12.8$ mm.
7. Base is a parallelogram for which base is 53.7 cm and height is 17.7 cm; $h = 23.5$ cm.
8. Base is a parallelogram for which base is 1.37 yd and height is 1.13 yd; $h = 1.27$ yd.

In Exercises 9 through 16, determine (a) the lateral areas and (b) the total areas of the prisms with the given dimensions.

9. Rectangular solid: $l = 16.3$ ft, $w = 11.5$ ft, $h = 14.3$ ft
10. Rectangular solid: $l = 72.0$ mm, $w = 32.3$ mm, $h = 10.3$ mm
11. Cube: $e = 0.820$ m
12. Cube: $e = 2.73$ ft
13. Quadrilateral base for which $B = 3270 \text{ in.}^2$, $p = 238$ in.; $h = 51.8$ in.
14. Polygon base for which $B = 7360 \text{ cm}^2$, $p = 421$ cm; $h = 42.9$ cm
15. Base is a right triangle with legs 4.25 cm and 6.50 cm; $h = 3.10$ cm
16. Base is a right triangle with legs 17.0 in. and 10.5 in.; $h = 3.25$ in.

In Exercises 17 through 24, determine the volumes of the cylinders with the given values of r, d (diameter of base), and h.

17. $r = 20.0$ cm; $h = 15.0$ cm

18. $r = 7.00$ in.; $h = 4.00$ in.

19. $r = 15.0$ ft; $h = 3.60$ ft

20. $r = 1.58$ m; $h = 8.48$ m

21. $d = 366$ mm; $h = 140$ mm

22. $d = 0.634$ yd; $h = 0.156$ yd

23. $d = 24.2$ in.; $h = 32.3$ in.

24. $d = 22.0$ cm; $h = 12.2$ cm

In Exercises 25 through 32, determine (a) the lateral areas and (b) the total areas of the cylinders with the given values of r, d, and h.

25. $r = 300$ mm; $h = 120$ mm

26. $r = 60.0$ ft; $h = 80.0$ ft

27. $r = 8.20$ in.; $h = 2.40$ in.

28. $r = 2.30$ m; $h = 1.10$ m

29. $d = 24.0$ cm; $h = 8.50$ cm

30. $d = 3.80$ in.; $h = 7.50$ in.

31. $d = 86.4$ ft; $h = 12.4$ ft

32. $d = 84.2$ mm; $h = 123$ mm

In Exercises 33 through 40, solve the given problems. All involve prisms or cylinders.

33. How many cubic meters of concrete are needed for a driveway 20.0 m long, 2.75 m wide, and 0.100 m thick?

34. A swimming pool is 50.0 ft wide, 80.0 ft long, 3.00 ft deep at one end, and 8.00 ft deep at the other end. How many cubic feet of water will it hold? (Assume that the slope on the bottom is constant.)

35. A rectangular box is to be used to store radioactive materials. The inside of the box is 12.0 in. long, 9.00 in. wide, and 9.00 in. deep. What is the area of sheet lead which must be used to line the inside of the box?

36. A glass prism used in the study of optics has a right triangular base. The legs of the right triangle are 3.00 cm and 4.00 cm. The prism is 8.50 cm high. What is the total surface area of the prism?

37. What is the weight of 1 km of copper wire which is 0.500 cm in diameter? The density of copper is 8.90 g/cm^3.

38. A cylindrical grain storage container 82.0 ft high has a radius of 24.3 ft. One bushel of grain occupies about 1.24 ft^3. How many bushels can be stored in the container?

39. At 25¢ per square yard, how much would it cost to paint the outside of the storage container of Exercise 38? (The bottom is not to be painted.)

40. A lawn roller is a cylinder 96.0 cm long and 30.0 cm in radius. What area is rolled in one complete revolution of the roller?

13-6 SOLID GEOMETRIC FIGURES: PYRAMIDS, CONES, SPHERES

The solid geometric figures discussed in the previous section have parallel and equal bases. In this section, we discuss other types of solids. Applications of these figures are found in architecture, engineering, and astronomy.

The base of a **pyramid** *is a polygon. The other faces, the* **lateral faces,** *are triangles which meet at a common point, the* **vertex.** *The edges where the lateral faces meet are the* **lateral edges.** *The* **height,** *or* **altitude,** *of the pyramid is the perpendicular distance from the vertex to the base.* These are illustrated in Fig. 13-75. *We shall consider only* **regular pyramids**—*those whose bases are regular polygons and whose lateral faces are congruent isosceles triangles.*

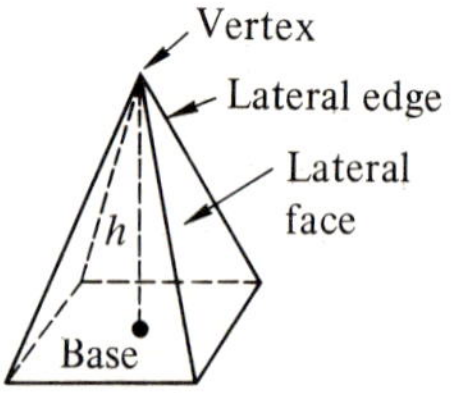

Figure 13-75

If we compare the volumes of a prism and a pyramid of the same base and height, we can see that the volume of the pyramid is considerably less than that of the prism (see Fig. 13-76). Although we cannot prove it here, *the volume of the pyramid is exactly one-third of the volume of the prism.*

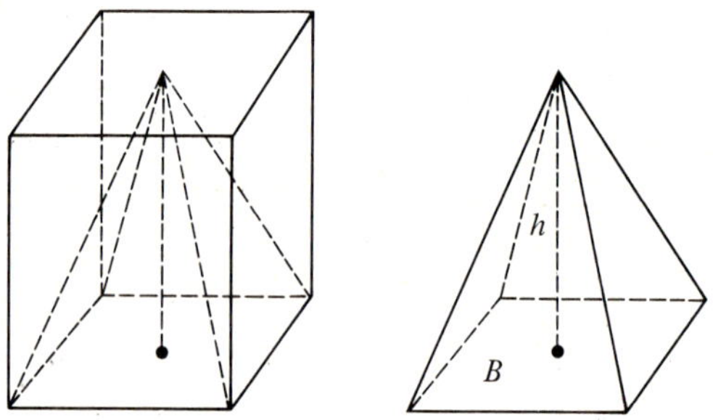

Figure 13-76

Thus,

$$V = \frac{1}{3} Bh \tag{13-15}$$

is the formula for the volume of a pyramid of base area B and height h.

Example A The base of a pyramid is a square 12.0 cm on a side. The height of the pyramid is 18.5 cm. To find the volume, we must note that the base area B is $(12.0\text{ cm})^2$. Thus,

$$
\begin{aligned}
V &= \frac{1}{3}(12.0)^2(18.5) \\
&= 888\text{ cm}^3
\end{aligned}
$$

In a regular pyramid the lateral faces are congruent triangles. The altitude, or height, of each of these triangles is the **slant height** *s of the pyramid,* as shown in Fig. 13-77. Therefore, the area of each triangle is one-half the product of the slant height and the base of the triangle. The sum of the bases of the triangles is the perimeter of the base of the pyramid. Thus, *the* **lateral area,** *which is the sum of the areas of the lateral faces of the pyramid, is*

$$\boxed{L = \frac{1}{2}ps} \tag{13-16}$$

Here p is the perimeter of the base and s is the slant height.

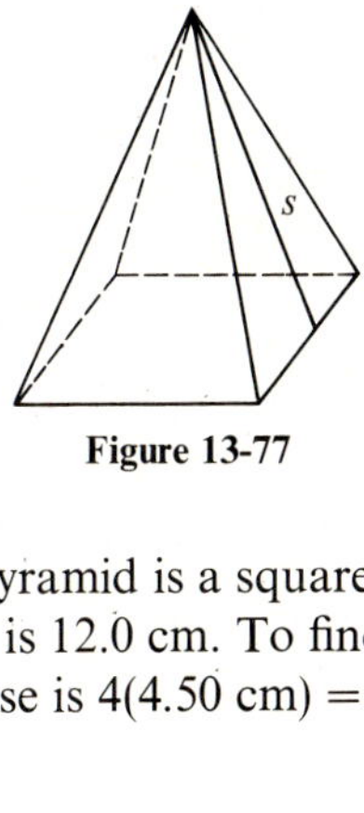

Figure 13-77

Example B The base of a pyramid is a square of side 4.50 cm, and the slant height of the pyramid is 12.0 cm. To find the lateral area, we note that the perimeter of the base is 4(4.50 cm) = 18.0 cm. Thus,

$$
\begin{aligned}
L &= \frac{1}{2}(18.0)(12.0) \\
&= 108\text{ cm}^2
\end{aligned}
$$

A **conical surface** *is generated by rotating a straight line about one of its points. A* **cone** *is a solid geometric figure bounded by the conical surface and a plane which cuts the surface.* Although many types of cones may be generated, one type is of particular importance. *It is the* **right circular cone,** *which is generated by rotating a right triangle about one of its legs.* Thus, the **base** of the cone is a circle and the **slant height** is the hypotenuse of the right triangle (see Fig. 13-78). The **height,** or **altitude,** of the cone is one leg of the right triangle. The radius of the base is the length of the other leg. We shall restrict our attention to right circular cones, and whenever we use "cone," we mean right circular cone.

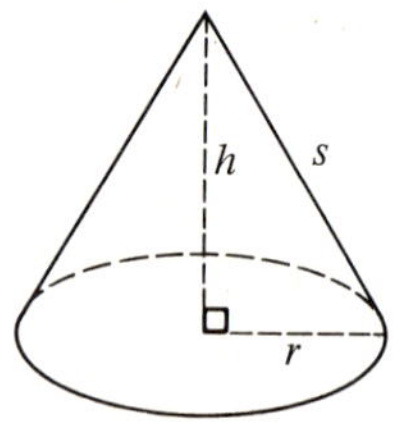

Figure 13-78

As with the pyramid and prism, if we compare the volumes of a cone and cylinder with the same base and height, we see that the volume of the cone is considerably less than that of the cylinder. In fact, the volume of the cone is exactly one-third of the volume of the cylinder, just as with the pyramid and prism. Thus, *the volume of a right circular cone is*

$$\boxed{V = \frac{1}{3}\pi r^2 h} \qquad (13\text{-}17)$$

Here r is the radius of the base and h is the height of the cone.

Example C The volume of a cone for which $r = 5.50$ in. and $h = 3.75$ in. is

$$V = \frac{1}{3}\pi(5.50)^2(3.75)$$

$$= \frac{1}{3}(3.14)(30.25)(3.75)$$

$$= 119 \text{ in.}^3$$

It is reasonable that the lateral area of a cone follows the same basic formula as that of a pyramid. Recalling that the slant height of a cone is the hypotenuse of the right triangle with legs r and h, and that the perimeter of a circle is its circumference, we have

$$L = \frac{1}{2}ps = \frac{1}{2}(2\pi r)s$$

or

$$L = \pi r s \tag{13-18}$$

as the formula for the lateral area of a cone. Here r is the radius of the base and s is the slant height.

Example D A tent is in the shape of a cone with a radius of 3.20 m and height of 3.60 m. What is the **surface area** (lateral area) of the tent?

We must first find the slant height s. This is done by use of the Pythagorean theorem. As noted above, the radius and height are the legs of a right triangle with the slant height as the hypotenuse (see Fig. 13-79). Thus,

$$\begin{aligned} s^2 &= r^2 + h^2 \\ &= (3.20)^2 + (3.60)^2 \\ &= 10.24 + 12.96 = 23.2 \end{aligned}$$

or

$$s = 4.82 \text{ m}$$

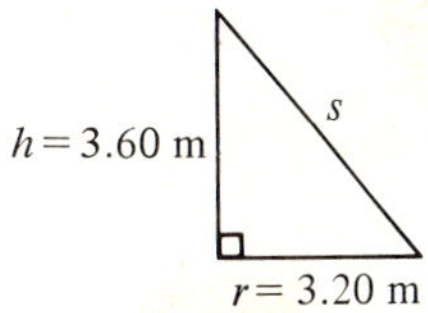

Figure 13-79

Thus, the lateral area is

$$\begin{aligned} L &= \pi(3.20)(4.82) = (3.14)(3.20)(4.82) \\ &= 48.4 \text{ m}^2 \end{aligned}$$

The final solid figure we shall consider is the **sphere.** *A sphere is generated by rotating a circle about a diameter. The* **radius** *of the sphere is a line segment joining the center and a point on the sphere. The* **diameter** *of the sphere is a line segment through the center and having its endpoints on the sphere. The intersection of a plane and a sphere is a circle.* If the plane of intersection contains the center, the circle of intersection is a **great circle.** Other circles in planes not containing the center are **small circles.** These are illustrated in Fig. 13-80.

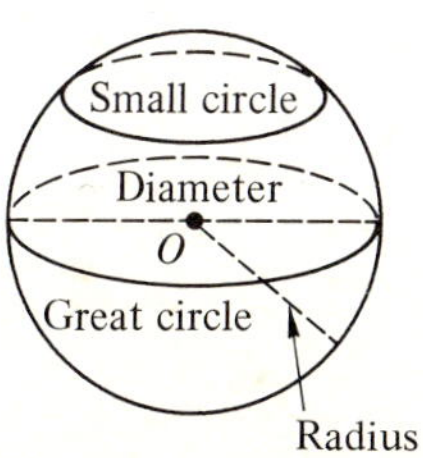

Figure 13-80

Due to the close relationship between the circle and the sphere, the volume and the surface area of the sphere are expressed in terms of π. Although we cannot prove it here, *the volume of a sphere is given by*

$$V = \frac{4}{3}\pi r^3 \tag{13-19}$$

Here r is the radius of the sphere. Also, *the total surface area of a sphere is given by*

$$A = 4\pi r^2 \tag{13-20}$$

Example E The radius of the earth (which is approximately a sphere) is about 3960 mi. The volume of the earth is

$$V = \frac{4}{3}\pi(3960)^3 = 2.60 \times 10^{11}\ \text{mi}^3$$

The area of the earth is

$$A = 4\pi(3960)^2 = 1.97 \times 10^8\ \text{mi}^2$$

Example F A grain storage container is in the shape of a cylinder surmounted by a **hemisphere** (half a sphere). See Fig. 13-81. Given that the radius of the cylinder is 40.0 ft and its height is 120 ft, find the volume of the container.

Figure 13-81

The volume of the container is the volume of the cylinder plus the volume of the hemisphere. By the construction we see that the radius of the hemisphere is the same as the radius of the cylinder. Thus,

$$\begin{aligned} V &= \pi r^2 h + \frac{1}{2}\left(\frac{4}{3}\pi r^3\right) = \pi r^2 h + \frac{2}{3}\pi r^3 \\ &= (3.14)(40.0)^2(120) + \frac{2}{3}(3.14)(40.0)^3 \\ &= 603{,}000 + 134{,}000 = 737{,}000\ \text{ft}^3 \end{aligned}$$

Exercises 13-6

In Exercises 1 through 4, find the volumes of the pyramids with the given bases and heights.

1. Base is a polygon with $B = 3600 \text{ ft}^2$; $h = 45.0$ ft.
2. Base is a polygon with $B = 7850 \text{ cm}^2$; $h = 38.4$ cm.
3. Base is a square of side 25.0 mm; $h = 4.60$ mm.
4. Base is a square of side 8.50 ft; $h = 4.85$ ft.

In Exercises 5 through 8, find the lateral areas of the pyramids with the given bases and slant heights (or heights).

5. Base is a polygon with $p = 18.0$ yd; $s = 8.50$ yd.
6. Base is a polygon with $p = 880$ cm; $s = 350$ cm.
7. Base is a square of side 12.0 m; $h = 8.00$ m.
8. Base is a square of side 30.0 in.; $h = 36.0$ in.

In Exercises 9 through 12, find the volumes of the cones with the given base radii (or diameters) and heights.

9. $r = 10.0$ ft; $h = 14.0$ ft
10. $r = 25.0$ cm; $h = 40.0$ cm
11. $d = 62.8$ cm; $h = 26.3$ cm
12. $d = 17.8$ ft; $h = 22.3$ ft

In Exercises 13 through 16, find the lateral areas of the cones with the given base radii (or diameters) and slant heights (or heights).

13. $r = 7.00$ in.; $s = 9.00$ in.
14. $d = 2.80$ m; $s = 1.80$ m
15. $r = 45.0$ cm; $h = 24.0$ cm
16. $d = 17.8$ ft; $h = 10.5$ ft

In Exercises 17 through 20, find the volumes of the spheres with the given radii or diameters.

17. $r = 3.00$ ft
18. $r = 20.0$ cm
19. $d = 220$ mm
20. $d = 36.2$ ft

In Exercises 21 through 24, find the surface areas of the spheres with the given radii or diameters.

21. $r = 30.0$ in.
22. $r = 60.0$ m
23. $d = 346$ mm
24. $d = 62.4$ in.

In Exercises 25 through 40, solve the given problems involving pyramids, cones, and spheres.

25. When built, the Great Pyramid of Egypt had a square base approximately 250 yd on a side. Its height was about 160 yd. What was its volume?
26. A steel wedge is made in the shape of a pyramid with a square base, 3.50 cm on a side, and a height of 8.25 cm. Given that the density of steel is 7.80 g/cm^3, what is the mass of the wedge?
27. A sheet metal container is in the shape of a pyramid (inverted) with a square base. The side of the square is 12.0 cm and the depth of the container is 8.75 cm. At a cost of 75¢ per square centimeter, what is the cost of the material of the container (no base)?
28. A tent has the shape of a pyramid on a square base. The side of the square is 9.25 ft and the height of the pyramid is 10.5 ft. What is the area of the canvas needed for the four sides and the floor?

29. A conical cistern 10.0 ft high has a radius at the top of 6.00 ft. Water weighs 62.4 lb/ft^3. How many pounds of water does the cistern hold?

30. A pile of sand is in the shape of a cone. The diameter of the base is 12.0 m and its height is 2.50 m. What is the volume of sand in the pile?

31. A paper cup is in the shape of a cone with radius 1.80 in. and height 3.50 in. What is the surface area of the cup?

32. A conical funnel is 8.50 cm deep and 10.0 cm in diameter. What is the surface area of the funnel? (Neglect the opening of the funnel.)

33. What is the weight of a gold sphere 6.00 in. in diameter? The density of gold is 0.697 lb/in^3.

34. The circumference of a basketball is about 29.8 in. What is its volume?

35. The diameter of the moon is about 3480 km. What is the surface area of the moon? Compare the radius of the moon with that of the earth (See Example E—unit conversion is necessary.), and then compare the areas.

36. A lampshade is in the shape of a hemisphere of diameter 36.0 cm. What is the surface area of the shade?

37. A certain rivet is shaped as shown in Fig. 13-82 (a conical part on a cylindrical part). Find the volume of the rivet shown.

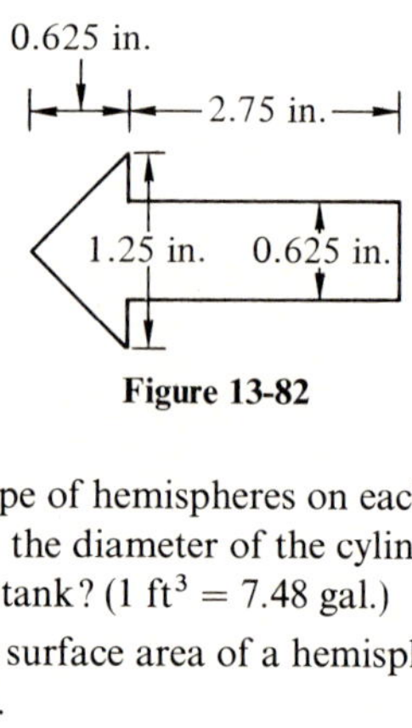

Figure 13-82

38. An oil storage tank is in the shape of hemispheres on each end of a cylinder. The total length of the tank is 45.0 ft, and the diameter of the cylinder (or sphere) is 12.5 ft. What is the volume, in gallons, of the tank? (1 ft^3 = 7.48 gal.)

39. Derive the formula for the total surface area of a hemispherical volume of radius r (curved surface and flat surface).

40. Derive the formula for the volume formed by placing a hemisphere of radius r on a cone of radius r and height h.

13-7 REASONING AND PROOF IN MATHEMATICS

Throughout the development of the material in this text, we have defined many terms and have developed many basic mathematical operations. We based our conclusions on "reasonable," although informal and intuitive, statements. Many of these "reasonable" statements were unproved, and no attempt was made to *prove* them.

Example A In Chapter 3, when we were discussing the solution of an equation, we stated that

> "if the same number is added to each side (of an equation), the two sides are still equal."

Certainly, if two quantities are equal and equal numbers are added to each, the results should be equal. This is a "reasonable" statement, but simply stating it does not *prove* it.

In Chapter 6, when discussing fractions, we stated the fundamental principle of fractions. A large part of the material on fractions is based on this principle, although no attempt was made to *prove* it.

In a more rigorous approach to mathematics, precise lines of reasoning are set up and followed. Although we have followed appropriate reasoning processes, we have been informal. However, bad reasoning, which often leads to fallacious results, must be avoided. In the following example, we discuss five basic errors in reasoning that often occur.

Example B

1. *Reasoning by analogy between situations which are alike in many respects but not in all respects is not logical.* For example, all **equilateral** (all sides equal) triangles are **equiangular** (all angles equal). Concluding that therefore all equilateral quadrilaterals are also equiangular is false. The rhombus is not equiangular. Properties of triangles and quadrilaterals differ in many respects.

2. *Assuming as true the statement that is being proved is a common error.* One commits this error if, for example, in proving two lines to be perpendicular, one uses the right angles involved before proving them to be right angles.

3. *Reasoning by using facts that are irrelevant to the question is fallacious.* For example, if each of the angles in a given polygon equals 108°, the conclusion that a particular angle in any other polygon is 108° is not valid unless it is known that it is in a similar polygon. The first part of the statement has no bearing on the conclusion.

4. *Assuming that which is true in one or more cases is true in all cases is illogical.* For example, when we note that a pair of lines meet at a point and thereupon conclude that any pair of lines will meet at a point, we are making an assumption that is not necessarily correct. In Chapter 12, we saw that two lines do not necessarily meet at a point.

5. *"Jumping to a conclusion" is a common type of error in reasoning.* Stating that two triangles are congruent by proving that corresponding angles are equal illustrates this point. The proof must proceed further by showing that at least one pair of corresponding sides are equal.

Sets, along with simple **circle diagrams,** are often useful as a means of testing a particular line of reasoning. This testing is done by letting the points within a given circle represent the elements of a specified set. Example C illustrates the use of such a circle diagram for testing a line of reasoning in a nonmathematical problem. It should be realized that proper reasoning is important in all fields.

Example C Let us assume that a dog is an animal with four legs and that your pet has four legs. A conclusion is then drawn that your pet is a dog.

To test this conclusion, we note that the first part of the assumption deals with dogs and four-legged animals. Therefore, we let A represent the set of all four-legged animals and D the set of all dogs. This means that set D is a subset of set A. Your pet p is a member of the set of four-legged animals. Therefore, it must lie within the circle A in Figure 13-83, but there is nothing to indicate that it must be within circle D as well. Therefore, the conclusion is not necessarily correct, and the reasoning which led to the conclusion that your pet is a dog is not correct.

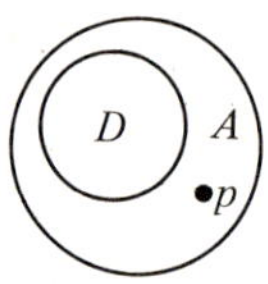

Figure 13-83

If we assume that a dog is an animal with four legs and that your pet is a dog, it follows properly that your pet has four legs. (Note that the assumption that a dog is an animal with four legs removes from consideration any dogs which may not have four legs.)

One important method of reasoning is called **inductive reasoning.** *This method proceeds from the specific case to the general case.* If there are enough examples of a particular occurrence and no counterexamples are known, the general occurrence is then accepted. This method is used in physics and in science generally. It is also essentially the method used in psychology, where a statistical study is used to predict what is most probable. Inductive reasoning can lead to incorrect results, and cases can be cited in the history of science where this has happened. However, scientists usually collect as many data as is reasonably possible before drawing any conclusions, and much of the progress in science has stemmed from experimentation and inductive reasoning.

Another important method of reasoning is that of **deductive reasoning.** *In a manner of speaking, this method proceeds from the general to the specific. In mathematics, this is the accepted method of proof.*

Before describing the method of a deductive proof, we must introduce certain forms. *A* **postulate** (*or* **axiom**) *is a statement which is accepted without* *proof. A* **theorem** *is a statement that must be proved before it is accepted. A* **definition** *explains the meaning of a word in terms of previously defined and accepted words. A* **premise** *is information (assumed true for the theorem) given in a theorem, from which we try to* **prove** *a particular* **conclusion,** *which is also stated in the theorem.*

As a means of illustration, and for future use in this section, we shall now present several important postulates.

Postulates

1. *If equals are added to equals, the sums are equal.*
2. *If equals are subtracted from equals, the remainders are equal.*
3. *If equals are multiplied by equals, the products are equal.*
4. *If equals are divided by equals (not zero), the quotients are equal.*
5. *The whole is equal to the sum of its parts.*
6. *A quantity may be substituted for its equal.*
7. *Quantities equal to the same or equal quantities are equal to each other.*
8. *Only one straight line passes through two points.*
9. *Through a given point, one and only one straight line can be drawn parallel to a given line.*
10. *A straight line cannot intersect a circle at more than two points.*

Since we have presented numerous definitions throughout this text, we shall not give any examples here, although we shall refer to certain definitions in the proofs that follow.

By means of Example D, we shall illustrate a theorem, the premise, the conclusion to be proved, and the general method of a deductive proof.

Example D

Theorem. *If two lines intersect, the vertical angles formed are equal.*

The theorem is the above statement, and it is our task to show that it is true. The premise is that two lines intersect. This is given information and may be accepted without proof. The conclusion that must be proved is that the vertical angles are equal. (We stated that vertical angles are equal in Section 13-1, although we never actually proved it.) The only acceptable reasons for particular steps in the proof are the premises, postulates, definitions, and previously proven theorems.

We now draw a figure to represent that which we are to prove. In Fig. 13-84, lines AB and CD intersect at E. We shall prove that vertical angles AED and BEC are equal. This is formally done in the following manner.

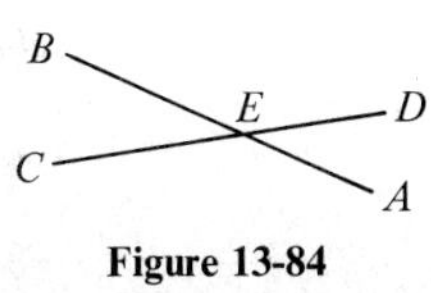

Figure 13-84

Given: Lines AB and CD intersect at E.
Prove: $\angle AED = \angle BEC$

Statement	Reason
1. Lines AB and CD intersect at E.	1. Given
2. $\angle AED + \angle DEB = 180°$	2. Definition of straight angle
3. $\angle AED = 180° - \angle DEB$	3. Postulate 2
4. $\angle BEC + \angle DEB = 180°$	4. Definition of straight angle
5. $\angle BEC = 180° - \angle DEB$	5. Postulate 2
6. $\therefore \angle AED = \angle BEC$	6. Postulate 7

($\therefore$ means "therefore.")

The solution to an algebraic equation follows a deductive line of reasoning. The reasoning follows the same pattern as that used in proving a theorem. The given information (premise) is the equation, and the conclusion is the value of the unknown that is found.

Example E Solve for x: $4x + 5 = 12$.

Statement	Reason
1. $4x + 5 = 12$	1. Given
2. $4x = 7$	2. Postulate 2
3. $\therefore x = \frac{7}{4}$	3. Postulate 4

Examples D and E illustrate the **direct method** of a deductive proof. Some theorems are more readily proved by means of the **indirect method.** *In this method, all admissible possibilities are stated and that which is false is assumed true. This assumption is then shown to lead to a contradiction of a known fact.* Example F illustrates a theorem for which this type of proof is useful, although we shall only discuss the procedure that would be used.

Example F

Theorem. *If two angles of a triangle are unequal, the sides opposite those angles are unequal in the same order* (see Fig. 13-85).

Given: $\triangle ABC$ with $\angle B > \angle A$.
Prove: $AC > BC$.

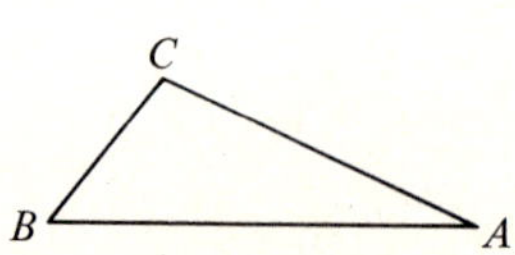

Figure 13-85

We note that there are three possibilities with respect to the relative size of sides AC and BC. They are $AC = BC$, $AC < BC$, and $AC > BC$. The proof proceeds to show that the first two are not possible, and therefore the third must be true. (We shall not show the actual proof of this theorem.)

If a proposed theorem is not true in general, a method of disproving it is by showing a **counterexample.** *That is, a case is demonstrated for which the proposed theorem does not hold. This is sufficient, for any valid theorem must hold in* **all** *cases.*

Example G A student notes in all the cases he tries that the square of a number is greater than the number itself. He therefore theorizes that $a^2 > a$ for any value of a. However, his friend then points out that $(\frac{1}{2})^2 = \frac{1}{4}$ and that $\frac{1}{4} < \frac{1}{2}$. Since his theorem is not true for $a = \frac{1}{2}$, it is not a true theorem. (Actually, a^2 is less than a for any positive a which is less than 1.)

Exercises 13-7

In Exercises 1 through 12, state the type of incorrect reasoning as identified in Example B. (Even though a conclusion is correct, the reasoning leading to it may be incorrect.)

1. If a triangle is both equilateral and equiangular, the sum of its angles is 180°.
2. If the midpoints of the sides of a quadrilateral are joined in order, a parallelogram is formed. Therefore, if the midpoints of any polygon are joined in order, a parallelogram is formed.
3. The diagonals of a rectangle are equal. Therefore, the diagonals of a parallelogram are equal.
4. In an isosceles triangle the altitude bisects the vertex angle. Therefore, the base angles are equal.
5. In $\triangle ABC$, $AB = BC$. Therefore, $\angle ADB = \angle CDB$ (see Fig. 13-86).

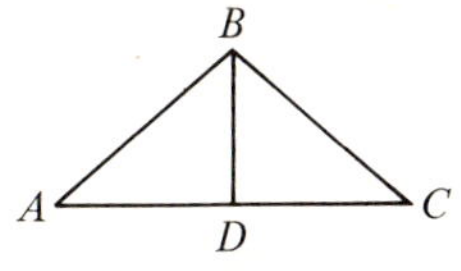

Figure 13-86

6. The integers 3, 5, and 7 are prime. Therefore, all odd integers are prime.
7. In quadrilateral $ABCD$, $AB \parallel CD$. Therefore, $BC \parallel DA$, and $ABCD$ is a parallelogram.
8. In a square all sides are equal and all angles are equal. Therefore, the perimeter is equal to the sum of the four sides.
9. A man with a gun in his hand is standing by a woman who has been shot. Therefore, he has killed her.

10. Since none of the presidents of the United States has been a woman, it is not possible that a woman will ever be president.
11. All rich people drive big cars. You drive a big car. Therefore, you are rich.
12. If all students who do not study fail, and John failed, it follows that he did not study.

In Exercises 13 through 16, draw an appropriate circle diagram to determine whether the following statements follow logical reasoning.

13. Since the diagonals of a rectangle bisect each other, the diagonals of a square bisect each other.
14. If the base angles of a triangle are equal, the triangle is equilateral.
15. During periods of recession, unemployment is high; unemployment is now high. Therefore, we are now in a period of recession.
16. If honest men do not steal and Jim is honest, he will not steal.

In Exercises 17 through 22, use deductive proofs to prove the given theorems. In the steps of the proof, use only the postulates listed in this chapter, any appropriate definitions, and those premises given in the theorem.

17. If two angles are both adjacent and complementary, their exterior sides are perpendicular to each other.
18. If alternate interior angles are equal, the sum of the angles of a triangle is 180°. (*Hint:* Draw a triangle with a line passing through one vertex parallel to the opposite base. Use Postulate 9.)
19. If the sum of the angles of a triangle is 180°, the sum of the angles of a quadrilateral is 360°. (Draw a quadrilateral and one of its diagonals.)
20. If $x + y = 20$ and $y = 8$, then $x = 12$.
21. If $5x + 9 = 39$, then $x = 6$.
22. If the sum of two variables is 18 and one of them is 5, then the other is 13.

In Exercises 23 through 26, solve the given equations for x. Use deductive reasoning, giving an appropriate reason for each step.

23. $x + 6 = 8$
24. $2x - 7 = 19$
25. $\frac{2}{3}x + 4 = 18$
26. $\frac{x}{4} + 3y = 8$

In Exercises 27 and 28, prove the given algebraic theorems by an indirect proof. If the conclusion is that $x = a$, assume that $x > a$ and then $x < a$, and show that these assumptions are false.

27. If $3x = 9$, then $x = 3$.
28. If $x + 4 = 9$, then $x = 5$.
29. Explain how one would use indirect reasoning to solve a murder mystery.
30. Explain how one could use indirect reasoning to diagnose the cause of an illness.

In Exercises 31 and 32, prove the given "theorems" to be incorrect by stating a counterexample.

31. If any number is doubled, the result is greater than the number.
32. If two sides and an angle of one triangle are respectively equal to the corresponding sides and angle of another triangle, the triangles are similar.

In Exercises 33 and 34, devise an appropriate theorem based upon the indicated circle diagram.

33. The diagram of Fig. 13-87

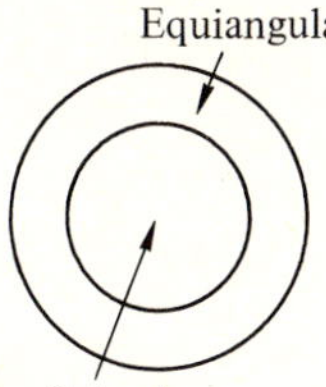

Figure 13-87

34. The diagram of Fig. 13-88

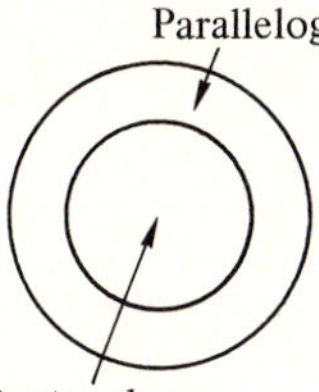

Figure 13-88

13-8 SUMMARY

Many of the basic concepts of geometry have been discussed in this chapter. We identified many different kinds of angles and introduced some of the basic features of polygons, including perimeter and area. In addition, we discussed the basic features of the circle, including its circumference, which is found from $c = \pi d$ or $c = 2\pi r$. The primary emphasis was placed on the important types of triangles and quadrilaterals.

In Section 13-4, we discussed the basic features of similar triangles and briefly mentioned congruent triangles. In Sections 13-5 and 13-6, we considered the surface areas and volumes of the more important solid figures.

Throughout the chapter, many important formulas were given. Below is a summary of these important formulas.

Circumference of circle (radius form):	$c = 2\pi r$	(13-1)
Circumference of circle (diameter form):	$c = \pi d$	(13-2)
Area of parallelogram or rectangle:	$A = bh$	(13-3)
Area of square:	$A = s^2$	(13-4)
Area of triangle:	$A = \frac{bh}{2}$	(13-5)
Area of trapezoid:	$A = \frac{h(b_1 + b_2)}{2}$	(13-6)
Area of circle:	$A = \pi r^2$	(13-7)
Volume of rectangular solid:	$V = lwh$	(13-8)
Volume of prism:	$V = Bh$	(13-9)
Surface area of rectangular solid:	$A = 2lw + 2lh + 2wh$	(13-10)

Surface area of cube:	$A = 6e^2$	(13-11)
Volume of right circular cylinder:	$V = \pi r^2 h$	(13-12)
Lateral area of cylinder:	$L = 2\pi rh$	(13-13)
Surface area of right circular cylinder:	$A = 2\pi r^2 + 2\pi rh$	(13-14)
Volume of pyramid:	$V = \frac{1}{3} Bh$	(13-15)
Lateral area of regular pyramid:	$L = \frac{1}{2} ps$	(13-16)
Volume of right circular cone:	$V = \frac{1}{3} \pi r^2 h$	(13-17)
Lateral area of cone:	$L = \pi rs$	(13-18)
Volume of sphere:	$V = \frac{4}{3} \pi r^3$	(13-19)
Surface area of sphere:	$A = 4\pi r^2$	(13-20)

We concluded the chapter by introducing the method of formal mathematical proof, along with a brief analysis of the process of reasoning. Applications of reasoning and proof were drawn from geometry, algebra, and other areas. It could be seen that a geometric proof and an algebraic proof follow the same basic method, called the deductive method.

Many new terms were introduced in the chapter. The list below includes those which should now be familiar to the reader.

Acute angle
Adjacent angles
Alternate exterior angles
Alternate interior angles
Altitude
Angle
Angle bisectors
Arc
Area
Axiom
Base angles
Base of figure
Center of circle
Central angle
Centroid
Chord
Circle
Circumference
Complementary angles
Conclusion
Cone
Congruent polygons
Conical surface
Corresponding angles of || lines
Corresponding angles of ~ figures
Corresponding sides
Counterexample
Cube (geometric figure)
Cylindrical surface
Decagon
Decimal parts of a degree
Deductive reasoning
Definition
Degrees

Diagonal
Diameter
Direct method of proof
Dodecagon
Edge of polyhedron
Equiangular polygon
Equilateral polygon
Face of polyhedron
Great circle (of sphere)
Height of figure
Hexagon
Hypotenuse
Indirect method of proof
Inductive reasoning
Interior angles
Lateral area
Lateral edge
Lateral face
Legs of right triangle
Line
Measure of angle
Median
Midpoint
Minutes (angle measurement)
Obtuse angle
Parallel
Parallelogram
Pentagon
Perimeter
Perpendicular
Plane
Point
Polyhedron
Polygon
Postulate
Premise
Prism
Properties of similar triangles
Proportional sides
Protractor
Pyramid
Quadrilateral
Radius
Rectangle
Rectangular solid
Regular pyramid
Rhombus
Right angle
Right circular cone
Right circular cylinder
Right prism
Scale drawing
Secant
Seconds (angle measurement)
Similar triangles
Slant height
Small circle (of sphere)
Sphere
Square (geometric figure)
Straight angle
Supplementary angles
Surface area
Tangent
Theorem
Total area
Transversal
Trapezoid
Triangle
    equilateral
    isosceles
    right
    scalene
Vertex of angle
Vertical angles
Vertices of figure
Volume

You should review the above terms until you understand all of them. The following exercises will help in this review, as well as in reviewing the other concepts of the chapter.

Exercises 13-8

In Exercises 1 through 4, use Fig. 13-89 and identify the indicated angles.

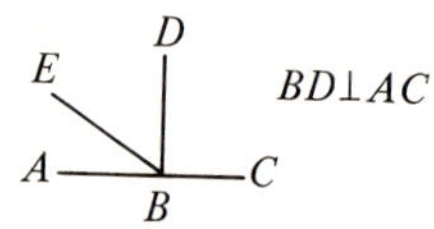

Figure 13-89

1. An obtuse angle
2. Two acute angles
3. The complement of $\angle ABE$
4. The supplement of $\angle ABE$

In Exercises 5 through 8, use Fig. 13-90 and identify or evaluate the following.

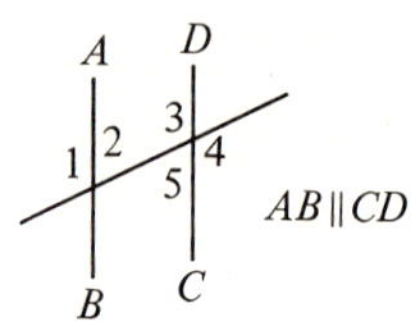

Figure 13-90

5. A pair of alternate interior angles
6. A pair of corresponding angles
7. $\angle 2 + \angle 3 = ?$
8. $\angle 1 + \angle 5 = ?$

In Exercises 9 through 20, use the indicated figures. Determine the indicated angles.

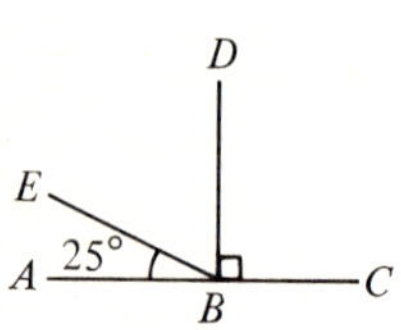

Figure 13-91

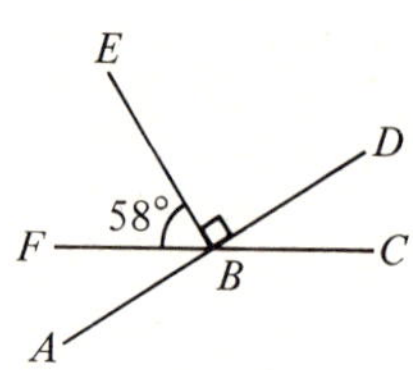

Figure 13-92

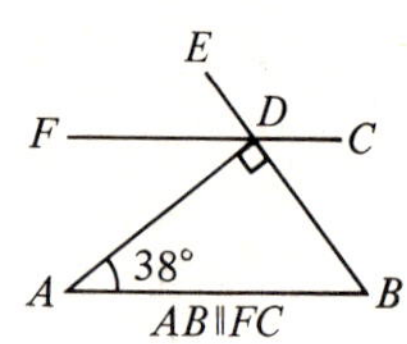

Figure 13-93

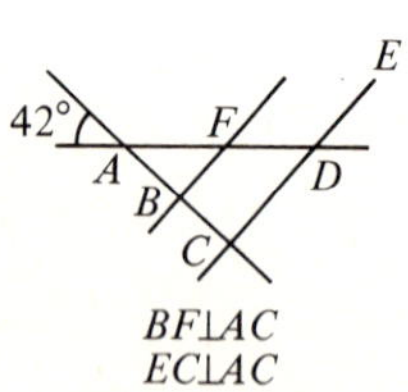

Figure 13-94

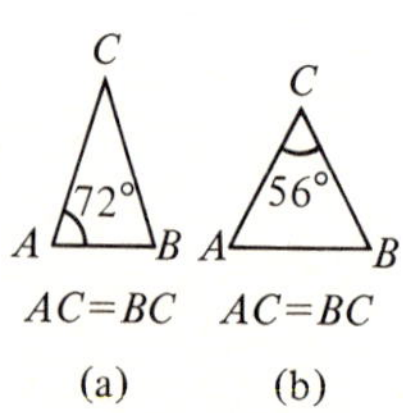

Figure 13-95

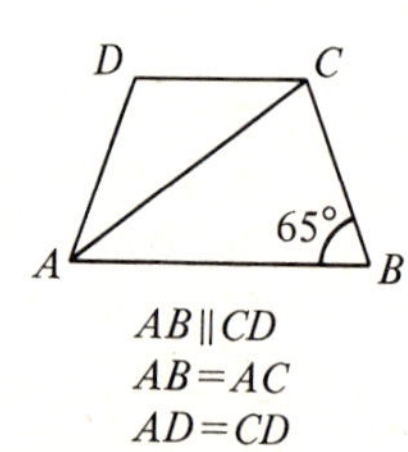

Figure 13-96

9. $\angle EBD$ in Fig. 13-91
10. $\angle EBC$ in Fig. 13-91
11. $\angle CBD$ in Fig. 13-92
12. $\angle ABC$ in Fig. 13-92
13. $\angle BDC$ in Fig. 13-93
14. $\angle CDE$ in Fig. 13-93
15. $\angle BFD$ in Fig. 13-94
16. $\angle FDE$ in Fig. 13-94
17. $\angle C$ in Fig. 13-95a
18. $\angle B$ in Fig. 13-95b
19. $\angle DCA$ in Fig. 13-96
20. $\angle ADC$ in Fig. 13-96

In Exercises 21 through 24, use Fig. 13-97. Determine the indicated arcs (denoted by ⌒*) and angles. Angles CAB and CBA are* **inscribed angles**. *An inscribed angle equals one-half of its intercepted arc.*

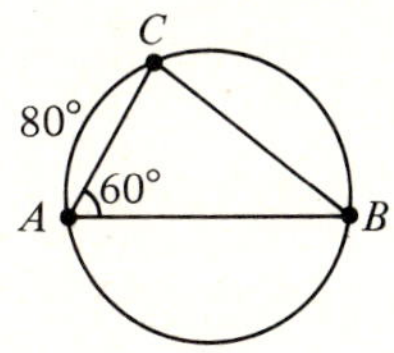

Figure 13-97

21. $\overset{\frown}{BC}$ **22.** $\overset{\frown}{AB}$ **23.** $\angle ABC$ **24.** $\angle ACB$

In Exercises 25 through 28, use Fig. 13-98. Line CT is tangent to the circle with center at O. Determine the indicated angles. (See Exercises 21 through 24.) A tangent line is perpendicular to a diameter at the point of contact.

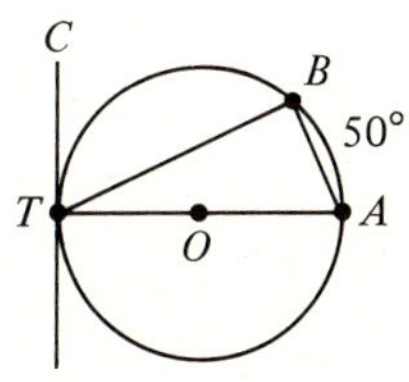

Figure 13-98

25. $\angle BTA$ **26.** $\angle TAB$ **27.** $\angle BTC$ **28.** $\angle ABT$

In Exercises 29 through 44, find the perimeters of the indicated geometric figures.

29. Triangle: $a = 17.5$ in., $b = 13.8$ in., $c = 8.9$ in.

30. Quadrilateral: $a = 22.4$ cm, $b = 68.5$ cm, $c = 37.3$ cm, $d = 29.9$ cm

31. Equilateral triangle: $s = 8.5$ mm

32. Equilateral triangle: $s = 14.9$ ft

33. Isosceles triangle: equal sides of 0.38 yd, third side of 0.53 yd

34. Isosceles triangle: equal sides of 8.13 mm, third side of 7.09 mm

35. Square: $s = 6.8$ m

36. Rhombus: side of 15.2 in,

37. Parallelogram: $a = 692$ ft, $b = 207$ ft

38. Parallelogram: $a = 7.8$ m, $b = 6.2$ m

39. Rectangle: $l = 96$ cm, $w = 43$ cm

40. Rectangle: $l = 108$ in., $w = 92$ in.

41. Circle: $r = 4.25$ ft

42. Circle: $d = 38.0$ cm

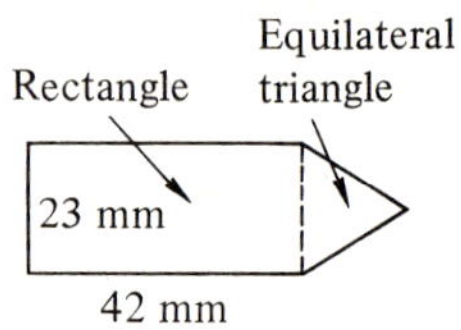

Figure 13-99

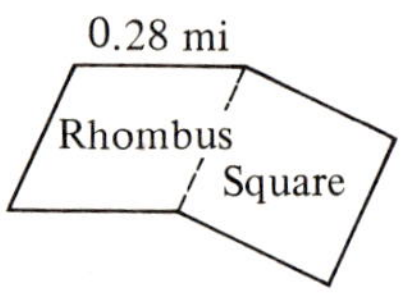

Figure 13-100

43. The figure in Fig. 13-99 (equilateral triangle attached to rectangle)

44. The figure in Fig. 13-100 (square and rhombus attached)

In Exercises 45 through 60, find the area of the indicated figures.

45. Rectangle: $l = 17.5$ cm, $w = 12.8$ cm

46. Square: $s = 94.0$ in.

47. Parallelogram: $b = 2.68$ yd, $h = 1.15$ yd

48. Parallelogram: $b = 172$ mm, $h = 85$ mm

49. Triangle: $b = 4.68$ cm, $h = 2.05$ cm

50. Triangle: $b = 68.0$ ft, $h = 43.0$ ft

51. Right triangle: legs are 3.25 in. and 1.88 in.

52. Right triangle: legs are 40.8 m and 19.5 m

53. Trapezoid: $h = 0.016$ km, $b_1 = 0.118$ km, $b_2 = 0.067$ km

54. Trapezoid: $h = 36.8$ in., $b_1 = 12.4$ in., $b_2 = 11.1$ in.

55. Circle: $r = 3.36$ in.

56. Circle: $r = 81.2$ cm

57. Circle: $d = 420$ mm

58. Circle: $d = 0.608$ ft

59. The figure in Fig. 13-99 if h of the triangle is 20.0 mm

60. The figure in Fig. 13-100 if h of the rhombus is 0.220 mi

In Exercises 61 and 62, solve for the unknown sides of the given triangle by use of an appropriate proportion.

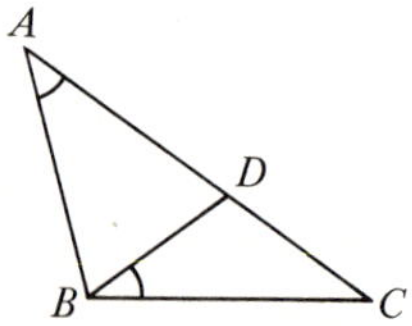

Figure 13-101

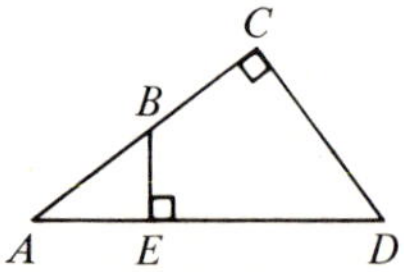

Figure 13-102

61. In Fig. 13-101, $AC = 12$, $BC = 8$, $BD = 5$. Find AB.

62. In Fig. 13-102, $AB = 4$, $BC = 4$, $CD = 6$. Find BE.

In Exercises 63 and 64, use Fig. 13-103 and find the indicated sides. Given: BF is a diameter, CE is tangent at D, AB = 2.0 in., *BC* = 4.0 in., *AE* = 18.0 in., *BF* ∥ *CE.*

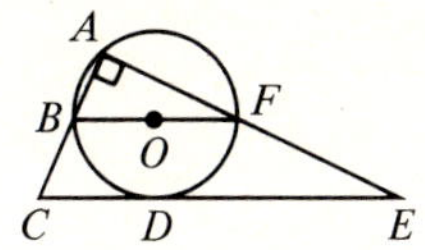

Figure 13-103

63. $BF = ?$ **64.** $CE = ?$

In Exercises 65 through 76, find the volumes of the indicated figures for the given values.

65. Prism: base area 40.0 ft^2, height 15.0 ft
66. Prism: base area 800 cm^2, height 50.0 cm
67. Prism: base is right triangle with legs 26.0 cm and 34.0 cm; height 14.0 cm
68. Prism: base is right triangle with legs 6.20 ft and 3.80 ft; height 8.50 ft
69. Cylinder: base radius 36.0 in., altitude 24.0 in.
70. Cylinder: base diameter 120 mm, height 58.0 mm
71. Pyramid: base area 3850 cm^2, height 125 cm
72. Pyramid: base is square with side 6.44 in., altitude 2.25 in.
73. Cone: base radius 18.2 ft, height 11.5 ft
74. Cone: base diameter 76.2 mm, height 22.1 mm
75. Sphere: radius 18.0 cm
76. Sphere: diameter 52.0 ft

In Exercises 77 through 88, find the lateral area or total area of the indicated figure for the given values.

77. Lateral area of prism: base perimeter 60.0 ft, height 15.0 ft
78. Total area of prism of Exercise 67
79. Total area of rectangular solid: length 2.00 m, width 1.50 m, height 1.20 m
80. Total area of cube of edge 3.5 yd
81. Lateral area of cylinder of Exercise 69
82. Total area of cylinder of Exercise 70
83. Lateral area of pyramid: base perimeter 240 cm, slant height 140 cm
84. Lateral area of pyramid of Exercise 72
85. Lateral area of cone of Exercise 73
86. Lateral area of cone of Exercise 74
87. Total area of sphere of Exercise 75
88. Total area of sphere of Exercise 76

In Exercises 89 through 126, solve the given problems.

89. What is the diagonal distance between corners of a rectangular room 12.5 ft wide and 17.0 ft long?

90. An observer is 550 m from the launch pad of a rocket. After the rocket has ascended vertically to a point 750 m from the observer, how far has it ascended?

91. In a certain pulley system, the center of pulley *A* is 180 cm above and 145 cm to the right of the center of pulley *B*. How far is it between the centers of the pulleys?

92. The base of a 20.0-ft ladder is 6.25 ft from the base of a vertical wall. How far up the wall does the ladder touch?

93. A loop of wire is in the shape shown in Fig. 13-104. (The two geometric figures are a semicircle and an isosceles right triangle.) Find the length of the wire loop.

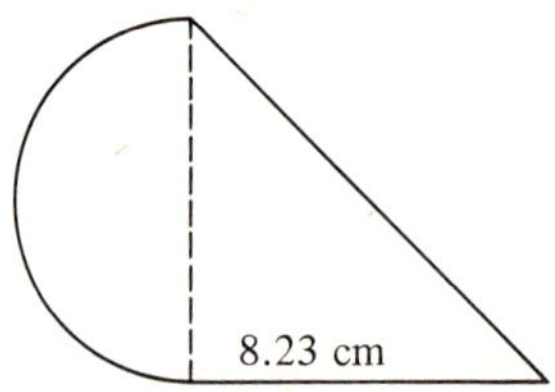

Figure 13-104

94. The hypotenuse of a right triangle is 24.0 m and one side is twice the other. Find the perimeter of the triangle.

95. One of the acute angles of a right triangle is three times the other. How many degrees are there in each angle?

96. In a given triangle, the second angle is three times the first, and the third angle is twice the first. How many degrees are there in each angle?

97. A tree and a telephone pole cast shadows as shown in Fig. 13-105. Find the height of the telephone pole.

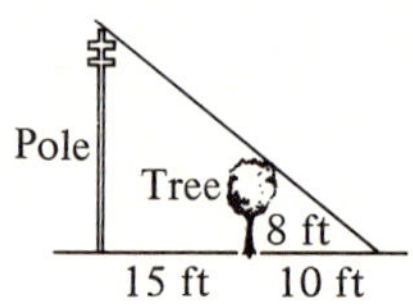

Figure 13-105

98. The diameter of the sun is 860,000 mi, the diameter of the earth is 7920 mi, and the distance from the earth to the sun (center to center) is 93,000,000 mi. What is the distance from the center of the earth to the end of the earth's shadow due to the rays from the sun?

99. A good approximation of the height of a tree can be made by following the procedure suggested in Fig. 13-106. By measuring *DE*, *AE*, and *BC* (use a ruler), the length of the tree $DE + EF$ can be found. Find the height of a tree if $AB = 50$ cm, $BC = 30$ cm, $AE = 2400$ cm, and $DE = 150$ cm.

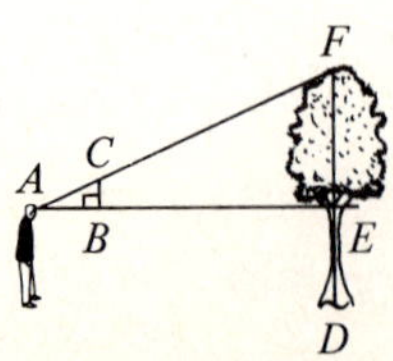

Figure 13-106

100. What is the length of the steel support in the structure shown in Fig. 13-107?

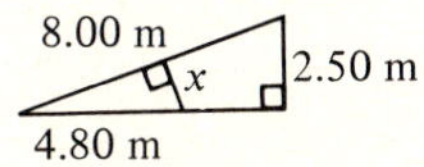

Figure 13-107

101. A drafting student is making a drawing with a scale of $2\frac{3}{4}$ in. = 5 ft. What distance on her drawing should be used to represent 18 ft 3 in.?

102. On the blueprint of a certain building, a certain hallway is 38.5 in. long. The scale is $1\frac{1}{8}$ in. = 6 ft. How long is the hallway?

103. A square frame is constructed in a circular rim such that the vertices of the square are in contact with the rim. What is the arc intercepted by adjacent sides of the square?

104. The pulley belt shown in Fig. 13-108 is in contact with the pulley along an arc of 220°. What is the indicated angle between sections of the belt? (See Exercises 25–28.)

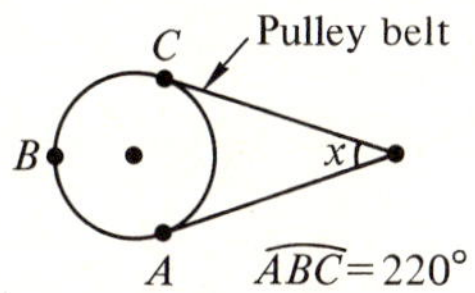

Figure 13-108

105. A vertical beam has a rectangular cross section 15.4 cm by 10.5 cm. The beam is 240 cm high. What is the lateral area of the beam?

106. A fish tank is 3.50 ft long, 2.75 ft wide, and 2.25 ft high. What is the surface area of the tank? (There is no top.)

107. The circumference of a cylindrical water tank is 152 ft, and its height is 48.5 ft. How many gallons of water can it hold? (1 ft^3 = 7.48 gal.)

108. How many meters of steel rods 0.500 cm in radius can be made from a steel ingot which is in the shape of a rectangular solid 6.00 cm by 6.00 cm by 24.0 cm? Assume no losses in manufacturing.

109. A hollow concrete conduit is 12.0 m long. The inside diameter is 54.0 cm and the outside diameter is 75.0 cm. Find the volume of concrete in the conduit.

110. How many cubic inches of metal are there in a length of pipe 15.5 ft long when the inside diameter is 8.00 in. and the metal is 0.500 in. thick?

111. What is the area of a label (covering the lateral surface) on a cylindrical can 3.00 in. in diameter and 4.25 in. high? Assume the ends just meet and do not overlap.

112. A furnace is built in the shape of a cylinder 1.40 m high and 0.950 m in diameter. How many square meters of insulation are required to cover the sides and top of the furnace?

113. What is the volume of a conical rod with a base diameter of 0.750 cm and a length of 48.0 cm?

114. What is the volume of darkness of the earth's shadow? Assume that it is a right circular cone with the base through the center of the earth. (see Exercise 98). Compare this with the volume of the earth.

115. An umbrella opens into the shape of a cone 3.00 ft in diameter and 0.500 ft deep. What is the surface area of the umbrella?

116. A sheet metal cover is in the shape of a cone with radius 17.0 cm and height 6.00 cm. What is the surface area of the cover?

117. A marble monument is in the shape of a pyramid with a square base 160 cm on a side. The height of the monument is 140 cm. What is the mass, in kilograms, of the monument? The density of marble is 2.70×10^4 kg/m^3.

118. A concrete base for a piece of machinery is in the shape of a *frustum* of a pyramid. (A frustum is formed by cutting off the top by a plane parallel to the base.) The top of the frustum is a rectangle 18.0 ft by 12.5 ft. The length of the base is 24.0 ft, and the depth of the frustum is 3.25 ft. How many cubic feet of concrete were used in making the base?

119. A wedge is in the shape of a prism with a right triangular base. The base has a width of 4.00 in. and a length of 14.5 in. What is the total surface area of the wedge, given that its altitude is 5.38 in.?

120. A hopper containing grain is in the shape of an inverted pyramid (with a square base) surmounted by a cube. The base of the pyramid is 3.25 m on a side, and the height of the pyramid is 2.50 m. What is the capacity of the hopper?

121. What is the radius of a 16.0-lb shot? The density of iron used in the shot (spherical) is 450 lb/ft^3.

122. A hollow metal sphere has an outer radius of 1.50 ft and an inner radius of 1.25 ft. What is the volume of the metal in the sphere?

123. How many square centimeters of material are used to make a tennis ball whose circumference is 21.0 cm?

124. A hot water tank is the shape of a right circular cylinder surmounted by a hemisphere. The total height of the tank is 6.75 ft, and the diameter of the base is 2.50 ft. How many gallons does the tank contain? (See Exercise 107.)

125. Derive the formula for the total surface area of a volume formed by placing a hemisphere of radius r on a cylinder of radius r and altitude h.

126. Derive the formula for the volume formed by placing a pyramid of square base (edge e) and altitude h on a cube of edge e.

In Exercises 127 through 130, determine whether the conclusion, derived through deductive reasoning and based upon the given premises, is correct.

127. If no triangle is a rectangle and no isosceles triangle is a rectangle, then every isosceles triangle is a triangle.

128. If no triangle is a rectangle and all squares are rectangles, then no square is a triangle.

129. If the first of all sets is the best of the set, then George Washington was the best president of the United States.

130. If the weather is good, Jim will go on a picnic. Therefore, if the weather is bad, Jim will not go on a picnic.

In Exercises 131 through 136, use deductive proofs to prove the given theorems. In the steps of the proofs, use only the postulates of Section 13-7, any appropriate definitions, and the given premises.

131. If alternate interior angles of parallel lines are equal, a diagonal of a parallelogram divides the parallelogram into two congruent triangles.

132. If $2x - 9 = 13$, then $x = 11$.

133. At night, a person hears a sound from the direction of the kitchen. Therefore, the noise did not come from outside the house. Is this an appropriate conclusion?

134. If two sides and an angle of one triangle are equal respectively to two sides and an angle of another triangle, the triangles are congruent. Is this theorem correct?

135. By use of an indirect proof, show that $x = 7$ satisfies the equation $2x + 1 = 15$.

136. Use a circle diagram to determine whether the following reasoning is correct. Sulfur dioxide is a major air pollutant. The air in a certain city is badly polluted. Therefore, there is sulfur dioxide in the air of the city.

14 TRIGONOMETRY

14-1 THE TRIGONOMETRIC RATIOS

The solution of a great many applied problems involves the use of triangles, especially right triangles. Problems that can be solved by the use of triangles include the determination of distances which cannot be measured directly, such as widths of rivers and distances between various points in the universe. Also, problems involving building construction, forces on objects, and airplane headings lend themselves readily to solution by use of triangles.

The branch of mathematics which develops the methods for measuring triangles and solving related problems is **trigonometry,** the literal meaning of which is "triangle measurement." Through the methods of trigonometry, applications in astronomy, navigation, surveying, architecture, and various fields of science may be solved. Trigonometry is one of the most practical and most applicable branches of mathematics.

In discussing similar triangles in Chapter 13, we noted that corresponding sides of similar triangles are proportional. From this it follows that: *The ratio of one side to another side in a triangle is the same as the ratio of corresponding sides in any triangle which is similar.* For example, in a right triangle the ratio of the side opposite an angle to the hypotenuse is the same as the ratio of the side opposite the corresponding angle to the hypotenuse in a similar right triangle. This is further illustrated in the following example.

Example A The two right triangles in Fig. 14-1 are similar, although the distances represented are of very different magnitudes. Since they are similar, we know that

$$\frac{13}{2600} = \frac{5}{1000}$$

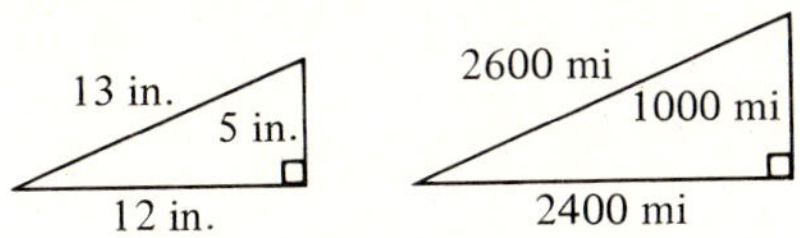

Figure 14-1

However, we can also see that the ratio of the shortest side to the hypotenuse is the same in each. That is,

$$\frac{5}{13} = \frac{1000}{2600}$$

Ratios of other pairs of corresponding sides can also be set up and shown to be equal.

We shall now define the **trigonometric ratios.** These definitions are based on two properties: *Corresponding angles of similar right triangles are equal, and the ratio of two sides of a right triangle is the same as the ratio of the two corresponding sides in any similar right triangle.*

In Fig. 14-2, we define the following trigonometric ratios:

$$\begin{aligned}\text{Sine of angle } A &= \frac{\text{side opposite angle } A}{\text{hypotenuse}}\\ \text{Cosine of angle } A &= \frac{\text{side adjacent to angle } A}{\text{hypotenuse}}\\ \text{Tangent of angle } A &= \frac{\text{side opposite angle } A}{\text{side adjacent to angle } A}\end{aligned} \tag{14-1}$$

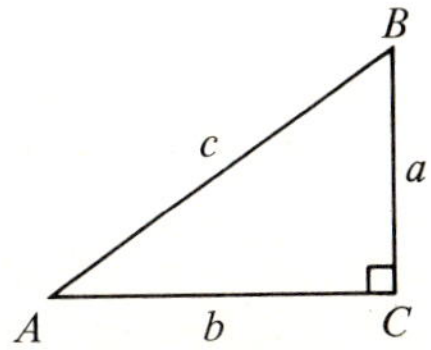

Figure 14-2

Example B In the triangle in Fig. 14-3, we have the following values for the trigonometric ratios:

$$\text{Sine of angle } A = \frac{8}{17}$$

$$\text{Cosine of angle } A = \frac{15}{17}$$

$$\text{Tangent of angle } A = \frac{8}{15}$$

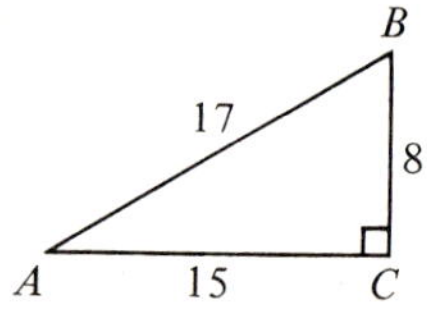

Figure 14-3

For convenience of notation, the names of the ratios are usually abbreviated and the sides are designated by the letters shown in Fig. 14-2. Therefore, we write:

$$\sin A = \frac{a}{c} \qquad \cos A = \frac{b}{c} \qquad \tan A = \frac{a}{b} \tag{14-2}$$

The definitions given in Eq. 14-1 can be used to find the trigonometric ratios of any acute angle in a right triangle. That is:

$$\sin B = \frac{\text{side opposite angle } B}{\text{hypotenuse}}$$

$$\cos B = \frac{\text{side adjacent angle } B}{\text{hypotenuse}}$$

$$\tan B = \frac{\text{side opposite angle } B}{\text{side adjacent angle } B}$$

Using the abbreviations for the trigonometric ratios and the sides designated as in Fig. 14-2, we have

$$\sin B = \frac{b}{c} \qquad \cos B = \frac{a}{c} \qquad \tan B = \frac{b}{a} \tag{14-3}$$

Example C In the triangle shown in Fig. 14-3, we have

$$\sin A = \frac{8}{17} \qquad \cos A = \frac{15}{17} \qquad \tan A = \frac{8}{15}$$

$$\sin B = \frac{15}{17} \qquad \cos B = \frac{8}{17} \qquad \tan B = \frac{15}{8}$$

In a given right triangle, if two of the sides are known, the third may be found by use of the Pythagorean theorem. Then, in turn, the trigonometric ratios may be found for the angles. The following example illustrates the method.

Example D In Fig. 14-4, one side and the hypotenuse of the triangle are given. To find the other side, a, we use the Pythagorean theorem, which gives

$$a = \sqrt{11^2 - 7^2} = \sqrt{121 - 49} = \sqrt{72} = 8.485$$

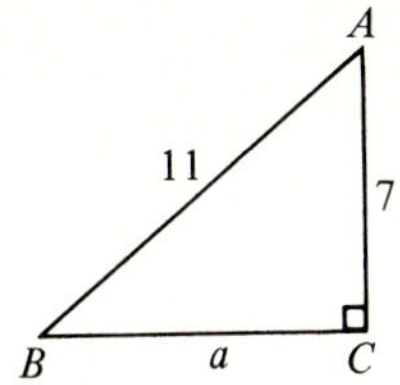

Figure 14-4

We are now in a position to find any of the ratios associated with the triangle. For example,

$$\sin A = \frac{8.485}{11} = 0.771 \quad \text{and} \quad \tan B = \frac{7}{8.485} = 0.825$$

In any right triangle there are six possible ratios which may be set up. So far, we have named three of them. The others are called the **cotangent, secant,** and **cosecant.** Referring to Fig. 14-2 and using the abbreviations for these ratios, we make the following definitions:

$$\boxed{\cot A = \frac{b}{a} \qquad \sec A = \frac{c}{b} \qquad \csc A = \frac{c}{a}} \tag{14-4}$$

Using the general definitions implied in Eqs. (14-4), we can find the trigonometric ratios for any acute angle in a right triangle.

Example E Referring to Fig. 14-3, we have

$$\cot A = \frac{15}{8} \qquad \sec A = \frac{17}{15} \qquad \csc A = \frac{17}{8}$$

$$\cot B = \frac{8}{15} \qquad \sec B = \frac{17}{8} \qquad \csc B = \frac{17}{15}$$

If one of the trigonometric ratios for a given angle is known, it is possible to determine the values of the other five. The following example illustrates how this may be done.

Example F If we know that the sine of an angle is $\frac{2}{3}$, we know that the ratio of the side opposite that angle to the hypotenuse is 2 to 3. Therefore, for purposes of finding the other ratios, we may assume that the lengths of these two sides are 2 units and 3 units. For convenience, we may draw a triangle with these two sides, as shown in Fig. 14-5. The third side may be found by use of the Pythagorean theorem. Therefore, calling this side x, we have

$$x = \sqrt{3^2 - 2^2} = \sqrt{5}$$

We now determine the other trigonometric ratios of the angle A, whose sine $\frac{2}{3}$ is given. They are

$$\cos A = \frac{\sqrt{5}}{3}$$

$$\tan A = \frac{2}{\sqrt{5}} = \frac{2\sqrt{5}}{5} \qquad \cot A = \frac{\sqrt{5}}{2}$$

$$\sec A = \frac{3}{\sqrt{5}} = \frac{3\sqrt{5}}{5} \qquad \csc A = \frac{3}{2}$$

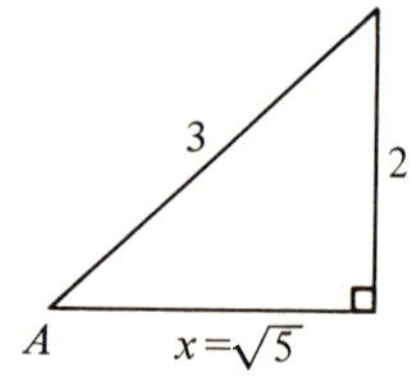

Figure 14-5

Decimal equivalents can also be easily determined.

Example G Given that $\tan A = 0.748$, we know that the ratio of the side opposite angle A to the side adjacent to it is 0.748. This means that we may assume a triangle of sides 0.748 and 1, as shown in Fig. 14-6. Using the Pythagorean theorem to find the hypotenuse c, we have

$$c = \sqrt{1^2 + 0.748^2} = \sqrt{1.560} = 1.25$$

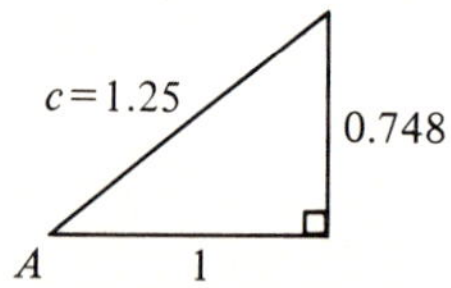

Figure 14-6

Therefore, the other trigonometric ratios for this angle are

$$\sin A = \frac{0.748}{1.25} = 0.598 \qquad \cos A = \frac{1}{1.25} = 0.800$$

$$\cot A = \frac{1}{0.748} = 1.34 \qquad \sec A = \frac{1.25}{1} = 1.25$$

$$\csc A = \frac{1.25}{0.748} = 1.67$$

Exercises 14-1

In Exercises 1 through 4, find the indicated trigonometric ratios in fractional form from Fig. 14-7.

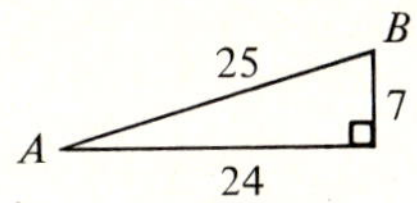

Figure 14-7

1. $\sin A$, $\tan A$, $\cos B$

2. $\cos A$, $\sin B$, $\cot A$

3. $\cot B$, $\sec A$, $\tan B$

4. $\sec B$, $\csc A$, $\csc B$

In Exercises 5 through 8, find the indicated trigonometric ratios in fractional form from Fig. 14-8.

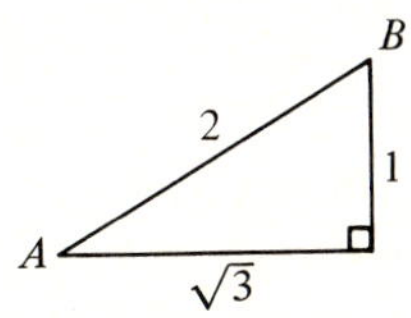

Figure 14-8

5. $\sin A$, $\sec B$, $\cot A$

6. $\csc A$, $\sin B$, $\cot B$

7. $\tan A$, $\cos B$, $\sec A$

8. $\tan B$, $\csc B$, $\cos A$

In Exercises 9 through 12, find the indicated trigonometric ratios in decimal form from Fig. 14-9.

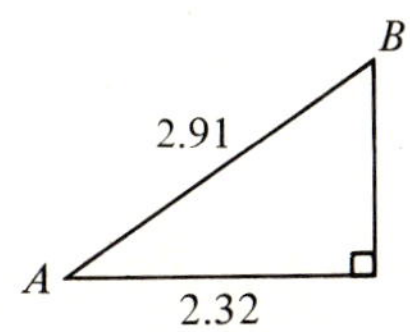

Figure 14-9

9. $\cos A$, $\tan A$, $\csc B$

10. $\sin B$, $\sec A$, $\cot B$

11. $\sin A$, $\cos B$, $\cot A$

12. $\csc A$, $\tan B$, $\sec B$

In Exercises 13 through 20, determine the indicated trigonometric ratios. The listed sides are those shown in Fig. 14-10.

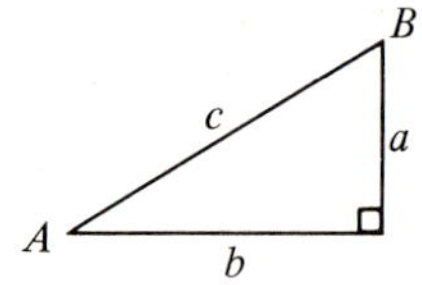

Figure 14-10

13. $a = 3$, $b = 4$. Find sin A and tan B.

14. $a = 5$, $c = 13$. Find cos A and csc B.

15. $b = 9$, $c = 41$. Find cot A and cos B.

16. $a = 8$, $c = 19$. Find sin A and sec B.

17. $a = 1$, $b = 1$. Find tan A and sin B.

18. $a = 2$, $c = 4$. Find sec A and tan B.

19. $b = 14$, $c = 23$. Find csc A and cos B.

20. $a = 132$, $b = 75$. Find cos A and cot B.

In Exercises 21 through 28, using the given trigonometric ratios, find the indicated trigonometric ratios.

21. $\tan A = 1$. Find sin A.

22. $\sin A = \frac{1}{2}$. Find cos A.

23. $\cos A = 0.70$. Find csc A.

24. $\sec A = 1.6$. Find sin A.

25. $\cot A = 0.563$. Find cos A.

26. $\csc A = 2.64$. Find tan A.

27. $\sin A = 0.720$. Find tan A.

28. $\tan A = 0.350$. Find cos A.

In Exercises 29 through 34, solve the given problems.

29. From the definitions of the trigonometric ratios, it can be seen that csc A is the reciprocal of sin A. What ratio is the reciprocal of cos A? of cot A?

30. From the definitions of the trigonometric ratios, it can be seen that $\sin A = \cos B$. What ratio associated with angle B equals tan A? csc A?

31. State the definitions of the cotangent, secant, and cosecant in terms of adjacent side, opposite side, and hypotenuse.

32. Draw three right triangles: (a) with sides 3 in., 4 in., and 5 in.; (b) with sides 6 cm, 8 cm, and 10 cm; (c) with sides 4.5 in., 6.0 in., and 7.5 in. For each triangle determine the sine and tangent of the smallest angle. What is the relationship of the three triangles? What is true of the trigonometric ratios found?

33. In Fig. 14-10, if $a = 3$ and $b = 4$, what is the value of $(\sin A)^2 + (\cos A)^2$?

34. In Fig. 14-10, given that $a = 3$ and $b = 4$, calculate the values of sin A, cos A, and tan A. Then show that $\sin A/\cos A = \tan A$.

14-2 VALUES OF THE TRIGONOMETRIC RATIOS

In the first section of this chapter, we defined the trigonometric ratios of an angle, but we did not mention the size of the angle. We did note, however, that the ratio of two sides of a right triangle is the same as the ratio of the two corresponding sides in any similar right triangle. Therefore, for an angle of a particular size, a given trigonometric ratio has a particular value.

Thus, if an angle has a specified number of degrees, there is a specific set of values of trigonometric ratios. The actual method used to determine these values for the purpose of setting up tables requires the use of more advanced mathematics than we can discuss here. However, by using certain basic geometric properties, we can establish the values of the trigonometric ratios for certain angles.

A basic geometric fact is that *in a 30°-60°-90° triangle, the side opposite the 30° angle is one-half the hypotenuse* (see Fig. 14-11a). We can easily verify this statement by referring to the equilateral triangle shown in Fig. 14-11b, which has been divided into two smaller congruent triangles by the altitude. Each of the smaller triangles has angles of 30° (at top), 60°, and 90°.

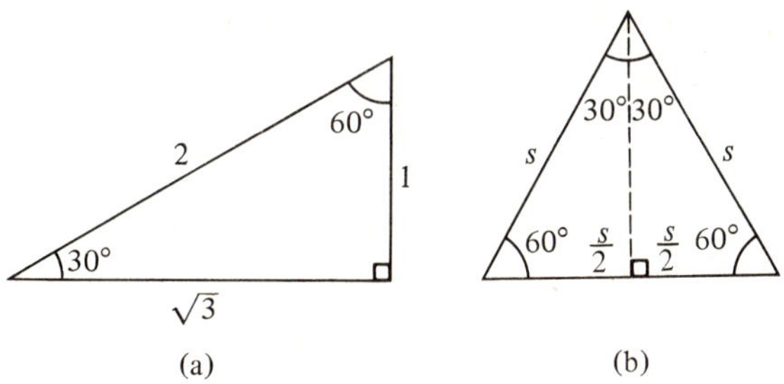

Figure 14-11

In Fig. 14-11a, the hypotenuse has been given the value 2, the side opposite the 30° angle has been given the value 1, and from the Pythagorean theorem we determine that the third side has the value $\sqrt{3}$. Therefore, for the 30° angle and the 60° angle, the trigonometric ratios may be established as shown in this table:

Angle	sin	cos	tan	cot	sec	csc
30°	$\frac{1}{2}$	$\frac{\sqrt{3}}{2}$	$\frac{\sqrt{3}}{3}$	$\sqrt{3}$	$\frac{2\sqrt{3}}{3}$	2
60°	$\frac{\sqrt{3}}{2}$	$\frac{1}{2}$	$\sqrt{3}$	$\frac{\sqrt{3}}{3}$	2	$\frac{2\sqrt{3}}{3}$

The following example illustrates the use of another geometric property to establish the values of the trigonometric ratios of 45°.

Example A To find the trigonometric ratios for 45°, we construct an isosceles right triangle, as shown in Fig. 14-12. Since the triangle is isosceles, both acute angles are 45° and the legs are equal. Therefore, each is given the value 1. From the Pythagorean theorem, we find that the hypotenuse is $\sqrt{2}$.

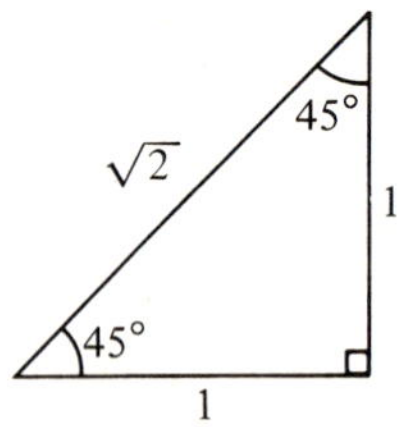

Figure 14-12

Therefore, we have

$$\sin 45^\circ = \frac{\sqrt{2}}{2} \qquad \cos 45^\circ = \frac{\sqrt{2}}{2} \qquad \tan 45^\circ = 1$$

$$\cot 45^\circ = 1 \qquad \sec 45^\circ = \sqrt{2} \qquad \csc 45^\circ = \sqrt{2}$$

We refer the reader to Example C, Section 8-4.

Combining the values from the preceding table for 30° and 60° and the values from Example A for 45°, we can set up the following short table of values of the trigonometric ratios (in decimal form):

Angle	sin	cos	tan	cot	sec	csc
30°	0.500	0.866	0.577	1.732	1.155	2.000
45°	0.707	0.707	1.000	1.000	1.414	1.414
60°	0.866	0.500	1.732	0.577	2.000	1.155

A more complete table, showing the values of the trigonometric ratios (not including the secant and cosecant) for each 0.1° from 0° to 90°, is given in Table 3 of Appendix C. It will be noted that from 0° to 45° the values of the angle are found in the left column, and from 45° to 90° they are found in the right column. This arrangement is chosen because the values of each **cofunction** (cosine and cotangent) from 0° to 45° or from 45° to 90° are the same as those of the corresponding functions (sine and tangent) from 45° to 90° or from 0° to 45°. The reason for this is brought out in the following example.

Example B In the right triangle shown in Fig. 14-13, the two acute angles add up to 90°. Since $\sin A = a/c$ and $\cos B = a/c$, we see that $\sin A = \cos B$.

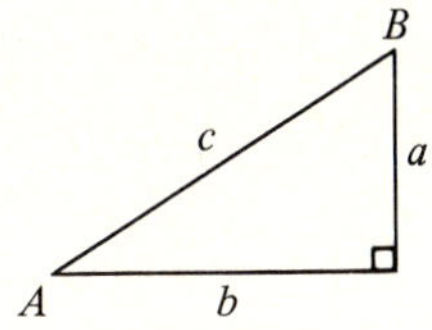

Figure 14-13

Also, since $A + B = 90°$,

$$B = 90° - A \quad \text{or} \quad \sin A = \cos(90° - A)$$

If $A = 32°$, for example, then

$$\sin 32° = \cos 58°$$

Therefore, in the table, all the values are included for the ratios from 0° to 45°, and it is necessary only to relabel the columns (as is done at the bottom) to obtain the values for the ratios from 45° to 90°.

The following examples illustrate how the table of trigonometric ratios is used to find values of the ratios and to find the angle when one of the ratios is given.

Example C To find sin 17.0° from Table 3, we look in the column labeled "sin" (at the top) to the right of 17.0, which is in the degrees column. Here we find the number 0.2924. Thus, $\sin 17.0° = 0.2924$.

To find cot 35.2°, we look in the column labeled "cot" (at the top) to the right of 35.2 of the degrees column. Here we find 1.418, which means that $\cot 35.2° = 1.418$.

Example D To find cos 82.0° from Table 3, we must look *above* "cos," which is at the bottom of the column. Also, we must look to the *left* of 82.0, which appears in the right-hand degrees column. Therefore, in the "cos" column (labeled at the bottom) and to the left of 82.0 in the degrees column, we find 0.1392. Thus, $\cos 82.0° = 0.1392$.

To find tan 56.8°, we must look *above* "tan" and to the *left* of 56.8 in the degrees column. Here we find 1.528. Thus, $\tan 56.8° = 1.528$.

Example E Given that $\sin \alpha = 0.6293$, find α. (The Greek letter alpha is often used to designate angles, as are many of the Greek letters.) First we look for 0.6293 in one of the "sin" columns. Finding this value *under* "sin," we then look to the *left*-hand degrees column and determine that $\alpha = 39.0°$.

Given that $\tan \alpha = 3.625$, find α to the nearest 0.1°. We look for 3.625 in the columns labeled "tan" and find that the nearest number to 3.625 is 3.630. Since 3.630 is *above* "tan," we must look to the *right*-hand degrees column. Here we find 74.6. Therefore, $\alpha = 74.6°$.

If the electronic calculator is used, it may not be necessary to refer to the table for the values of the trigonometric ratios or to use interpolation (see Section 11-6) to obtain additional accuracy. It should be noted that in this regard the calculator is another source from which these values are available and does not eliminate the need to understand how these values are used.

Before using a calculator, read the handbook to determine just how values are found and operations performed. We do note here that the trigonometric ratios are found by use of the [SIN], [COS], and [TAN] keys. To find the angle when the ratio is known, we use the [SIN^{-1}], [ARCSIN], or [INV-SIN] keys (or corresponding keys for the cosine or tangent). The symbols SIN^{-1}, ARCSIN, and INV-SIN mean "the angle whose sine is."

If two sides of a right triangle are known, using the definitions of the trigonometric ratios along with the table allows us to determine the angles of the triangle. Consider the following example.

Example F For the triangle in Fig. 14-14, find α.

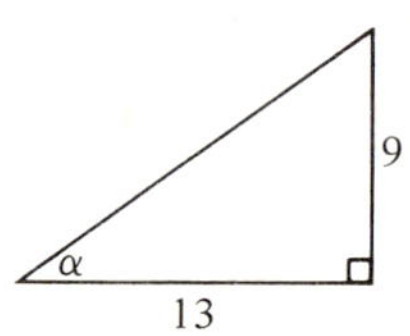

Figure 14-14

Since we know the two legs of the triangle, we can directly determine the value of the tangent of α. Thus,

$$\tan \alpha = \frac{9}{13} = 0.6923$$

which means that $\alpha = 34.7°$ (to the nearest 0.1°), since 0.6923 is closer to 0.6924 than any other value in the tangent column of Table 3. On a calculator, we would divide 9 by 13 and then use the [TAN^{-1}] or [ARCTAN] key.

Exercises 14-2

In Exercises 1 through 16, determine the value of the indicated trigonometric ratio from Table 3.

1. sin 39.0°	**2.** cos 18.0°	**3.** tan 41.5°	**4.** cot 22.3°
5. cot 61.2°	**6.** tan 75.1°	**7.** cos 52.3°	**8.** sin 49.9°
9. cot 81.6°	**10.** tan 57.7°	**11.** sin 44.2°	**12.** cos 7.6°
13. cos 10.5°	**14.** sin 30.9°	**15.** tan 77.0°	**16.** cot 46.1°

In Exercises 17 through 32, determine the value of the angle α *to the nearest* 0.1° *from Table 3.*

17. $\tan\alpha = 0.2493$

18. $\sin\alpha = 0.6157$

19. $\cos\alpha = 0.7478$

20. $\tan\alpha = 0.2295$

21. $\cot\alpha = 0.5985$

22. $\cos\alpha = 0.3380$

23. $\sin\alpha = 0.9949$

24. $\cos\alpha = 0.3629$

25. $\tan\alpha = 1.021$

26. $\cot\alpha = 0.8330$

27. $\cot\alpha = 2.081$

28. $\tan\alpha = 0.1694$

29. $\sin\alpha = 0.9797$

30. $\cos\alpha = 0.0680$

31. $\tan\alpha = 1.791$

32. $\cot\alpha = 0.7166$

In Exercises 33 through 36, for the given trigonometric ratio find the indicated trigonometric ratio from Table 3.

33. $\sin\alpha = 0.6428$; $\cot\alpha$

34. $\cos\alpha = 0.5736$; $\tan\alpha$

35. $\tan\alpha = 1.036$; $\cos\alpha$

36. $\cot\alpha = 0.1051$; $\sin\alpha$

In Exercises 37 through 44, find, to the nearest 0.1°, *angle* α *in Fig. 14-15 for the given sides of the triangle.*

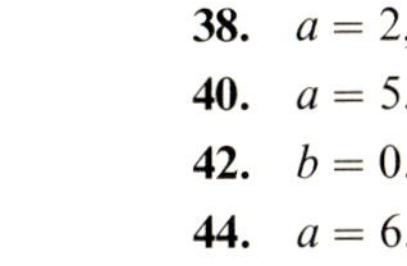

Figure 14-15

37. $a = 3, c = 5$

38. $a = 2, b = 5$

39. $b = 5.3, c = 7.1$

40. $a = 5.6, b = 1.9$

41. $a = 16.0, c = 21.3$

42. $b = 0.076, c = 0.152$

43. $a = 8700, b = 960$

44. $a = 6.75, c = 56.2$

In Exercises 45 through 48, draw a right triangle, including the indicated angle (use a protractor). Draw the side adjacent to the angle 10 cm long. By measuring the other sides and using the definitions of the ratios, verify the values of the ratios with the values from the table.

45. $\sin 32°$ **46.** $\tan 56°$ **47.** $\cos 13°$ **48.** $\cot 78°$

In Exercises 49 and 50, solve the given problems.

49. When a light ray enters glass from the air, it bends somewhat toward a line perpendicular to the surface. The *index of refraction* of a medium (in this case glass) is defined as

$$n = \frac{\sin i}{\sin r}$$

where i is the angle between the perpendicular and the ray in air and r is the angle between the perpendicular and the ray in the medium. A typical case for glass is $i = 59.0°$ and $r = 34.5°$. Determine the index of refraction for this case.

50. One end of a 25.0-m metal rod lies on a flat surface; the rod makes an angle of 17.0° with the surface. The length of the shadow of the rod due to a light shining vertically down on it is (25.0)(cos 17.0°). Find the length of the shadow.

14-3 ELEMENTARY APPLICATIONS

In this section, we shall investigate, by means of examples and exercises, many of the applications of the trigonometric ratios. These include indirect measurement, which was mentioned in the first section of this chapter. First, however, we shall briefly consider the general idea of **solving a triangle.**

In every triangle there are three angles and three sides. If three of these six parts are known, the other three can be determined, so long as at least one of the known parts is a side. *By solving a triangle, we mean finding the unknown parts of the triangle by using the known parts so that the result is that all six parts are known.*

In a right triangle we must know only two parts other than the right angle in order to solve it. The following two examples illustrate solving right triangles.

Example A Given that the hypotenuse of a right triangle is 16.0 and that one of the acute angles is 35.0°, find the other acute angle and the two sides. See Fig. 14-16.

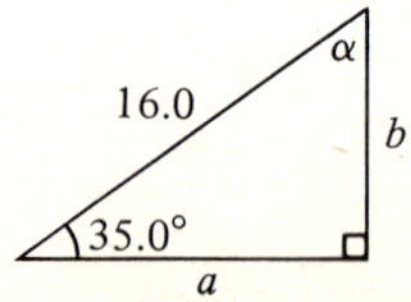

Figure 14-16

Since the angles of a triangle contain a total of 180° and the right angle contains 90°, the two acute angles must total 90°. Therefore, the unknown angle α must be 55.0°.

Next, since $\sin 35.0° = \dfrac{b}{16.0}$, we have

$$b = 16.0 \sin 35.0° \quad \text{or} \quad b = 16.0(0.5736) = 9.18$$

Also, since $\cos 35° = \dfrac{a}{16.0}$, we have

$$a = 16.0 \cos 35.0° = 16.0(0.8192) = 13.1$$

Therefore, the required angle is 55.0° and the required sides are 9.18 and 13.1.

Example B In a right triangle, the two legs are 5.00 and 8.00. Find the hypotenuse and the two acute angles. See Fig. 14-17.

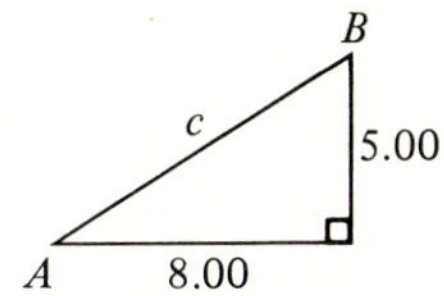

Figure 14-17

From the figure, we see that $\tan A = \frac{5.00}{8.00}$, which means that $\tan A =$ 0.6250. From the table, we then determine that $\angle A = 32.0°$ (to the nearest 0.1°). This in turn means that $\angle B = 58.0°$.

We can find side c by the Pythagorean theorem or by using a ratio of one of the angles just determined. The Pythagorean theorem is preferable, since it does not involve derived numbers, which could conceivably be in error. Thus,

$$c = \sqrt{(5.00)^2 + (8.00)^2} = \sqrt{89.0} = 9.43$$

Therefore, the required angles are 32.0° and 58.0° and the required side is 9.43.

The application of the trigonometric ratios is essentially the same as solving triangles, although it is usually one specific part of the triangle that we are asked to determine. The following five examples illustrate some of the basic applications.

Example C A ladder 22.0 ft long leans against the wall of a house. It makes an angle of 50.0° with the ground. How high on the house does the ladder reach? See Fig. 14-18.

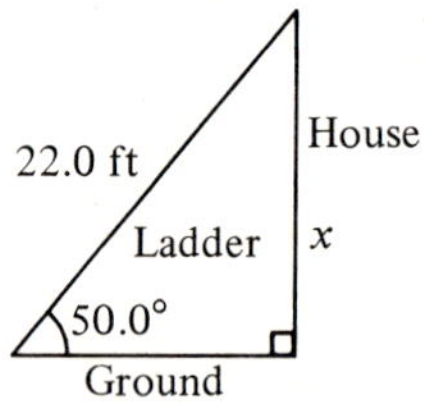

Figure 14-18

Since the side opposite the known angle is to be determined and the hypotenuse is known, the solution is most easily found by use of the sine. Thus,

$$\sin 50.0^\circ = \frac{x}{22.0}$$

$$x = 22.0 \sin 50.0^\circ = 22.0(0.7660)$$

$$= 16.9 \text{ ft}$$

Example D A guy wire attached to the top of a telephone pole is 10.5 m long. The grounded end of the wire is 4.75 m from the base of the pole. What is the angle between the wire and the ground? See Fig. 14-19.

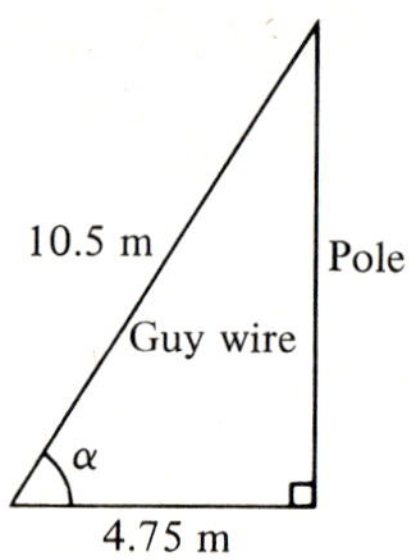

Figure 14-19

Since we know the hypotenuse and the adjacent side of the desired angle, we can easily find the solution by using the cosine. Thus,

$$\cos \alpha = \frac{4.75}{10.5} = 0.4524$$

From Table 3, we find that the cosine value of 0.4524 is for the angle 63.1° Thus,

$$\alpha = 63.1^\circ$$

Example E A lighthouse is 190 ft high. The **angle of elevation** (*the angle between the horizontal and the line of sight, upward with respect to the observer*) from a point on the shore to the top of the lighthouse is 21.3°. Find the distance from the point on shore to the base of the lighthouse. See Fig. 14-20.

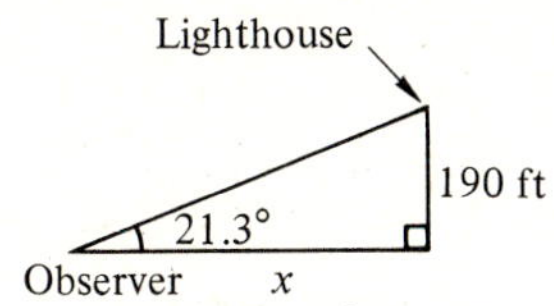

Figure 14-20

Since the known side is that which is opposite to the known angle, and the side to be determined is adjacent, the solution can be completed by using the tangent or the cotangent. We shall use the cotangent:

$$\cot 21.3^\circ = \frac{x}{190}$$

$$x = 190 \cot 21.3^\circ = 190(2.565)$$

$$= 487 \text{ ft}$$

Example F The pilot in an airplane checks the **angle of depression** (*the angle between the horizontal and the line of sight, downward with respect to the observer*) to the control tower of an airport to be 28.0°. If the plane is 10,000 m high, what is its horizontal distance from the tower? See Fig. 14-21.

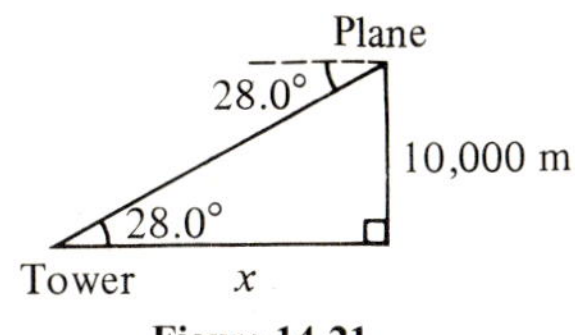

Figure 14-21

The angle of depression referred to in the statement of the problem is the upper 28.0° angle in Fig. 14-21. Since the alternate interior angles are equal, however, the angle marked at the tower is also 28.0°. Since opposite and adjacent sides are involved, we can use tangent or cotangent. Here we shall use the tangent. Thus,

$$\tan 28.0^\circ = \frac{10,000}{x}$$

$$x = \frac{10,000}{0.5317} = 18,800 \text{ m}$$

Example G A television antenna is on the roof of a building. From a point on the ground 36.0 ft from the building, the angles of elevation of the top and the bottom of the antenna are 51.0° and 42.0°, respectively. How tall is the antenna?

In Fig. 14-22, we let x represent the distance from the top of the building to the ground and y represent the distance from the top of the antenna to the ground. Therefore,

$$\frac{x}{36.0} = \tan 42.0^\circ$$

$$x = 36.0(\tan 42.0^\circ) = 36.0(0.9004)$$

$$= 32.4 \text{ ft}$$

and

$$\frac{y}{36.0} = \tan 51.0^\circ$$

$$y = 36.0(\tan 51.0^\circ) = 36.0(1.235)$$

$$= 44.5 \text{ ft}$$

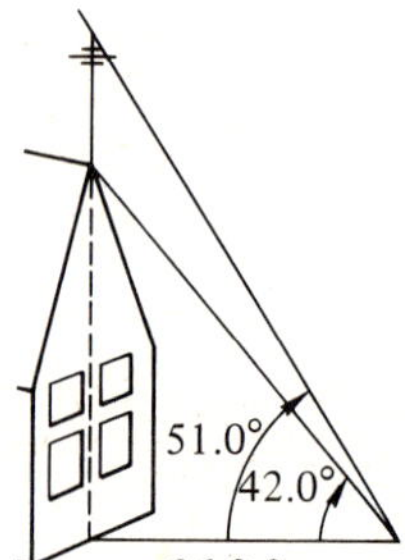

Figure 14-22

The length of the antenna is the difference of these distances, or 12.1 ft.

Exercises 14-3

In Exercises 1 through 16, solve the triangles with the given parts. The parts are indicated in Fig. 14-23. (Angles are indicated only by the appropriate capital letter.)

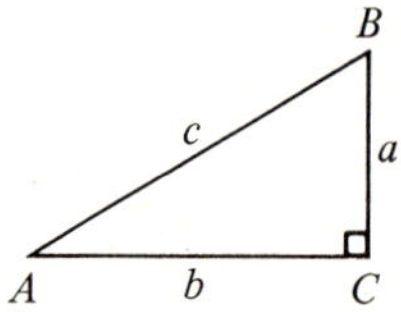

Figure 14-23

1. $A = 30.0^\circ, a = 12.0$

2. $A = 45.0^\circ, b = 16.0$

3. $B = 56.3^\circ, c = 22.5$

4. $B = 17.1^\circ, a = 15.7$

5. $A = 76.8^\circ, c = 31.4$

6. $B = 35.7^\circ, b = 1.45$

7. $a = 0.650, c = 1.35$

8. $a = 4.70, b = 7.40$

9. $b = 5.80, c = 45.0$

10. $a = 734, b = 129$

11. $a = 9.72, c = 10.8$

12. $b = 0.195, c = 0.321$

13. $A = 7.0^\circ, b = 15.3$

14. $B = 84.5^\circ, c = 1730$

15. $a = 65.1, b = 98.3$

16. $b = 1.89, c = 7.14$

In Exercises 17 through 42, solve the given problems by finding the appropriate part of a triangle.

17. The angle of elevation to the top of a statue is 10.0° from a point 165 ft from the base of the statue. Find the height of the statue.

18. The angle of elevation of the sun is 51.3° at the time a tree casts a shadow 23.7 m long. Find the height of the tree.

19. A robot is on the surface of Mars. The angle of depression from a camera in the robot to a rock on the surface of Mars is 13.3°. The camera is 196 cm above the surface. How far is the camera from the rock?

20. A guy wire whose grounded end is 16.0 ft from the pole it supports makes an angle of 56.0° with the ground. How long is the wire?

21. A roadway rises 150 ft for every 2000 ft along the road. Find the angle of inclination of the roadway.

22. A searchlight is 520 ft from a building. The beam lights up the building at a point 150 ft above the ground. What angle does the beam make with the ground?

23. The bottom of a picture on a wall is on the same level as a person's eye. The picture is 135 cm high, and the angle of elevation to the top of the picture is 23.0°. How far is the person's eye from the bottom of the picture?

24. Along the shore of a river, from a rock at a height of 300 ft above the river, the angle of depression of the closest point on the opposite shore is 12.0°. What is the distance across the river from the base of the height to the closest point on the opposite shore?

25. A point near the top of the Leaning Tower of Pisa is about 50.5 m from a point at the base of the tower (measured along the tower). This top point is also directly above a point on the ground 4.25 m from the same base point. What angle does the tower make with the ground?

26. A roof rafter is 5.25 m long (neglecting the overhang), and its upper end is 1.70 m above the lower end. Find the angle between the rafter and the horizontal.

27. A ship is traveling toward a port from the west when it changes course by 18.0° to the north. It then travels 23.0 mi until it is due north of the port. How far was it from port when it turned?

28. A draftsman sets the legs of a pair of dividers so that the angle between them is 36.0°. What is the distance between the points if each leg is 4.75 in. long?

29. A person is building a swimming pool 12.5 m long. The depth at one end is to be 1.00 m and the depth at the other end is to be 2.50 m. Find, to the nearest 0.1°, the slope of the bottom of the pool.

30. A shelf is supported by a straight 65.0-cm support attached to the wall at a point 53.5 cm below the bottom of the shelf. What angle does the support make with the wall?

31. The total angle through which a pendulum 42.5 in. long swings is 6.2°. Find the horizontal distance between the extreme positions of the pendulum.

32. A television antenna is 600 ft high. If the angle between the guy wires (attached at the top) and the antenna is 55.0°, how long are the guy wires?

33. An observer in a helicopter 800 ft above the ground notes that the angle of depression of an object is 26.0°. How far from directly below the helicopter is this object?

34. One way of finding the distance from the earth to the moon is indicated in Fig. 14-24. From point P the moon is directly overhead, and from point Q the moon is just visible. Both points are on the equator. The angle at E is about 89.0°. Given that the radius of the earth is 6360 km, how far is it to the moon (PM)?

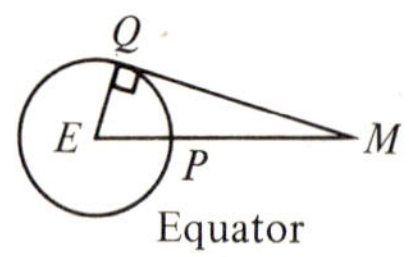

Figure 14-24

35. An astronaut observes two cities which are 45.0 km apart; the angle between the lines of sight is 10.6°. Given that he is the same distance from each, how high above the surface of the earth is he?

36. A jet cruising at 610 km/h climbs at an angle of 12.0°. What is its gain in altitude in 3 min?

37. What is the distance between two points on a wheel 38.0° apart on the circumference, given that the diameter of the wheel is 28.0 in.?

38. Eight bolt holes (spaced equally) are drilled on the circumference of a circle whose diameter is 16.6 in. What is the center-to-center distance between holes?

39. A surveyor wishes to measure the width of a large lake. She sights a point B on the opposite side of the lake from point A. She then measures off 200 ft from point A to point C, such that $CA \perp AB$. She then determines that $\angle ABC = 8.5°$. How wide is the lake?

40. An astronaut circling the moon at an altitude of 160 km notes that the angle of depression of the horizon is 23.8°. What is the radius of the moon?

41. The angle of depression of a fire noticed directly north of a 24.5-m fire tower is 5.5°. The angle of depression of a stream running east to west, and also north of the tower, is 13.4°. How far is the fire from the stream?

42. From an airplane 5000 ft above the surface of the water, a pilot observes two boats directly ahead. The angles of depression are 20.0° and 12.0°. How far apart are the boats?

14-4 SUMMARY

This chapter has been devoted to a brief introduction to trigonometry. The definitions of the trigonometric ratios in terms of the sides of a right triangle (more general definitions are given in a more complete treatment) were presented in Eqs. (14-1), (14-2), (14-3), and (14-4). We saw that any of the ratios has only one value for a particular angle due to a property of similar triangles, which states that corresponding sides are proportional.

In the second section, we explained the use of Table 3 for finding values of the trigonometric ratios, and we introduced the electronic calculator.

The third section was devoted to solving triangles and some of the elementary applications of trigonometry. These sections emphasized the importance of the Pythagorean theorem.

The terminology of the chapter included the following:

Angle of depression	Secant
Angle of elevation	Sine
Cofunction	Solving a triangle
Cosecant	Tangent
Cosine	Trigonometric ratios
Cotangent	Trigonometry

All of these should now be familiar, and the following set of exercises is intended to provide a review of the concepts of the chapter.

Exercises 14-4

In Exercises 1 through 4, find the indicated trigonometric ratios in fractional form for the angles of the triangle in Fig. 14-25.

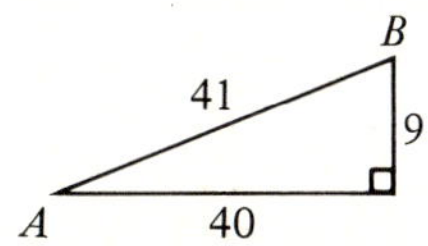

Figure 14-25

1. $\sin A$, $\cos B$ **2.** $\tan B$, $\cos A$ **3.** $\sec B$, $\tan A$ **4.** $\cot A$, $\csc B$

In Exercises 5 through 8, find the indicated trigonometric ratios in decimal form for the angles of the triangle in Fig. 14-26

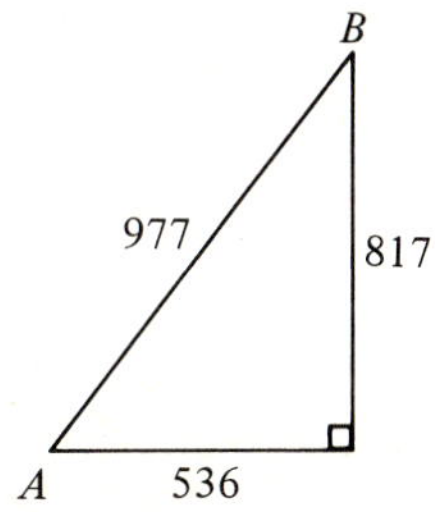

Figure 14-26

5. $\cos A$, $\tan B$ **6.** $\cot B$, $\sin A$ **7.** $\sin B$, $\tan A$ **8.** $\sec A$, $\csc B$

In Exercises 9 through 16, determine the indicated trigonometric ratios. The listed sides are those shown in Fig. 14-27. Do not use Table 3.

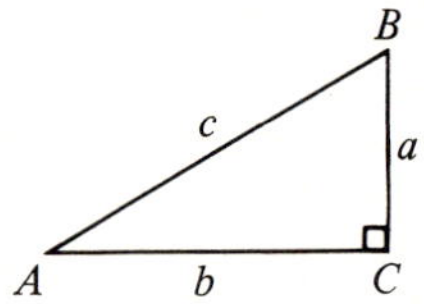

Figure 14-27

9. $a = 12.0$, $b = 5.00$. Find $\cos A$ and $\tan B$.

10. $b = 9.00$, $c = 15.0$. Find $\sec A$ and $\cot B$.

11. $a = 8.70$, $b = 2.30$. Find $\tan A$ and $\cos B$.

12. $a = 125$, $c = 148$. Find $\sin A$ and $\sec B$.

13. $a = 0.0890$, $c = 0.0980$. Find $\cot A$ and $\sin B$.

14. $b = 1670$, $c = 4200$. Find $\csc A$ and $\csc B$.

15. $b = 0.0943$, $c = 0.105$. Find $\cos A$ and $\tan B$.

16. $a = 8.60$, $b = 7.90$. Find $\sin A$ and $\cot B$.

In Exercises 17 through 24, use Table 3 to find, to the nearest 0.1°, *the indicated angle. The parts of the triangle are those shown in Fig. 14-27.*

17. $a = 8.00$, $c = 13.0$. Find A.

18. $b = 51.0$, $c = 58.0$. Find A.

19. $a = 5.60$, $b = 1.30$. Find B.

20. $a = 0.780$, $c = 2.40$. Find B.

21. $b = 3420$, $c = 7200$. Find B.

22. $a = 0.00670$, $b = 0.0156$. Find A.

23. $a = 4910$, $b = 3650$. Find A.

24. $b = 0.860$, $c = 0.915$. Find B.

In Exercises 25 through 32, solve the triangles with the given parts. The given parts are shown in Fig. 14-27.

25. $A = 21.0°$, $c = 6.93$

26. $B = 18.0°$, $a = 0.360$

27. $B = 32.7°$, $b = 45.9$

28. $B = 7.4°$, $c = 1890$

29. $a = 8.70$, $b = 5.20$

30. $a = 0.00760$, $b = 0.0120$

31. $a = 97.0$, $c = 108$

32. $b = 17.4$, $c = 54.0$

In Exercises 33 through 50, solve the problems.

33. When the angle of elevation of the sun is 40.0°, what is the length of the shadow of a tree 65.0 ft tall?

34. The distance from a point on the shore of a lake to an island is 3500 ft. From a point directly above the shore on a cliff the angle of depression to the island is 15.0°. How high is the cliff?

35. The distance from point B to point C along the shore of a river is 500 m. Point A is directly across from point C on the opposite shore. Given that $\angle ABC$ is 21.9°, how wide is the river?

36. A guy wire supporting a television tower is grounded 250 m from the base of the tower. Given that the wire makes an angle of 29.0° with the ground, how high up on the tower is it attached?

37. The span of a roof is 30.0 ft. Its rise is 8.00 ft at the center of the span. What is the angle the roof makes with the horizontal?

38. A roadway rises 85.0 ft for every 1000 ft along the road. Determine the angle of inclination of the roadway.

39. A jet climbs at an angle of 35.0° while traveling at 600 km/h. How long will it take to climb to an altitude of 10,000 m?

40. An observer 3000 ft from the launch pad of a rocket measures the angle of inclination to the rocket soon after liftoff to be 65.0°. How high is the rocket, assuming it has moved vertically?

41. The level of a pond is 14.5 m above the level of a nearby stream. A straight 75.0-m drainage pipe extends from the pond surface to a point 3.0 m above the stream. What angle does the pipe make with the horizontal?

42. A lakefront property is shown in Fig. 14-28. What is the angle θ between the shoreline and the property line?

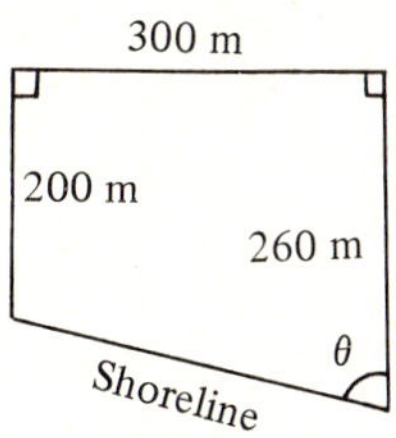

Figure 14-28

43. Two laser beams from the same source are transmitted to two points 100 m apart on a flat surface. One beam is perpendicular to the surface and the other makes an angle of 87.5° with the surface. How far from the surface is the source?

44. A person walks 3.25 km to the west from an intersection of roads. She then turns and walks 2.50 km in a direction 20.0° north of west. What is her displacement from the intersection?

45. The side of an equilateral triangle is 24.0 in. How long is its altitude?

46. A certain machine part is a regular hexagon (six sides) 0.844 in. on a side. What is the distance across from one side to the opposite side?

47. The altitude of an equilateral triangle is 6.93 cm. How long is its side?

48. In determining the height h of a building which is 220 m distant, a surveyor may use the equation $h = 220 \tan\theta$, where θ is the angle of elevation to the top of the building. What should θ be if $h = 130$ m?

49. Given that A and B are acute angles, is $\cos A > \cos B$ if $A > B$? Is $\tan A > \tan B$ if $A > B$?

50. Given that A and B are acute angles, is $\sin A > \sin B$ if $A > B$? Is $\cot A > \cot B$ if $A > B$?

Appendix A
REVIEW TOPICS FROM ARITHMETIC

A-1 BASIC OPERATIONS WITH WHOLE NUMBERS

In this appendix, a brief discussion of the basic operations of arithmetic is included for purposes of review and reference. These basic operations not only are essential for performing various types of computations, but they are fundamental to the development of the mathematics presented in the text.

The process of finding the total number of objects in two different sets of these objects is **addition.** When we are finding the **sum,** *which is the result of addition*, we must take into account positional notation (see Section 1-7) and add only those numbers with the same positional value. In this way, we are adding **like quantities.**

In order to perform addition accurately and with reasonable speed, it is essential to know the basic sums through $9 + 9 = 18$ without hesitation. Learning the basic sums well takes practice. Also, in the process of addition, when the sum of one column exceeds 9, it is necessary to "carry" a number to the next column.

Example A The addition of the given numbers is indicated to the right. The small numbers indicate the number being carried. Since the sum of the numbers in the ones column is 26, a 6 is placed in the sum, and the 2 is carried. The remaining digits in the sum are determined in a similar manner.

$$\begin{array}{r} {\scriptstyle 1\,3\,2} \\ 386 \\ 475 \\ 569 \\ 496 \\ \hline 1926 \end{array}$$

When we have to determine how much greater one number is than another, we use **subtraction,** *which is the inverse of addition. Subtraction consists of reducing one number* (*the* **minuend**) *by the number being subtracted* (*the* **subtrahend**) *to obtain the* **difference.** If the digit in the subtrahend is greater than the digit in the minuend for any column, it is necessary to "borrow" from the next column.

Example B If we wish to subtract 29 from 73, we find that, in the ones column, we are to reduce 3 by 9. When considering whole numbers, we cannot do this. However, if we "borrow" 10 ones from the tens column of 73, the subtraction amounts to subtracting 2 tens + 9 ones from 6 tens + 13 ones. Thus, we have

$$\begin{array}{rl} 73 & \text{minuend} \\ -29 & \text{subtrahend} \\ \hline 44 & \text{difference} \end{array} \quad \text{or} \quad \begin{array}{r} \overset{6}{7}{}^{1}3 \\ -29 \\ \hline 44 \end{array}$$

The basic arithmetic operation in which we find the total number of objects in several sets, each set containing the same number of objects, is called **multiplication.** In this sense, we see that multiplication is a process of adding equal numbers. *The number that is multiplied is called the* **multiplicand,** *the number of times it is taken is the* **multiplier,** *and the result is the* **product.** The multiplicand and multiplier are also called **factors** of the product.

In order to perform multiplication accurately and with reasonable speed, it is necessary, as with addition, to know the basic products through $9 \times 9 = 81$ without hesitation. In the process of multiplying, it is necessary to align the numbers being multiplied properly and to "carry" properly from one position to another.

Example C The product of 26 and 124 would appear as follows:

$$\begin{array}{rl} 124 & \text{multiplicand} \\ \times\ 26 & \text{multiplier} \\ \hline 744 & \\ 248 & \\ \hline 3224 & \text{product} \end{array}$$

To find the product, we must multiply 124 by 6. We begin by multiplying 4 by 6 and obtaining 24. The 4 is placed below the 6 and the 2 is carried and then added to the product of 2 and 6, obtaining 14. This 4 is placed in the next position for the product and the 1 is carried, and so on. The product of 124 and 2 is moved one place to the left to account for positional notation. Addition then gives us the total product.

Just as subtraction is the inverse process of addition, **division** *is the inverse process of multiplication.* Therefore, we can look upon division as a process of repeated subtraction. For example, if we subtract 2 from 10 five times, the result is zero. This means that 10 divided by 2 is 5. Here 10 is the **dividend,** 2 is the **divisor,** and 5 is the **quotient.** A common notation is $10 \div 2 = 5$, where $\div$ means "divided by." In Example D, the process of division is reviewed.

Example D In order to divide 3288 by 24, or $3288 \div 24$, we set up the problem as follows:

```
             137   quotient
divisor 24 ) 3288  dividend
             24              (24 × 1)
             ---
              88
              72             (24 × 3)
             ---
              168
              168            (24 × 7)
              ---
```

We note first that 24 will not divide into 3 but that it will divide once into 32. Thus, a 1 is placed in the quotient, and the product of 1 and 24 is placed below the 32. Now 24 is subtracted from 32, obtaining 8, and the 8 is then carried down from the dividend. We now note that 24 divides 3 times into 88 but not 4 times. Thus, the 3 is placed in the quotient, and so on.

In many divisions, the divisor will not divide exactly into the dividend. In these cases, there is a **remainder** in the answer. Example E illustrates a division with a remainder.

Example E Divide 5286 by 25.

```
       211
25 ) 5286
     50      (25 × 2)
     --
      28
      25     (25 × 1)
      --
       36
       25    (25 × 1)
       --
       11    (remainder)
```

Since 11 is smaller than 25, the divisor, and all of the digits of the dividend have been used, the division process is discontinued and 11 becomes the remainder.

Exercises A-1

In Exercises 1 through 12, add the given numbers.

1.	**2.**	**3.**	**4.**
28	83	726	521
37	98	97	879
96	46	813	106

5.	**6.**	**7.**	**8.**
826	278	3973	9806
992	446	5868	3791
809	915	9039	986
67	548	5792	7372

9.	**10.**	**11.**	**12.**
3873	989	30964	87657
9295	3216	9877	93984
4082	4807	92286	57609
399	736	5547	8726
7646	9297	965	92875

In Exercises 13 through 20, subtract the lower number from the upper number.

13.	**14.**	**15.**	**16.**
873	921	8305	2006
292	224	7356	1197

17.	**18.**	**19.**	**20.**
36047	10906	290078	872110
26249	9928	194396	682324

In Exercises 21 through 28, perform the indicated multiplications.

21. 27×259 **22.** 36×564 **23.** 329×7105 **24.** 915×3794

25.	**26.**	**27.**	**28.**
3879	9276	52438	97877
× 4081	× 6814	× 4794	× 6985

In Exercises 29 through 36, perform the indicated divisions.

29. $3\overline{)729}$ **30.** $81\overline{)2997}$ **31.** $54\overline{)17604}$ **32.** $29\overline{)14674}$

33. $52664 \div 37$ **34.** $59685 \div 28$ **35.** $607964 \div 387$ **36.** $920337 \div 726$

A-2 OPERATIONS WITH FRACTIONS

We shall now consider the basic concepts and operations on arithmetic fractions. Some of the principles discussed here are also included in Chapter 6, where algebraic fractions are considered.

In general, *a* **fraction** *is the indicated division of one number by another.* It is also possible to interpret this definition as a certain number of equal parts of a given unit or of a given group. For example, the fraction $\frac{5}{8}$ is the indicated division of 5 by 8, where 5 is the **numerator** and 8 is the **denominator.** It is also possible to interpret the fraction $\frac{5}{8}$ as referring to 5 of 8 equal segments of a line.

If the numerator of a fraction is numerically less than the denominator, the fraction is called a **proper fraction.** *If the numerator equals or is numerically greater than the denominator, the fraction is called an* **improper fraction.** Since an improper fraction in which the numerator is greater than the denominator represents a number numerically greater than one, it is often convenient to use *a* **mixed number,** *which is a whole number plus a proper fraction*, to represent the same number.

Example A The fraction $\frac{4}{9}$ is a proper fraction, whereas the fraction $\frac{9}{4}$ is an improper fraction. The mixed number $2\frac{1}{4}$, meaning $2 + \frac{1}{4}$, equals $\frac{9}{4}$.

To convert $\frac{9}{4}$ to a mixed number, we divide 9 by 4. The result is 2 with a remainder of 1. Thus, $\frac{9}{4} = 2\frac{1}{4}$.

To convert $2\frac{1}{4}$ to an improper fraction, we multiply 2×4 to get 8, which is then added to 1 to obtain 9. This 9 is the required numerator of the improper fraction. Thus, $2\frac{1}{4} = \frac{9}{4}$.

Two fractions are called **equivalent** *if the numerator and denominator of one of the fractions can be multiplied, or divided, by the same number (not zero) in order to obtain the other fraction. A fraction is said to be* **reduced** *if it is changed to a fraction in which the numerator and denominator are numerically smaller. A fraction is converted to* **higher terms** *if the resulting numerator and denominator are numerically larger.*

Among most important operations performed on a fraction is reducing it to **lowest terms** (*or* **simplest form**). This means that *the resulting numerator and denominator are not both divisible by any whole number other than one.*

Example B The fraction $\frac{6}{16}$ can be converted to $\frac{30}{80}$ by multiplying the numerator and denominator by 5. Therefore, $\frac{6}{16} = \frac{30}{80}$.

The fraction $\frac{18}{24}$ can be reduced by dividing the numerator and denominator by 2. This gives the fraction $\frac{9}{12}$. However, this is not the simplest form since both 9 and 12 are divisible by 3.

We find the simplest form of $\frac{18}{24}$ by dividing the numerator and denominator by 6. This gives us $\frac{18}{24} = \frac{3}{4}$, where $\frac{3}{4}$ is $\frac{18}{24}$ in lowest terms.

Other illustrations of the simplest form of fractions are as follow:

$$\frac{10}{18} = \frac{5}{9} \qquad \frac{18}{21} = \frac{6}{7} \qquad \frac{45}{50} = \frac{9}{10} \qquad \frac{7}{42} = \frac{1}{6} \qquad \frac{152}{96} = \frac{19}{12}$$

In order to add fractions in which the denominators are the same, we place the sum of the numerators over the common denominator. In the same way, when we subtract fractions, we place the difference of the numerators over the common denominator. Considering the meaning of a fraction as a number of equal parts of a given unit, we see that we are combining like quantities.

Example C

a. $\frac{2}{9} + \frac{5}{9} = \frac{2+5}{9} = \frac{7}{9}$

b. $\frac{5}{11} - \frac{3}{11} = \frac{5-3}{11} = \frac{2}{11}$

c. $\frac{3}{8} + \frac{1}{8} = \frac{4}{8} = \frac{1}{2}$

Note that in illustration (c) the result was put in the simplest form of $\frac{1}{2}$. The final result should always be put in simplest form.

If the fractions being added or subtracted do not have the same denominators, it is necessary to convert each of them to a form in which the denominators are equal. After this has been completed, the fractions can then be combined. Although any proper common denominator can be used, any simplification required in the final result will be minimized if this denominator is the least possible value. This form is known as the **lowest common denominator** for these fractions. In general, *the lowest common denominator of a set of fractions is the smallest number that is evenly divisible by all of the denominators of the set of fractions.*

Example D In many of the simpler cases, the lowest common denominator can be determined by inspection. For example, if we wish to add $\frac{1}{2} + \frac{3}{8}$, we can see that the lowest common denominator is 8 because both 2 and 8 divide evenly into 8. Thus,

$$\frac{1}{2} + \frac{3}{8} = \frac{4}{8} + \frac{3}{8} = \frac{4+3}{8} = \frac{7}{8}$$

If we cannot readily determine the lowest common denominator by inspection, find all the **prime factors** in each denominator. *A* **prime number** *is a whole number that is divisible evenly only by itself and* 1. (The first few prime integers are 2, 3, 5, 7, 11, and 13.) *The lowest common denominator is the product of all the different prime factors, each taken the greatest number of times it occurs in any one of the denominators.*

Example E When we want to find the sum $\frac{5}{12} + \frac{7}{8}$, we must find the prime factors of 12 and 8. Since $12 = 2 \times 2 \times 3$ and $8 = 2 \times 2 \times 2$, we see that the only different prime factors are 2 and 3. Since 2 occurs twice in 12 and three times in 8, it is taken three times in the least common denominator. Since 3 occurs once in 12 and does not occur in 8, it is taken once. Therefore, the lowest common denominator is

$$2 \times 2 \times 2 \times 3 = 24$$

This means that

$$\frac{5}{12} + \frac{7}{8} = \frac{10}{24} + \frac{21}{24} = \frac{31}{24}$$

Example F Combine: $\dfrac{2}{15} + \dfrac{11}{27} - \dfrac{7}{50}$

First we must determine the prime factors of 15, 27, and 50. Therefore,

$$15 = 3 \times 5 \qquad 27 = 3 \times 3 \times 3 \qquad 50 = 2 \times 5 \times 5$$

We now observe that the necessary prime factors for the lowest common denominator are 2, 3, and 5. The greatest number of times each appears is once for 2 (in 50), three times for 3 (in 27), and twice for 5 (in 50). Accordingly, the lowest common denominator is

$$2 \times 3 \times 3 \times 3 \times 5 \times 5 = 1350$$

Therefore, we have

$$\begin{aligned}\frac{2}{15} + \frac{11}{27} - \frac{7}{50} &= \frac{2(90)}{15(90)} + \frac{11(50)}{27(50)} - \frac{7(27)}{50(27)} \\ &= \frac{180}{1350} + \frac{550}{1350} - \frac{189}{1350} = \frac{180 + 550 - 189}{1350} \\ &= \frac{730 - 189}{1350} = \frac{541}{1350}\end{aligned}$$

The product of two fractions is the fraction whose numerator is the product of the numerators and whose denominator is the product of the denominators. If the resulting fraction is not in simplest form, it should be reduced to this form. This reduction is best accomplished by first only *indicating* the multiplications of numerators and denominators and then dividing out any common factors before actually doing any multiplication.

Example G

a. $\frac{5}{7} \times \frac{2}{3} = \frac{5 \times 2}{7 \times 3} = \frac{10}{21}$

b. $\frac{4}{5} \times \frac{3}{8} = \frac{4 \times 3}{5 \times 8} = \frac{1 \times 3}{5 \times 2} = \frac{3}{10}$ (Divide 4 and 8 by 4.)

c. $\frac{16}{15} \times \frac{5}{12} = \frac{16 \times 5}{15 \times 12} = \frac{4 \times 1}{3 \times 3} = \frac{4}{9}$ (Divide 16 and 12 by 4, and divide 5 and 15 by 5.)

d. $4 \times \frac{3}{7} = \frac{4}{1} \times \frac{3}{7} = \frac{4 \times 3}{1 \times 7} = \frac{12}{7}$

We note that, in illustration (a), there were no common factors in the numerator and the denominator. However, in illustrations (b) and (c), the common factors were divided out before the multiplications were actually performed. In illustration (d), we treated 4 as $\frac{4}{1}$ before proceeding.

In order to divide a number by a fraction, invert the divisor and multiply. By **invert,** we mean interchange the numerator and denominator of the divisor.

Example H

a. $\frac{3}{7} \div \frac{4}{9} = \frac{3}{7} \times \frac{9}{4} = \frac{27}{28}$

b. $\dfrac{\frac{8}{5}}{\frac{4}{15}} = \frac{8}{5} \times \frac{15}{4} = \frac{8 \times 15}{5 \times 4} = \frac{2 \times 3}{1 \times 1} = 6$

Exercises A-2

In Exercises 1 through 20, reduce the given fractions to lowest terms.

1. $\frac{3}{9}$ **2.** $\frac{4}{16}$ **3.** $\frac{12}{26}$ **4.** $\frac{18}{15}$

5. $\frac{18}{30}$ **6.** $\frac{25}{35}$ **7.** $\frac{54}{63}$ **8.** $\frac{5}{40}$

9. $\frac{36}{132}$ **10.** $\frac{42}{120}$ **11.** $\frac{104}{72}$ **12.** $\frac{36}{135}$

13. $\frac{154}{70}$ **14.** $\frac{240}{135}$ **15.** $\frac{204}{228}$ **16.** $\frac{162}{189}$

17. $\frac{221}{351}$ **18.** $\frac{132}{385}$ **19.** $\frac{220}{43}$ **20.** $\frac{76}{51}$

In Exercises 21 through 56, perform the indicated operations, expressing each result in simplest form.

21. $\frac{1}{5}+\frac{3}{5}$
22. $\frac{2}{11}+\frac{5}{11}$
23. $\frac{4}{9}+\frac{7}{9}+\frac{1}{9}$
24. $2\frac{1}{16}+\frac{9}{16}-\frac{3}{16}$
25. $\frac{1}{2}+2\frac{1}{4}$
26. $\frac{2}{3}-\frac{1}{6}$
27. $\frac{7}{12}-\frac{1}{4}$
28. $\frac{13}{18}-\frac{1}{2}$
29. $\frac{23}{6}-\frac{3}{8}$
30. $\frac{22}{9}+\frac{5}{12}$
31. $\frac{11}{20}+\frac{5}{8}$
32. $\frac{19}{28}-\frac{5}{24}$
33. $\frac{3}{8}+\frac{7}{12}+\frac{1}{18}$
34. $\frac{29}{9}+\frac{7}{15}-\frac{4}{25}$
35. $\frac{1}{3}+\frac{23}{14}-\frac{5}{21}$
36. $\frac{26}{27}-\frac{7}{18}+\frac{1}{10}$
37. $\frac{2}{7}\times\frac{3}{11}$
38. $\frac{7}{8}\times\frac{3}{5}$
39. $3\times\frac{2}{5}$
40. $\frac{2}{7}\times 1\frac{2}{3}$
41. $\frac{7}{8}\div\frac{5}{6}$
42. $\frac{7}{4}\div\frac{2}{5}$
43. $\frac{5}{9}\div 3$
44. $2\frac{1}{2}\div\frac{8}{3}$
45. $\frac{8}{9}\times\frac{5}{16}$
46. $\frac{5}{12}\times\frac{3}{7}$
47. $\frac{7}{3}\times\frac{9}{14}$
48. $\frac{22}{25}\times\frac{15}{33}$
49. $\frac{8}{15}\div\frac{12}{35}$
50. $\frac{21}{44}\div\frac{28}{33}$
51. $\frac{8}{17}\div 4$
52. $\frac{39}{35}\div\frac{13}{21}$
53. $\frac{3}{5}\times\frac{15}{7}\times\frac{14}{9}$
54. $\frac{2}{7}\times\frac{19}{4}\times\frac{21}{38}$
55. $\left(\frac{3}{4}\times\frac{28}{27}\right)\div\frac{35}{6}$
56. $\frac{18}{25}\div\left(\frac{17}{5}\div\frac{34}{15}\right)$

A-3 DECIMALS AND PERCENT

Fractions whose denominators are 10, 100, 1000, *etc., are called* **decimal fractions.** For example, $\frac{7}{10}$ and $\frac{193}{10{,}000}$ are decimal fractions. Making further use of positional notation (see Section 1-7), we place a **decimal point** to the right of the one's position and let the digit in the first position to the right stand for the number of tenths, the digit in the second position to the right stand for the number of hundredths, and so on. *Numbers written in this form are called* **decimals.**

Example A The meaning of the decimal 6352.1879 is shown to the right. We may also show the meaning of this decimal number as

thousands	*hundreds*	*tens*	*ones*		*tenths*	*hundredths*	*thousandths*	*ten-thousandths*
6	3	5	2	.	1	8	7	9

$$6(1000)+3(100)+5(10)+2(1)+\frac{1}{10}+\frac{8}{100}+\frac{7}{1000}+\frac{9}{10000}$$

The meaning of this decimal can also be shown in terms of powers of ten, but this form requires the use of zero and negative exponents as developed in Section 4-3. Using them, we have

$$6352.1879 = 6(10^3) + 3(10^2) + 5(10^1) + 2(10^0) + 1(10^{-1}) + 8(10^{-2}) + 7(10^{-3}) + 9(10^{-4})$$

Since a fraction is the indicated division of one number by another, we can change a fraction to an equivalent decimal by division. We place a decimal point and additional zeros to the right of the one's position of the numerator and then perform the division.

Example B In order to change $\frac{5}{8}$ into an equivalent decimal, we perform the following division.

```
     .625
8 ) 5.000
    4 8
    ---
      20
      16
      --
       40
       40
       --
```

Therefore, $\frac{5}{8} = 0.625$.

When we convert a fraction to a decimal, the division does not always come out even regardless of the number of places to the right of the decimal point. In cases like this, we generally **round off** *the result of the division.* See Section 11-4 for a discussion of rounding off.

In order to change a decimal to a fractional form, we first write it in its decimal fraction form and then simplify it.

Example C

$$0.5 = \frac{5}{10} = \frac{1}{2} \qquad 0.125 = \frac{125}{1000} = \frac{1}{8}$$

$$0.00164 = \frac{164}{100000} = \frac{41}{25000}$$

When we consider the addition and subtraction of decimals, we see that the principles are the same as those that were developed for whole numbers. *Since only like quantities can be added or subtracted, we must then add tenths to tenths, hundredths to hundredths, and so on. Therefore, when we add and subtract decimals, we place the decimal points directly below each other and perform the addition or subtraction as with whole numbers.*

Example D

$$\begin{array}{r} 326.49 \\ 98.362 \\ +5937.8 \\ \hline 6362.652 \end{array} \qquad \begin{array}{r} 7862.472 \\ -794.56 \\ \hline 7067.912 \end{array}$$

When we multiply one decimal by another, the number of decimal places (to the right of the decimal point) in the product is the sum of the number of decimal places in the numbers being multiplied. Also, it is not necessary to line up decimal points as we do in addition and subtraction. The numbers are multiplied as if they were whole numbers, and the decimal point is then properly positioned in the product.

Example E

$$\begin{array}{rl} 5.307 & \text{(3 places)} \\ 2.63 & \text{(2 places)} \\ \hline 15921 & \\ 31842 & \\ 10614 & \\ \hline 13.95741 & \text{(5 places)} \end{array}$$

As with the other basic operations, the division of one decimal by another is very similar to the division of natural numbers. *Before the division is actually performed, however, the decimal point in the divisor is moved to the right a sufficient number of places to make the divisor a whole number. The decimal point in the dividend is then moved the same number of places.* (This is equivalent to multiplying the numerator and the denominator of the fraction form of the division by the same number.) *The decimal point in the quotient is directly above that of the dividend.*

Example F When dividing 74.362 by 3.26, we have the following setup.

```
              22.8
3.26. ) 74.36.2
         65 2
         -----
          9 16
          6 52
          -----
          2 642
          2 608
          -----
             34
```

Closely related to decimals is the topic of **percent.** *The word "percent" means "per hundred" or "hundredths." The familiar % symbol is used to denote percent.*

Example G 3% means $\frac{3}{100}$ or 3(0.01). Thus, 3% = 0.03.

$$25\% = \frac{25}{100} = 0.25 \qquad 300\% = \frac{300}{100} = 3.00 \qquad 0.4\% = \frac{0.4}{100} = 0.004$$

Using the meaning of percent allows us to convert percents to decimals, decimals to percents, percents to fractions, and fractions to percents. These conversions are illustrated in Example H.

Example H

$$5\% = \frac{5}{100} = 0.05 \qquad \text{(percent to decimal)}$$

$$0.45 = \frac{45}{100} = 45\% \qquad \text{(decimal to percent)}$$

$$0.6\% = \frac{0.6}{100} = \frac{6}{1000} = \frac{3}{500} \qquad \text{(percent to fraction)}$$

$$\frac{15}{12} = 1.25 = \frac{125}{100} = 125\% \qquad \text{(fraction to percent)}$$

Note that a percent can be changed to a decimal by moving the decimal point two places to the left and omitting the % sign. Similarly, a decimal can be changed to a percent by moving the decimal point two places to the right and inserting the % sign. When we change a percent to a fraction, we reduce the resulting fraction to its lowest terms. When we change a fraction to percent, we carry out the division (rounding off if necessary) and express the resulting decimal as a percent.

In most cases in which percentage is used in calculations, it is converted to decimal form. *Here we find the* **percentage** *of a given number, called the* **base,** *by multiplying it by the* **rate,** *which is the percent expressed as a decimal. We may express this in the* **percentage formula.**

percentage = rate × base

The percentage formula may also be used in the forms

$$\text{rate} = \frac{\text{percentage}}{\text{base}} \quad \text{or} \quad \text{base} = \frac{\text{percentage}}{\text{rate}}$$

depending on what information is given and what is required.

Example I A state has a sales tax of $3\frac{1}{2}\%$. What is the sales tax on an automobile costing \$3000?

Here we are given the rate, $3\frac{1}{2}\% = 0.035$, and the base, \$3000. Thus, we find the percentage, which in this case is the sales tax.

$$0.035 \times 3000 = \$105$$

Therefore, the sales tax on \$3000 is \$105.

Exercises A-3

In Exercises 1 through 8, change the given fractions to decimals. Where necessary, round off the result to the accuracy indicated.

1. $\frac{2}{5}$ **2.** $\frac{3}{25}$ **3.** $\frac{7}{8}$ **4.** $\frac{9}{16}$

5. $\frac{47}{27}$ (tenths) **6.** $\frac{882}{67}$ (tenths)

7. $\frac{4}{19}$ (hundredths) **8.** $\frac{29}{426}$ (thousandths)

In Exercises 9 through 12, change the given decimals into equivalent fractions in simplest form.

9. 0.42 **10.** 0.002 **11.** 5.34 **12.** 0.0084

In Exercises 13 through 20, perform the indicated operations. In Exercises 19 and 20, round off the result to tenths.

13. $18.046 + 1924.3 + 874.91$ **14.** $0.046 + 19.35 + 186.6942$

15. $18.623 - 9.86$ **16.** $0.8694 - 0.0996$

17. 2.36×5.932 **18.** 37.4×8.207

19. $32.6 \div 2.6$ **20.** $186 \div 0.87$

In Exercises 21 through 24, change the given percentages to equivalent decimals.

21. 8% **22.** 78% **23.** 236% **24.** 0.3%

In Exercises 25 through 28, change the given decimals to percents.

25. 0.27 **26.** 0.09 **27.** 21.6 **28.** 0.0064

In Exercises 29 through 32, change the given percents to equivalent fractions.

29. 30% **30.** 48% **31.** 0.8% **32.** 120%

In Exercises 33 through 36, change the given fractions to percents. Round off to the nearest tenth where necessary.

33. $\frac{3}{5}$ **34.** $\frac{7}{20}$ **35.** $\frac{4}{7}$ **36.** $\frac{16}{11}$

In Exercises 37 through 46, solve the given problems.

37. Find 20% of 65.
38. Find 2.6% of 230.
39. What percent of 72 is 18?
40. What percent of 250 is 5?
41. 3.6 is 25% of what number?
42. 7 is 1.75% of what number?
43. The sales tax in a certain city is 4%. What is the tax in this city on an article costing $378?
44. A tire costing $18.00 is to be sold at a 15% discount. What is the new price?
45. Of 520 lb of a certain alloy, 182 lb are zinc. What percent of the alloy is zinc?
46. The minimum distance from the earth of an artificial satellite is about 46% of its maximum distance from the earth. If its minimum distance is about 1020 mi, what is its maximum distance to the nearest mile?

Appendix B
UNITS OF MEASUREMENT: THE METRIC SYSTEM

As we mentioned in Section 1-2, many of the numbers we use represent some type of measurement. *Such numbers are called* **denominate numbers,** *and associated with these denominate numbers are* **units of measurement.** In order for the number or calculation to be meaningful, we must know these units.

Example A If we measure the length of a piece of metal pipe to be 12, we must also know whether it is being measured in inches, feet, meters, or some other unit of length.

If the time for one rotation of a wheel is measured to be 0.02, we must also know whether the unit of time being used is a second, minute, hour, or some other unit.

Certain universally accepted **base units** *are used to measure fundamental quantities.* The units for numerous other quantities are expressed in terms of the base units. Fundamental quantities for which base units are defined are (1) length, (2) mass or force, depending on the system of units being used, (3) time, (4) electric current, (5) temperature, (6) amount of substance, and (7) luminous intensity. *Other units, referred to as* **derived units,** *are expressible in terms of the units for these quantities.*

Example B The distance that a plane travels in flying from city A to city B can be measured in a unit of length, for example the mile. The time of flight can be measured in a unit of time, for example the hour. The average speed of the plane during the flight is then measured in miles per hour. Therefore, speed is measured in terms of units of length and time.

The unit for electric charge, the *coulomb*, is defined as one ampere second (the unit of electric current times the unit of time). Therefore, electric charge is defined in terms of the base units of electric current and time.

Here we see that some derived units are given special names and others are not. The coulomb is a derived unit with a special name, but there is no special name given to the unit of speed.

Although two basic systems of units, the **metric system** and the **British system,** are in use today, nearly every country in the world is either using the metric system or in the process of converting to it. This includes the United States and all the other countries which now use or previously used the British system. In the United States the conversion is presently voluntary and is coordinated by a government committee. Most major United States industrial firms will have completed the conversion for their products by the mid-1980s. In most of the other countries now converting, timetables for completing various phases of the conversion have been set up, and some phases are now completed.

For the present, both systems are of importance, although the metric system will eventually be used almost universally. For that reason, some of the exercises and examples in this book have metric units and others have British units.

Although more than one system has been developed in which metric units are used, the **International System of Units (SI)** is now accepted. This system, established in 1960, uses some different definitions for base units than the previously developed metric units. The measurement of the base units is more accessible, however, and the differences are very slight. Therefore, when we refer to the metric system, we are using SI units.

In the British system, the base unit of length is the **foot** and that of force is the **pound.** In the metric system, the base unit of length is the **meter** (spelled *metre* outside the United States) and that of mass is the **kilogram.** Here we see a difference in definition which causes some difficulty when units are converted from one system to the other. That is, a base unit in the British system is a unit of force, and a base unit in the metric system is a unit of mass.

The distinction between mass and force is very significant in physics. Along with the difference in definitions of base units of the systems, it causes difficulty when discussing the weight of an object. **Weight,** *which is the force with which an object is attracted to the earth, is different from mass, which is a measure of the amount of material in an object.* Although they are different quantities, mass and weight are closely related. Near the surface of the earth the weight of an object is directly proportional to its mass. However, at great distances from the earth the weight of an object will be essentially zero whereas its mass does not change.

Since force and mass are different, it is not strictly correct to convert pounds to kilograms. However, since the pound is the base unit in the British system, and the kilogram is the base unit in the metric system, *at the earth's surface 1 kg corresponds to 2.21 lb, in the sense that the gravitational force on a mass of 1 kg is 2.21 lb.* Therefore, when designating units for weight, we use pounds in the British system.

In the metric system, kilograms are commonly used for weight. This usage is considered acceptable even though it is mass that is actually being denoted. The metric unit of force, the *newton*, is also used for weight. Thus, it is preferable to specify *the mass of an object in kilograms and the force of gravity on an object in newtons.* Thereby we avoid using the word "weight" unless its meaning is completely clear.

As for the other base units, both systems use the **second** as the base unit for time. In the SI system the **ampere** is defined as the base unit of electric current, and this unit can also be used in the British system. As for temperature, **degrees Fahrenheit** are used with the British system and **degrees Celsius** (formerly centigrade) are used with the metric system. (Actually, the **kelvin** is defined as the base unit of thermodynamic temperature, although for temperature intervals one kelvin equals one degree Celsius.) In the SI system, the base unit for the amount of a substance is the **mole** and the base unit of luminous intensity is the **candela.**

The base units in either system are not always of a convenient size for many types of measurements. Therefore, other units of more convenient size are used in each system. When designating a unit of measurement, it is also convenient to use a **symbol** rather than the complete name of the unit.

Example C Long distances are not conveniently measured in feet, and therefore we generally use miles. For short distances, inches are more convenient and more commonly used.

In denoting feet, the commonly used symbol is ft. Inches are denoted by in. and miles by mi.

It is the relationship among units within each system which makes the metric system more convenient to use. *Units of different magnitudes differ by a power of 10, and therefore only a shift of a decimal point is necessary when changing units.* No specific arrangement exists in the British system. Also, the metric system uses a specific set of prefixes to denote the multiple being used. Table B-1 shows some of the common prefixes with their symbols and meanings.

TABLE B-1. METRIC PREFIXES

Prefix	Symbol	Factor by Which Unit Is Multiplied	Meaning
mega	M	$1{,}000{,}000 = 10^6$	million
kilo	k	$1000 = 10^3$	thousand
centi	c	$0.01 = \frac{1}{10^2}$	hundredth
milli	m	$0.001 = \frac{1}{10^3}$	thousandth
micro	μ	$0.000001 = \frac{1}{10^6}$	millionth

Example D The symbol for the meter is m. Therefore, since a kilometer equals one thousand meters, we show this as $1\text{ km} = 10^3\text{ m}$ or $1\text{ km} = 1000\text{ m}$. In the same way, we have the following relationships, with meanings shown on the left and proper symbols on the right:

One megameter equals one million meters: $1\text{ Mm} = 10^6\text{ m}$

One centimeter equals one one-hundredth meter: $1\text{ cm} = 0.01\text{ m}$

One millimeter equals one one-thousandth meter: $1\text{ mm} = 0.001\text{ m}$

Table B-2 lists some of the units used most commonly in the metric system, British system, or both. Also shown are the symbols and some of the basic relationships.

TABLE B-2. METRIC AND BRITISH UNITS

	Metric			British		
Quantity	Unit	Symbol		Unit	Symbol	
Length	meter	m		foot	ft	
	centimeter	cm	1 cm = 0.01 m	yard	yd	1 yd = 3 ft
	millimeter	mm	1 mm = 0.001 m	inch	in.	1 in. = $\frac{1}{12}$ ft
	kilometer	km	1 km = 1000 m	mile	mi	1 mi = 5280 ft
Mass, force	kilogram	kg		pound	lb	
	gram	g	1 g = 0.001 kg	ounce	oz	1 oz = $\frac{1}{16}$ lb
	milligram	mg	1 mg = 0.001 g	short ton	t	1 t = 2000 lb
Capacity	liter	L		quart	qt	
	milliliter	mL	1 mL = 0.001 L	pint	pt	1 pt = $\frac{1}{2}$ qt
	kiloliter	kL	1 kL = 1000 L	gallon	gal	1 gal = 4 qt
Temperature	degrees Celsius	°C		degrees Fahrenheit	°F	
Time (both systems)	second	s				
	minute	min	1 min = 60 s			
	hour	h	1 h = 3600 s			
Electrical units (both systems)	ampere	A	current			
	volt	V	potential (voltage)			
	ohm	Ω	resistance			
	coulomb	C	charge			
	watt	W	power			

Note: The SI symbol for liter is l, but since this symbol may be confused with the numeral 1, the symbol L is recognized for use in the U.S. and Canada. Also, ℓ is recognized in several countries.

The metric prefixes are used with the various units of the metric system, as in Example D. Example E further illustrates the use of units, prefixes, and symbols.

Example E

Unit	**Symbol**	**Meaning**
centiliter	cL	1 cL = 0.01 L
millisecond	ms	1 ms = 0.001 s
microvolt	μV	1 μV = 0.000001 V
megacoulomb	MC	1 MC = 1,000,000 C

As mentioned earlier, the derived units for other quantities are expressible in the base units. For example, the units of area, volume, and speed are expressed in terms of the base units. The units for energy and pressure have special names, but they may be expressed in terms of the base units.

Example F We find the area of a rectangle by multiplying one length by another. Considering the meaning of exponents, we can think of area as being measured in units of length $\times$ length $=$ $(\text{length})^2$. Therefore, area is measured in square feet (ft^2), square miles (mi^2), square meters (m^2), and so on.

To find the volume of a rectangular solid, we multiply the length by the width by the depth. This means that volume is measured in units of length $\times$ length $\times$ length $=$ $(\text{length})^3$. Thus, we measure volume in cubic feet (ft^3), cubic centimeters (cm^3), and so on.

Density, which is weight (or mass) per unit volume, is measured in lb/ft^3 or kg/m^3, for example.

If we have the product of two different units, we show that the units are to be multiplied by placing a raised dot between the symbols. To show the unit kilogram meter (simply a space between unit names indicates product), we use $kg \cdot m$.

When designating units of area or volume, where square or cubic units are used, we shall use exponents in the designation. As in Example F, we use m^2 rather than sq m and $in.^3$ rather than cu in.

If a unit is used in the SI system or in both systems, we use the SI symbol for the unit. Standard British symbols are used for the units in that system only. However, in writing numbers we use the notation which is still prevalent. For long numbers, we use commas to separate groupings of three. (In the SI system spaces, not commas, are used.) For example, we write twenty-five thousand as 25,000 (not as 25 000, which is the SI style). It is expected, however, that the SI style will eventually become universal.

When we are working with denominate numbers, it is sometimes necessary to change from one set of units to another. *A change within a system is called a* **reduction.** *A change from one system to another is called a* **conversion.**

Example G To find the cost of floor covering for a room, it is necessary to know the area of the room, and this is generally given in square feet. Floor-covering cost is often given in cost per square yard. Therefore, it is necessary to reduce square feet to square yards.

Distances in most parts of Europe are given in kilometers. To determine equivalent distances in miles, one has to convert kilometers to miles.

For purposes of changing units, Table B-3 gives some basic reduction and conversion factors. Other reduction factors are given in Table B-2.

TABLE B-3. REDUCTION AND CONVERSION FACTORS

Reduction Factors	
144 in.^{2} = 1 ft^2	100 mm^2 = 1 cm^2
9 ft^2 = 1 yd^2	10,000 cm^2 = 1 m^2
1728 in.^{3} = 1 ft^3	1000 mm^3 = 1 cm^3
27 ft^3 = 1 yd^3	1000 cm^3 = 1 L
Conversion factors	
1 in. = 2.54 cm	1 ft^3 = 28.32 L
1 m = 39.37 in.	1 pt = 473.2 cm^3
1 mi = 1.609 km	1 L = 1.057 qt
1 lb = 453.6 g	
1 kg = 2.205 lb	

To change a given number of one set of units into another set of units, we *perform multiplications and divisions with the units themselves*. These computations are essentially the same as the ones we do when working with fractions.

Example H If we had a number representing feet per second to be multiplied by another number representing seconds per minute, as far as the units are concerned we have

$$\frac{\text{ft}}{\text{s}} \times \frac{\text{s}}{\text{min}} = \frac{\text{ft} \times \not{\text{s}}}{\not{\text{s}} \times \text{min}} = \frac{\text{ft}}{\text{min}}$$

This means that the final result would be in feet per minute.

In changing a number from one set of units to another set of units, we use reduction and conversion factors from Tables B-2 and B-3 and apply the principle illustrated in Example H for operating with the units themselves. The convenient way to use the values in the tables is in the form of fractions. Since the given values are equal to each other, their quotient is 1. For Example, since 1 in. = 2.54 cm,

$$\frac{1\text{ in.}}{2.54\text{ cm}} = 1 \quad \text{or} \quad \frac{2.54\text{ cm}}{1\text{ in.}} = 1$$

since each represents the division of a certain length by itself. Multiplying a given quantity by 1 does not change its value. The following examples illustrate reduction and conversion of units.

Example I Change 30 mi/h to feet per second.

$$30 \text{ mi/h} = \left(30 \frac{\cancel{\text{mi}}}{\cancel{\text{h}}}\right)\left(\frac{5280 \text{ ft}}{1 \cancel{\text{mi}}}\right)\left(\frac{1 \cancel{\text{h}}}{60 \cancel{\text{min}}}\right)\left(\frac{1 \cancel{\text{min}}}{60 \text{ s}}\right)$$

$$= \frac{(30)(5280) \text{ ft}}{(60)(60) \text{ s}} = 44 \text{ ft/s}$$

Note that the only units remaining are those that were required.

Example J Reduce 575 g/cm^3 to kilograms per cubic meter.

$$575 \frac{\text{g}}{\text{cm}^3} = \left(575 \frac{\text{g}}{\text{cm}^3}\right)\left(\frac{100 \text{ cm}}{1 \text{ m}}\right)^3\left(\frac{1 \text{ kg}}{1000 \text{ g}}\right)$$

$$= \left(575 \frac{\cancel{\text{g}}}{\cancel{\text{cm}^3}}\right)\left(\frac{\overset{1000}{\cancel{1000000}} \cancel{\text{cm}^3}}{1 \text{ m}^3}\right)\left(\frac{1 \text{ kg}}{\cancel{1000} \cancel{\text{g}}}\right)$$

$$= 575 \times 1000 \frac{\text{kg}}{\text{m}^3} = 575{,}000 \frac{\text{kg}}{\text{m}^3}$$

Example K Convert 62.8 lb/in.^{2} to kilograms per square meter.

$$62.8 \text{ lb/in.}^2 = \left(62.8 \frac{\text{lb}}{\text{in.}^2}\right)\left(\frac{1 \text{ kg}}{2.205 \text{ lb}}\right)\left(\frac{1 \text{ in.}}{2.54 \text{ cm}}\right)^2\left(\frac{100 \text{ cm}}{1 \text{ m}}\right)^2$$

$$= \left(62.8 \frac{\cancel{\text{lb}}}{\cancel{\text{in.}^2}}\right)\left(\frac{1 \text{ kg}}{2.205 \cancel{\text{lb}}}\right)\left(\frac{1 \cancel{\text{in.}^2}}{2.54^2 \cancel{\text{cm}^2}}\right)\left(\frac{10000 \cancel{\text{cm}^2}}{1 \text{ m}^2}\right)$$

$$= \frac{(62.8)(10000)}{(2.205)(2.54^2)} \frac{\text{kg}}{\text{m}^2} = 44{,}150 \text{ kg/m}^2$$

Exercises for Appendix B

In the following exercises, use only the values given in Tables B-2 and B-3. Where applicable, round off to the accuracy noted.

1. How many yards are there in 1 mi?
2. How many ounces are there in 1 t?
3. How many centimeters are there in 1 km?
4. How many milligrams are there in 1 kg?
5. Show that 1 ft^2 = 144 in.^{2}
6. Show that 1 cm^3 = 1000 mm^3.
7. Convert 5 in. to centimeters.
8. Convert 6 kg to pounds.
9. Reduce 8 gal to pints.
10. Reduce 20 kL to milliliters.

11. Convert 10 qt to liters (round off to hundredths).
12. Convert 18 cm to inches (round off to tenths).
13. Reduce 39.6 cm^2 to square meters.
14. Reduce 0.087 L to cubic centimeters.
15. Convert 829 $in.^3$ to liters (round off to hundredths).
16. Convert 0.068 kL to cubic feet (round off to tenths).
17. An Atlas rocket weighed 260,000 lb at takeoff. How many tons is this?
18. An airplane is flying at 37,000 ft. What is its altitude in miles? (Round off to hundredths.)
19. The speedometer of a car is calibrated to kilometers per hour. If the speedometer of such a car reads 60, how fast in miles per hour is the car traveling? (Round off to units.)
20. The acceleration due to gravity is about 980 cm/s^2. Convert this to feet per second squared (round off to tenths).
21. The speed of sound is about 1130 ft/s. Change this speed to miles per hour (round off to units).
22. The density of water is about 62.4 lb/ft^3. Convert this to kilograms per cubic meter (round off to units).
23. The average density of the earth is about 5.52 g/cm^3. Express this in kilograms per cubic meter (round off to units).
24. The moon travels about 1,500,000 mi in about 28 d in one rotation about the earth. Express its velocity in feet per second (round off to tens). (d is the symbol for day.)
25. One horsepower is defined as 550 ft·lb/s. Express this in kilogram centimeters per second (round off to tens).
26. A unit used to measure the flow of water is cubic foot per minute. Convert 1 ft^3/min to liters per second (round off to hundredths).
27. The metric unit for force is the *newton* (N), where 1 N = 0.225 lb. Convert 250 ft·lb into newton meters (round off to units).
28. The metric unit for pressure is the *pascal* (Pa), where 1 Pa = 1 N/m^2. Convert 14.7 $lb/in.^2$ (atmospheric pressure) to pascals (round off to thousands). (See Exercise 27.)

Appendix C
TABLES

TABLE 1. SQUARES AND SQUARE ROOTS

N	N^2	$\sqrt{N}$	$\sqrt{10N}$	N	N^2	$\sqrt{N}$	$\sqrt{10N}$
1.0	1.00	1.000	3.162	5.5	30.25	2.345	7.416
1.1	1.21	1.049	3.317	5.6	31.36	2.366	7.483
1.2	1.44	1.095	3.464	5.7	32.49	2.387	7.550
1.3	1.69	1.140	3.606	5.8	33.64	2.408	7.616
1.4	1.96	1.183	3.742	5.9	34.81	2.429	7.681
1.5	2.25	1.225	3.873	6.0	36.00	2.449	7.746
1.6	2.56	1.265	4.000	6.1	37.21	2.470	7.810
1.7	2.89	1.304	4.123	6.2	38.44	2.490	7.874
1.8	3.24	1.342	4.243	6.3	39.69	2.510	7.937
1.9	3.61	1.378	4.359	6.4	40.96	2.530	8.000
2.0	4.00	1.414	4.472	6.5	42.25	2.550	8.062
2.1	4.41	1.449	4.583	6.6	43.56	2.569	8.124
2.2	4.84	1.483	4.690	6.7	44.89	2.588	8.185
2.3	5.29	1.517	4.796	6.8	46.24	2.608	8.246
2.4	5.76	1.549	4.899	6.9	47.61	2.627	8.307
2.5	6.25	1.581	5.000	7.0	49.00	2.646	8.367
2.6	6.76	1.612	5.099	7.1	50.41	2.665	8.426
2.7	7.29	1.643	5.196	7.2	51.84	2.683	8.485
2.8	7.84	1.673	5.292	7.3	53.29	2.702	8.544
2.9	8.41	1.703	5.385	7.4	54.76	2.720	8.602
3.0	9.00	1.732	5.477	7.5	56.25	2.739	8.660
3.1	9.61	1.761	5.568	7.6	57.76	2.757	8.718
3.2	10.24	1.789	5.657	7.7	59.29	2.775	8.775
3.3	10.89	1.817	5.745	7.8	60.84	2.793	8.832
3.4	11.56	1.844	5.831	7.9	62.41	2.811	8.888
3.5	12.25	1.871	5.916	8.0	64.00	2.828	8.944
3.6	12.96	1.897	6.000	8.1	65.61	2.846	9.000
3.7	13.69	1.924	6.083	8.2	67.24	2.864	9.055
3.8	14.44	1.949	6.164	8.3	68.89	2.881	9.110
3.9	15.21	1.975	6.245	8.4	70.56	2.898	9.165
4.0	16.00	2.000	6.325	8.5	72.25	2.915	9.220
4.1	16.81	2.025	6.403	8.6	73.96	2.933	9.274
4.2	17.64	2.049	6.481	8.7	75.69	2.950	9.327
4.3	18.49	2.074	6.557	8.8	77.44	2.966	9.381
4.4	19.36	2.098	6.633	8.9	79.21	2.983	9.434
4.5	20.25	2.121	6.708	9.0	81.00	3.000	9.487
4.6	21.16	2.145	6.782	9.1	82.81	3.017	9.539
4.7	22.09	2.168	6.856	9.2	84.64	3.033	9.592
4.8	23.04	2.191	6.928	9.3	86.49	3.050	9.644
4.9	24.01	2.214	7.000	9.4	88.36	3.066	9.695
5.0	25.00	2.236	7.071	9.5	90.25	3.082	9.747
5.1	26.01	2.258	7.141	9.6	92.16	3.098	9.798
5.2	27.04	2.280	7.211	9.7	94.09	3.114	9.849
5.3	28.09	2.302	7.280	9.8	96.04	3.131	9.899
5.4	29.16	2.324	7.348	9.9	98.01	3.146	9.950

TABLE 2. FOUR-PLACE LOGARITHMS OF NUMBERS

N	0	1	2	3	4	5	6	7	8	9
10	0000	0043	0086	0128	0170	0212	0253	0294	0334	0374
11	0414	0453	0492	0531	0569	0607	0645	0682	0719	0755
12	0792	0828	0864	0899	0934	0969	1004	1038	1072	1106
13	1139	1173	1206	1239	1271	1303	1335	1367	1399	1430
14	1461	1492	1523	1553	1584	1614	1644	1673	1703	1732
15	1761	1790	1818	1847	1875	1903	1931	1959	1987	2014
16	2041	2068	2095	2122	2148	2175	2201	2227	2253	2279
17	2304	2330	2355	2380	2405	2430	2455	2480	2504	2529
18	2553	2577	2601	2625	2648	2672	2695	2718	2742	2765
19	2788	2810	2833	2856	2878	2900	2923	2945	2967	2989
20	3010	3032	3054	3075	3096	3118	3139	3160	3181	3201
21	3222	3243	3263	3284	3304	3324	3345	3365	3385	3404
22	3424	3444	3464	3483	3502	3522	3541	3560	3579	3598
23	3617	3636	3655	3674	3692	3711	3729	3747	3766	3784
24	3802	3820	3838	3856	3874	3892	3909	3927	3945	3962
25	3979	3997	4014	4031	4048	4065	4082	4099	4116	4133
26	4150	4166	4183	4200	4216	4232	4249	4265	4281	4298
27	4314	4330	4346	4362	4378	4393	4409	4425	4440	4456
28	4472	4487	4502	4518	4533	4548	4564	4579	4594	4609
29	4624	4639	4654	4669	4683	4698	4713	4728	4742	4757
30	4771	4786	4800	4814	4829	4843	4857	4871	4886	4900
31	4914	4928	4942	4955	4969	4983	4997	5011	5024	5038
32	5051	5065	5079	5092	5105	5119	5132	5145	5159	5172
33	5185	5198	5211	5224	5237	5250	5263	5276	5289	5302
34	5315	5328	5340	5353	5366	5378	5391	5403	5416	5428
35	5441	5453	5465	5478	5490	5502	5514	5527	5539	5551
36	5563	5575	5587	5599	5611	5623	5635	5647	5658	5670
37	5682	5694	5705	5717	5729	5740	5752	5763	5775	5786
38	5798	5809	5821	5832	5843	5855	5866	5877	5888	5899
39	5911	5922	5933	5944	5955	5966	5977	5988	5999	6010
40	6021	6031	6042	6053	6064	6075	6085	6096	6107	6117
41	6128	6138	6149	6160	6170	6180	6191	6201	6212	6222
42	6232	6243	6253	6263	6274	6284	6294	6304	6314	6325
43	6335	6345	6355	6365	6375	6385	6395	6405	6414	6425
44	6435	6444	6454	6464	6474	6484	6493	6503	6513	6522
45	6532	6542	6551	6561	6571	6580	6590	6599	6609	6618
46	6628	6637	6646	6656	6665	6675	6684	6693	6702	6712
47	6721	6730	6739	6749	6758	6767	6776	6785	6794	6803
48	6812	6821	6830	6839	6848	6857	6866	6875	6884	6893
49	6902	6911	6920	6928	6937	6946	6955	6964	6972	6981
50	6990	6998	7007	7016	7024	7033	7042	7050	7059	7067
51	7076	7084	7093	7101	7110	7118	7126	7135	7143	7152
52	7160	7168	7177	7185	7193	7202	7210	7218	7226	7235
53	7243	7251	7259	7267	7275	7284	7292	7300	7308	7316
54	7324	7332	7340	7348	7356	7364	7372	7380	7388	7396

TABLE 2 (*Continued*)

N	0	1	2	3	4	5	6	7	8	9
55	7404	7412	7419	7427	7435	7443	7451	7459	7466	7474
56	7482	7490	7497	7505	7513	7520	7528	7536	7543	7551
57	7559	7566	7574	7582	7589	7597	7604	7612	7619	7627
58	7634	7642	7649	7657	7664	7672	7679	7686	7694	7701
59	7709	7716	7723	7731	7738	7745	7752	7760	7767	7774
60	7782	7789	7796	7803	7810	7818	7825	7832	7839	7846
61	7853	7860	7868	7875	7882	7889	7896	7903	7910	7917
62	7924	7931	7938	7945	7952	7959	7966	7973	7980	7987
63	7993	8000	8007	8014	8021	8028	8035	8041	8048	8055
64	8062	8069	8075	8082	8089	8096	8102	8109	8116	8122
65	8129	8136	8142	8149	8156	8162	8169	8176	8182	8189
66	8195	8202	8209	8215	8222	8228	8235	8241	8248	8254
67	8261	8267	8274	8280	8287	8293	8299	8306	8312	8319
68	8325	8331	8338	8344	8351	8357	8363	8370	8376	8382
69	8388	8395	8401	8407	8414	8420	8426	8432	8439	8445
70	8451	8457	8463	8470	8476	8482	8488	8494	8500	8506
71	8513	8519	8525	8531	8537	8543	8549	8555	8561	8567
72	8573	8579	8585	8591	8597	8603	8609	8615	8621	8627
73	8633	8639	8645	8651	8657	8663	8669	8675	8681	8686
74	8692	8698	8704	8710	8716	8722	8727	8733	8739	8745
75	8751	8756	8762	8768	8774	8779	8785	8791	8797	8802
76	8808	8814	8820	8825	8831	8837	8842	8848	8854	8859
77	8865	8871	8876	8882	8887	8893	8899	8904	8910	8915
78	8921	8927	8932	8938	8943	8949	8954	8960	8965	8971
79	8976	8982	8987	8993	8998	9004	9009	9015	9020	9025
80	9031	9036	9042	9047	9053	9058	9063	9069	9074	9079
81	9085	9090	9096	9101	9106	9112	9117	9122	9128	9133
82	9138	9143	9149	9154	9159	9165	9170	9175	9180	9186
83	9191	9196	9201	9206	9212	9217	9222	9227	9232	9238
84	9243	9248	9253	9258	9263	9269	9274	9279	9284	9289
85	9294	9299	9304	9309	9315	9320	9325	9330	9335	9340
86	9345	9350	9355	9360	9365	9370	9375	9380	9385	9390
87	9395	9400	9405	9410	9415	9420	9425	9430	9435	9440
88	9445	9450	9455	9460	9465	9469	9474	9479	9484	9489
89	9494	9499	9504	9509	9513	9518	9523	9528	9533	9538
90	9542	9547	9552	9557	9562	9566	9571	9576	9581	9586
91	9590	9595	9600	9605	9609	9614	9619	9624	9628	9633
92	9638	9643	9647	9652	9657	9661	9666	9671	9675	9680
93	9685	9689	9694	9699	9703	9708	9713	9717	9722	9727
94	9731	9736	9741	9745	9750	9754	9759	9763	9768	9773
95	9777	9782	9786	9791	9795	9800	9805	9809	9814	9818
96	9823	9827	9832	9836	9841	9845	9850	9854	9859	9863
97	9868	9872	9877	9881	9886	9890	9894	9899	9903	9908
98	9912	9917	9921	9926	9930	9934	9939	9943	9948	9952
99	9956	9961	9965	9969	9974	9978	9983	9987	9991	9996

TABLE 3. FOUR-PLACE VALUES OF TRIGONOMETRIC FUNCTIONS

Angle θ in Degrees and Tenths

Degrees	Sin θ	Cos θ	Tan θ	Cot θ		Degrees	Sin θ	Cos θ	Tan θ	Cot θ	
0.0	0.0000	1.0000	0.0000	—	90.0	5.0	0.0872	0.9962	0.0875	11.43	85.0
.1	0.0017	1.0000	0.0017	573.0	.9	.1	0.0889	0.9960	0.0892	11.20	.9
.2	0.0035	1.0000	0.0035	286.4	.8	.2	0.0906	0.9959	0.0910	10.99	.8
.3	0.0052	1.0000	0.0052	191.0	.7	.3	0.0924	0.9957	0.0928	10.78	.7
.4	0.0070	1.0000	0.0070	143.2	.6	.4	0.0941	0.9956	0.0945	10.58	.6
.5	0.0087	1.0000	0.0087	114.6	.5	.5	0.0958	0.9954	0.0963	10.39	.5
.6	0.0105	0.9999	0.0105	95.49	.4	.6	0.0976	0.9952	0.0981	10.20	.4
.7	0.0122	0.9999	0.0122	81.85	.3	.7	0.0993	0.9951	0.0998	10.02	.3
.8	0.0140	0.9999	0.0140	71.62	.2	.8	0.1011	0.9949	0.1016	9.845	.2
.9	0.0157	0.9999	0.0157	63.66	.1	.9	0.1028	0.9947	0.1033	9.677	.1
1.0	0.0175	0.9998	0.0175	57.29	89.0	6.0	0.1045	0.9945	0.1051	9.514	84.0
.1	0.0192	0.9998	0.0192	52.08	.9	.1	0.1063	0.9943	0.1069	9.357	.9
.2	0.0209	0.9998	0.0209	47.74	.8	.2	0.1080	0.9942	0.1086	9.205	.8
.3	0.0227	0.9997	0.0227	44.07	.7	.3	0.1097	0.9940	0.1104	9.058	.7
.4	0.0244	0.9997	0.0244	40.92	.6	.4	0.1115	0.9938	0.1122	8.915	.6
.5	0.0262	0.9997	0.0262	38.19	.5	.5	0.1132	0.9936	0.1139	8.777	.5
.6	0.0279	0.9996	0.0279	35.80	.4	.6	0.1149	0.9934	0.1157	8.643	.4
.7	0.0297	0.9996	0.0297	33.69	.3	.7	0.1167	0.9932	0.1175	8.513	.3
.8	0.0314	0.9995	0.0314	31.82	.2	.8	0.1184	0.9930	0.1192	8.386	.2
.9	0.0332	0.9995	0.0332	30.14	.1	.9	0.1201	0.9928	0.1210	8.264	.1
2.0	0.0349	0.9994	0.0349	28.64	88.0	7.0	0.1219	0.9925	0.1228	8.144	83.0
.1	0.0366	0.9993	0.0367	27.27	.9	.1	0.1236	0.9923	0.1246	8.028	.9
.2	0.0384	0.9993	0.0384	26.03	.8	.2	0.1253	0.9921	0.1263	7.916	.8
.3	0.0401	0.9992	0.0402	24.90	.7	.3	0.1271	0.9919	0.1281	7.806	.7
.4	0.0419	0.9991	0.0419	23.86	.6	.4	0.1288	0.9917	0.1299	7.700	.6
.5	0.0436	0.9990	0.0437	22.90	.5	.5	0.1305	0.9914	0.1317	7.596	.5
.6	0.0454	0.9990	0.0454	22.02	.4	.6	0.1323	0.9912	0.1334	7.495	.4
.7	0.0471	0.9989	0.0472	21.20	.3	.7	0.1340	0.9910	0.1352	7.396	.3
.8	0.0488	0.9988	0.0489	20.45	.2	.8	0.1357	0.9907	0.1370	7.300	.2
.9	0.0506	0.9987	0.0507	19.74	.1	.9	0.1374	0.9905	0.1388	7.207	.1
3.0	0.0523	0.9986	0.0524	19.08	87.0	8.0	0.1392	0.9903	0.1405	7.115	82.0
.1	0.0541	0.9985	0.0542	18.46	.9	.1	0.1409	0.9900	0.1423	7.026	.9
.2	0.0558	0.9984	0.0559	17.89	.8	.2	0.1426	0.9898	0.1441	6.940	.8
.3	0.0576	0.9983	0.0577	17.34	.7	.3	0.1444	0.9895	0.1459	6.855	.7
.4	0.0593	0.9982	0.0594	16.83	.6	.4	0.1461	0.9893	0.1477	6.772	.6
.5	0.0610	0.9981	0.0612	16.35	.5	.5	0.1478	0.9890	0.1495	6.691	.5
.6	0.0628	0.9980	0.0629	15.89	.4	.6	0.1495	0.9888	0.1512	6.612	.4
.7	0.0645	0.9979	0.0647	15.46	.3	.7	0.1513	0.9885	0.1530	6.535	.3
.8	0.0663	0.9978	0.0664	15.06	.2	.8	0.1530	0.9882	0.1548	6.460	.2
.9	0.0680	0.9977	0.0682	14.67	.1	.9	0.1547	0.9880	0.1566	6.386	.1
4.0	0.0698	0.9976	0.0699	14.30	86.0	9.0	0.1564	0.9877	0.1584	6.314	81.0
.1	0.0715	0.9974	0.0717	13.95	.9	.1	0.1582	0.9874	0.1602	6.243	.9
.2	0.0732	0.9973	0.0734	13.62	.8	.2	0.1599	0.9871	0.1620	6.174	.8
.3	0.0750	0.9972	0.0752	13.30	.7	.3	0.1616	0.9869	0.1638	6.107	.7
.4	0.0767	0.9971	0.0769	13.00	.6	.4	0.1633	0.9866	0.1655	6.041	.6
.5	0.0785	0.9969	0.0787	12.71	.5	.5	0.1650	0.9863	0.1673	5.976	.5
.6	0.0802	0.9968	0.0805	12.43	.4	.6	0.1668	0.9860	0.1691	5.912	.4
.7	0.0819	0.9966	0.0822	12.16	.3	.7	0.1685	0.9857	0.1709	5.850	.3
.8	0.0837	0.9965	0.0840	11.91	.2	.8	0.1702	0.9854	0.1727	5.789	.2
.9	0.0854	0.9963	0.0857	11.66	.1	.9	0.1719	0.9851	0.1745	5.730	.1
5.0	0.0872	0.9962	0.0875	11.43	85.0	10.0	0.1736	0.9848	0.1763	5.671	80.0
	Cos θ	Sin θ	Cot θ	Tan θ	Degrees		Cos θ	Sin θ	Cot θ	Tan θ	Degrees

TABLE 3 (*Continued*)

Degrees	Sin θ	Cos θ	Tan θ	Cot θ	
10.0	0.1736	0.9848	0.1763	5.671	80.0
.1	0.1754	0.9845	0.1781	5.614	.9
.2	0.1771	0.9842	0.1799	5.558	.8
.3	0.1788	0.9839	0.1817	5.503	.7
.4	0.1805	0.9836	0.1835	5.449	.6
.5	0.1822	0.9833	0.1853	5.396	.5
.6	0.1840	0.9829	0.1871	5.343	.4
.7	0.1857	0.9826	0.1890	5.292	.3
.8	0.1874	0.9823	0.1908	5.242	.2
.9	0.1891	0.9820	0.1926	5.193	.1
11.0	0.1908	0.9816	0.1944	5.145	79.0
.1	0.1925	0.9813	0.1962	5.097	.9
.2	0.1942	0.9810	0.1980	5.050	.8
.3	0.1959	0.9806	0.1998	5.005	.7
.4	0.1977	0.9803	0.2016	4.959	.6
.5	0.1994	0.9799	0.2035	4.915	.5
.6	0.2011	0.9796	0.2053	4.872	.4
.7	0.2028	0.9792	0.2071	4.829	.3
.8	0.2045	0.9789	0.2089	4.787	.2
.9	0.2062	0.9785	0.2107	4.745	.1
12.0	0.2079	0.9781	0.2126	4.705	78.0
.1	0.2096	0.9778	0.2144	4.665	.9
.2	0.2113	0.9774	0.2162	4.625	.8
.3	0.2130	0.9770	0.2180	4.586	.7
.4	0.2147	0.9767	0.2199	4.548	.6
.5	0.2164	0.9763	0.2217	4.511	.5
.6	0.2181	0.9759	0.2235	4.474	.4
.7	0.2198	0.9755	0.2254	4.437	.3
.8	0.2215	0.9751	0.2272	4.402	.2
.9	0.2233	0.9748	0.2290	4.366	.1
13.0	0.2250	0.9744	0.2309	4.331	77.0
.1	0.2267	0.9740	0.2327	4.297	.9
.2	0.2284	0.9736	0.2345	4.264	.8
.3	0.2300	0.9732	0.2364	4.230	.7
.4	0.2317	0.9728	0.2382	4.198	.6
.5	0.2334	0.9724	0.2401	4.165	.5
.6	0.2351	0.9720	0.2419	4.134	.4
.7	0.2368	0.9715	0.2438	4.102	.3
.8	0.2385	0.9711	0.2456	4.071	.2
.9	0.2402	0.9707	0.2475	4.041	.1
14.0	0.2419	0.9703	0.2493	4.011	76.0
.1	0.2436	0.9699	0.2512	3.981	.9
.2	0.2453	0.9694	0.2530	3.952	.8
.3	0.2470	0.9690	0.2549	3.923	.7
.4	0.2487	0.9686	0.2568	3.895	.6
.5	0.2504	0.9681	0.2586	3.867	.5
.6	0.2521	0.9677	0.2605	3.839	.4
.7	0.2538	0.9673	0.2623	3.812	.3
.8	0.2554	0.9668	0.2642	3.785	.2
.9	0.2571	0.9664	0.2661	3.758	.1
15.0	0.2588	0.9659	0.2679	3.732	75.0
	Cos θ	Sin θ	Cot θ	Tan θ	Degrees

Degrees	Sin θ	Cos θ	Tan θ	Cot θ	
15.0	0.2588	0.9659	0.2679	3.732	75.0
.1	0.2605	0.9655	0.2698	3.706	.9
.2	0.2622	0.9650	0.2717	3.681	.8
.3	0.2639	0.9646	0.2736	3.655	.7
.4	0.2656	0.9641	0.2754	3.630	.6
.5	0.2672	0.9636	0.2773	3.606	.5
.6	0.2689	0.9632	0.2792	3.582	.4
.7	0.2706	0.9627	0.2811	3.558	.3
.8	0.2723	0.9622	0.2830	3.534	.2
.9	0.2740	0.9617	0.2849	3.511	.1
16.0	0.2756	0.9613	0.2867	3.487	74.0
.1	0.2773	0.9608	0.2886	3.465	.9
.2	0.2790	0.9603	0.2905	3.442	.8
.3	0.2807	0.9598	0.2924	3.420	.7
.4	0.2823	0.9593	0.2943	3.398	.6
.5	0.2840	0.9588	0.2962	3.376	.5
.6	0.2857	0.9583	0.2981	3.354	.4
.7	0.2874	0.9578	0.3000	3.333	.3
.8	0.2890	0.9573	0.3019	3.312	.2
.9	0.2907	0.9568	0.3038	3.291	.1
17.0	0.2924	0.9563	0.3057	3.271	73.0
.1	0.2940	0.9558	0.3076	3.251	.9
.2	0.2957	0.9553	0.3096	3.230	.8
.3	0.2974	0.9548	0.3115	3.211	.7
.4	0.2990	0.9542	0.3134	3.191	.6
.5	0.3007	0.9537	0.3153	3.172	.5
.6	0.3024	0.9532	0.3172	3.152	.4
.7	0.3040	0.9527	0.3191	3.133	.3
.8	0.3057	0.9521	0.3211	3.115	.2
.9	0.3074	0.9516	0.3230	3.096	.1
18.0	0.3090	0.9511	0.3249	3.078	72.0
.1	0.3107	0.9505	0.3269	3.060	.9
.2	0.3123	0.9500	0.3288	3.042	.8
.3	0.3140	0.9494	0.3307	3.024	.7
.4	0.3156	0.9489	0.3327	3.006	.6
.5	0.3173	0.9483	0.3346	2.989	.5
.6	0.3190	0.9478	0.3365	2.971	.4
.7	0.3206	0.9472	0.3385	2.954	.3
.8	0.3223	0.9466	0.3404	2.937	.2
.9	0.3239	0.9461	0.3424	2.921	.1
19.0	0.3256	0.9455	0.3443	2.904	71.0
.1	0.3272	0.9449	0.3463	2.888	.9
.2	0.3289	0.9444	0.3482	2.872	.8
.3	0.3305	0.9438	0.3502	2.856	.7
.4	0.3322	0.9432	0.3522	2.840	.6
.5	0.3338	0.9426	0.3541	2.824	.5
.6	0.3355	0.9421	0.3561	2.808	.4
.7	0.3371	0.9415	0.3581	2.793	.3
.8	0.3387	0.9409	0.3600	2.778	.2
.9	0.3404	0.9403	0.3620	2.762	.1
20.0	0.3420	0.9397	0.3640	2.747	70.0
	Cos θ	Sin θ	Cot θ	Tan θ	Degrees

TABLE 3 (*Continued*)

Degrees	Sin θ	Cos θ	Tan θ	Cot θ		Degrees	Sin θ	Cos θ	Tan θ	Cot θ	
20.0	0.3420	0.9397	0.3640	2.747	70.0	25.0	0.4226	0.9063	0.4663	2.145	65.0
.1	0.3437	0.9391	0.3659	2.733	.9	.1	0.4242	0.9056	0.4684	2.135	.9
.2	0.3453	0.9385	0.3679	2.718	.8	.2	0.4258	0.9048	0.4706	2.125	.8
.3	0.3469	0.9379	0.3699	2.703	.7	.3	0.4274	0.9041	0.4727	2.116	.7
.4	0.3486	0.9373	0.3719	2.689	.6	.4	0.4289	0.9033	0.4748	2.106	.6
.5	0.3502	0.9367	0.3739	2.675	.5	.5	0.4305	0.9026	0.4770	2.097	.5
.6	0.3518	0.9361	0.3759	2.660	.4	.6	0.4321	0.9018	0.4791	2.087	.4
.7	0.3535	0.9354	0.3779	2.646	.3	.7	0.4337	0.9011	0.4813	2.078	.3
.8	0.3551	0.9348	0.3799	2.633	.2	.8	0.4352	0.9003	0.4834	2.069	.2
.9	0.3567	0.9342	0.3819	2.619	.1	.9	0.4368	0.8996	0.4856	2.059	.1
21.0	0.3584	0.9336	0.3839	2.605	69.0	26.0	0.4384	0.8988	0.4877	2.050	64.0
.1	0.3600	0.9330	0.3859	2.592	.9	.1	0.4399	0.8980	0.4899	2.041	.9
.2	0.3616	0.9323	0.3879	2.578	.8	.2	0.4415	0.8973	0.4921	2.032	.8
.3	0.3633	0.9317	0.3899	2.565	.7	.3	0.4431	0.8965	0.4942	2.023	.7
.4	0.3649	0.9311	0.3919	2.552	.6	.4	0.4446	0.8957	0.4964	2.014	.6
.5	0.3665	0.9304	0.3939	2.539	.5	.5	0.4462	0.8949	0.4986	2.006	.5
.6	0.3681	0.9298	0.3959	2.526	.4	.6	0.4478	0.8942	0.5008	1.997	.4
.7	0.3697	0.9291	0.3979	2.513	.3	.7	0.4493	0.8934	0.5029	1.988	.3
.8	0.3714	0.9285	0.4000	2.500	.2	.8	0.4509	0.8926	0.5051	1.980	.2
.9	0.3730	0.9278	0.4020	2.488	.1	.9	0.4524	0.8918	0.5073	1.971	.1
22.0	0.3746	0.9272	0.4040	2.475	68.0	27.0	0.4540	0.8910	0.5095	1.963	63.0
.1	0.3762	0.9265	0.4061	2.463	.9	.1	0.4555	0.8902	0.5117	1.954	.9
.2	0.3778	0.9259	0.4081	2.450	.8	.2	0.4571	0.8894	0.5139	1.946	.8
.3	0.3795	0.9252	0.4101	2.438	.7	.3	0.4586	0.8886	0.5161	1.937	.7
.4	0.3811	0.9245	0.4122	2.426	.6	.4	0.4602	0.8878	0.5184	1.929	.6
.5	0.3827	0.9239	0.4142	2.414	.5	.5	0.4617	0.8870	0.5206	1.921	.5
.6	0.3843	0.9232	0.4163	2.402	.4	.6	0.4633	0.8862	0.5228	1.913	.4
.7	0.3859	0.9225	0.4183	2.391	.3	.7	0.4648	0.8854	0.5250	1.905	.3
.8	0.3875	0.9219	0.4204	2.379	.2	.8	0.4664	0.8846	0.5272	1.897	.2
.9	0.3891	0.9212	0.4224	2.367	.1	.9	0.4679	0.8838	0.5295	1.889	.1
23.0	0.3907	0.9205	0.4245	2.356	67.0	28.0	0.4695	0.8829	0.5317	1.881	62.0
.1	0.3923	0.9198	0.4265	2.344	.9	.1	0.4710	0.8821	0.5340	1.873	.9
.2	0.3939	0.9191	0.4286	2.333	.8	.2	0.4726	0.8813	0.5362	1.865	.8
.3	0.3955	0.9184	0.4307	2.322	.7	.3	0.4741	0.8805	0.5384	1.857	.7
.4	0.3971	0.9178	0.4327	2.311	.6	.4	0.4756	0.8796	0.5407	1.849	.6
.5	0.3987	0.9171	0.4348	2.300	.5	.5	0.4772	0.8788	0.5430	1.842	.5
.6	0.4003	0.9164	0.4369	2.289	.4	.6	0.4787	0.8780	0.5452	1.834	.4
.7	0.4019	0.9157	0.4390	2.278	.3	.7	0.4802	0.8771	0.5475	1.827	.3
.8	0.4035	0.9150	0.4411	2.267	.2	.8	0.4818	0.8763	0.5498	1.819	.2
.9	0.4051	0.9143	0.4431	2.257	.1	.9	0.4833	0.8755	0.5520	1.811	.1
24.0	0.4067	0.9135	0.4452	2.246	66.0	29.0	0.4848	0.8746	0.5543	1.804	61.0
.1	0.4083	0.9128	0.4473	2.236	.9	.1	0.4863	0.8738	0.5566	1.797	.9
.2	0.4099	0.9121	0.4494	2.225	.8	.2	0.4879	0.8729	0.5589	1.789	.8
.3	0.4115	0.9114	0.4515	2.215	.7	.3	0.4894	0.8721	0.5612	1.782	.7
.4	0.4131	0.9107	0.4536	2.204	.6	.4	0.4909	0.8712	0.5635	1.775	.6
.5	0.4147	0.9100	0.4557	2.194	.5	.5	0.4924	0.8704	0.5658	1.767	.5
.6	0.4163	0.9092	0.4578	2.184	.4	.6	0.4939	0.8695	0.5681	1.760	.4
.7	0.4179	0.9085	0.4599	2.174	.3	.7	0.4955	0.8686	0.5704	1.753	.3
.8	0.4195	0.9078	0.4621	2.164	.2	.8	0.4970	0.8678	0.5727	1.746	.2
.9	0.4210	0.9070	0.4642	2.154	.1	.9	0.4985	0.8669	0.5750	1.739	.1
25.0	0.4226	0.9063	0.4663	2.145	65.0	30.0	0.5000	0.8660	0.5774	1.732	60.0
	Cos θ	Sin θ	Cot θ	Tan θ	Degrees		Cos θ	Sin θ	Cot θ	Tan θ	Degrees

TABLE 3 (*Continued*)

Degrees	Sin θ	Cos θ	Tan θ	Cot θ		Degrees	Sin θ	Cos θ	Tan θ	Cot θ	
30.0	0.5000	0.8660	0.5774	1.732	60.0	35.0	0.5736	0.8192	0.7002	1.428	55.0
.1	0.5015	0.8652	0.5797	1.725	.9	.1	0.5750	0.8181	0.7028	1.423	.9
.2	0.5030	0.8643	0.5820	1.718	.8	.2	0.5764	0.8171	0.7054	1.418	.8
.3	0.5045	0.8634	0.5844	1.711	.7	.3	0.5779	0.8161	0.7080	1.412	.7
.4	0.5060	0.8625	0.5867	1.704	.6	.4	0.5793	0.8151	0.7107	1.407	.6
.5	0.5075	0.8616	0.5890	1.698	.5	.5	0.5807	0.8141	0.7133	1.402	.5
.6	0.5090	0.8607	0.5914	1.691	.4	.6	0.5821	0.8131	0.7159	1.397	.4
.7	0.5105	0.8599	0.5938	1.684	.3	.7	0.5835	0.8121	0.7186	1.392	.3
.8	0.5120	0.8590	0.5961	1.678	.2	.8	0.5850	0.8111	0.7212	1.387	.2
.9	0.5135	0.8581	0.5985	1.671	.1	.9	0.5864	0.8100	0.7239	1.381	.1
31.0	0.5150	0.8572	0.6009	1.664	59.0	36.0	0.5878	0.8090	0.7265	1.376	54.0
.1	0.5165	0.8563	0.6032	1.658	.9	.1	0.5892	0.8080	0.7292	1.371	.9
.2	0.5180	0.8554	0.6056	1.651	.8	.2	0.5906	0.8070	0.7319	1.366	.8
.3	0.5195	0.8545	0.6080	1.645	.7	.3	0.5920	0.8059	0.7346	1.361	.7
.4	0.5210	0.8536	0.6104	1.638	.6	.4	0.5934	0.8049	0.7373	1.356	.6
.5	0.5225	0.8526	0.6128	1.632	.5	.5	0.5948	0.8039	0.7400	1.351	.5
.6	0.5240	0.8517	0.6152	1.625	.4	.6	0.5962	0.8028	0.7427	1.347	.4
.7	0.5255	0.8508	0.6176	1.619	.3	.7	0.5976	0.8018	0.7454	1.342	.3
.8	0.5270	0.8499	0.6200	1.613	.2	.8	0.5990	0.8007	0.7481	1.337	.2
.9	0.5284	0.8490	0.6224	1.607	.1	.9	0.6004	0.7997	0.7508	1.332	.1
32.0	0.5299	0.8480	0.6249	1.600	58.0	37.0	0.6018	0.7986	0.7536	1.327	53.0
.1	0.5314	0.8471	0.6273	1.594	.9	.1	0.6032	0.7976	0.7563	1.322	.9
.2	0.5329	0.8462	0.6297	1.588	.8	.2	0.6046	0.7965	0.7590	1.317	.8
.3	0.5344	0.8453	0.6322	1.582	.7	.3	0.6060	0.7955	0.7618	1.313	.7
.4	0.5358	0.8443	0.6346	1.576	.6	.4	0.6074	0.7944	0.7646	1.308	.6
.5	0.5373	0.8434	0.6371	1.570	.5	.5	0.6088	0.7934	0.7673	1.303	.5
.6	0.5388	0.8425	0.6395	1.564	.4	.6	0.6101	0.7923	0.7701	1.299	.4
.7	0.5402	0.8415	0.6420	1.558	.3	.7	0.6115	0.7912	0.7729	1.294	.3
.8	0.5417	0.8406	0.6445	1.552	.2	.8	0.6129	0.7902	0.7757	1.289	.2
.9	0.5432	0.8396	0.6469	1.546	.1	.9	0.6143	0.7891	0.7785	1.285	.1
33.0	0.5446	0.8387	0.6494	1.540	57.0	38.0	0.6157	0.7880	0.7813	1.280	52.0
.1	0.5461	0.8377	0.6519	1.534	.9	.1	0.6170	0.7869	0.7841	1.275	.9
.2	0.5476	0.8368	0.6544	1.528	.8	.2	0.6184	0.7859	0.7869	1.271	.8
.3	0.5490	0.8358	0.6569	1.522	.7	.3	0.6198	0.7848	0.7898	1.266	.7
.4	0.5505	0.8348	0.6594	1.517	.6	.4	0.6211	0.7837	0.7926	1.262	.6
.5	0.5519	0.8339	0.6619	1.511	.5	.5	0.6225	0.7826	0.7954	1.257	.5
.6	0.5534	0.8329	0.6644	1.505	.4	.6	0.6239	0.7815	0.7983	1.253	.4
.7	0.5548	0.8320	0.6669	1.499	.3	.7	0.6252	0.7804	0.8012	1.248	.3
.8	0.5563	0.8310	0.6694	1.494	.2	.8	0.6266	0.7793	0.8040	1.244	.2
.9	0.5577	0.8300	0.6720	1.488	.1	.9	0.6280	0.7782	0.8069	1.239	.1
34.0	0.5592	0.8290	0.6745	1.483	56.0	39.0	0.6293	0.7771	0.8098	1.235	51.0
.1	0.5606	0.8281	0.6771	1.477	.9	.1	0.6307	0.7760	0.8127	1.230	.9
.2	0.5621	0.8271	0.6796	1.471	.8	.2	0.6320	0.7749	0.8156	1.226	.8
.3	0.5635	0.8261	0.6822	1.466	.7	.3	0.6334	0.7738	0.8185	1.222	.7
.4	0.5650	0.8251	0.6847	1.460	.6	.4	0.6347	0.7727	0.8214	1.217	.6
.5	0.5664	0.8241	0.6873	1.455	.5	.5	0.6361	0.7716	0.8243	1.213	.5
.6	0.5678	0.8231	0.6899	1.450	.4	.6	0.6374	0.7705	0.8273	1.209	.4
.7	0.5693	0.8221	0.6924	1.444	.3	.7	0.6388	0.7694	0.8302	1.205	.3
.8	0.5707	0.8211	0.6950	1.439	.2	.8	0.6401	0.7683	0.8332	1.200	.2
.9	0.5721	0.8202	0.6976	1.433	.1	.9	0.6414	0.7672	0.8361	1.196	.1
35.0	0.5736	0.8192	0.7002	1.428	55.0	40.0	0.6428	0.7660	0.8391	1.192	50.0
	Cos θ	Sin θ	Cot θ	Tan θ	Degrees		Cos θ	Sin θ	Cot θ	Tan θ	Degrees

TABLE 3 (*Continued*)

Degrees	Sin θ	Cos θ	Tan θ	Cot θ	
40.0	0.6428	0.7660	0.8391	1.192	50.0
.1	0.6441	0.7649	0.8421	1.188	.9
.2	0.6455	0.7638	0.8451	1.183	.8
.3	0.6468	0.7627	0.8481	1.179	.7
.4	0.6481	0.7615	0.8511	1.175	.6
.5	0.6494	0.7604	0.8541	1.171	.5
.6	0.6508	0.7593	0.8571	1.167	.4
.7	0.6521	0.7581	0.8601	1.163	.3
.8	0.6534	0.7570	0.8632	1.159	.2
.9	0.6547	0.7559	0.8662	1.154	.1
41.0	0.6561	0.7547	0.8693	1.150	49.0
.1	0.6574	0.7536	0.8724	1.146	.9
.2	0.6587	0.7524	0.8754	1.142	.8
.3	0.6600	0.7513	0.8785	1.138	.7
.4	0.6613	0.7501	0.8816	1.134	.6
.5	0.6626	0.7490	0.8847	1.130	.5
.6	0.6639	0.7478	0.8878	1.126	.4
.7	0.6652	0.7466	0.8910	1.122	.3
.8	0.6665	0.7455	0.8941	1.118	.2
.9	0.6678	0.7443	0.8972	1.115	.1
42.0	0.6691	0.7431	0.9004	1.111	48.0
.1	0.6704	0.7420	0.9036	1.107	.9
.2	0.6717	0.7408	0.9067	1.103	.8
.3	0.6730	0.7396	0.9099	1.099	.7
.4	0.6743	0.7385	0.9131	1.095	.6
.5	0.6756	0.7373	0.9163	1.091	.5
.6	0.6769	0.7361	0.9195	1.087	.4
.7	0.6782	0.7349	0.9228	1.084	.3
.8	0.6794	0.7337	0.9260	1.080	.2
.9	0.6807	0.7325	0.9293	1.076	.1
43.0	0.6820	0.7314	0.9325	1.072	47.0
.1	0.6833	0.7302	0.9358	1.069	.9
.2	0.6845	0.7290	0.9391	1.065	.8
.3	0.6858	0.7278	0.9424	1.061	.7
.4	0.6871	0.7266	0.9457	1.057	.6
.5	0.6884	0.7254	0.9490	1.054	.5
.6	0.6896	0.7242	0.9523	1.050	.4
.7	0.6909	0.7230	0.9556	1.046	.3
.8	0.6921	0.7218	0.9590	1.043	.2
.9	0.6934	0.7206	0.9623	1.039	.1
44.0	0.6947	0.7193	0.9657	1.036	46.0
.1	0.6959	0.7181	0.9691	1.032	.9
.2	0.6972	0.7169	0.9725	1.028	.8
.3	0.6984	0.7157	0.9759	1.025	.7
.4	0.6997	0.7145	0.9793	1.021	.6
.5	0.7009	0.7133	0.9827	1.018	.5
.6	0.7022	0.7120	0.9861	1.014	.4
.7	0.7034	0.7108	0.9896	1.011	.3
.8	0.7046	0.7096	0.9930	1.007	.2
.9	0.7059	0.7083	0.9965	1.003	.1
45.0	0.7071	0.7071	1.0000	1.000	45.0
	Cos θ	**Sin θ**	**Cot θ**	**Tan θ**	**Degrees**

TABLE 4. NATURAL LOGARITHMS OF NUMBERS

N	$\ln N$	N	$\ln N$	N	$\ln N$
0.0		4.5	1.5041	9.0	2.1972
0.1	7.6974–10	4.6	1.5261	9.1	2.2083
0.2	8.3906–10	4.7	1.5476	9.2	2.2192
0.3	8.7960–10	4.8	1.5686	9.3	2.2300
0.4	9.0837–10	4.9	1.5892	9.4	2.2407
0.5	9.3069–10	5.0	1.6094	9.5	2.2513
0.6	9.4892–10	5.1	1.6292	9.6	2.2618
0.7	9.6433–10	5.2	1.6487	9.7	2.2721
0.8	9.7769–10	5.3	1.6677	9.8	2.2824
0.9	9.8946–10	5.4	1.6864	9.9	2.2925
1.0	0.0000	5.5	1.7047	10	2.3026
1.1	0.0953	5.6	1.7228	11	2.3979
1.2	0.1823	5.7	1.7405	12	2.4849
1.3	0.2624	5.8	1.7579	13	2.5649
1.4	0.3365	5.9	1.7750	14	2.6391
1.5	0.4055	6.0	1.7918	15	2.7081
1.6	0.4700	6.1	1.8083	16	2.7726
1.7	0.5306	6.2	1.8245	17	2.8332
1.8	0.5878	6.3	1.8405	18	2.8904
1.9	0.6419	6.4	1.8563	19	2.9444
2.0	0.6931	6.5	1.8718	20	2.9957
2.1	0.7419	6.6	1.8871	25	3.2189
2.2	0.7885	6.7	1.9021	30	3.4012
2.3	0.8329	6.8	1.9169	35	3.5553
2.4	0.8755	6.9	1.9315	40	3.6889
2.5	0.9163	7.0	1.9459	45	3.8067
2.6	0.9555	7.1	1.9601	50	3.9120
2.7	0.9933	7.2	1.9741	55	4.0073
2.8	1.0296	7.3	1.9879	60	4.0943
2.9	1.0647	7.4	2.0015	65	4.1744
3.0	1.0986	7.5	2.0149	70	4.2485
3.1	1.1314	7.6	2.0281	75	4.3175
3.2	1.1632	7.7	2.0412	80	4.3820
3.3	1.1939	7.8	2.0541	85	4.4427
3.4	1.2238	7.9	2.0669	90	4.4998
3.5	1.2528	8.0	2.0794	95	4.5539
3.6	1.2809	8.1	2.0919	100	4.6052
3.7	1.3083	8.2	2.1041		
3.8	1.3350	8.3	2.1163		
3.9	1.3610	8.4	2.1282		
4.0	1.3863	8.5	2.1401		
4.1	1.4110	8.6	2.1518		
4.2	1.4351	8.7	2.1633		
4.3	1.4586	8.8	2.1748		
4.4	1.4816	8.9	2.1861		

ANSWERS TO ODD-NUMBER EXERCISES

Exercises 1-2, Page 5

1. Add the number 7 to the number 2

3. Add the number $+9$ to the number $+5$

5. Add the number -3 to the number -4

7. Subtract the number $+2$ from the number -5

9. -7; $+7$

11. $-\frac{1}{2}$; $+\frac{1}{3}$

13 and **15.**

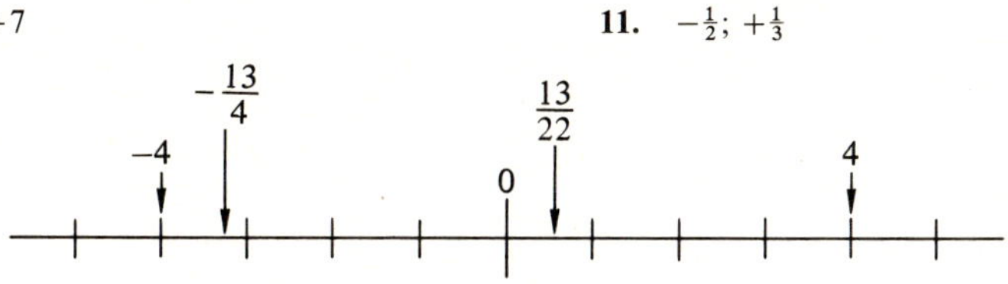

17. 4 **19.** -2 **21.** $6 > 2$ **23.** $0 < 4$ **25.** $-3 > -7$ **27.** $\frac{2}{3} > \frac{1}{2}$

29. $0.8 < 1.2$ **31.** $|-9| = |9|$ **33.** 6; 6 **35.** $\frac{6}{7}$; $\frac{8}{5}$ **37.** -4 **39.** -35

41. -35 **43.** -10

Exercises 1-3, Page 10

1. Comm. **3.** Assoc. **5.** $+11$ **7.** -15 **9.** $+3$ **11.** -8 **13.** $+7$

15. -3 **17.** -10 **19.** $+16$ **21.** $+1$ **23.** -6 **25.** -8 **27.** $+1$

29. $+6$ **31.** -17 **33.** -2 **35.** -16 **37.** (a) -5°F (b) -15°F

39. -55 m **41.** 120 ft/s

43. $(-2) - (-7) = +5$; $(-7) - (-2) = -5$; $(-2) - (-7) \neq (-7) - (-2)$

Exercises 1-4, Page 15

1. Assoc. **3.** Comm. **5.** +80 **7.** −63 **9.** −84 **11.** +30 **13.** −56
15. +240 **17.** +360 **19.** 0 **21.** −168 **23.** +36; +36 **25.** +49; −49
27. −1024 **29.** +1; −1 **31.** −128; −128 **33.** −50 mm **35.** +7000 ft

Exercises 1-5, Page 20

1. +4 **3.** −9 **5.** −11 **7.** +15 **9.** 0 **11.** Undefined **13.** −16
15. −2 **17.** +3 **19.** −6 **21.** −16 **23.** −12 **25.** −5 **27.** +9
29. −15 **31.** −8 **33.** −21 **35.** +32 **37.** −250 ft/min **39.** −$2500
41. $\frac{+8}{-2} = -4$; $\frac{-2}{+8} = -\frac{1}{4}$; $\frac{+8}{-2} \neq \frac{-2}{+8}$

Exercises 1-6, Page 24

1. {1, 2, 3, 4, 5, 6, 7}
3. {−4, −3, −2}
5. {11, 12, 13, . . .}
7. $\{-2, -\frac{1}{2}, \frac{11}{3}\}$
9. No; Yes
11. {−2, −1, 0, 1, 3, 5, 6, 7, 9, 10}
13. {3, 6}
15. {0, 1, 5, 6}
17. $E \subseteq P$
19. $P \cup O$
21. R
23. {0, 1, 2}, {0}, {1}, {2}, {0, 1} {0, 2}, {1, 2}, { }

Exercises 1-7 Page 28

1. $3(10)^2 + 5(10) + 2$
3. $8(10)^3 + 6(10)^2 + 1(10) + 7$
5. (a) 1110_{two} (b) 24_{five}
7. (a) 1000101_{two} (b) 234_{five}
9. 14 **11.** 21 **13.** 17 **15.** 284 **17.** 1100_{two} **19.** 242_{five}
21. 1, 2, 3, 10, 11, 12, 13, 20, 21, 22, 23, 30, 31, 32, 33, 100, 101, 102, 103, 110.
23. 1032_{four}
25. 1, 2, 3, 4, 5, 6, 7, 8, 9, t, e, 10, 11, 12, 13, 14, 15, 16, 17, 18, 19, 1t, 1e, 20, 21.
27. $20t_{twelve}$ **29.** 122_{five} **31.** 2313_{four} **33.** 134_{ten} **35.** 356_{eight} **37.** 2312_{four}
39. 63_{eight}

Exercises 1-8, Page 32

1. Assoc. of add.
3. Comm. of mult.
5. Comm. of add.
7. Assoc. of mult.
9. $-3 < +1$
11. $|-2| > -3$
13.

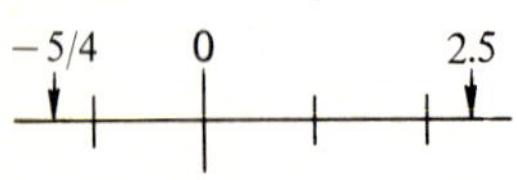

15. $\frac{2}{3}$; $\frac{1}{4}$

17. −2 **19.** −13 **21.** −63 **23.** +9 **25.** +10 **27.** +19 **29.** +96
31. −60 **33.** +10 **35.** −6 **37.** +256; −256 **39.** −125; −125 **41.** +5
43. +12 **45.** 0 **47.** −5 **49.** +10 **51.** −7 **53.** $\{0, \frac{11}{3}\}$ **55.** No, Yes
57. 93 **59.** 1212_{five} **61.** 242 **63.** 10101_{two} **65.** +1; −1 **67.** −10 s
69. +18 **71.** −4 **73.** 396°C **75.** −20 ft/s **77.** +14,778 ft **79.** +$125
81. −6°C **83.** {. . . , 1, 2, 3} **85.** $\{\frac{1}{3}, \frac{1}{2}, \frac{2}{3}, 1, \frac{3}{2}, 2, 3\}$ **89.** No; Yes

Exercises 2-1, Page 39

1. ab **3.** x^2 **5.** πr^2 **7.** a^3b^2 **9.** b, c **11.** $7, p, q, r$ **13.** i, i, R
15. a, b, c, c, c **17.** 6 **19.** 2π **21.** $4\pi e$ **23.** mw^2 **25.** $x = 2y$ **27.** $i = 12f$
29. $i = 1000m$ **31.** $V = 7.5lwd$ **33.** $N = 8tn^3$ **35.** $s = \frac{1}{2}gt^2$ **37.** $A = s^2$
39. $C = s^3c$ **41.** -54 **43.** 1,800,000 km **45.** 1.5 V **47.** 192 in. **49.** 90 gal
51. 48 J

Exercises 2-2, Page 44

1. $a^2, -3ab, 7b$ **3.** $3, -6ab, 7a, -8/b$
5. $x, 2x$ **7.** $3x^3, -5x^3, -x^3$
9. $-2mn, -mn$ **11.** $-ky^2, 2ky^2$
13. $6a(a - x)$ **15.** $x^2(a - x)(a + x)$
17. $\frac{2}{5}$ **19.** $\dfrac{6}{a - b}$
21. 6 **23.** 24
25. $p = 2l + 2w$ **27.** $w = 100 + 5t^2$
29. $A = 2\pi r(r + h)$ **31.** $A = \dfrac{a + b + c + d + e}{5}$
33. $T = 2\pi\sqrt{l/g}$ **35.** $C = \frac{5}{9}(F - 32)$
37. 20 in. **39.** 75.4 m^2 **41.** \$620 **43.** 79 **45.** 5 ft

Exercises 2-3, Page 50

1. $3a - b$ **3.** $7a - 4b^2$ **5.** $5s + 3t$ **7.** $5x + 5y$ **9.** $6a - 3b$ **11.** $7x/2$
13. $1/2xy$ **15.** $7a - 15b$ **17.** $12x + y$ **19.** $4x$ **21.** $5rs - 15r$ **23.** $5abc - c^2 - b$
25. $5ax - 3$ **27.** $5x - 10y$ **29.** $9x - 1000$ **31.** $20a + 10b + 5$ **33.** $100t + 110$
35. $2xy + 5y + 15$ **37.** 86 **39.** -250 **41.** 215 cents **43.** 385 km **45.** 6650
47. 846

Exercises 2-4, Page 53

1. $3x + 2y$ **3.** $6a^2 - 2ax$ **5.** $3x + y$ **7.** $-ac + c$ **9.** $2a - b$
11. $3a + 11b - ab$ **13.** $2ax + 2a$ **15.** $2a - ax$ **17.** $-xy + 5x$ **19.** $3ab + a - 2b$
21. $3 + 5a + ax$ **23.** $13a + 3b$ **25.** $5a + 4b$ **27.** $4ax + 2ay - 5xy$ **29.** $6a^2$
31. $12x$ **33.** $l = w + 24$ **35.** $E = mc^2$ **37.** $p = \dfrac{st}{t^2 + 1}$ **39.** $V = 40n + 145$
41. $d = 6at + 8a$ **43.** $N = \dfrac{x + y}{t}$ **45.** $2a + b$ **47.** $6a + 5b + 2$ **49.** 98 **51.** 910
53. 12% **55.** 195 ft **57.** $V = 41p + 180$, 836 cents **59.** 3 ft, 1 yd

Exercises 3-1, Page 61

1. $\{5\}$ **3.** $\{3\}$ **5.** $\{8\}$ **7.** $\{35\}$ **9.** $\{-4\}$ **11.** $\{9\}$ **13.** $\{-2\}$ **15.** $\{3\}$
17. $\{-6\}$ **19.** $\{-1\}$ **21.** $\{8\}$ **23.** $\{-12\}$ **25.** $\{\frac{1}{2}\}$ **27.** $\{\frac{1}{3}\}$
29. $x - 6 = 7$, $\{13\}$ **31.** $3x = 33$, $\{11\}$ **33.** $5x + 8 = 3x$, $\{-4\}$
35. $4(x + 3) = 20 + 2x$, $\{4\}$ **37.** $\{\ \}$, $\{2\}$

Exercises 3-2, Page 66

1. $M = CR/100$
3. $v_1 = v_2 - at$
5. $T = PV/R$
7. $N = Vt/k$
9. $m = \dfrac{2E}{v^2}$
11. $C = \dfrac{5F - 160}{9}$
13. $r = \dfrac{A - P}{Pt}$
15. $k = \dfrac{a + PV^2}{V}$
17. $P = \dfrac{R_1 + ak - R_2}{k}$
19. $H = \dfrac{wL}{R(w + L)}$
21. $d = a - bc$
23. $x = \dfrac{f - 3y}{a}$
25. $x = \dfrac{3y - 2ay}{2a}$
27. $a = 2b + 4$
29. $b = \dfrac{x_1 - x_2 - 3a}{a}$
31. $R_2 = 2R_3 - R_1$
33. $z = \dfrac{3y + 6 - 7ay}{7a}$
35. $y = \dfrac{x - 3a - ax}{a}$
37. $y = n + 100$, $n = y - 100$
39. $A = \dfrac{a + b + c}{3}$, $c = 3A - a - b$
41. $10x + 25(x + 2) + 50(x - 1) = y$, $x = y/85$

Exercises 3-3, Page 70

1. (a) $8 > 6$ (b) $-5 < 1$
3. (a) $3 > -6$ (b) $1 < 2$
5. (a) $2 > 0$ (b) $-2 > -3$
7. $-1 > -6$
9. $\{x|x > 12\}$
11. $\{x|x < 5\}$
13. $\{x|x > 18\}$
15. $\{x|x < -6\}$
17. $\{x|x < 4\}$
19. $\{x|x < -1\}$
21. $\{x|x < -4\}$
23. $\{x|x < -2\}$
25. $\{x|x > -1\}$
27. $\{x|x > 6\}$
29. $\{x|x > 2\}$
31. Less than \$87,500
33. More than 50 mi/h

Exercises 3-4, Page 74

1. $\{x|x \le 8\}$ **3.** $\{x|x \ge 5\}$

5. $\{x|x \le 8\}$ **7.** $\{x|x \ge 6\}$

9. $\{x|x \ge 3\}$ **11.** $\{x|x \ge -\frac{1}{2}\}$

13. $\{x|x \ge -5\}$ **15.** $\{x|x \ge \frac{3}{5}\}$

17. 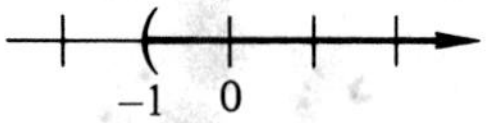

19. 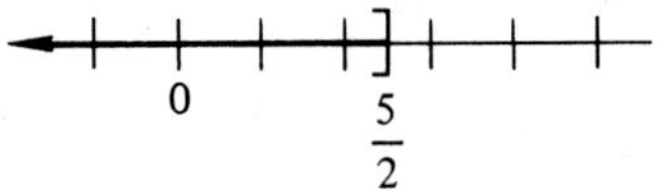

21. $\{x|x > -2\}$ 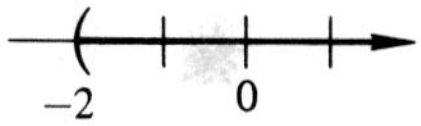

23. $\{x|x \ge 2\}$ 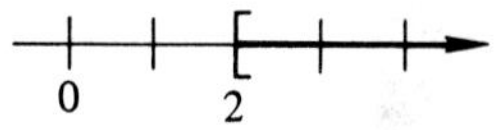

25. $\{x|x \ge -1\}$ 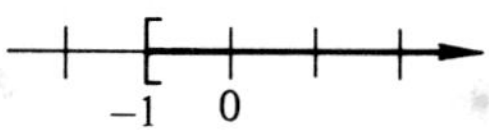

27. $\{x|x < \frac{13}{5}\}$ 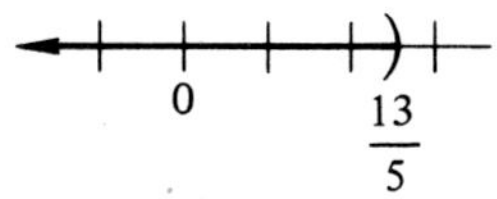

29. $\{x|x \ge -6\}$ **31.** At least \$75

Exercises 3-5, Page 79

1. $\frac{12}{5}; \frac{3}{23}$ **3.** $\frac{7}{1}; \frac{1}{6}$ **5.** $\frac{15}{4}$ **7.** $\frac{3}{10}$ **9.** $\frac{1}{6}$ **11.** $\frac{4}{9}$ **13.** 2 m/s **15.** $\frac{2}{9}$ lb/ft^3

17. $\frac{5}{4}$ **19.** $\frac{6}{7}$ **21.** $\frac{15}{7}$ **23.** $\frac{3}{8}$ **25.** $8/x = 2/y$ **27.** $27/m = 3/n$ **29.** 14

31. $\frac{35}{2}$ **33.** 4540 g **35.** $\frac{100}{3}$ in. **37.** 6 m; 9 m

39. 10,000,000 gal of regular; 4,000,000 gal of lead-free **41.** 15 in; 20 in.

Exercises 3-6, Page 85

1. $y = kt$ **3.** $y = ks^2$ **5.** $t = k/y$ **7.** $y = kst$ **9.** $y = ks/t$ **11.** $x = kyz/t^2$

13. $y = 5s$ **15.** $s = 2t^3$ **17.** $u = 272/d^2$ **19.** $y = 9x/t$ **21.** $d = 20s/3\sqrt{v}$ **23.** 16

25. $\frac{32}{5}$ **27.** 36 **29.** $\frac{72}{5}$ **31.** $\frac{50}{27}$ **33.** $v = 32t$ **35.** 175 bc/h **37.** 10 in.

39. 400 units **41.** 480 gal **43.** $I = 400/d^2$ **45.** $P = RT/V$

Exercises 3-7, Page 88

1. $\{8\}$ **3.** $\{9\}$ **5.** $\{-4\}$ **7.** $\{4\}$ **9.** $\{-3\}$ **11.** $\{\frac{1}{2}\}$ **13.** $P = S - C - E$

15. $g = wa/F$ **17.** $M_1 = \dfrac{PT - M_2V_2}{V_1}$ **19.** $D = \dfrac{2L_1T + Ld}{L}$ **21.** $y = \dfrac{ax + 2a}{3}$

23. $c = \dfrac{ax + ab - bx}{b}$ **25.** $\{x|x < 9\}$ **27.** $\{x|x \ge 5\}$ **29.** $\{x|x < -3\}$

31. $\{x|x \le -4\}$ **33.** $\{x|x < -2\}$ **35.** $\{x|x < 3\}$ **37.** $\{x|x \ge 3\}$ **39.** $\{x|x \le 4\}$

41. $y = kz^2$ **43.** $t = ks/x^4$ **45.** $w = k(mn)^3$ **47.** $y = k/(pq)^2$ **49.** $x = k(mn)^2$

51. $w = k\sqrt{rt}/s$ **53.** $\frac{18}{5}$ **55.** $\frac{9}{2}$ **57.** (a) $\frac{2}{3}$ (b) $\frac{2}{3}$ **59.** (a) $\frac{105}{2}$ (b) $\frac{105}{2}$

61. $x = yz/9$ **63.** 30 **65.** $\{3\}$ **67.** $\{-8\}$ **69.** $\{x|x < -3\}$ **71.** $\{x|x \ge -3\}$

73. $\frac{7}{3}$ **75.** $\frac{21}{2}$ **77.** $R = 435 - D$ **79.** Less than \$22 **81.** 46,000 cans **83.** 9 lb

85. 3.33 s **87.** $T_2 = \dfrac{Ht + kAT_1}{kA}$

Exercises 4-1, Page 95

1. $3s - xy - a$ **3.** $9y - 3x - 4a - 9xy$ **5.** $3x + 2xy$ **7.** $-3x^2 + 3xy - 2s$
9. $2a + 3$ **11.** $-4 - 4x$ **13.** $2s - 1$ **15.** $-9 + 4y$ **17.** $10x + 1$ **19.** $2s + 2$
21. $6x - 12$ **23.** $7t - 5x - 5p^2 + 9$ **25.** $\{3\}$ **27.** $\{1\}$ **29.** $\{\frac{13}{10}\}$ **31.** $\{\frac{6}{5}\}$
33. -1 **35.** $v = \dfrac{mv_O - I}{m}$

Exercises 4-2, Page 99

1. x^{10} **3.** y^8 **5.** x^{21} **7.** y^{15} **9.** $-6a^3x^2y$ **11.** $-45a^2ct^7$ **13.** $-84r^2s^2t^4$
15. $-8s^3t^9x^3$ **17.** $2x^3 + 8ax$ **19.** $6a^2x - 3a^4$ **21.** $-2s^2tx + 2st^3y$
23. $-3x^3y^2 + 9ax^2y^7$ **25.** $x^2 + 2x - 3$ **27.** $s^2 + s - 6$ **29.** $2x^2 - x - 6$
31. $2x^2 + x - 1$ **33.** $10v^2 + 17v + 3$ **35.** $a^2 - 3ax + 2x^2$ **37.** $6a^2 + ac - 2c^2$
39. $2x^2 + 18x - 5tx - 45t$ **41.** $4a^2 - 81p^2y^2$ **43.** $2a^3 - 5a^2 - 13a - 5$
45. $a^2 + 2axy - 4ax - 2x^2y + 3x^2$ **47.** $x^3 - 3x^2 - 10x + 24$ **49.** $x^2 - 4x + 4$
51. $x^2 + 4xy + 4y^2$ **53.** $x^3 + 3x^2 + 3x + 1$ **55.** $y^3 - 12cy^2 + 48c^2y - 64c^3$
57. $(x + 2)(x - 1) = x^2 + x - 2$ **59.** $(6 + 4)(6 - 4) = 4(9) - 16; (10)(2) = 36 - 16; 20 = 20$

Exercises 4-3, Page 105

1. x^3 **3.** $1/p^{10}$ **5.** a **7.** $1/a^9$ **9.** $1/b^4$ **11.** a^4 **13.** 1 **15.** 1
17. 1 **19.** $2a$ **21.** $\frac{1}{2}$ **23.** $x - 1$ **25.** $4n^3$ **27.** $-ax$ **29.** $2at^4/c^2$
31. $-x/4r^2$ **33.** $2/b$ **35.** $b/2$ **37.** $3b/a$ **39.** $1/9b^2$ **41.** $7t^2u/6$
43. $2x/y^2z^3$ **45.** $1/ax$ **47.** $2/c^8$ **49.** $6/x^2$ **51.** $6x^8$ **53.** $1/a^2b^4$ **55.** $a^2/9$
57. y^2/x^6 **59.** y^5/a^8x^5 **61.** $2ay$ **63.** $p = 10/3^x$
65. 0.0000000000000000000000000009 g

Exercises 4-4, Page 110

1. $6b + 5$ **3.** $3m - 1$ **5.** $-a^2x^2 + ax$ **7.** $y^2 - 2xy^3$ **9.** $bc - ab^3c^4 - 2a$
11. $-ab^2 + 2a^2b^3 + 1 + b$ **13.** $x - 3$ **15.** $2x + 1$ **17.** $3x - 6 + \dfrac{23}{x + 4}$ **19.** $x + 3$
21. $4y - 1$ **23.** $8x - 1$ **25.** $9x + 1$ **27.** $x^2 - x - 3$ **29.** $2x^2 - x + 3$
31. $x^2 - 4xy - 2y^2$ **33.** $2x^3 - x^2 + 2x + 1 - \dfrac{2}{3x - 1}$ **35.** $a - 3b$ **37.** $a - 1$
39. $a - 2b - 3c$ **41.** $a + 5$ **43.** $c - 8$ **45.** $\dfrac{4x^2 + 6x}{2x} = 2x + 3$ lb
47. $10r - 2 + \dfrac{2.4}{6r + 1.2}$

Exercises 4-5, Page 113

1. $3a - x$ **3.** $9x - 13y$ **5.** $-6a^3b^6$ **7.** $56x^4y^9z^8$ **9.** $-4ax^3$ **11.** $2x^5/3y^3$

13. $1/ay^2$ **15.** $1/4ab^2$ **17.** $\dfrac{1}{x^2(2x + y)}$ **19.** $4y - 2b$ **21.** $5x - 2a$ **23.** $5b - 10$

25. $-2a^3x + 2a^2t$ **27.** $2x^2 + x - 21$ **29.** $6a^2 - 11ab - 10b^2$ **31.** $6b^2 - 12ab + 6a^2$

33. $14b^2n + 5abn - a^2n$ **35.** $-2x^3 + 6x^2 + 4x - 16$ **37.** $-2y^3 + 3x^3$

39. $h - 3j^2 - 6h^3j^3$ **41.** $x + 4$ **43.** $x^2 - x + 1 - \dfrac{11}{2x + 3}$ **45.** $3x^2 + 2x + 2$

47. $2x^2 - x - 4$ **49.** $\{7\}$ **51.** $a^2 + 1$ **53.** $3x + 13$ units **55.** $36x - 4x^2$

57. $mv_2^2 - mv_1^2$ **59.** $lh + 2lhat + lha^2t^2$ **61.** $pV + pb + aV + ab$ **63.** $1 - p^2 - r + rp$

65. $1 - \dfrac{Q_2}{Q_1}$ **67.** $a(1 + r^2)(1 + r)$

Exercises 5-1, Page 118

1. 2, 3, 5, 7 **3.** 19, 29 **5.** 41, 43, 47 **7.** 61, 67 **9.** (2)(5) **11.** (2)(2)(7)

13. (2)(5)(5) **15.** (2)(2)(2)(11) **17.** (2)(2)(5)(7) **19.** (2)(3)(5)(7) **21.** $2, x - 1, x + 3$

23. $3, 3, c, c, ax - b, cx + d$ **25.** Yes **27.** No, $(x + 2)^2 = x^2 + 4x + 4$ **29.** 3

31. $x + 3$ **33.** Yes **35.** $P \subseteq I$ **37.** No

Exercises 5-2, Page 121

1. $5(x + y)$ **3.** $7(a^2 - 2bc)$ **5.** $a(a + 2)$ **7.** $2x(x - 2)$ **9.** $3a(b - 1)$

11. $2p(2 - 3q)$ **13.** $3ab(a + 3)$ **15.** $acf(abc - 4)$ **17.** $2(x + y - z)$

19. $5(x^2 + 3xy - 4y^3)$ **21.** $2x(3x + 2y - 4)$ **23.** $4pq(3q - 2 - 7q^2)$

25. $7a^2b^2(5ab^2c^2 + 2b^3c^3 - 3a)$ **27.** $3a(2ab - 1 + 3b^2 - 4ab^2)$ **29.** $2pr(12kp - k^2r + 4 + 16p)$

31. $4z(7xy - 14x^2y^2z + 16x^2z^2 + 9y^2)$ **33.** $2(lw + lh + hw)$ **35.** $Rv(1 + v + v^2 + v^3 + v^4 + v^5)$

Exercises 5-3, Page 123

1. $9, 25, 4x^2$ **3.** a^2b^4 **5.** $x^2 - y^2$ **7.** $4a^2 - b^2$ **9.** $49a^2x^4 - p^6$

11. $x^2 + 4x + 4$ **13.** 399 **15.** 39900 **17.** $(a + h)(a - h)$ **19.** $(4 + x)(4 - x)$

21. $(2x^2 + y)(2x^2 - y)$ **23.** $(3y + v)(3y - v)$ **25.** $(5w + u)(5w - u)$ **27.** $(ab + 3y)(ab - 3y)$

29. $(9x^2 + 2y^3)(9x^2 - 2y^3)$ **31.** $9(v^2 + 3w^3)(v^2 - 3w^3)$ **33.** $25(2x^3 + z^2)(2x^3 - z^2)$

35. $4(x + 5y)(x - 5y)$ **37.** $(x^2 + 1)(x + 1)(x - 1)$ **39.** $4(x^2 + 9y^2)$ **41.** $m(v_1 + v_2)(v_1 - v_2)$

43. 1591 in.^{2} **45.** $x(x + y)(x - y)$

Exercises 5-4, Page 129

1. $(x + 2)(x + 1)$ **3.** $(x + 4)(x - 3)$ **5.** $(y - 5)(y + 1)$ **7.** $(x + 5)^2$

9. $(2q + 1)(q + 5)$ **11.** $(3x + 1)(x - 3)$ **13.** $(5c - 1)(c + 7)$ **15.** Cannot be factored

17. $(2s - 3t)(s - 5t)$ **19.** $(5x + 2)(x + 3)$ **21.** $(2x - 1)(2x - 3)$ **23.** $(6q + 1)(2q + 3)$

25. $(6t - 5u)(t + 2u)$ **27.** $(2x + 3)(4x - 3)$ **29.** $(4x - 3)(x + 6)$ **31.** $(4n - 5)(2n + 3)$

33. $2(x - 3)(x - 8)$ **35.** $2(2x - 3z)(x + 2z)$ **37.** $2x(x + 1)(x + 2)$ **39.** $a(5x - y)(2x + 5y)$

41. $(x + 4)(x - 10)$ **43.** $3(s + 60)(s - 50)$ **45.** $16(t + 2)(t - 10)$ **47.** $4x(x - 6)^2$

Exercises 5-5, Page 133

1. $(x + y)(x^2 - xy + y^2)$

3. $(2x - y)(4x^2 + 2xy + y^2)$

5. $(4x - 5)(16x^2 + 20x + 25)$

7. $(ab + 1)(a^2b^2 - ab + 1)$

9. $(2x - 3y)(4x^2 + 6xy + 9y^2)$

11. $2(2x + 3y)(4x^2 - 6xy + 9y^2)$

13. $4(x - 5)(x^2 + 5x + 25)$

15. Cannot be factored

17. $(4 + y)(x + y)$

19. $(x - 4v)(4b + 7)$

21. $(c + d)(2 + d)$

23. $(3 + b)(a - b)$

25. $(x + 2)(x + 2)(x - 2)$

27. $(k - p)(n + q)(n - q)$

29. $(y^2 - 6)(y - 8)$

31. $(c - d)(2 + d)$

33. $(a - b + 1)(a - b - 1)$

35. $(x + 3y + 2)(x + 3y - 2)$

37. $(y + x - 3)(y - x + 3)$

39. $(2m + 2n + 1)(2m - 2n - 1)$

41. $(a - b)(a^2 + ab + b^2) = a^3 - b^3$

43. $(x - 5y)(x^2 + 5xy + 25y^2)$

Exercises 5-6, Page 134

1. $5(a - c)$

3. $3a(a + 2)$

5. $4ab(3a + 1)$

7. $3xyz(1 + 6y)$

9. $(2x + y)(2x - y)$

11. $(4y^2 + x)(4y^2 - x)$

13. $(x + 1)^2$

15. $(x - 6)(x - 1)$

17. $(x - 10)(x - 2)$

19. $a(x^2 + 3ax - a^2)$

21. $2nm(m^2 - 2nm + 3n^2)$

23. $4t^2(p^3 - 3t^2 - 1 + a)$

25. $(4rs + 3y)(4rs - 3y)$

27. Cannot be factored

29. $(2x + 7)(x + 1)$

31. $(2t - 3)(7t + 1)$

33. $(2x + 1)(2x - 9)$

35. $(3x + 1)^2$

37. $(x + 2y)(x + y)$

39. $(5c - d)(2c + 5d)$

41. $(8x + 7)(11x - 12)$

43. $16y^3(3 - y)(1 - y)$

45. $5(x^2 + 5)(x^2 - 5)$

47. $2(x + 3y)(x - 3y)$

49. $4(x^2 + 9)$

51. $8x^4y^2(xy + 2)(xy - 2)$

53. $3a(x + 4)(x - 3)$

55. $3r(6r - 13s)(3r + 4s)$

57. $(4x^2 + 1)(2x - 1)(2x + 1)$

59. $4(3x^2 + 1)(3x^2 - 1)$

61. $(xy + pq)(x^2y^2 - xypq + p^2q^2)$

63. $(2x - yz)(4x^2 + 2xyz + y^2z^2)$

65. $(2a^2 + b)(4a^4 - 2a^2b + b^2)$

67. $(2a - 3)(a + b)$

69. $(t - 1)(2t + s)$

71. $2(x + y)(4 + y)$

73. $4(a - b + 3)(a - b - 3)$

75. $3(1 + x - y)(1 - x + y)$

77. $P(N + 2)$

79. $at(2 - t)$

81. $(x - 5)(x - 3)$

83. $(x - 4)(2x + 13)$

85. $(T + 530)(T - 10)$

87. $T(T_b + T_p) = T_bT_p,\ T = \dfrac{T_bT_p}{T_b + T_p}$

Exercises 6-1, Page 141

1. $\frac{4}{6}, \frac{10}{15}$ **3.** $\dfrac{6a^3x}{2a^2b}, \dfrac{3a^2bx}{ab^2}$

5. $\dfrac{x^2-2xy+y^2}{x^2-y^2}, \dfrac{x^2-y^2}{x^2+2xy+y^2}$ **7.** $\dfrac{12b-12-3b^2}{b^2-4b+4}; \dfrac{12-3b^2}{b^2-4}$

9. $\frac{4}{7}$ **11.** $2a/3a^2$ **13.** $\dfrac{2x-1}{x+1}$ **15.** $\dfrac{2a-1}{2}$ **17.** $\frac{2}{9}$ **19.** $\frac{4}{15}$ **21.** $\frac{1}{3}$ **23.** $\frac{1}{3}$

25. $ab/4$ **27.** $2t/7r^2s$ **29.** $\dfrac{x-4}{x+4}$ **31.** $\dfrac{(x+1)(x+2)}{2(x+3)}$ **33.** $\dfrac{x+1}{x-1}$ **35.** $\dfrac{x}{x+2}$

37. $\dfrac{3x-2}{4x+3}$ **39.** $\dfrac{x+3y}{3y}$ **41.** $\dfrac{1-3x}{3x+1}$ **43.** $\dfrac{5-x}{2+x}$ **45.** $\frac{3}{7}$ **47.** -1

Exercises 6-2, Page 146

1. $1/8n; 13s$ **3.** $3b/a; a/3b$ **5.** $\frac{2}{15}$ **7.** $rt/12$ **9.** $\frac{81}{256}$ **11.** $a^{10}/32x^5$

13. $x^5y^{10}/32$ **15.** $\frac{24}{7}$ **17.** $x/8m$ **19.** $y/45x$ **21.** $a^4/2b^5$ **23.** $\dfrac{a+3b}{a+b}$

25. $\dfrac{5x(x-y)}{6}$ **27.** $\dfrac{5b-2}{10}$ **29.** $\dfrac{(x+3)(x-3)}{(x-2)(x-4)}$ **31.** $\dfrac{2x+3}{2x+1}$ **33.** 3

35. $\dfrac{3(a-b)}{(a-2b)(a+b)}$ **37.** $\dfrac{(s+2)(s+7)}{(s-12)(s+11)}$ **39.** $\dfrac{(x-1)^2}{3x+2}$ **41.** $\dfrac{2a(a+b)}{(2a+b)(a-b)}$

43. $\dfrac{1}{2x+5}$ **45.** $\dfrac{512(3x+4y)(3x-4y)}{b}$ **47.** $4000/x$

Exercises 6-3, Page 152

1. 12 **3.** 90 **5.** 420 **7.** 750 **9.** $5a$ **11.** $15a^2$ **13.** $15s^2$ **15.** $84k^2$

17. $96a^3b^3$ **19.** $60a^2cx^3$ **21.** $x^2(x+1)$ **23.** $2a^2(a-3)$ **25.** $8x(x-1)$

27. $9x(x-1)^2$ **29.** $6ax(a-3)$

31. $3x^2(x-1)(x+1)$ **33.** $6x(x+y)(x-y)$

35. $(a+b)(a-2b)(a+2b)$ **37.** $12(x-3)(x+3)^2$

39. $2(x-3y)(x+3y)(3x+2y)$

Exercises 6-4, Page 157

1. $\frac{20}{36}-\frac{21}{36}$ **3.** $\dfrac{5b}{abx}+\dfrac{a}{abx}-\dfrac{4bx}{abx}$

5. $\dfrac{8x}{2x^2(x-1)}-\dfrac{3}{2x^2(x-1)}$

7. $\dfrac{x(x+2)^2}{2(x-2)(x+2)^2}+\dfrac{10(x+2)}{2(x-2)(x+2)^2}-\dfrac{6x(x-2)}{2(x-2)(x+2)^2}$

9. $\frac{1}{40}$ **11.** $\dfrac{23x-15}{30x}$

13. $19/10x$ **15.** $\dfrac{8y-b}{by^2}$

17. $\dfrac{2xy+5x^2-3}{x^2y}$ **19.** $\dfrac{42yz-15xz+2xy}{12xyz}$

21. $\dfrac{13x + 5}{x^2(2x + 1)}$ 23. $\dfrac{2y + 9x + 6}{y^2(3x + 2)}$

25. $\dfrac{5y^2 + 3y + 36}{15(y - 3)(y + 3)}$ 27. $\dfrac{2x^2 + 3x + 9}{4(x - 3)(x + 3)}$

29. $\dfrac{y^2 - 6y - 3}{3(y - 3)(y + 3)}$ 31. $-\dfrac{x + 6}{2(2x + 3)}$

33. $\dfrac{(3x - 14)(x - 1)}{(2 - 3x)(2 + 3x)}$ 35. $\dfrac{-5x^2 + 15x + 4}{(x - 2)(x + 2)}$

37. $\dfrac{7p^2 + 4p + 8q - 175q^2}{8(p - 5q)^2}$ 39. $\dfrac{-57p^2 + 47pq + 80q^2 + 30p^2q + 20pq^2}{10p(3p + 2q)(3p - 5q)}$

41. $\dfrac{56}{(t^2 + 4)(t + 2)(t - 2)}$ 43. $\dfrac{y + 2x}{y - 2x}$

45. $\frac{35}{82}$ 47. $\dfrac{2}{M + 1}$

Exercises 6-5, Page 162

1. $\{4\}$ 3. $\{-3\}$ 5. $\{10\}$ 7. $\{-\frac{3}{4}\}$ 9. $1 - 3b$ 11. $4a + 2$ 13. $\dfrac{2a - 4}{3a^2}$

15. $-16/7b$ 17. $\{2\}$ 19. $\{\frac{7}{3}\}$ 21. $\{\frac{3}{4}\}$ 23. $\{\ \}$, No solution 25. $\dfrac{b}{1 - 3b}$

27. $\dfrac{3}{2n - 4}$ 29. $\{\frac{52}{11}\}$ 31. $Ptr + P$

Exercises 6-6, Page 164

1. $\frac{10}{21}$ 3. $y/6x$ 5. $a/3bc^2$ 7. $\dfrac{2a - b}{a - 2b}$ 9. $\dfrac{a}{x + y}$ 11. $4a/5b$ 13. $\dfrac{p + 1}{p + 3}$

15. $\dfrac{2(y - 5)}{3 - 2y}$ 17. $\frac{33}{112}$ 19. $\frac{93}{32}$ 21. $\frac{2}{15}$ 23. $15y/4x$ 25. b^2/ac^2 27. $\dfrac{10b - 3a}{5a^2b}$

29. $\dfrac{10cd - 6 + c}{2c^2d}$ 31. $\dfrac{2}{x(x + 1)}$

33. $\dfrac{x - 5}{4}$ 35. $(3x + 2y)(y - 2x)$

37. $\dfrac{5x + 9}{x^2(x + 3)}$ 39. $\dfrac{x^2 - 4x + 3}{x^2 - 4x + 4}$

41. $\dfrac{-4x^2 - 12x - 9}{(2x + 5)(x + 3)(x - 3)}$ 43. $\frac{2}{5}$

45. $\dfrac{x + 1}{2 - x}$ 47. $\dfrac{2y + 5x}{2y - 5x}$ 49. $\{-4\}$ 51. $\{5\}$ 53. $\dfrac{2a + 1}{4}$ 55. $\dfrac{a + 6}{4(b - a)}$

57. $\{-\frac{2}{5}\}$ 59. $\{\ \}$, No solution 61. $4\frac{6}{7}$ hours 63. $\dfrac{(LC\omega^2 - 1)^2}{C^2\omega^2}$

65. $\dfrac{g_2^2 - 2g_1g_2 + g_1^2}{g_2^2}$ 67. $\dfrac{3r - h}{12r^3}$

69. $\dfrac{8A + 4w^2 + \pi w^2}{4w}$ 71. $\dfrac{kh - Uk}{Uh}$

Exercises 7-1, Page 170

1. Let x = width of rectangle; $x + 10$ = length of rectangle.
   Equation: $140 = 2(x + 10) + 2x$.

3. Let x = length of third side of triangle; $x - 2$ = length of one of two equal sides of triangle.
   Equation: $29 = x + 2(x - 2)$.

5. Let x = one part; $87 - x$ = other part.
   Equation: $2x = 87 - x$.

7. Let x = amount he saves; $10x$ = amount he spends.
   Equation: $13{,}200 = x + 10x$.

9. Let x = number of quarters; $21 - x$ = number of nickels.
   Equation: $25x + 5(21 - x) = 265$.

11. Let x = part invested at 5%; $2000 - x$ = part invested at 6%.
    Equation: $0.05x + 0.06(2000 - x) = 105$.

13. Let x = amount student pays; $x + 100$ = amount state pays; $x + 100$ = amount county pays.
    Equation: $x + (x + 100) + (x + 100) = 2300$.

15. Let x = number of papers printed on a weekday: $x + 50{,}000$ = number of papers printed on Sunday.
    Equation: $6x + x + 50{,}000 = 1{,}100{,}000$.

17. Let x = smaller integer; $x + 1$ = larger integer.
    Equation: $x + (x + 1) = 31$.

19. Let x = smallest integer; $x + 1$ and $x + 2$ are the other integers.
    Equation: $x + (x + 1) + (x + 2) = 156$.

21. Let x = rate of one car; $x + 10$ = rate of other car.
    Equation: $8x + 8(x + 10) = 800$.

23. Let x = distance by car; $15x - 40$ = distance by plane.
    Equation: $x + (15x - 40) = 200$.

Exercises 7-2, Page 175

1. 18, 75 **3.** 527,000 in New England; 782,000 in the Middle Atlantic States **5.** 7 in. by 13 in.
7. 270 cm **9.** \$3.50 for his regular job and \$2.50 for his part-time job. **11.** \$720
13. 43, 44 **15.** 44, 46, 48 **17.** 3 **19.** \$4,550 **21.** 36 mi/h, 44 mi/h **23.** 6 h
25. 22.5 L **27.** 8 lb at \$2.00, 12 lb at \$3.20

Completed solutions for Section 7-1

1. 30 in. by 40 in. **3.** 9 cm, 9 cm, and 11 cm **5.** 29, 58 **7.** \$1,200
9. 8 quarters, 13 nickels **11.** \$1,500 **13.** \$700 **15.** 150,000 daily; 200,000 Sunday
17. 15, 16 **19.** 51, 52, 53 **21.** 45 mi/h, 55 mi/h **23.** 15 km

Exercises 7-3, Page 177

1. 46 on the first nine, 41 on the second nine **3.** Washington, 69,000 mi^2; Idaho, 84,000 mi^2; Montana, 147,000 mi^2
5. 91 **7.** $16\frac{1}{3}$ cm **9.** 4 in. **11.** 52 nickels, 27 dimes, 1 quarter **13.** 24, 26, 27
15. 12 cm by 18 cm **17.** 4500 ft. **19.** 72 years **21.** 8, 57 **23.** 54 in. by 36 in.
25. $\frac{12}{7}$ days **27.** 15 m **29.** $7\frac{11}{12}$ hours **31.** 16 km **33.** $\frac{20}{3}$ qt **35.** 4.21 cm^3

Exercises 8-1, Page 183

1. 6 **3.** -13 **5.** 0.5 **7.** -0.2 **9.** 20 **11.** -40 **13.** 2 **15.** -2
17. -5 **19.** 0.5 **21.** 2 **23.** 3 **25.** 2 **27.** $2i$ **29.** $-20i$ **31.** $0.7i$
33. -25 **35.** -12 **37.** rational: $\sqrt{4}$, $\sqrt[3]{8}$, $\sqrt[3]{-8}$; irrational: $\sqrt{3}$; imaginary: $\sqrt{-8}$
39. rational: $\sqrt[4]{0.0001}$, $\sqrt{0.16}$; irrational: $\sqrt[3]{100}$, $\sqrt[5]{64}$; imaginary: $\sqrt{-1}$ **41.** π s **43.** 8 m

Exercises 8-2, Page 188

1. $2\sqrt{2}$ **3.** $3\sqrt{3}$ **5.** $3\sqrt{7}$ **7.** $5\sqrt{6}$ **9.** $6\sqrt{3}$ **11.** $8\sqrt{3}$ **13.** $\sqrt{2}/2$
15. $3\sqrt{2}/4$ **17.** $c\sqrt{a}$ **19.** $ab\sqrt{a}$ **21.** $\sqrt{2a}/2$ **23.** $3\sqrt{3ab}/b$ **25.** $2ac\sqrt{bc}$
27. $4x^2z^2\sqrt{5yz}$ **29.** $b\sqrt{3a}/6$ **31.** $x\sqrt{10y}/5a^4$ **33.** 7.071 **35.** 25.69 **37.** 130.4
39. 0.6325 **41.** 0.09644 **43.** 0.816 **45.** $3\sqrt[3]{2}$ **47.** $2a\sqrt[3]{a}$ **49.** $2a^2\sqrt[4]{a}$
51. $2x\sqrt[5]{2x^2}$ **53.** 25.3 cm **55.** $\sqrt{LC}/2\pi LC$

Exercises 8-3, Page 192

1. $3\sqrt{3}$ **3.** $4\sqrt{7}+\sqrt{5}$ **5.** $-5\sqrt{15}$ **7.** $6\sqrt{2}-2\sqrt{3}$ **9.** $5\sqrt{3}$ **11.** $9\sqrt{10}$
13. $9\sqrt{2}$ **15.** $\sqrt{7}$ **17.** $-4\sqrt{3}-\sqrt{2}$ **19.** $5\sqrt[3]{2}-2\sqrt[3]{3}$ **21.** $2\sqrt[5]{2}$ **23.** $4\sqrt{a}$
25. $3\sqrt{2a}+4a\sqrt{2a}$ **27.** $7a\sqrt{2}-2\sqrt{3a}$ **29.** $4bc\sqrt[4]{a^2c}$ **31.** $\sqrt{21}-9\sqrt{2}$ **33.** 26
35. $a\sqrt{b}+3a\sqrt{c}$ **37.** $1+\sqrt{6}$ **39.** $-17-3\sqrt{15}$ **41.** $2a-\sqrt{ac}-15c$ **43.** $\dfrac{2+\sqrt{6}}{2}$
45. $\dfrac{7-\sqrt{21}}{4}$ **47.** $\dfrac{4+3\sqrt{2}}{2}$ **49.** $\dfrac{a-2\sqrt{ab}}{a-4b}$ **51.** $75\sqrt{2}$ ft. **53.** $\dfrac{13-2\sqrt{30}}{7}$

Exercises 8-4, Page 197

1. 5 **3.** 8 **5.** $4\sqrt{7}$ **7.** $15\sqrt{3}$ **9.** $2\sqrt{6}$ **11.** 17 **13.** 9 **15.** 12
17. 13 **19.** $3\sqrt{3}$ **21.** No **23.** No **25.** 21.9 in. **27.** 601 ft **29.** 56.2 m
31. 28.3 ft **33.** 17.5 ft **35.** 29.7 km **37.** 15.7 ft **39.** 36.8 ft

Exercises 8-5, Page 202

1. $\sqrt{3}$ **3.** $\sqrt[4]{x}$ **5.** $\sqrt[5]{a^2}$ or $(\sqrt[5]{a})^2$ **7.** $\sqrt[3]{x^5}$ or $(\sqrt[3]{x})^5$ **9.** 3 **11.** 4 **13.** 2
15. -2 **17.** 8 **19.** 16 **21.** 27 **23.** 4 **25.** $\frac{1}{6}$ **27.** $\frac{1}{2}$ **29.** $\frac{1}{2}$ **31.** 18
33. 2 **35.** 2 **37.** $\frac{1}{48}$ **39.** $-\frac{1}{14}$ **41.** $a^{3/2}$ **43.** $a^{3/4}b$ **45.** $x^{29/15}$
47. $x^{5/6}$ **49.** $x = A^{1/2}$ **51.** 24 atm

Exercises 8-6, Page 204

1. $\frac{1}{5}$ **3.** 8 **5.** $\frac{1}{6}$ **7.** 13 **9.** 6 **11.** -4 **13.** $\frac{1}{3}$ **15.** $-\frac{3}{11}$ **17.** 10
19. 10 **21.** 343 **23.** 1331 **25.** $\frac{8}{9}$ **27.** 324 **29.** $9i$ **31.** $-0.8i$
33. $3b/a^2$ **35.** m^4/n^2 **37.** $2x^5/3y^3$ **39.** $-x^2y^3/a^2$ **41.** $a^{7/12}$ **43.** $a^{7/6}$
45. $x^{1/2}/y$ **47.** $s^{2/3}t^{7/3}$ **49.** 7.211 **51.** 83.07 **53.** 15.49 **55.** 13.42
57. 0.6856 **59.** 0.2449 **61.** $2\sqrt{10}$ **63.** $8\sqrt{2}$ **65.** $6\sqrt{2}$ **67.** $6\sqrt{3}$ **69.** $\sqrt{5}/5$
71. $2\sqrt{7}/7$ **73.** $2a$ **75.** $5b\sqrt{5c}$ **77.** $\sqrt{6a}/a$ **79.** $2\sqrt{21a}/3a$ **81.** $2a\sqrt[3]{2}$
83. $2a\sqrt[5]{2a^3}$ **85.** $-\sqrt{7}$ **87.** $5\sqrt{7}-2\sqrt{6}$ **89.** $\sqrt{70}-5\sqrt{10}$ **91.** $(6-3a)\sqrt{2}$
93. $-6\sqrt{3}$ **95.** $2\sqrt{15}-30$ **97.** $a\sqrt{b}-3\sqrt{5ab}$ **99.** $27-5\sqrt{30}$ **101.** $81-17\sqrt{42}$
103. $2a-5\sqrt{ab}-3b$ **105.** $\dfrac{\sqrt{10}+2}{3}$ **107.** $5\sqrt{2}-7$ **109.** 15 **111.** 48
113. 20; Pythagorean numbers **115.** 33; Pythagorean triplet **117.** $\sqrt{abd}/b$
119. $5\sqrt{14}/9 = 2.08$; 0.00208 in. **121.** $\frac{2}{11}$ ft **123.** 5 m **125.** $a^{2/5}/b^{2/5}$

Exercises 9-1, Page 210

1. $x^2-7x-4=0$; $a=1, b=-7, c=-4$ **3.** Not quadratic
5. $x^2+4x+4=0$; $a=1, b=4, c=4$ **7.** $7x^2-x=0$; $a=7, b=-1, c=0$

9. Not quadratic **11.** $-6x^2 + 3x - 1 = 0;\ a = -6, b = 3, c = -1$ **13.** $\{-1, 2\}$
15. $\{\frac{1}{2}, 1\}$ **17.** $\{3, -4\}$ **19.** $3x^2 - 10x + 8 = 0$; None of values in solution set
21. $s^2 - 4s + 4 = 0$; 2 is a member of solution set **23.** $n^2 - 9 = 0;\ \{3, -3\}$
25. $x^2 - x - 42 = 0$ **27.** $t^2 - 6t + 9 = 0;\ (3)^2 - 6(3) + 9 = 0$

Exercises 9-2, Page 215

1. $\{2, -2\}$ **3.** $\{3, 2\}$ **5.** $\{7, -1\}$ **7.** $\{\frac{1}{2}, -2\}$ **9.** $\{-1, \frac{9}{2}\}$ **11.** $\{\frac{1}{5}, 4\}$
13. $\{3, -4\}$ **15.** $\{\frac{2}{3}, -\frac{1}{2}\}$ **17.** $\{\frac{5}{2}, 6\}$ **19.** $\{-\frac{7}{2}, \frac{1}{2}\}$ **21.** $\{-1, \frac{3}{4}\}$
23. $\{3\}$ (Double root) **25.** $\{-2, \frac{5}{6}\}$ **27.** $\{-6, 0\}$ **29.** $\{-\frac{5}{3}\}$ (Double root)
31. $\{2a, -2a\}$ **33.** $x^2 - 2x + 1 = 0$ **35.** $y^2 + 6y + 8 = 0$ **37.** $v^2 + v - 42 = 0$
39. $30w^2 + w - 1 = 0$ **41.** 4 **43.** 20 m by 30 m **45.** 8 in. **47.** $x = 2L$
49. 50 cm by 50 cm

Exercises 9-3, Page 220

1. $\{3, -2\}$ **3.** $\{4, 3\}$ **5.** $\{\frac{1}{2}, -1\}$ **7.** $\left\{\frac{-5 + \sqrt{13}}{2}, \frac{-5 - \sqrt{13}}{2}\right\}$ **9.** $\{\frac{1}{2}, \frac{3}{2}\}$
11. $\{\frac{3}{2}, -\frac{3}{2}\}$ **13.** $\{\frac{7}{2}, -\frac{1}{2}\}$ **15.** $\left\{\frac{-1 + i\sqrt{7}}{2}, \frac{-1 - i\sqrt{7}}{2}\right\}$ **17.** $\left\{\frac{-1 + \sqrt{33}}{4}, \frac{-1 - \sqrt{33}}{4}\right\}$
19. $\{\frac{5}{3}, 0\}$ **21.** $\left\{\frac{3 + i\sqrt{55}}{4}, \frac{3 - i\sqrt{55}}{4}\right\}$ **23.** $\{0, \frac{10}{3}\}$ **25.** $\left\{\frac{5 + \sqrt{57}}{2}, \frac{5 - \sqrt{57}}{2}\right\}$
27. $\{\frac{1}{2}, 7\}$ **29.** $\{1 + \sqrt{7}, 1 - \sqrt{7}\}$ **31.** $\left\{-\frac{1}{3a}, \frac{-3}{2a}\right\}$ **33.** $\{-3, 1\}$ **35.** $\{3, 1\}$
37. $\{-1, -\frac{1}{2}\}$ **39.** 4 ft, 7 ft **41.** 0.94 s, 5.31 s
43. $I = \frac{E + \sqrt{E^2 - 4PR}}{2R}, I = \frac{E - \sqrt{E^2 - 4PR}}{2R}$

Exercises 9-4, Page 221

1. $\{-6, -1\}$ **3.** $\{1, \frac{5}{2}\}$ **5.** $\{-\frac{1}{6}, 6\}$ **7.** $\{\frac{3}{4}\}$ **9.** $\{0, -\frac{7}{18}\}$ **11.** $\{10, -11\}$
13. $\{4, 1\}$ **15.** $\{1, \frac{5}{6}\}$ **17.** $\{1\}$ **19.** $\{\frac{5}{3}, -\frac{5}{3}\}$ **21.** $\left\{\frac{-1 + i\sqrt{3}}{2}, \frac{-1 - i\sqrt{3}}{2}\right\}$
23. $\left\{\frac{5 + \sqrt{17}}{4}, \frac{5 - \sqrt{17}}{4}\right\}$ **25.** $\{3 + \sqrt{15}, 3 - \sqrt{15}\}$ **27.** $\left\{\frac{-2 + i\sqrt{2}}{2}, \frac{-2 - i\sqrt{2}}{2}\right\}$
29. $\{2i, -2i\}$ **31.** $\{3, -3\}$ **33.** $\{-1 + i, -1 - i\}$ **35.** $\{3, -\frac{1}{2}\}$
37. $\{2 + \sqrt{3}, 2 - \sqrt{3}\}$ **39.** $\left\{\frac{1 + i\sqrt{7}}{4}, \frac{1 - i\sqrt{7}}{4}\right\}$ **41.** $x^2 - 4x + 3 = 0$
43. $x^2 - 3x - 18 = 0$ **45.** $6p^2 - p - 1 = 0$ **47.** $t^2 - 4t = 0$ **49.** 13, 7 or $-13, -7$
51. 5 and 20 **53.** 5 by 4 **55.** 8 m **57.** $t = \frac{-v + \sqrt{v^2 + 2as}}{a}$

Exercises 10-1, Page 226

1. y dependent, x independent **3.** p dependent, V independent **5.** Multiplication by 3 **7.** Square r and subtract from 2 **9.** $f(x) = 5 - x$ **11.** $f(t) = t^2 - 3t$ **13.** 0, 1, −1 **15.** 7, −5 **17.** $-2, -\frac{5}{4}$ **19.** −1, 8 **21.** $a^2 - 2a^4, \dfrac{a-2}{a^2}$ **23.** $\frac{1}{2}$, Undefined **25.** 2, 6 **27.** 2, 2 **29.** Function **31.** Function **33.** $p = 4s$ **35.** $t = 0.02n$ **37.** $s = 3v$ **39.** $i = V/5$ **41.** $F = 0.25 + 0.50\,d$ (d in km) **43.** $N = b^2 + 3b + 2$

Exercises 10-2, Page 230

1.

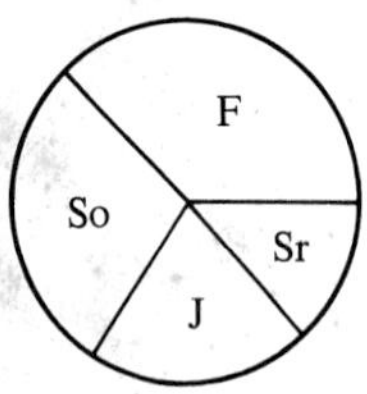

3.

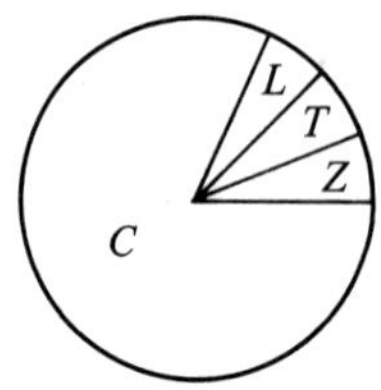

5.

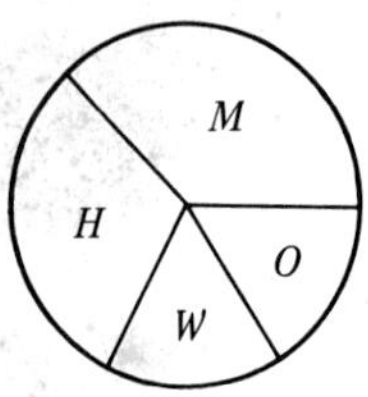

7.

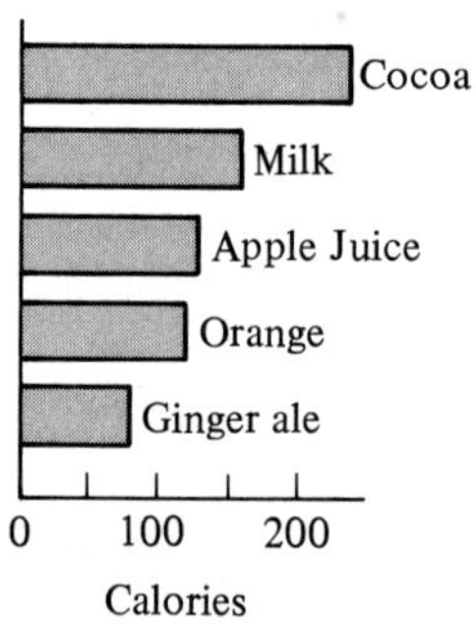

9.

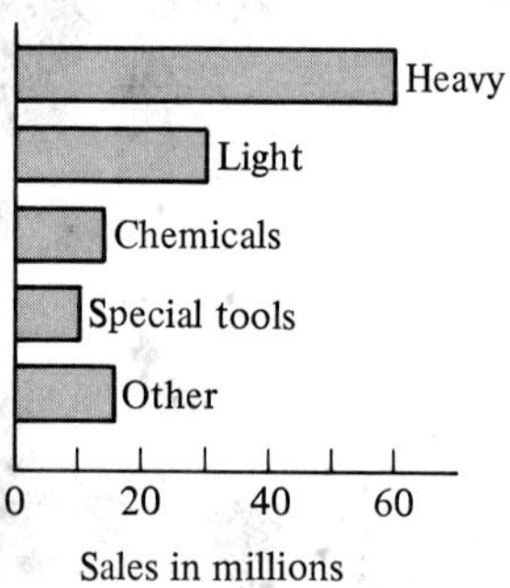

11.

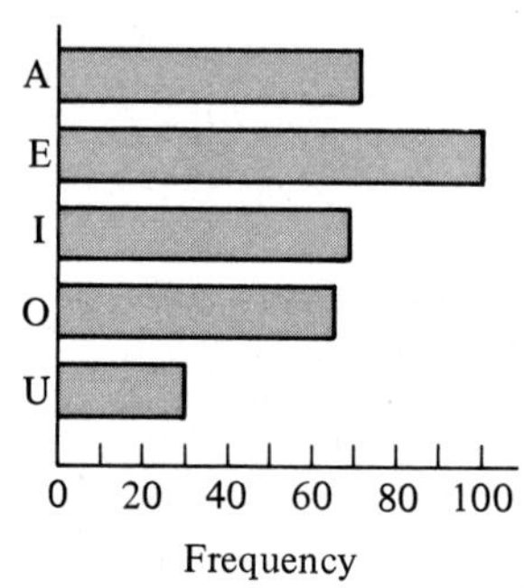

13.

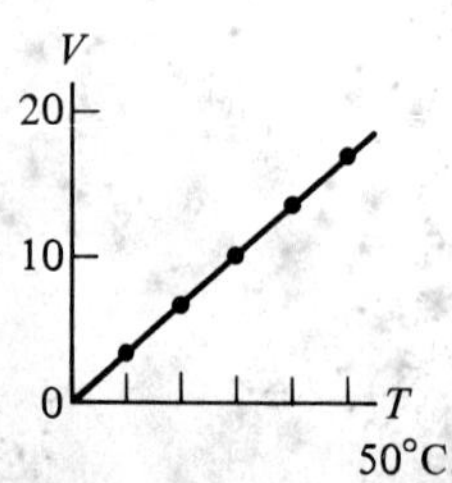

15.

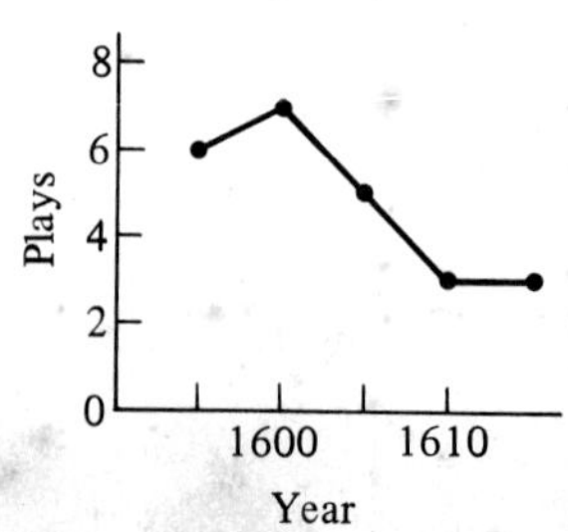

17. 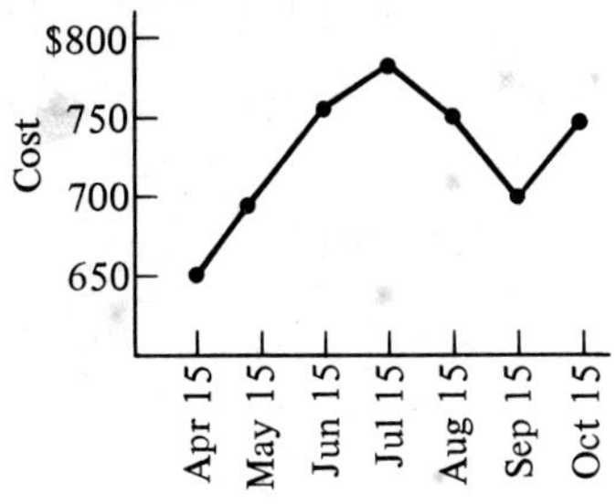

19. (Approximate)
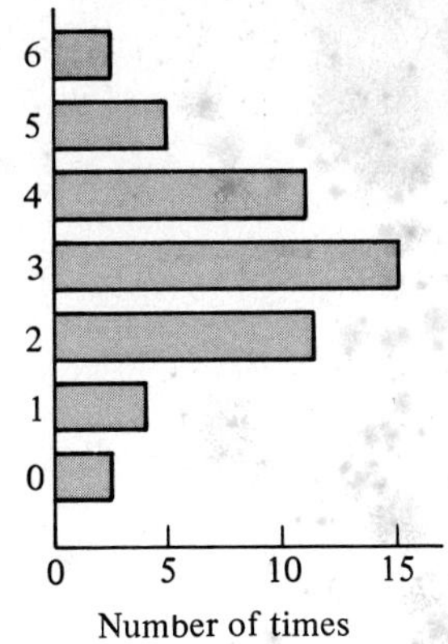

21. No specific solution

Exercises 10-3, Page 234

1. $A(2, 1)$, $B(-2, 3)$

3. $E(4, 0)$, $F(-\frac{3}{2}, \frac{1}{2})$

5. 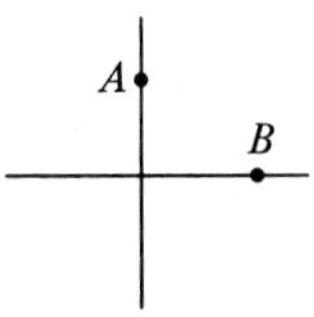

7. 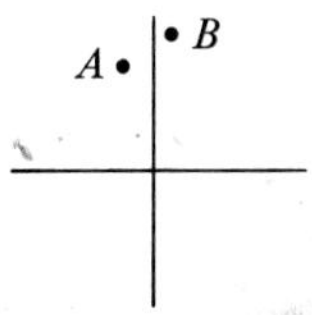

9. 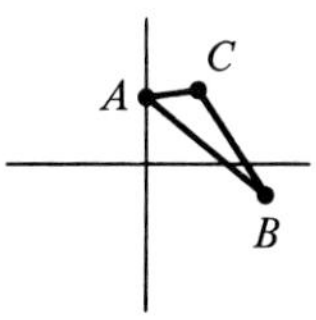

11. 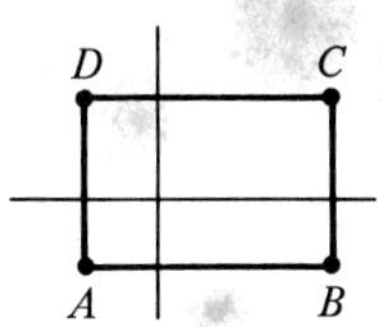

13. 0 **15.** I and III **17.** IV **19.** (0, 0), (4, 0), (0, 4), (4, 4) **21.** (1, 2)

23. Circle, center at (0, 0), radius of 3

Exercises 10-4, Page 240

1. 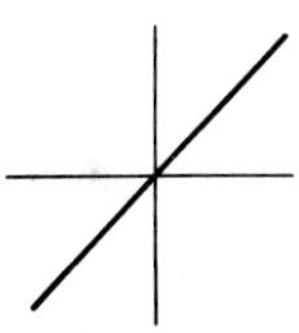

3. 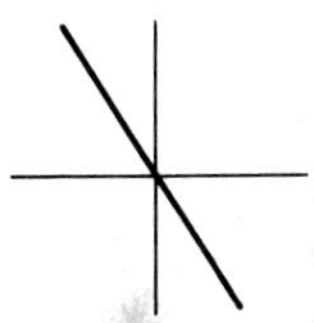

5. 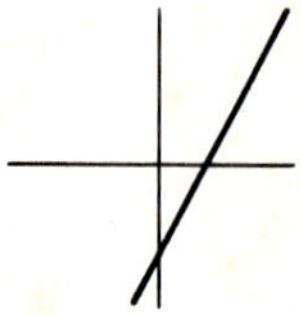

7. 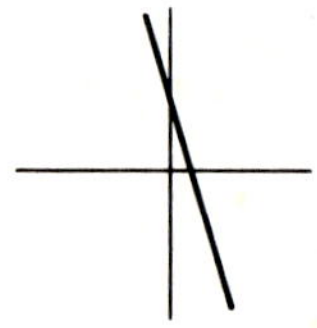

9. 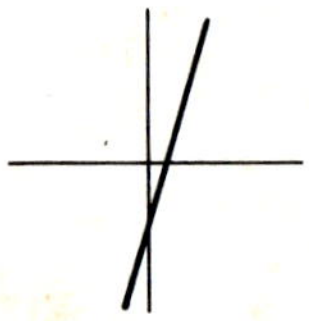

11. 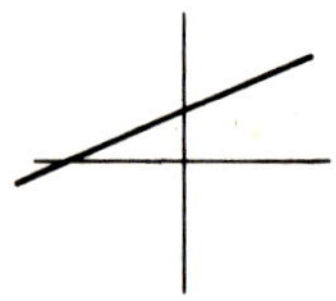

13. 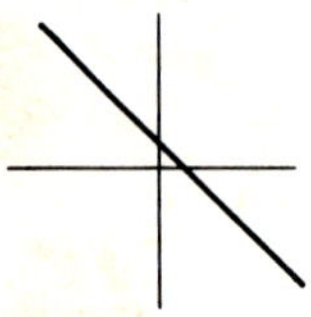

15. 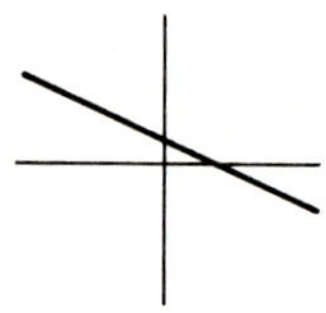

17. 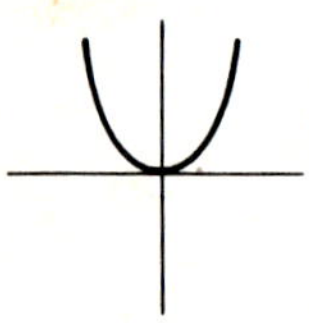

19. 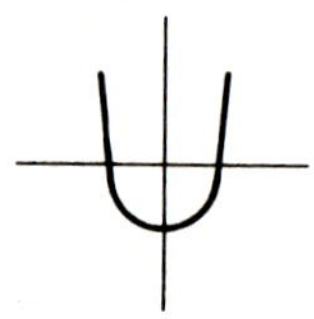

21. 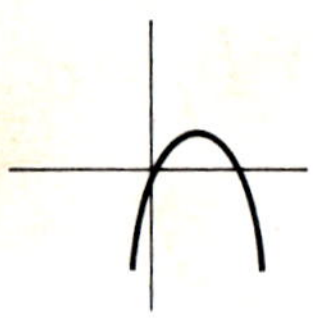

23. 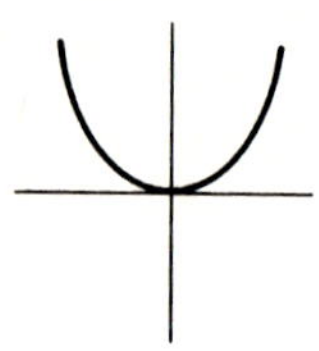

25. 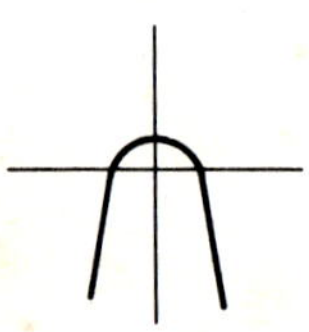

27. 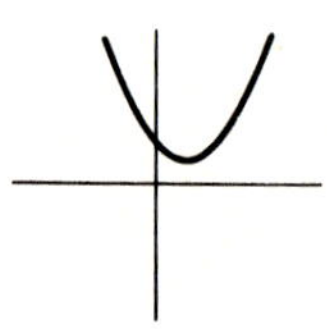

29. 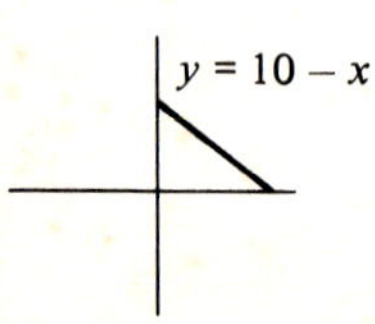

31. $p = 1.5x - 200$

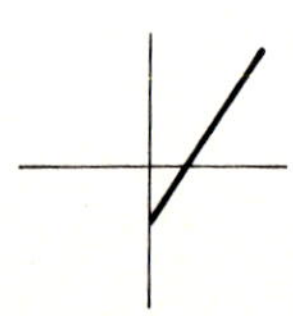

33. $f(d) = \pi d^2/4$

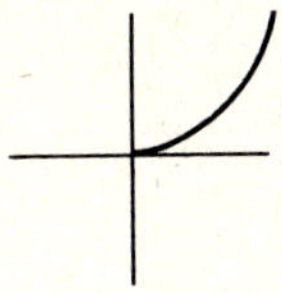

35. $d = x - x^2$

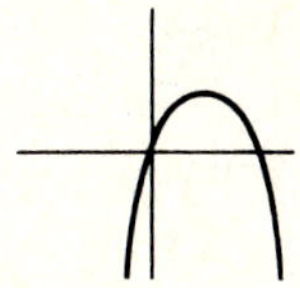

37. Relation

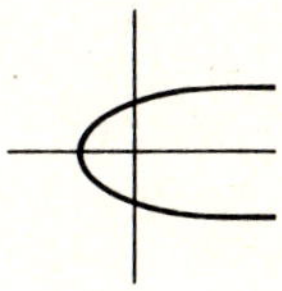

39. Function

41. $\{1.7, -1.2\}$

43. $\{0.7, -7.7\}$

Exercises 10-5, Page 244

1.

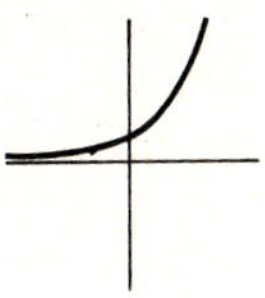

3.

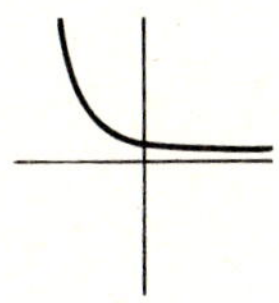

5.

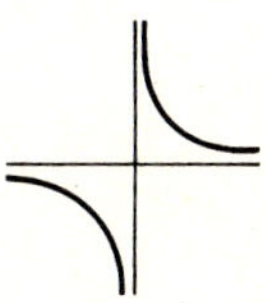

7.

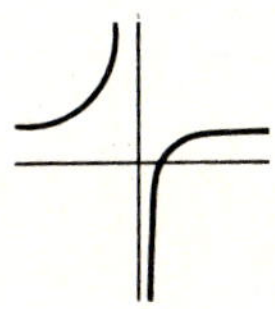

9.

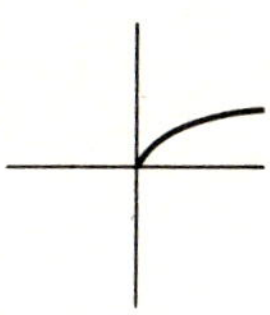

11.

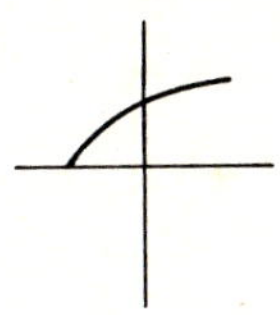

13.

15.

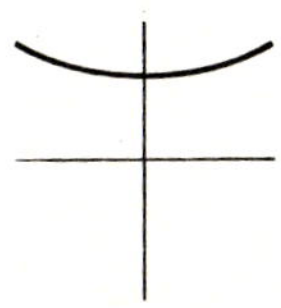

17. 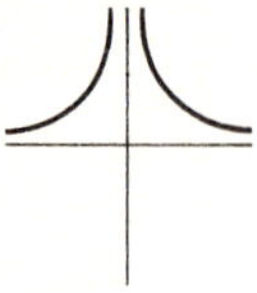

19. 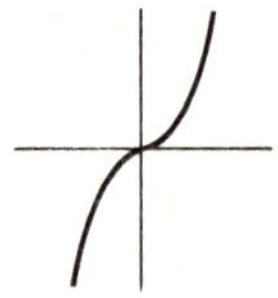

21. 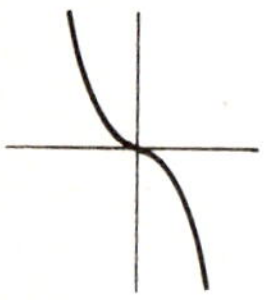

23. 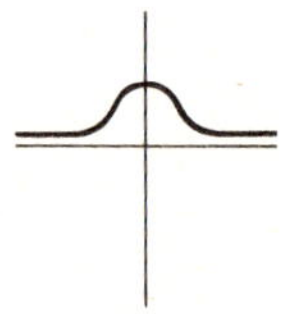

25. 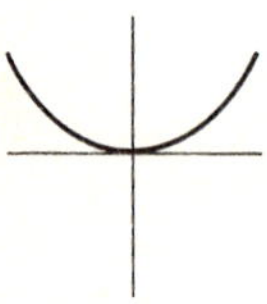

27. 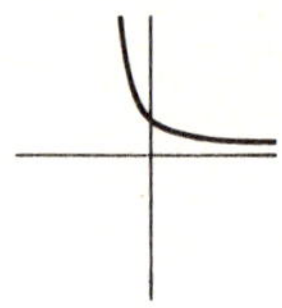

29. 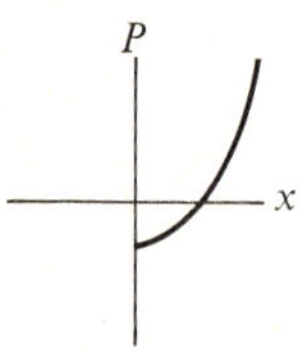

31. 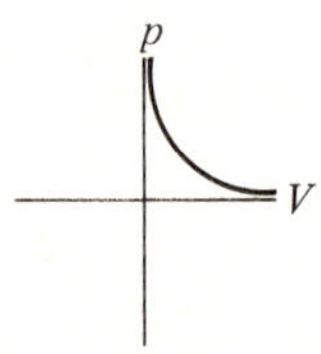

33. 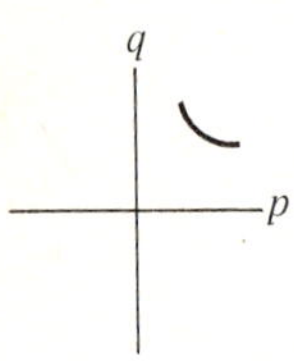

Exercises 10-6, Page 249

1. 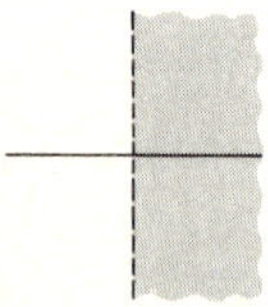

3. 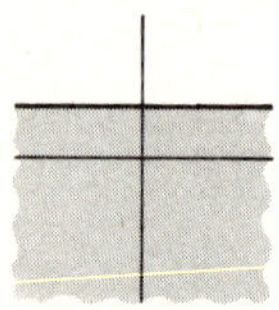

5. 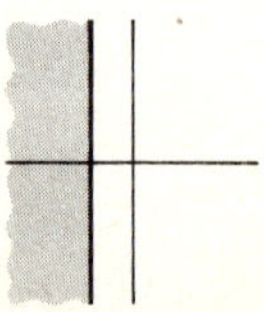

7. 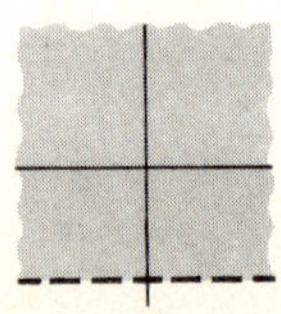

9. 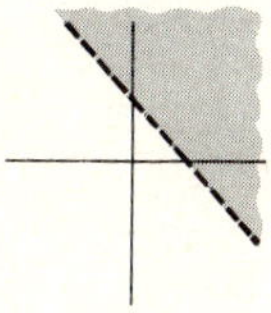

11. 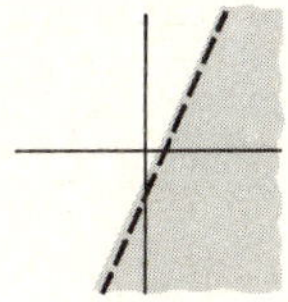

13. 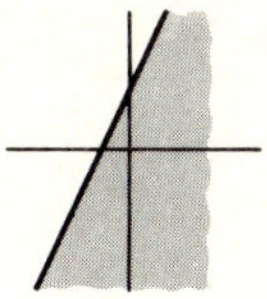

15. 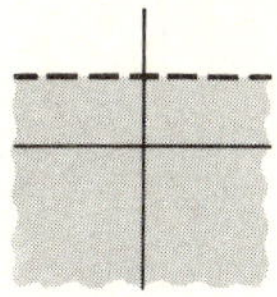

17. 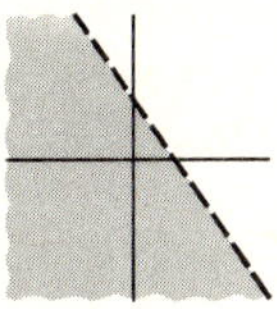

19. 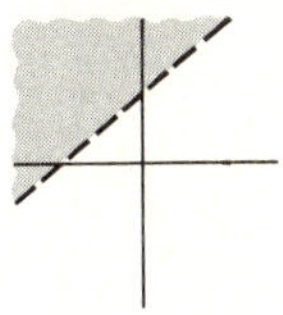

21. 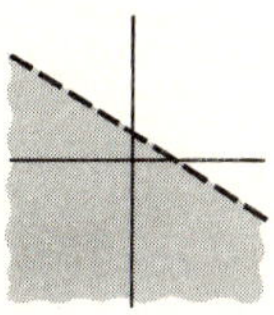

23. 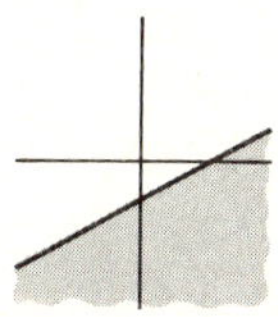

25. 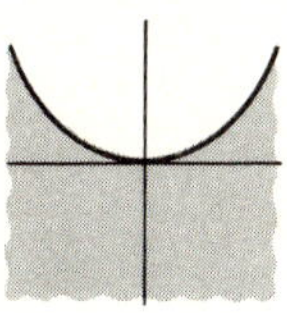

27. 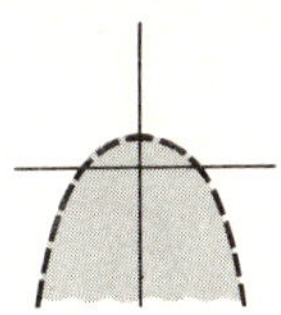

29. $s \geq 31$ km/h

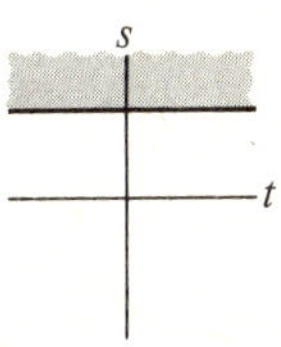

31. $y < 2x + 6$

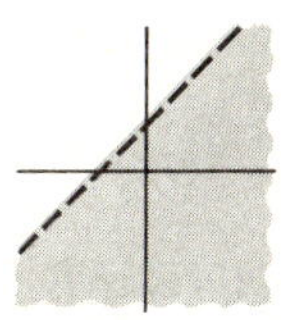

Exercises 10-7, Page 254

1. $\frac{3}{2}$ **3.** -1 **5.** $\frac{2}{3}$ **7.** $-\frac{3}{2}$ **9.** $m = 3, b = 1$ **11.** $m = -1, b = 6$

13. $m = \frac{5}{3}, b = 0$ **15.** $m = -4, b = 0$ **17.** $m = 1, b = 3$ **19.** $m = \frac{1}{3}, b = -\frac{1}{2}$

21. $m = 1, b = -9$

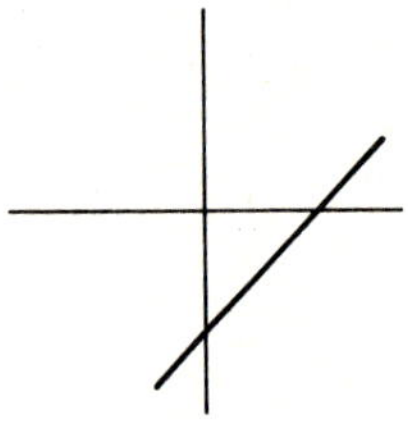

23. $m = -\frac{3}{2}, b = 3$

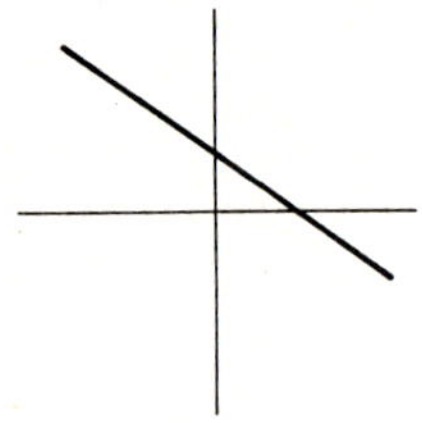

25. $y = 4x - 7$ **27.** $y = -5x - 6$ **29.** $y = -x + 2$ **31.** $y = x - 3$ **33.** 20 ft

35. $C = 2n + 400$; The fixed cost is the y-intercept and the cost per unit is the slope.

Exercises 10-8, Page 260

1. $f^{-1}(x) = x - 4$

3. $f^{-1}(x) = \dfrac{x + 1}{3}$

5. $f^{-1}(x) = \dfrac{7 - x}{5}$

7. $f^{-1}(x) = \dfrac{2x + 3}{2}$

9. $f^{-1}(x) = 1/x$

11. $f^{-1}(x) = \sqrt[3]{x}$

13. $f^{-1}(x) = x + 1$

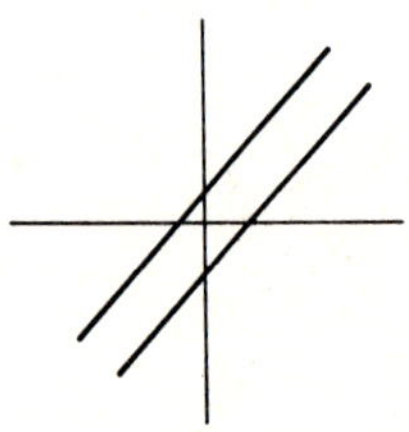

15. $f^{-1}(x) = \dfrac{x - 3}{4}$

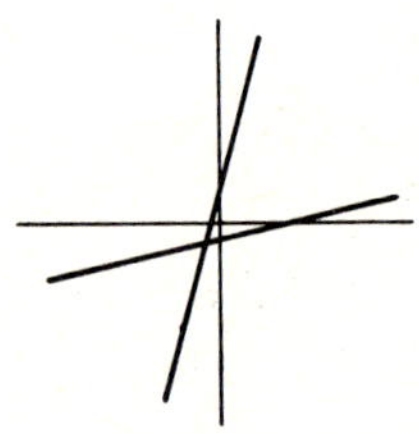

17. $f^{-1}(x) = -\dfrac{8x + 1}{2}$

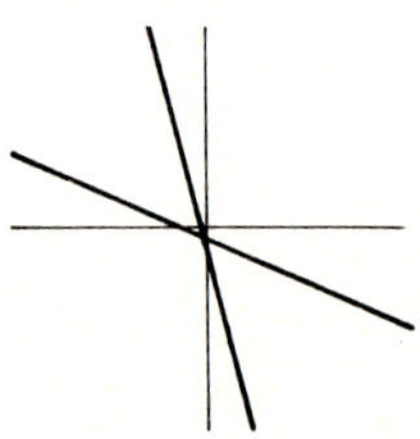

19. $f^{-1}(x) = -2/x$

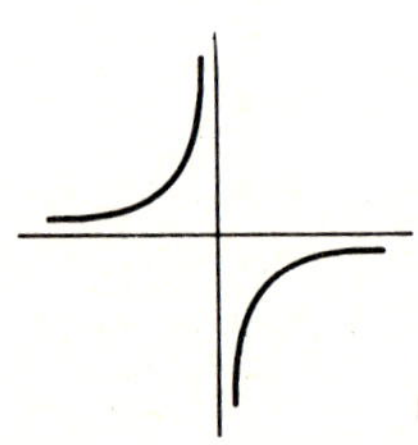

21. $f^{-1}(f(x)) = (x + 4) - 4 = x$

23. $f^{-1}(f(x)) = \dfrac{(3x - 1) + 1}{3} = x$

25. $f^{-1}(f(x)) = \dfrac{2\left(\dfrac{2x - 3}{2}\right) + 3}{2} = x$

27. $f^{-1}(f(x)) = \dfrac{1}{(1/x)} = x$

29. For example: $f(-1) = f(1) = 2$

31. For example: $f(-2) = f(0) = f(2) = 0$

33. $p = f(s) = 4s$; $f^{-1}(s) = s/4$

35. $f^{-1}(x) = \dfrac{x + 50}{10}$; They can use it to find the number of items they must sell in order to earn a given profit.

Exercises 10-9, Page 262

1. $1, \frac{1}{2}$ **3.** $-20, -11$ **5.** $-9, -16$ **7.** $0, -r^3/8$ **9.** $10, 2w^4 + 2$

11. $-2, 8q^3 - 3$

13.

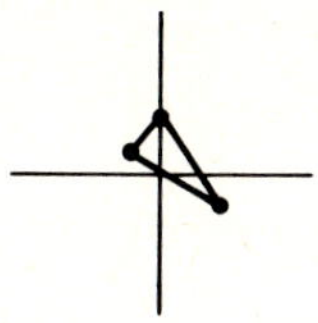

15.

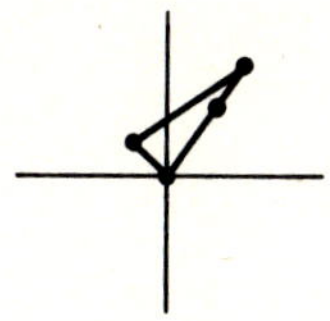

17.

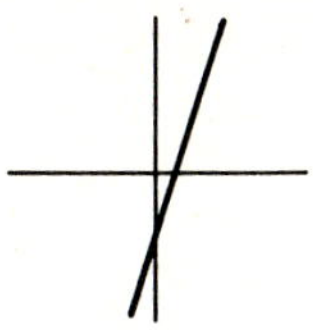

19.

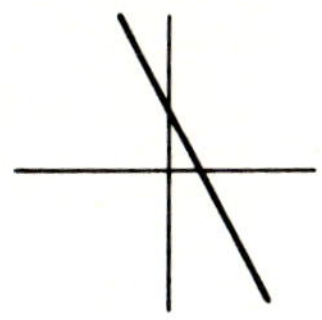

21.

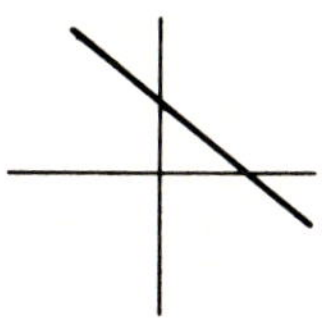

23.

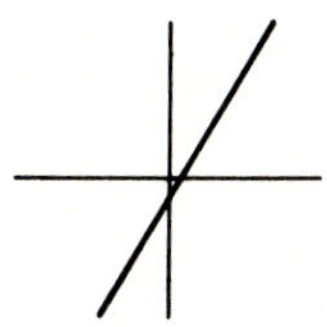

25.

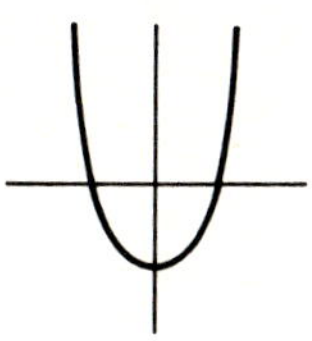

27.

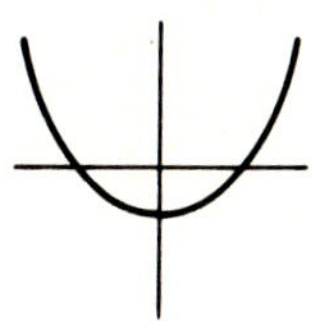

29.

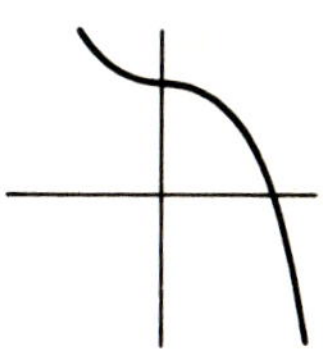

31.

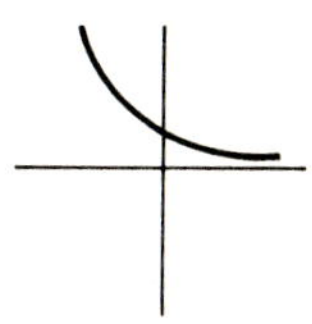

33.

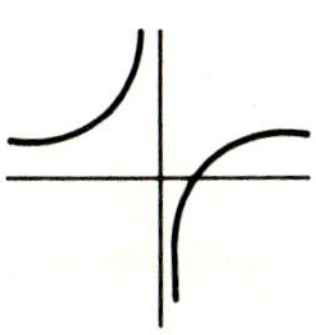

35.

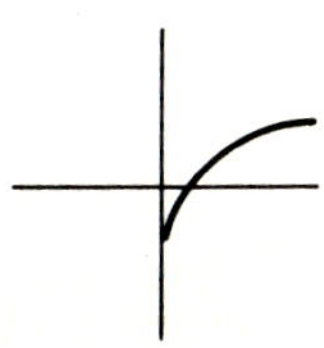

37. $y = 4x$

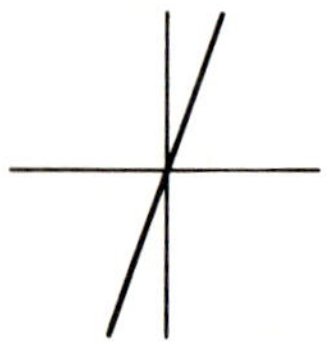

39. $y = \dfrac{17 - 27x}{16}$

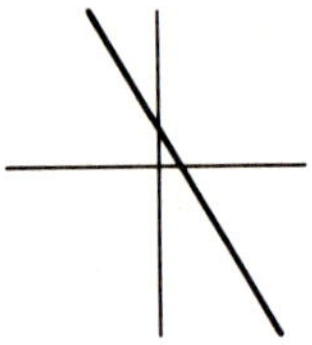

41. $m = -\frac{2}{3}, b = -5$

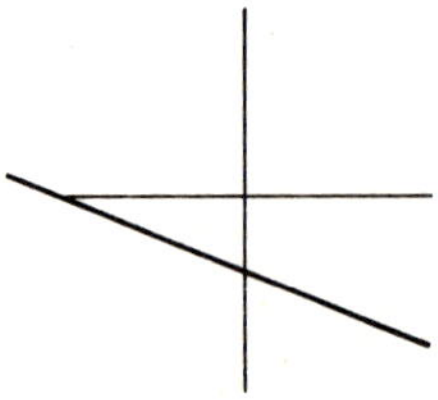

43. $m = \frac{3}{2}, b = 1$

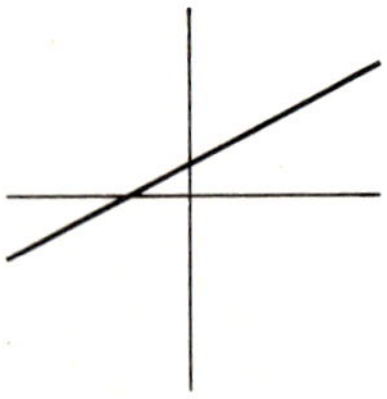

45.

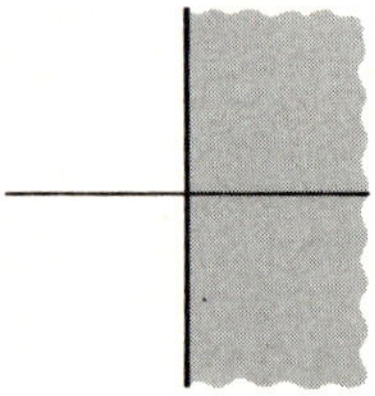

47.

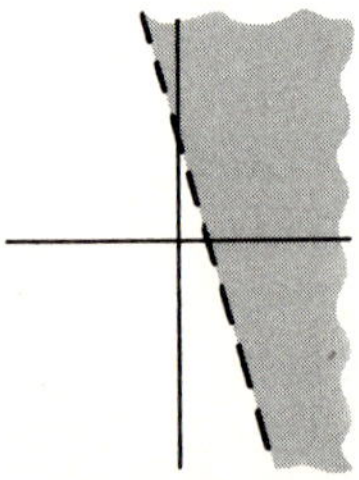

49.

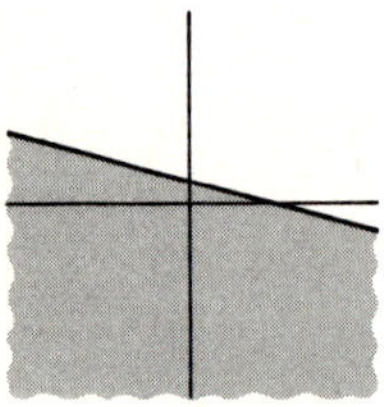

51.

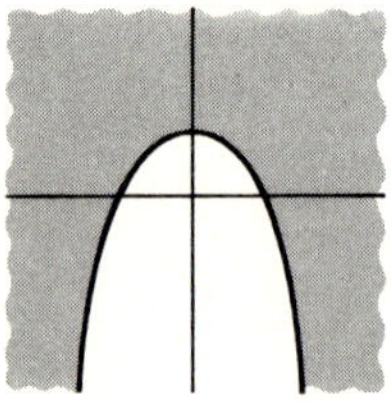

53. It has an inverse.

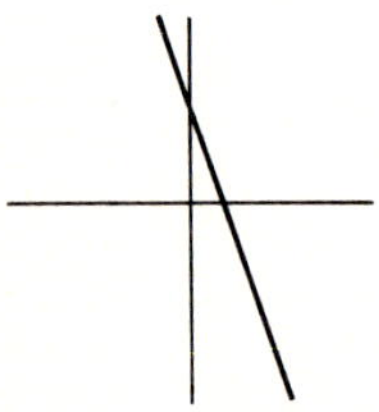

55. It has no inverse.

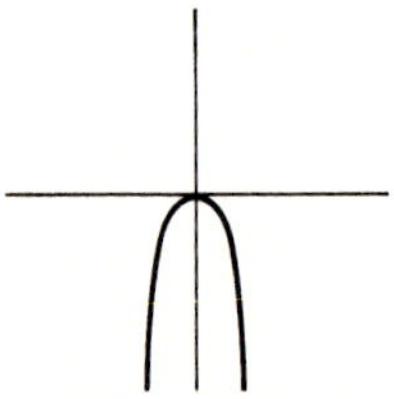

57. $f^{-1}(x) = 2x + 6$;
$f^{-1}(f(x)) = 2(\frac{1}{2}x - 3) + 6 = x$

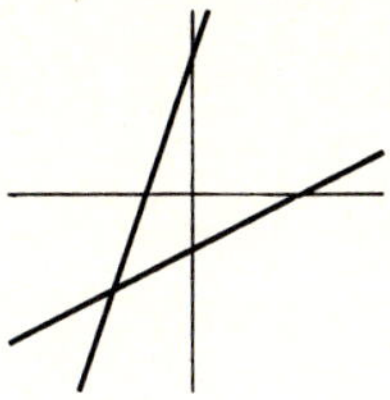

59. $f^{-1}(x) = \dfrac{11 - 3x}{3}$;

$$f^{-1}(f(x)) = \frac{11 - 3\left(\dfrac{-3x + 11}{3}\right)}{3} = x$$

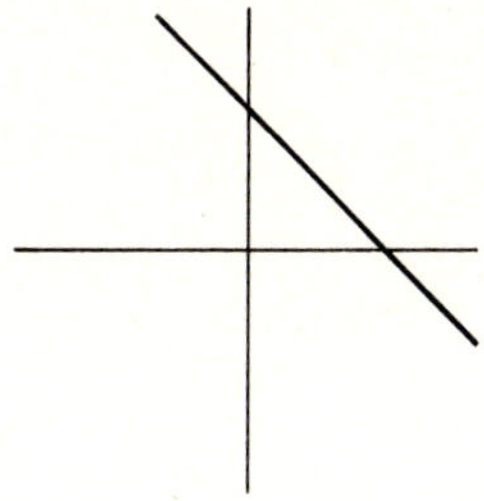

61. $f^{-1}(x) = -3/x$; $f^{-1}(f(x)) = -\dfrac{3}{(-3/x)} = x$

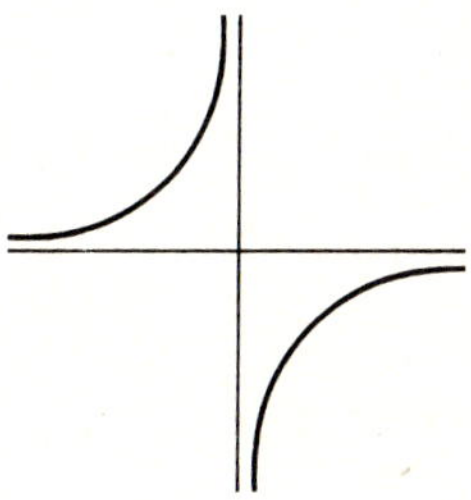

63. $f^{-1}(x) = \sqrt[3]{x - 1}$;

$f^{-1}(f(x)) = \sqrt[3]{(x^3 + 1) - 1} = x$

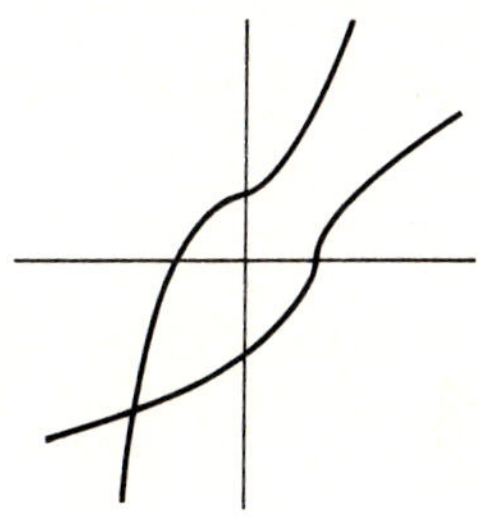

65.

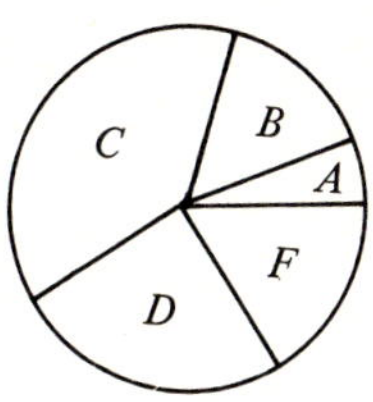

67.

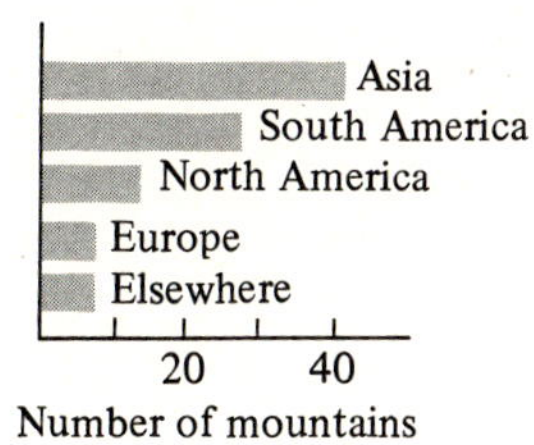

69.

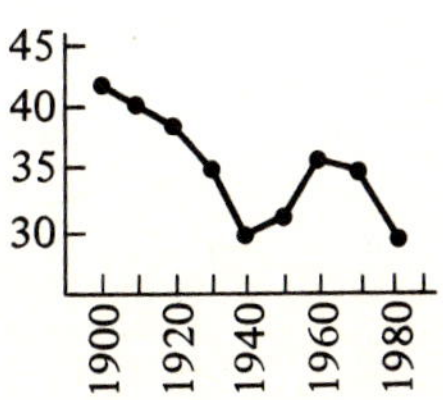

71. $p = 2x + 8$

73. $IQ = 20M/3$

75. $V = 2A$

77. $T = 160/I^2$

79. $h = \frac{2}{3}w$

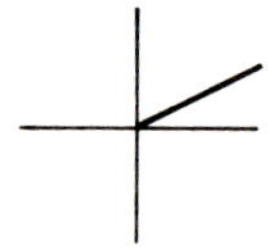

81. $N = 2x^2$

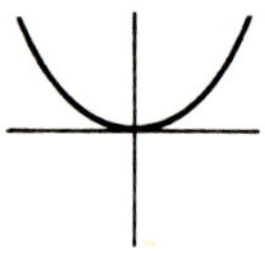

83. $A = 400w - w^2$

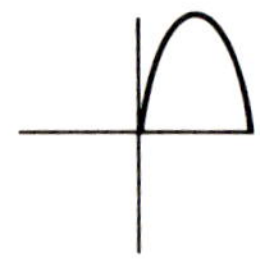

Exercises 11-1, Page 268

1. $\log_3 27 = 3$ **3.** $\log_4 256 = 4$ **5.** $\log_4(\frac{1}{16}) = -2$ **7.** $\log_2(\frac{1}{64}) = -6$ **9.** $\log_8 2 = \frac{1}{3}$
11. $\log_{1/4}(\frac{1}{16}) = 2$ **13.** $81 = 3^4$ **15.** $9 = 9^1$ **17.** $5 = 25^{1/2}$ **19.** $3 = 243^{1/5}$
21. $0.1 = 10^{-1}$ **23.** $16 = (0.5)^{-4}$ **25.** 2 **27.** -2 **29.** 343 **31.** $\frac{1}{4}$ **33.** 9
35. $\frac{1}{64}$ **37.** 0.2 **39.** -3 **41.** $t = \log_2(N/1000)$ **43.** $N = N_o e^{-kt}$

Exercises 11-2, Page 272

1.

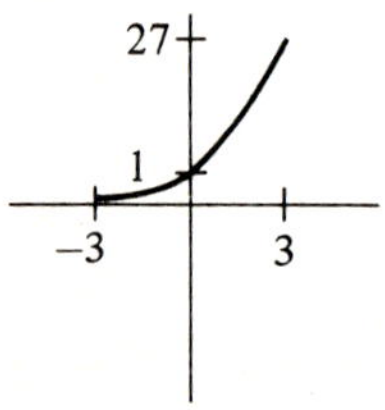

3.

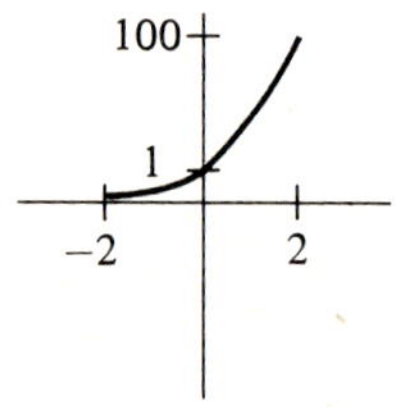

5.

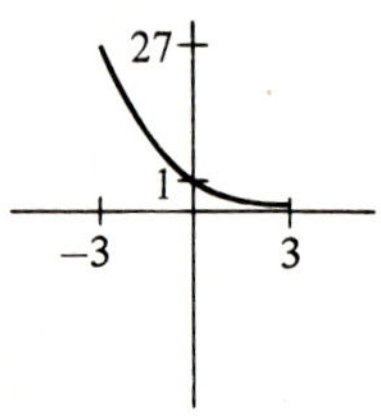

7.

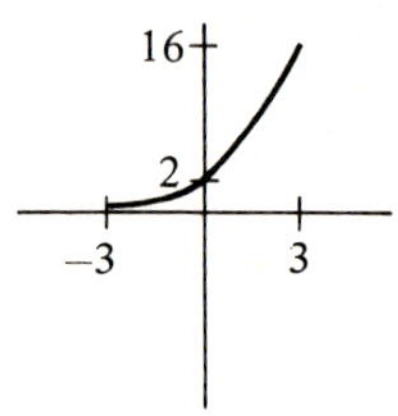

9.

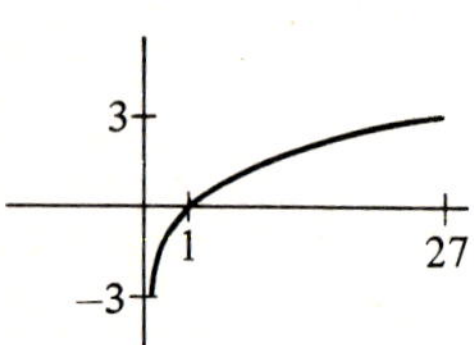

11.

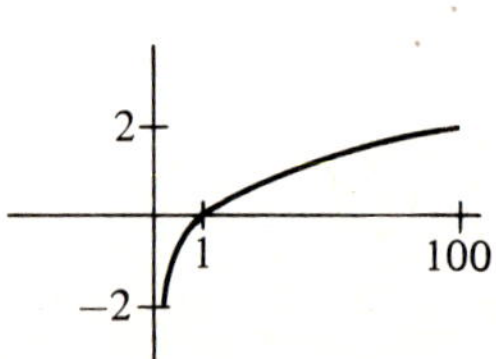

13.

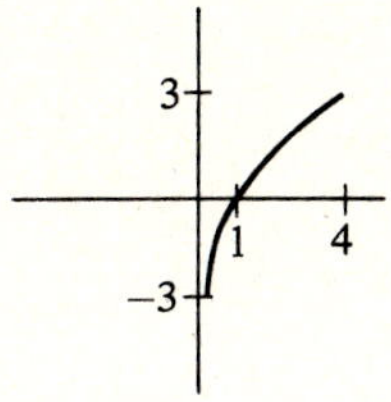

15.

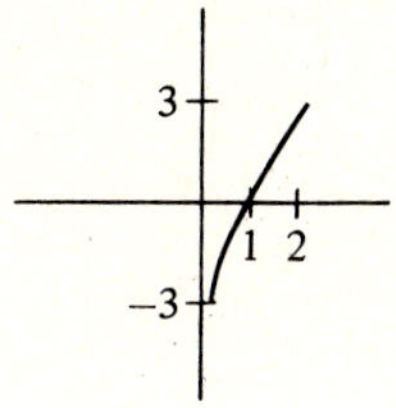

17. They are inverse functions.

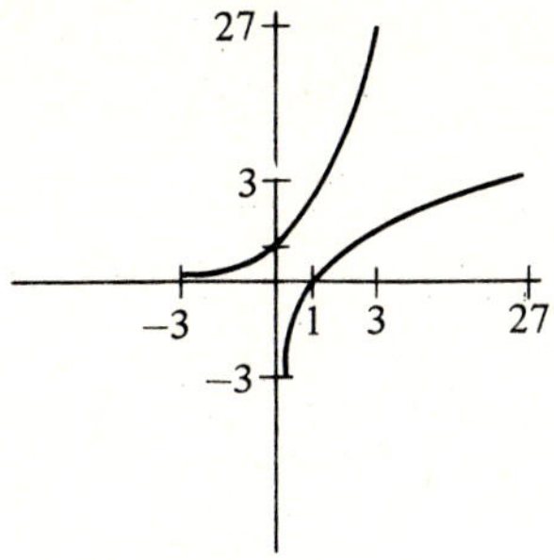

19.

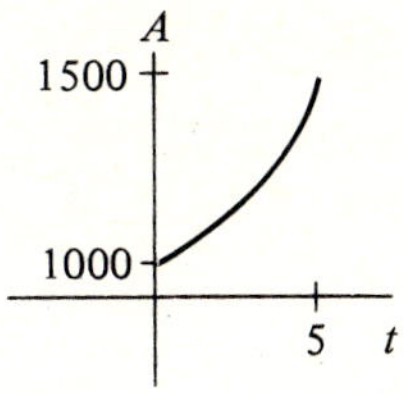

Exercises 11-3, Page 276

1. $\log_5 x + \log_5 y$ **3.** $\log_7 5 - \log_7 a$

5. $3 \log_2 a$ **7.** $\log_6 a + \log_6 b + \log_6 c$

9. $\frac{1}{4} \log_5 y$ **11.** $\frac{1}{2} \log_2 x - 2 \log_2 a$

13. $\log_b ac$ **15.** $\log_5 3$ **17.** $\log_b x^{3/2}$ **19.** $\log_e 4n^3$ **21.** -5 **23.** 2.5 **25.** $\frac{1}{2}$

27. $\frac{3}{4}$ **29.** $2 + \log_3 2$ **31.** $-1 - \log_2 3$ **33.** $\frac{1}{2}(1 + \log_3 2)$ **35.** $4 + 3 \log_2 3$

37. $3 + \log_{10} 3$ **39.** $-2 + 3 \log_{10} 3$ **41.** 0.602 **43.** -0.301 **45.**

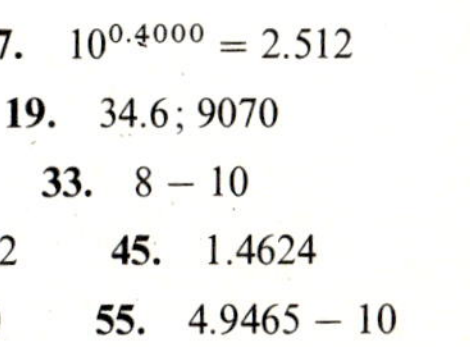

Exercises 11-4, Page 284

1. $10^{0.6667} = 4.642$ **3.** $10^{0.9000} = 7.943$ **5.** $10^{0.1667} = 1.468$ **7.** $10^{0.4000} = 2.512$

9. 3 **11.** 0.3010 **13.** 4; 2 **15.** 1; 1 **17.** 7.86; 0.957 **19.** 34.6; 9070

21. 1 **23.** 3 **25.** 0 **27.** 9 − 10 **29.** 7 − 10 **31.** 1 **33.** 8 − 10

35. 8 − 10 **37.** 0.8407 **39.** 0.7007 **41.** 3.7634 **43.** 2.0212 **45.** 1.4624

47. 2.9440 **49.** 8.8075 − 10 **51.** 9.4814 − 10 **53.** 6.9360 − 10 **55.** 4.9465 − 10

57. 6.1271 **59.** 41.7782 **61.** 100 dB, 1000 times as great

Exercises 11-5, Page 290

1. 9.52 **3.** 76.5 **5.** 19,100 **7.** 115,000 **9.** 2.65 **11.** 1.36 **13.** 0.0198

15. 12.9 **17.** 41.6 **19.** 41.1 **21.** 0.282 **23.** 5,090,000 **25.** 5.18 **27.** 3.85

29. 2.38 **31.** 0.770 **33.** 1.99 **35.** 7.20 **37.** 3.74 **39.** 4.12 **41.** 20.1

43. 221 **45.** 23.4 m^2 **47.** 112 **49.** 17.6 cm **51.** \$148

Exercises 11-6, Page 294

1. 0.8415 **3.** 9.5007 − 10 **5.** 1.4641 **7.** 3.4972 **9.** 4.372 **11.** 87.34
13. 287.4 **15.** 0.00002862 **17.** 33.02 **19.** 139,200 **21.** 4638 **23.** $(2.532)(10^{-4})$
25. $47,440,000 **27.** 9.632 cm **29.** $(1.971)(10^8)$ mi^2 **31.** 1474 m/s

Exercises 11-7, Page 300

1. $\frac{1}{3}$ **3.** −1 **5.** 2.1282 **7.** 9.8946 − 10 **9.** 5.2983 **11.** 3.9890 **13.** 1.1786
15. 7.1866 − 10 **17.** 0.3646 **19.** 5.8377 **21.** 0.9745 **23.** 6.5118 **25.** 3.8522
27. 9.8374 − 10 **29.** 702 years

Exercises 11-8, Page 302

1. 10,000 **3.** 6 **5.** 6 **7.** $\frac{1}{5}$ **9.** $\log_3 2 + \log_3 x$ **11.** $2\log_3 t$ **13.** $2 + \log_2 7$
15. $1 + \frac{1}{2}\log_4 3$ **17.**

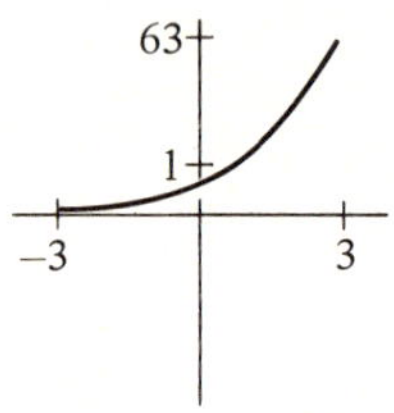

19.

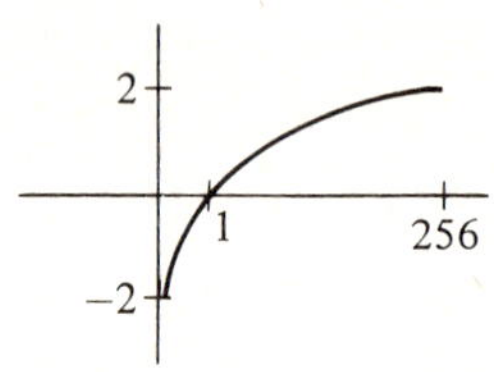

21. 34.8 **23.** $(1.31)(10^{-4})$ **25.** 584.4 **27.** 16.3
29. $(7.67)(10^{-4})$ **31.** 9.718 **33.** 1260 **35.** $(3.35)(10^7)$ **37.** 1463 **39.** 4.81
41. 1.05 **43.** 36.17 **45.** 693 **47.** 35.2 **49.** 2.1815 **51.** 0.7276 **53.** 1270 m^2
55. 84,200,000 **57.** 39.8% **59.** 499 s **61.** 736.0 **63.** 1370 **65.** 4420

Exercises 12-1, Page 307

1. Let x = larger number; y = smaller number.
Equations: $x + y = 19$; $x - y = 11$.

3. Let x = number of boys; y = number of girls.
Equations: $x + y = 500$; $x - y = 74$

5. Let d = number of dimes; q = number of quarters.
Equations: $0.10d + 0.25q = 3.50$; $d = q$.

7. Let h = cost of home insurance; c = cost of car insurance.
Equations: $h + c = 490$; $c = 2h + 10$.

9. Let x = length of longer part; y = length of shorter part.
Equations: $x + y = 12$; $x - y = \frac{3}{2}$.

11. Let x = amount earned in the winter; y = amount earned in the summer.
Equations: $x + y = 850$; $x = 3y + 70$.

13. Let x = amount invested at 6%; y = amount invested at 7%.
Equations: $x + y = 5000$; $0.06x + 0.07y = 330$.

15. Let x = speed of faster train; y = speed of slower train.
Equations: $x - y = 5$; $5x + 5y = 425$.

17. Let r = smaller resistance; R = larger resistance.
Equations: $r + R = 14$; $3r + 5R = 60$.

19. Let F = number of Fahrenheit degrees; C = number of Celsius degrees.
Equations: $F = 32 + \frac{9}{5}C$; $F = C$.

21. Let y = number of pennants won by Yankees; x = number of pennants won by other teams. Equations: $y = 2x - 1$; $x + y = 44$.

23. Let u = units digit; t = tens digit. Equations: $u + t = 9$; $(10u + t) = (10t + u) + 9$.

Exercises 12-2, Page 312

1. $x = 4.0, y = 0.0$

3. $x = 1.3, y = 3.3$

5. $r = 2.0, s = 1.0$

7. $x = 2.0, y = 2.0$

9. Dependent

11. $x = -2.0, y = 1.5$

13. $a = -1.9, b = -0.5$

15. $p = 2.4, q = 1.1$

17. $x = 2.5, y = -0.9$

19. Dependent

21. $x = -10.0, y = 6.0$

23. $x = 0.8, y = 3.6$

25. Let x = larger number, y = smaller number; $x + y = 18$, $x - y = 6$; $x = 12$, $y = 6$.

27. Let l = length, w = width; $2l + 2w = 28$, $l - w = 4$; $l = 9$ in., $w = 5$ in.

29. Let r_1 = rate (in mi/h) he rows in still water; r_2 = rate (in mi/h) of stream; $2(r_1 + r_2) = 10$, $4(r_1 - r_2) = 8$; 3.5 mi/h in still water, 1.5 mi/h is the rate of the stream.

Exercises 12-3, Page 316

1. $x = 6, y = 11$

3. $x = \frac{10}{3}, y = \frac{4}{3}$

5. $u = 7, v = 3$

7. $x = -1, y = 2$

9. $x = 2, y = 1$

11. $x = 0, y = 0$

13. $x = 6, y = 0$

15. $u = -\frac{6}{17}, k = \frac{20}{17}$

17. $x = \frac{81}{10}, y = \frac{63}{10}$

19. $x = -2, y = -1$

21. $r = 4, s = -2$

23. $x = \frac{32}{31}, k = -\frac{69}{31}$

25. Let x = larger number, y = smaller number; $x - 2y = 3$, $x + y = 12$; $x = 9$, $y = 3$.

27. Let g = number of field goals, f = number of foul shots; $2g + f = 90$, $g = 2f$; $g = 36$, $f = 18$.

29. Let x = weight in air, y = weight in water;

$$\frac{x}{x - y} = 10, \; x + y = 30; \; x = \tfrac{300}{19} \text{ lb}$$

Exercises 12-4, Page 321

1. $x = 3, y = 1$

3. $x = 6, y = 4$

5. $m = 3, n = 6$

7. $d = -1, t = 4$

9. $x = -\frac{21}{11}, n = -\frac{6}{11}$

11. $x = 9, y = 2$

13. Inconsistent

15. $a = 1, b = 2$

17. $p = \frac{3}{2}, q = -\frac{1}{12}$

19. $x = -14, y = -3$

21. $m = -4, n = 6$

23. $x = 1, y = 1$

25. Let x = larger number, y = smaller number; $x + y = 28$, $3y = x + 16$; $x = 17$, $y = 11$.

27. Let W = number in Wyoming, I = number in Idaho; $3W = 2I - 19$, $2W = I + 2$; $W = 23$, $I = 44$.

29. Let x = amount invested at 6%; y = amount invested at 8%; $x + y = 6000$; $0.06x + 0.08y = 400$; $x = \$4,000$, $y = \$2,000$

Exercises 12-5, Page 323

1. $x = \frac{12}{5}, y = \frac{6}{5}$
3. $x = 4, y = 4$
5. $x = 4, y = 4$
7. $p = \frac{18}{5}, q = \frac{2}{5}$
9. $u = 1, v = -1$
11. $a = 12, b = -3$
13. $x = -9, y = 11$
15. $y = 1, z = -\frac{1}{2}$
17. Dependent
19. $x = -\frac{7}{3}, y = 2$
21. $x = \frac{43}{19}, y = -\frac{22}{19}$
23. $s = \frac{94}{107}, t = -\frac{22}{107}$
25. $x = -1, y = 5$
27. $x = 12, y = -24$
29. $r = \frac{1}{3}, s = \frac{1}{5}$
31. $x = 100, y = -1$
33. $x = \frac{115}{28}, y = \frac{265}{14}$
35. $x = 1.0, y = 0.3$
37. $x = 1.6, y = -1.2$
39. $u = 2.0, v = -6.0$
41. Inconsistent
43. 4 and 15
45. 287 boys, 213 girls
47. 10 dimes, 10 quarters
49. \$160 for home, \$330 for car
51. $\frac{27}{4}$ cm from one end
53. \$655 in winter, \$195 in summer
55. \$2000 at 6%, \$3,000 at 7%
57. 45 mi/h, 40 mi/h
59. 9 ohms, 5 ohms
61. $-40°$
63. Yankees won 29 and others won 15
65. 45
67. 24 and 14
69. \$3000 at 6%, \$4000 at 7%
71. $i_1 = -\frac{3}{22}$A, $i_2 = -\frac{39}{110}$A
73. $T_2 = 25.0$ lb, $T_3 = 43.3$ lb
75. 7 dimes, 8 nickels
77. \$2,800 at 5%, \$2,200 at 6%
79. 49°C and 97°C
81. \$4.50/h and \$3.00/h
83. 45 mi/h, 55 mi/h
85. 34
87. Empty set

Exercises 13-1, Page 331

1. 25° **3.** 102° **5.** 156.25° **7.** 67.1° **9.** $\angle ADC$ **11.** $\triangle BCD$
13. $\angle DOA$, $\angle COE$
15. $\angle COB$, $\angle BOA$, $\angle AOF$, $\angle FOE$
17. $\angle COB$ and $\angle BOA$, $\angle BOA$ and $\angle AOF$, $\angle AOF$ and $\angle FOE$
19. $\angle COB$ and $\angle BOA$, $\angle BOA$ and $\angle AOF$, $\angle AOF$ and $\angle FOE$
21. $\angle 3$ and $\angle 4$
23. $\angle 2$ and $\angle 4$, $\angle 3$ and $\angle 5$
25. 60° **27.** 60° **29.** 85° **31.** 140° **33.** 40° **35.** 90° **37.** 1 **39.** No.

Exercises 13-2, Page 338

1. 12 ft **3.** 123 mm **5.** 896 m **7.** 84.8 in. **9.** 193.8 mm **11.** 57.1 in.
13. 2.6 m **15.** 0.6 mi **17.** 168 cm **19.** 230.6 ft **21.** 94.8 cm **23.** 35 ft
25. 40 cm **27.** $p = 2s + a$ **29.** $p = \pi r + 2r$ **31.** $a = p - b - c$ **33.** $r = c/2\pi$
35. 42 ft **37.** \$9.00 **39.** 70 ft **41.** 25.1 ft **43.** Two **45.** 5
47. Perpendicular, bisecting
49. Radius is perpendicular to the tangent.
51. It is a semicircle.

Exercises 13-3, Page 344

1. 2,700 cm^2 **3.** 57.8 in.^{2} **5.** 2450 mm^2 **7.** 106 in.^{2} **9.** 0.240 cm^2 **11.** 238 ft^2
13. 0.000258 km^2 **15.** 0.717 ft^2 **17.** (a) 36.0 in. (b) 60.0 in.^{2}
19. (a) 60.0 cm (b) 225 cm^2 **21.** (a) 30.0 in. (b) 30.0 in.^{2}
23. (a) 26.6 in. (b) 35.0 in.^{2} **25.** 60.0 in.^{2} **27.** 22.2 cm^2 **29.** $A = bh_1 + \frac{1}{2}bh_2$
31. $A = \frac{1}{2}ar + \frac{1}{4}\pi r^2$ **33.** 4100 cm^2 **35.** 5880 lb **37.** 41.1 ft^2 **39.** 7.26 cm^2

Exercises 13-4, Page 351

1. Similar **3.** Neither **5.** $\angle E$; side EF **7.** $\angle U$; side ST **9.** 10 **11.** 12
13. 21 **15.** 20 **17.** 16 **19.** 8 **21.**

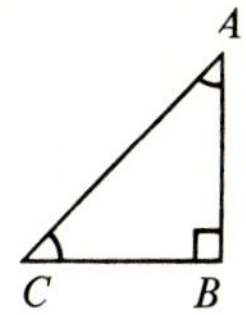

23. 3.3, 4.0

25. $\angle XKY = \angle NKF$, $\angle KXY = \angle KNF$, $\angle XYK = \angle NFK$
27. 8 in. **29.** 6.7 ft **31.** 3.5 m **33.** $2\frac{1}{4}$ in. **35.** 1200 km **37.** $\frac{16}{3}$ ft
39. 104 m

Exercises 13-5, Page 360

1. 6900 ft^3 **3.** 99,000 cm^3 **5.** 8.85 ft^3 **7.** 22,300 cm^3
9. (a) 795 ft^2 (b) 1170 ft^2 **11.** (a) 2.69 m^2 (b) 4.03 m^2
13. (a) 12,300 in.^{2} (b) 18,900 in.^{2} **15.** (a) 57.4 cm^2 (b) 85.0 cm^2
17. 18,800 cm^3 **19.** 2540 ft^3 **21.** 14,700,000 mm^3 **23.** 14,800 in.^{3}
25. (a) 226,000 mm^2 (b) 791,000 mm^2 **27.** (a) 124 in.^{2} (b) 546 in.^{2}
29. (a) 641 cm^2 (b) 1550 cm^2 **31.** (a) 3360 ft^2 (b) 15,100 ft^2 **33.** 5.50 m^3
35. 594 in.^{2} **37.** 175,000 g **39.** \$399

Exercises 13-6, Page 367

1. 54,000 ft^3 **3.** 958 mm^3 **5.** 76.5 yd^2 **7.** 240 m^2 **9.** 1470 ft^3 **11.** 27,100 cm^3
13. 198 in.^{2} **15.** 7210 cm^2 **17.** 113 ft^3 **19.** 5,570,000 mm^3 **21.** 11,300 in.^{2}
23. 376,000 mm^2 **25.** 3,330,000 yd^3 **27.** \$191 **29.** 23,500 lb **31.** 22.3 in.^{2}
33. 78.8 lb **35.** 38,000,000 km^2 **37.** 1.10 in.^{3} **39.** $3\pi r^2$

Exercises 13-7, Page 373

The answers listed for the odd-numbered problems 1 through 11 are the principal lines of false reasoning. However, others may be involved to some extent.

1. Irrelevancy **3.** Analogy
5. Jumping to conclusion **7.** Jumping to conclusion
9. Jumping to conclusion **11.** Assuming true for all cases
13. Logical **15.** Not logical
17. Use the given information and the definition of complementary angles to show that exterior sides form a right angle.

19. Use the division of the quadrilateral into two triangles and postulate 5 for the angles formed by the diagonal.

21. Use the given equation, postulate 2, and postulate 4.

23. Use the given equation and postulate 2.

25. Use the given equation, postulate 2, postulate 4, and postulate 3.

27. If $x > 3$, $3x > 9$; if $x < 3$, $3x < 9$; therefore, $x = 3$.

31. Zero or any negative number provides a counterexample.

33. All equilateral triangles are equiangular.

Exercises 13-8, Page 378

1. $\angle CBE$ **3.** $\angle EBD$ **5.** $\angle 2$ and $\angle 5$ **7.** 180° **9.** 65° **11.** 32° **13.** 52°

15. 132° **17.** 36° **19.** 50° **21.** 120° **23.** 40° **25.** 25° **27.** 65°

29. 40.2 in. **31.** 25.5 mm **33.** 1.29 yd **35.** 27.2 m **37.** 1798 ft **39.** 278 cm

41. 26.7 ft **43.** 153 mm **45.** 224 cm^2 **47.** 3.08 yd^2 **49.** 4.80 cm^2

51. 3.06 in.^{2} **53.** 0.00148 km^2 **55.** 35.4 in.^{2} **57.** 138,000 mm^2 **59.** 1200 mm^2

61. 7.5 **63.** 6.33 in. **65.** 600 ft^3 **67.** 6190 cm^3 **69.** 97,700 in.^{3}

71. 160,000 cm^3 **73.** 3990 ft^3 **75.** 24,400 cm^3 **77.** 900 ft^2 **79.** 14.4 m^2

81. 5430 in.^{2} **83.** 16,800 cm^2 **85.** 1230 ft^2 **87.** 4070 cm^2 **89.** 21.1 ft

91. 231 cm **93.** 32.8 cm **95.** 22.5°, 67.5° **97.** 20 ft **99.** 1590 cm

101. 10.0 in. **103.** 180° **105.** 12,400 cm^2 **107.** 667,000 gal **109.** 2.55 m^3

111. 40.0 in.^{2} **113.** 7.07 cm^3 **115.** 7.44 ft^2 **117.** 32,300 kg **119.** 238 in.^{2}

121. 0.204 ft **123.** 140 cm^2 **125.** $A = 3\pi r^2 + 2\pi rh$

127. Conclusion does not follow, although true.

129. Conclusion is correct.

131. Use the given information and the definition of a parallelogram to show corresponding sides and angles of triangles are equal.

133. No

135. If $x > 7$, $2x + 1 > 15$; if $x < 7$, $2x + 1 < 15$; therefore, $x = 7$.

Exercises 14-1, Page 391

1. $\frac{7}{25}, \frac{7}{24}, \frac{7}{25}$

3. $\frac{7}{24}, \frac{25}{24}, \frac{24}{7}$

5. $\frac{1}{2}, 2, \sqrt{3}$

7. $\sqrt{3}/3, \frac{1}{2}, 2\sqrt{3}/3$

9. 0.797, 0.757, 1.25

11. 0.604, 0.604, 1.32

13. $\frac{3}{5}, \frac{4}{3}$

15. $\frac{9}{40}, \frac{40}{41}$

17. $1, \sqrt{2}/2$

19. 1.26, 0.793

21. $\sqrt{2}/2$

23. 1.4

25. 0.490

27. 1.04

29. sec A, tan A

31. $\cot A = \dfrac{\text{adjacent side}}{\text{opposite side}}$, $\sec A = \dfrac{\text{hypotenuse}}{\text{adjacent side}}$, $\csc A = \dfrac{\text{hypotenuse}}{\text{opposite side}}$

33. 1

Exercises 14-2, Page 397

1. 0.6293 **3.** 0.8847 **5.** 0.5498 **7.** 0.6115 **9.** 0.1477 **11.** 0.6972

13. 0.9833 **15.** 4.331 **17.** 14.0° **19.** 41.6° **21.** 59.1° **23.** 84.2° **25.** 45.6°

27. 25.7° **29.** 78.4° **31.** 60.8° **33.** 1.192 **35.** 0.6947 **37.** 36.9° **39.** 41.7°

41. 48.7° **43.** 83.7° **45.** 0.53 **47.** 0.97 **49.** 1.51

Exercises 14-3, Page 40

1. $B = 60.0°, b = 20.8, c = 24.0$ **3.** $A = 33.7°, a = 12.5, b = 18.7$
5. $B = 13.2°, a = 30.6, b = 7.17$ **7.** $A = 28.8°, B = 61.2°, b = 1.18$
9. $A = 82.6°, B = 7.4°, a = 44.6$ **11.** $A = 64.2°, B = 25.8°, b = 4.71$
13. $B = 83.0°, a = 1.88, c = 15.4$ **15.** $A = 33.5°, B = 56.5°, c = 118$
17. 29.1 ft **19.** 852 cm **21.** 4.3° **23.** 318 cm **25.** 85.2° **27.** 21.9 mi
29. 6.8° **31.** 4.60 in. **33.** 1640 ft **35.** 243 km **37.** 9.12 in. **39.** 1340 ft
41. 151 m

Exercises 14-4, Page 405

1. $\frac{9}{41}, \frac{9}{41}$ **3.** $\frac{41}{9}, \frac{9}{40}$ **5.** 0.549, 0.656 **7.** 0.549, 1.52 **9.** 0.385, 0.417
11. 3.78, 0.967 **13.** 0.461, 0.418 **15.** 0.898, 2.04 **17.** 38.0° **19.** 13.1°
21. 28.4° **23.** 53.4°
25. $B = 69.0°, a = 2.48, b = 6.50$ **27.** $A = 57.3°, a = 71.5, c = 85.0$
29. $A = 59.1°, B = 30.9°, c = 10.1$ **31.** $A = 63.9°, B = 26.1°, b = 47.5$
33. 77.5 ft **35.** 201 m **37.** 28.1° **39.** 1.74 s **41.** 8.8° **43.** 2290 m
45. 20.8 in. **47.** 8.00 cm **49.** No; Yes

Exercises A-1, Page 411

1. 161 **3.** 1636 **5.** 2694 **7.** 24,672 **9.** 25,295 **11.** 139,639 **13.** 581
15. 949 **17.** 9,798 **19.** 95,682 **21.** 6,993 **23.** 2,337,545 **25.** 15,830,199
27. 251,387,772 **29.** 243 **31.** 326 **33.** 1423 rem. 13 **35.** 1570 rem. 374

Exercises A-2, Page 415

1. $\frac{1}{3}$ **3.** $\frac{6}{13}$ **5.** $\frac{3}{5}$ **7.** $\frac{6}{7}$ **9.** $\frac{3}{11}$ **11.** $\frac{13}{9}$ **13.** $\frac{11}{5}$ **15.** $\frac{17}{19}$ **17.** $\frac{17}{27}$
19. $\frac{220}{43}$ **21.** $\frac{4}{5}$ **23.** $\frac{4}{3}$ **25.** $\frac{11}{4}$ **27.** $\frac{1}{3}$ **29.** $\frac{83}{24}$ **31.** $\frac{47}{40}$ **33.** $\frac{73}{72}$
35. $\frac{73}{42}$ **37.** $\frac{6}{77}$ **39.** $\frac{6}{5}$ **41.** $\frac{21}{20}$ **43.** $\frac{5}{27}$ **45.** $\frac{5}{18}$ **47.** $\frac{3}{2}$ **49.** $\frac{14}{9}$
51. $\frac{2}{17}$ **53.** 2 **55.** $\frac{2}{15}$

Exercises A-3, Page 420

1. 0.4 **3.** 0.875 **5.** 1.7 **7.** 0.21 **9.** $\frac{21}{50}$ **11.** $\frac{267}{50}$ **13.** 2817.256
15. 8.763 **17.** 13.99952 **19.** 12.5 **21.** 0.08 **23.** 2.36 **25.** 27% **27.** 2160%
29. $\frac{3}{10}$ **31.** $\frac{1}{125}$ **33.** 60% **35.** 57.1% **37.** 13 **39.** 25% **41.** 14.4
43. \$15.12 **45.** 35%

Exercises for Appendix B, Page 429

1. 1760 yd/mi **3.** 100,000 cm/km **5.** (12 in./ft)(12 in./ft) = 144 $\text{in.}^2/\text{ft}^2$ **7.** 12.7 cm
9. 64 pt **11.** 9.46 L **13.** 0.00396 m^2 **15.** 13.6 L **17.** 130 t **19.** 37 mi/h
21. 770 mi/h **23.** 5520 kg/m^3 **25.** 7600 kg · cm/s **27.** 339 N · m

INDEX